METHODS IN MOLECULAR BIOLOGY™

Series Editor
John M. Walker
School of Life Sciences
University of Hertfordshire
Hatfield, Hertfordshire, AL10 9AB, UK

For other titles published in this series, go to
www.springer.com/series/7651

Computational Biology of Transcription Factor Binding

Edited by

Istvan Ladunga

Department of Statistics, University of Nebraska-Lincoln, Lincoln, NE, USA

Editor
Istvan Ladunga
Department of Statistics
University of Nebraska-Lincoln
1901 Vine St., E145 Beadle Center
Lincoln, NE 68588-0665, USA
sladunga@unl.edu

ISSN 1064-3745   e-ISSN 1940-6029
ISBN 978-1-60761-853-9   e-ISBN 978-1-60761-854-6
DOI 10.1007/978-1-60761-854-6
Springer New York Dordrecht Heidelberg London

Library of Congress Control Number: 2010934132

© Springer Science+Business Media, LLC 2010
All rights reserved. This work may not be translated or copied in whole or in part without the written permission of
the publisher (Humana Press, c/o Springer Science+Business Media, LLC, 233 Spring Street, New York, NY 10013,
USA), except for brief excerpts in connection with reviews or scholarly analysis. Use in connection with any form of
information storage and retrieval, electronic adaptation, computer software, or by similar or dissimilar methodology
now known or hereafter developed is forbidden.
The use in this publication of trade names, trademarks, service marks, and similar terms, even if they are not identified
as such, is not to be taken as an expression of opinion as to whether or not they are subject to proprietary rights.

Cover illustration: Crystal structure of Fis bound to 27 bp optimal binding sequence F2 from Stella, S., Cascio, D.,
Johnson, R.C. (2010) The shape of the DNA minor groove directs binding by the DNA-bending protein Fis. Genes
Dev. 24: 814–826.

Printed on acid-free paper

Humana Press is part of Springer Science+Business Media (www.springer.com)

Preface

Transcriptional regulation controls the basic processes of life. Its complex, dynamic, and hierarchical networks control the momentary availability of messenger RNAs for protein synthesis. Transcriptional regulation is key to cell division, development, tissue differentiation, and cancer as discussed in **Chapters 1** and **2**.

We have witnessed rapid, major developments at the intersection of computational biology, experimental technology, and statistics. A decade ago, researches were struggling with notoriously challenging predictions of isolated binding sites from low-throughput experiments. Now we can accurately predict *cis*-regulatory modules, conserved clusters of binding sites (**Chapters 13** and **15**), partly based on high-throughput chromatin immunoprecipitation experiments in which tens of millions of DNA segments are sequenced by massively parallel, next-generation sequencers (ChIP-seq, **Chapters 9**, **10**, and **11**). These spectacular developments have allowed for the genome-wide mappings of tens of thousands of transcription factor binding sites in yeast, bacteria, mammals, insects, worms, and plants.

Please also note the no less spectacular failures in many laboratories around the world. Having access to chromatin immunoprecipitation, next-generation sequencing, and software is no guarantee for success. The productive and creative use of computational and experimental tools requires a high-level understanding of the underlying biology, the technological characteristics, and the potential and limitation of statistical and computational solutions. This is the *raison d'être* of this volume, guiding scientists of all disciplines through the jungle of regulatory regions, ChIP-seq, about 200 motif discovery tools and others. As in previous volumes of the series *Methods in Molecular Biology^{TM}*, we help readers to understand the basic principles and give detailed guidance for the computational analyses and biological interpretations of transcription factor binding. We disclose critical practical information and caveats that may be missing from research publications. This volume serves not only computational biologists but experimentalists as well, who may want to understand better how to design and execute experiments and to communicate effectively with computational biologists, computer scientists, and statisticians. **Chapter 1** helps readers to find their way in the maze of resources by a high-level overview of the computational, biological, and some experimental solutions of transcription factor binding. **Chapter 1** highlights other units in this volume and discusses some of the issues not covered.

Why are there so many failed experiments and analyses? Consider, for an example, ChIP-seq, where background noise accounts for more than half of the sequencing reads. Potentially, this may lead to a vast array of false-positive observations. Careful investigators, however, can apply kernel-based density estimates and other background modeling and correction methods to find significantly enriched signals in such noisy observations (**Chapters 9** and **10**). Density estimates are followed by improved peak calling with controlled false discovery rate (**Chapter 10**). Another problem is that ChIP-seq peaks are tens to hundreds of times wider than the footprint of the transcription factor on the DNA. The highest peaks often come from amplification and sequencing bias,

v

vi   Preface

not from a bona fide biological signal (**Chapter 1**). These serious issues mandate the identification of shared, short, and variable DNA motifs, representations of variable binding sites, from moderate-to-low resolution ChIP-seq data using computational motif discovery algorithms. On the other hand, false negatives are also abundant. Consider the temporary nature of regulation, which responds to temporary environmental and internal stimuli. Therefore, a site is typically bound only at a fraction of time, easily missed by snapshot techniques like ChIP (**Chapter 24**). In order to reduce the number of false positives and negatives, motifs are trained by a wide spectrum of statistical learning methods. In spite of the diverse implementation of these tools, most of them stem from expectation maximization and Gibbs sampling (**Chapters 6, 7, and 11**) or support vector machines (**Chapter 13**). The trained tools can find binding sites missed by experiments in the predicted promoter regions (**Chapter 5**), all regulatory regions (**Chapter 4**), or in the whole genome.

In itself, de novo computational motif prediction is still not accurate enough (**Chapter 8**). Confidence levels can be increased greatly by integrating binding site locations with in vitro protein–DNA affinities (**Chapter 12**), evolutionary conserved regions (**Chapters 11, 14, and 18**), and transposable DNA elements that propagate binding sites through the genome (**Chapter 14**). Time-delayed co-expression as inferred from large compendia of gene expression experiments also indicates binding sites of shared transcription factors. This enormous wealth of information can be retrieved in computationally efficient ways from diverse databases including OregAnno (**Chapter 20**), Plant-TFDB (**Chapter 21**), cis-Lexicon (**Chapter 22**), and genome browsers (**Chapters 1, 10, and 22**).

The integrated observations and predictions help us to reconstruct complex, hierarchical, and dynamic transcriptional regulatory networks (**Chapters 23 and 24**). This task demands not only new experiments but also the re-annotation of existing experimental data and computational predictions and ongoing, major paradigm changes for all of us.

Istvan Ladunga

Contents

Preface . . . . *v*

Contributors . . . . *ix*

1. An Overview of the Computational Analyses and Discovery
of Transcription Factor Binding Sites 1
   Istvan Ladunga

2. Components and Mechanisms of Regulation of Gene Expression 23
   Alper Yilmaz and Erich Grotewold

3. Regulatory Regions in DNA: Promoters, Enhancers, Silencers, and Insulators . . . 33
   Jean-Jack M. Riethoven

4. Three-Dimensional Structures of DNA-Bound Transcriptional Regulators 43
   Tripti Shrivastava and Tahir H. Tahirov

5. Identification of Promoter Regions and Regulatory Sites 57
   Victor V. Solovyev, Ilham A. Shahmuradov, and Asaf A. Salamov

6. Motif Discovery Using Expectation Maximization and Gibbs' Sampling 85
   Gary D. Stormo

7. Probabilistic Approaches to Transcription Factor Binding Site Prediction 97
   *Stefan Posch, Jan Grau, André Gohr, Jens Keilwagen,
   and Ivo Grosse*

8. The Motif Tool Assessment Platform (MTAP) for Sequence-Based
Transcription Factor Binding Site Prediction Tools 121
   Daniel Quest and Hesham Ali

9. Computational Analysis of ChIP-seq Data 143
   Hongkai Ji

10. Probabilistic Peak Calling and Controlling False Discovery Rate
Estimations in Transcription Factor Binding Site Mapping from ChIP-seq 161
    Shuo Jiao, Cheryl P. Bailey, Shunpu Zhang, and Istvan Ladunga

11. Sequence Analysis of Chromatin Immunoprecipitation Data
for Transcription Factors 179
    Kenzie D. MacIsaac and Ernest Fraenkel

12. Inferring Protein–DNA Interaction Parameters from SELEX Experiments 195
    Marko Djordjevic

13. Kernel-Based Identification of Regulatory Modules 213
    Sebastian J. Schultheiss

vii

viii   Contents

14. Identification of Transcription Factor Binding Sites Derived
from Transposable Element Sequences Using ChIP-seq 225
Andrew B. Conley and I. King Jordan

15. Target Gene Identification via Nuclear Receptor Binding Site Prediction 241
Gabor Varga

16. Computing Chromosome Conformation 251
*James Fraser, Mathieu Rousseau, Mathieu Blanchette,
and Josée Dostie*

17. Large-Scale Identification and Analysis of C-Proteins 269
Valery Sorokin, Konstantin Severinov, and Mikhail S. Gelfand

18. Evolution of *cis*-Regulatory Sequences in *Drosophila* 283
Xin He and Saurabh Sinha

19. Regulating the Regulators: Modulators of Transcription Factor Activity 297
Logan Everett, Matthew Hansen, and Sridhar Hannenhalli

20. Annotating the Regulatory Genome 313
*Stephen B. Montgomery, Katayoon Kasaian, Steven J.M. Jones,
and Obi L. Griffith*

21. Computational Identification of Plant Transcription Factors
and the Construction of the PlantTFDB Database 351
*Kun He, An-Yuan Guo, Ge Gao, Qi-Hui Zhu, Xiao-Chuan Liu,
He Zhang, Xin Chen, Xiaocheng Gu, and Jingchu Luo*

22. Practical Computational Methods for Regulatory Genomics: A *cis*GRN-
Lexicon and *cis*GRN-Browser for Gene Regulatory Networks 369
Sorin Istrail, Ryan Tarpine, Kyle Schutter, and Derek Aguiar

23. Reconstructing Transcriptional Regulatory Networks Using Three-Way
Mutual Information and Bayesian Networks 401
Weijun Luo and Peter J. Woolf

24. Computational Methods for Analyzing Dynamic Regulatory Networks 419
Anthony Gitter, Yong Lu, and Ziv Bar-Joseph

Subject Index 443

Contributors

DEREK AGUIAR • *Department of Computer Science, Center for Computational Molecular Biology, Brown University, Providence, RI, USA*

HESHAM ALI • *Department of Pathology and Microbiology, College of Information Science & Technology, University of Nebraska, Omaha, NE, USA*

CHERYL P. BAILEY • *Department of Biochemistry, University of Nebraska-Lincoln, Lincoln, NE, USA*

ZIV BAR-JOSEPH • *Computer Science Department, Carnegie Mellon University, Pittsburgh, PA, USA*

MATHIEU BLANCHETTE • *McGill Centre for Bioinformatics, McGill University, Montréal, QC, Canada*

XIN CHEN • *National Laboratory of Protein Engineering and Plant Genetic Engineering, College of Life Sciences, Center for Bioinformatics, Peking University, Beijing, China*

ANDREW B. CONLEY • *School of Biology, Georgia Institute of Technology, Atlanta, GA, USA*

MARKO DJORDJEVIC • *Faculty of Biology, Department of Chemistry and Physics, University of Belgrade, Arkansas State University, Studentski try 16, 11000 Belgrade*

JOSÉE DOSTIE • *Department of Biochemistry, and Goodman Cancer Research Center, McGill University, Montréal, QC, Canada*

LOGAN EVERETT • *Department of Genetics, Penn Center for Bioinformatics, University of Pennsylvania, Philadelphia, PA, USA*

ERNEST FRAENKEL • *Computer Science and Artificial Intelligence Laboratory, Department of Biological Engineering, Massachusetts Institute of Technology, Cambridge, MA, USA*

JAMES FRASER • *Department of Biochemistry, and Goodman Cancer Research Center, McGill University, Montréal, QC, Canada*

GE GAO • *National Laboratory of Protein Engineering and Plant Genetic Engineering, College of Life Sciences, Center for Bioinformatics, Peking University, Beijing, China*

MIKHAIL S. GELFAND • *Bioinformatics, Institute for Information Transmission Problems RAS, Bolshoi Karetny 17, Moscow 127994, Russia*

ANTHONY GITTER • *Computer Science Department, Carnegie Mellon University, Pittsburgh, PA, USA*

ANDRÉ GOHR • *Leibniz Institute of Plant Biochemistry (IPB), Halle, Germany*

JAN GRAU • *Institute of Computer Science, Martin Luther University, Halle–Wittenberg, Germany*

OBI L. GRIFFITH • *Canada's Michael Smith Genome Sciences Centre, Vancouver, BC, Canada*

IVO GROSSE • *Institute of Computer Science, Martin Luther University, Halle–Wittenberg, Germany*

ERICH GROTEWOLD • *Plant Biotechnology Center and Department of Plant Cellular and Molecular Biology, The Ohio State University, Columbus, OH, USA*

XIAOCHENG GU • *National Laboratory of Protein Engineering and Plant Genetic Engineering, College of Life Sciences, Center for Bioinformatics, Peking University, Beijing, China*

AN-YUAN GUO • *National Laboratory of Protein Engineering and Plant Genetic Engineering, College of Life Sciences, Center for Bioinformatics, Peking University, Beijing, China*

SRIDHAR HANNENHALLI • *Department of Genetics, Penn Center for Bioinformatics, University of Pennsylvania, Philadelphia, PA, USA*

MATTHEW HANSEN • *Department of Genetics, Penn Center for Bioinformatics, University of Pennsylvania, Philadelphia, PA, USA*

KUN HE • *National Laboratory of Protein Engineering and Plant Genetic Engineering, College of Life Sciences, Center for Bioinformatics, Peking University, Beijing, China*

XIN HE • *Department of Computer Science, University of Illinois at Urbana-Champaign, Urbana, IL, USA*

SORIN ISTRAIL • *Department of Computer Science, Center for Computational Molecular Biology, Brown University, Providence, RI, USA*

HONGKAI JI • *Department of Biostatistics, The Johns Hopkins Bloomberg School of Public Health, Baltimore, MD, USA*

SHUO JIAO • *Institute Fred Hutchinson Cancer, Seattle, WA, USA*

STEVEN J.M. JONES • *Canada's Michael Smith Genome Sciences Centre, Vancouver, BC, Canada*

I. KING JORDAN • *School of Biology, Georgia Institute of Technology, Center for Bioinformatics and Computational Genomics, Atlanta, GA, USA*

KATAYOON KASAIAN • *Canada's Michael Smith Genome Sciences Centre, Vancouver, BC, Canada*

JENS KEILWAGEN • *Leibniz Institute of Plant Genetics and Crop Plant Research (IPK), Gatersleben, Germany*

ISTVAN LADUNGA • *Department of Statistics, University of Nebraska-Lincoln, Lincoln, NE, USA*

XIAO-CHUAN LIU • *National Laboratory of Protein Engineering and Plant Genetic Engineering, College of Life Sciences, Center for Bioinformatics, Peking University, Beijing, China*

YONG LU • *Department of Biological Chemistry and Molecular Pharmacology, Harvard Medical School, Boston, MA, USA*

JINGCHU LUO • *National Laboratory of Protein Engineering and Plant Genetic Engineering, College of Life Sciences, Center for Bioinformatics, Peking University, Beijing, China*

WEIJUN LUO • *Cold Spring Harbor Laboratory, Cold Spring Harbor, NY, USA*

KENZIE D. MACISAAC • *Department of Biological Engineering, Massachusetts Institute of Technology, Cambridge, MA, USA; Molecular Profiling Research Informatics, Merck Research Laboratories, Boston, MA, USA*

STEPHEN B. MONTGOMERY • *Wellcome Trust Sanger Institute, Cambridge, UK; Department of Genetic Medicine and Development, University of Geneva Medical School, Geneva, Switzerland*

STEFAN POSCH • *Institute of Computer Science, Martin Luther University, Halle–Wittenberg, Germany*

DANIEL QUEST • *Computational Biology Group, Biological Sciences Division, Oak Ridge National Laboratory, Oak Ridge, TN, USA*

JEAN-JACK M. RIETHOVEN • *Bioinformatics Core Research Facility, Center for Biotechnology and School for Biological Sciences, University of Nebraska-Lincoln, Lincoln, NE, USA*

MATHIEU ROUSSEAU • *McGill Centre for Bioinformatics, McGill University, Montréal, QC, Canada*

ASAF A. SALAMOV • *DOE Joint Genome Institute, Walnut Creek, CA, USA*

SEBASTIAN J. SCHULTHEISS • *Friedrich Miescher Laboratory of the Max Planck Society, Tübingen, Germany*

KYLE SCHUTTER • *Department of Biomedical Engineering, Center for Computational Molecular Biology, Brown University, Providence, RI, USA*

KONSTANTIN SEVERINOV • *Waksman Institute, Rutgers, The State University of New Jersey, New Brunswick, NJ, USA; Institute of Gene Biology and Institute of Molecular Genetics, Russian Academy of Sciences, Moscow, Russia*

ILHAM A. SHAHMURADOV • *Bioinformatics Laboratory, Institute of Botany, ANAS, Baku, Azerbaijan*

TRIPTI SHRIVASTAVA • *Eppley Institute for Research in Cancer and Allied Diseases, University of Nebraska Medical Center, Omaha, NE, USA*

SAURABH SINHA • *Department of Computer Science, University of Illinois at Urbana-Champaign, Urbana, IL, USA*

VICTOR V. SOLOVYEV • *Department of Computer Science, Royal Holloway, London, UK; Softberry Inc., University of London, Mount Kisco, NY, USA*

VALERY SOROKIN • *Faculty of Bioengineering and Bioinformatics, M.V. Lomonosov Moscow State University, Moscow, Russia*

GARY D. STORMO • *Department of Genetics, School of Medicine, Washington University, St. Louis, MO, USA*

TAHIR H. TAHIROV • *Eppley Institute for Research in Cancer and Allied Diseases, University of Nebraska Medical Center, Omaha, NE, USA*

RYAN TARPINE • *Department of Computer Science, Center for Computational Molecular Biology, Brown University, Providence, RI, USA*

GABOR VARGA • *Discovery Informatics, Lilly Research Labs, Eli Lilly and Company, Indianapolis, IN, USA*

PETER J. WOOLF • *Departments of Chemical and Biomedical Engineering and Bioinformatics Program, University of Michigan, Ann Arbor, MI, USA*

ALPER YILMAZ • *Plant Biotechnology Center and Department of Plant Cellular and Molecular Biology, The Ohio State University, Columbus, OH, USA*

HE ZHANG • *National Laboratory of Protein Engineering and Plant Genetic Engineering, College of Life Sciences, Center for Bioinformatics, Peking University, Beijing, China*

SHUNPU ZHANG • *Department of Statistics, University of Nebraska-Lincoln, Lincoln, NE, USA*

QI-HUI ZHU • *National Laboratory of Protein Engineering and Plant Genetic Engineering, College of Life Sciences, Center for Bioinformatics, Peking University, Beijing, China*

Chapter 1

An Overview of the Computational Analyses and Discovery of Transcription Factor Binding Sites

Istvan Ladunga

Abstract

Here we provide a pragmatic, high-level overview of the computational approaches and tools for the discovery of transcription factor binding sites. Unraveling transcription regulatory networks and their malfunctions such as cancer became feasible due to recent stellar progress in experimental techniques and computational analyses. While predictions of isolated sites still pose notorious challenges, *cis*-regulatory modules (clusters) of binding sites can now be identified with high accuracy. Further support comes from conserved DNA segments, co-regulation, transposable elements, nucleosomes, and three-dimensional chromosomal structures. We introduce computational tools for the analysis and interpretation of chromatin immunoprecipitation, next-generation sequencing, SELEX, and protein-binding microarray results. Because immunoprecipitation produces overly large DNA segments and well over half of the sequencing reads from constitute background noise, methods are presented for background correction, sequence read mapping, peak calling, false discovery rate estimation, and co-localization analyses. To discover short binding site motifs from extensive immunoprecipitation segments, we recommend algorithms and software based on expectation maximization and Gibbs sampling. Data integration using several databases further improves performance. Binding sites can be visualized in genomic and chromatin context using genome browsers. Binding site information, integrated with co-expression in large compendia of gene expression experiments, allows us to reveal complex transcriptional regulatory networks.

Key words: Transcription factor, transcription factor binding site, computational prediction, background correction, peak calling, chromatin immunoprecipitation, next-generation sequencing, ChIP-seq, protein-binding microarrays, transcriptional regulation, data integration.

1. Introduction

Transcriptional regulation affects the fundamental biological processes. By regulating cellular mRNA levels, it influences translation and the level of proteins. The sophistication of higher

I. Ladunga (ed.), *Computational Biology of Transcription Factor Binding*, Methods in Molecular Biology 674,
DOI 10.1007/978-1-60761-854-6_1, © Springer Science+Business Media, LLC 2010

eukaryotes resides primarily in the architecture and functioning of the regulatory networks, not in the number of proteins. In *Caenorhabditis elegans*, a relatively simple eukaryote, an adult hermaphrodite has 959 somatic cells and about 20,000 protein coding genes. Trillions of cells in a human individual carry only about twice the number of *C. elegans* genes. We are only beginning to understand the complex transcription regulatory networks and other mechanisms. It is vital to improve this understanding for curing regulatory malfunctions like cancer and autoimmune diseases. Although most of the regulators including transcription factors and microRNAs are known in human, the vast majority of their binding sites remain unexplored. The enormous variability of regulatory sites poses the most difficult challenge. For example, in the 12 half sites of the λ operators, as few as 2 of the 8 positions are conserved and most of the others are highly variable (1). The computational representation of such variable sequence sets – motifs – affects the performance of motif finder tools. Variability may also indicate differential DNA–TF affinity. Sites with higher affinity are expected to produce more transcripts than low-affinity regulatory sites (2). Fortunately, the prediction of promoter regions has matured (*see* **Chapter 5**).

Long and well-characterized motifs like those of p53 (3, 4) or PPARG (5, 6) are relatively easy to predict if some false negatives can be accepted. For shorter motifs, the naïve application of the over 200 published tools often provides somewhat inconsistent results [*see* **Chapter 8** and ref. (7)]. Such moderate performance mandates genome-wide experimental identification of binding sites samples. These samples have to represent motif variation to allow the training of prediction tools, which may find the rest of the sites including those unbound under the specific conditions of the experiment.

Obtaining representative samples is becoming increasingly affordable thanks to stellar progress at the intersection of biology, computational analyses/predictions, and experimental technology. This volume focuses on mapping the binding sites of transcription factors (TFs) to regulatory regions on the genome. TFs are regulatory proteins that bind to promoter, enhancer, and other DNA regions in a sequence-specific manner (**Chapters 2, 3, and 4**). TF binding affects the recruitment and dynamism of RNA polymerases and hence the transcription of genes. Most TFs provide control in one direction only: they either upregulate or downregulate the expression of a target gene, but not both. Certain other TFs, however, activate at low levels, but at high concentration, they repress the transcription of the same gene. C-proteins, for example, at low cellular concentration attach only to high-affinity sites and activate the target gene (**Chapter 17**). At high levels, C-proteins bind to the low-affinity sites as well, now inhibiting transcription. Such complex mechanisms are abundant

An Overview of the Computational Analyses and Discovery   3

in higher eukaryotes where most genes are regulated by multiple TFs. MicroRNAs (8), DNA methylation (9), and histone modifications (10, 11) also play major roles in transcriptional regulation.

A key message of this volume is that purist approaches, either "pure experiments" loosely associated with service-like computations or "pure algorithms" with marginal understanding of the biology and the technology, are equally elusive. Consider the most successful high-throughput experimental technique for the discovery of transcription factor binding sites (TFBS): chromatin immunoprecipitation (ChIP) combined with next-generation sequencing (ChIP-seq). Here tens of millions of sequencing reads are mapped onto the genome. Researchers have to correct for background noise and normalize between replicates. The background-corrected and normalized density distributions of reads allow calling peaks, significantly enriched regions that span over TF binding sites (TFBS, **Chapters 9**, **10**, and **11**). These 35–200 base pair wide peaks (*see* **Chapters 9**, **10**, and **11**) far exceed the 4–25 base pair footprints of TFs on the DNA (12, 13). Therefore locating the actual binding sites from (tens of) thousands of overly wide peaks requires computational discovery of shared binding site motifs.

2. Methods

2.1. Experimental Information

Computational binding site predictions or analyses invariably stem from some experimental information. Such observations include genomic, mRNA, and protein sequences, three-dimensional structures of DNA-bound TFs (**Chapter 4**), chromatin immunoprecipitation, Systematic Evolution of Ligands by EXponential enrichment [SELEX, **Chapter 12** and refs. (14–16)], protein-binding microarrays (17), co-expression of genes as calculated from compendia of gene expression experiments (**Chapters 23** and **24**), and DNAse I hypersensitive regions that indicate nucleosome-depleted, regulatory regions (18, 19).

Most information comes from high-throughput experiments at the cost of low resolution, significant background noise, and considerable systematic bias. Such undesirable features can be reduced by computational tools that take into account critical technological characteristics and biological issues. Abundant false positives and negatives can be reduced by motif analyses (**Chapter 6, 7, 8, 11**, and **13**), and integrating with Evolution of Ligands by EXponential enrichment (SELEX, **Chapter 12**), protein-binding microarrays, and co-expression results (**Chapters 23** and **24**). We also seek for a reasonable balance between false positives (overly permissive analyses) and false negatives (overly conservative settings). At the final steps, computational analyses converge to motif discovery and network reconstruction.

2.1.1. Perturbation Experiments: Mutagenesis and RNA Interference

Experimental mutagenesis of binding sites at regulatory regions and/or knocking out the TF genes provide for the most reliable TFBS localization (20) at the price of extremely low throughput. Knocking down TF genes by RNA interference (21) is a more economical solution but incomplete silencing could impair the results. Knockdown/knockout effects are evaluated by measuring the expression of target genes using PCR experiments.

2.1.2. Chromatin Immunoprecipitation (ChIP)

Chromatin Immunoprecipitation (ChIP) is the most powerful experimental technique for the in vivo mapping of DNA-associated proteins [**Chapters 9, 10, 11** and refs. (11, 22, 23)]. Essentially, proteins are cross-linked to their native genomic loci in vivo. Then cells are lysed and DNA is fragmented by sonication or shearing. Antibody-bound chromatin is immunoprecipitated and the extra DNA may be digested by micrococcal nuclease. Having reversed the cross-links, proteins are digested. DNA segments are either hybridized to promoter or tiling microarrays [ChIP-chip, (24)] or sequenced by ultra-high-throughput sequencing [ChIP-seq, (11)].

2.1.2.1. ChIP-chip

Immunoprecipitated, protein-free and size-selected DNA can be hybridized to genome-wide tiling or promoter microarray chips [ChIP-chip, (24)]. These microarrays differ from gene expression chips in that they span the whole genome or its selected parts like promoter regions or the ENCyclopedia of DNA Element (ENCODE) regions (25). ChIP-chip allowed the genome-wide mapping of TFBS in simple eukaryotes such as yeast (26, 27). In higher eukaryotes, the accuracy of ChIP-chip is compromised by intensive cross-hybridization between sample DNA and partially matching probes on the microarray chips. When millions of probes are crowded on a few chips, resolution (the spacing of probes on the genome) is sacrificed (28). Mapping chromatin around repetitive DNA elements that accounts for almost half of the mammalian genomes (29) would require longer probes and highly expensive microarrays. For less researched genomes, chip design and manufacturing may be economically unattractive.

2.1.2.2. ChIP and Next-Generation Sequencing: ChIP-seq

The above issues motivated researchers to sequence chromatin-immunoprecipitated DNA using next-generation sequencing (ChIP-seq, **Chapters 9, 10,** and **11**). ChIP-seq scales up well even to the most complex genomes. Since there is no need for species-specific microarrays, any genome can be sequenced. Cross-hybridization, the most burning issue with microarrays, is unknown in sequencing. ChIP-seq has a much finer resolution (25–200 bp) than ChIP-chip in large genomes (~200 bp).

The high resolution so much needed for distinguishing signal from background can be achieved at reasonable costs by using more fluid cells, each of them producing tens of millions of sequencing reads. The Illumina (formerly Solexa) Genome Analyzer platform (30) generates the highest coverage of 28–100 bases per sequencing read, while the Roche/454 instrument produces 250–400 base long reads at the price of much lower coverage (31). The Life Technologies (formerly Applied Biosystems, http://solid.appliedbiosystems.com) SOLiD platform is a compromise between the other two machines. Due to the short, 4–30 base pair footprint of TFs on the DNA (12, 13), long reads have no major advantages and most researchers opt for high coverage.

2.1.2.2.1. Base Calling in ChIP-seq

Vendors of sequencing platforms supply software packages that perform deterministic base calling. More real, probabilistic base calling can be achieved by Rolexa (32) or Alta-Cyclic (33), which are expected to increase the number of mappable sequencing reads as compared to the deterministic tools.

2.1.2.2.2. Mapping Sequencing Reads to the Genome

Mapping tens of millions of sequencing reads to a reference genome would take prohibitive time using traditional methods such as BLAST (34) or BLAT (35). Instead, the reference genome is represented as a suffix tree, and using the Burrows–Wheeler transformation, Bowtie (36) and related methods can map millions of reads in a few hours on a LINUX/UNIX computer (**Chapters 9** and **10**). A considerable proportion of the sequencing reads cannot be mapped unambiguously due to repetitive DNA sequences that make up ~46% of the human genome. Sequencing errors, to some extent, can be corrected by using read quality information. For a recent review of read mapping, confer Trapnell et al. (37).

2.1.2.2.3. Amplification and Sequencing Bias and Background Correction

ChIP produces very low amounts of DNA. Emerging technologies like the Helicos platform can sequence single molecules (38), but samples need to be amplified for the Illumina, ABI/Life Technologies' SOLiD and Roche instruments. Significant amplification bias has been observed (39). The considerable extent of amplification and sequencing bias is best studied in simple systems free from the complications of ChIP. Whole-genome sequencing is one such system where nonrepetitive DNA segments are expected to have equimolar concentrations in the sample. Significant departures from the uniform distribution in whole-genome sequencing reads over nonrepetitive DNA indicate bias in amplification, sequencing, and the accessibility of DNA. GC-rich regions tend to produce more sequencing tags than AT-rich segments (39–41). In whole-genome sequencing, nucleosome-depleted hence highly accessible gene boundaries produce significantly more sequencing tags than other genomic regions (41). Transcriptional activity increased the number of sequencing reads obtained: at the transcription start sites of highly expressed genes about four times more tags were obtained than at those of less expressed genes (41). This may be partly due to nucleosome depletion around the promoters of transcription (42), which makes the DNA more accessible for sequencing than nucleosome-rich regions. It is important to note that the highest ChIP-seq (or ChIP-chip) peaks frequently come not from bona fide TFBS but from the most accessible regions that also have positive amplification and sequencing bias (43).

2.1.2.2.4. Background Correction

According to Pepke et al. (43), ~60–90% of the sequencing reads come from background: most of the ChIP-DNA segments come from interactions other than the TF of interest. ChIP is limited by the specificity of the antibody used, a particularly serious issue with superclass/multigene family TFs. Four superclasses were proposed for TFs: leucine zippers (44), basic helix-loop-helix TFs (45), zinc fingers (46), and beta-scaffold factors with minor groove contacts (12). For a classification of regulatory proteins, *see* (12). Due to the moderate correlation of epitope and the actual DNA-binding residues, designing highly specific antibodies to multifamily TFs remains a major challenge (47). Also, antibodies may bind to other DNA-associated proteins including histones, chromatin remodeling enzymes, and chromatin scaffolding proteins. Antibody binding to untargeted proteins raises the issue of estimating background noise and false discovery rate (FDR) (**Chapters 9** and **10** and ref. (48)). Background noise estimations can be assisted by three major experimental approaches. In the first approach, chromatin is ligated with IgG or other non-specific antibody (11). In this type of control experiments, we measure the nonspecific binding of a general antibody to any part of the chromatin. Unfortunately, this approach cannot characterize the reaction of the selected TF-specific antibody with other members of the TF family. The second technique reverses the cross-links in vitro, which allows chromatin delocalization over the genome. The third type of control omits IP altogether and therefore assesses only the availability of genomic segments, their amplification, and sequencing biases. Note that each of these controls underestimates antibody binding to similar TFs. The similarity of epitope structures within large families may cause a drop in selectivity resulting in many false positives.

When no control experiment is available, statisticians develop models of background noise using theoretical distributions. Model-based analysis of ChIP-Seq (MACS) (49) estimates background using the Poisson distribution. The negative binomial distribution generates even better estimates (**Chapter 9**). Recently, most peak calling programs involve some kind of background correction.

An Overview of the Computational Analyses and Discovery 7

2.1.2.2.5. Peak Calling

ChIP-seq reads are enriched near the binding sites of the targeted TF as compared to genomic loci unbound by the TF of interest (**Chapter 10**, **Fig. 10.4**). When each strand of the ChIP-DNA fragments is sequenced from the 5′ end, the probability of polymerase detachment increases by progressing toward the 3′. Therefore the 5′ termini of both the Watson and the Crick strands are covered by more reads than their centers or 3′ ends. Two enrichment areas emerge, one upstream and another one downstream of the actual binding site. Typically, neither the highest point nor the center of the enrichment indicates exactly the binding site. Enrichment areas are somewhat irregular in shape and extend considerably wider than the actual binding site. Recognizing, merging, and calling the location of twin peaks is still a challenging problem as indicated by the dozens of diverse peak calling methods published. In **Chapter 9**, Hongkai Ji discusses modeling the background noise, peak calling, and as implemented in his CisGenome (22) tool. **Chapter 10** demonstrates kernel density estimates for peak calling and false discovery rate calculations as incorporated into the Quantitative Enrichment of Sequencing Tags software [QuEST, (50)]. For a detailed comparison of FindPeaks (51), SISSRs (52), USEq (53), PeakSeq (54), and dozens of other peak calling methods, we recommend Pepke et al.'s (43) comprehensive review.

2.1.2.2.6. False Discovery Rate (FDR)

Noise often exceeds signal coming from bona fide binding sites. Therefore calling peaks above any reasonable threshold unavoidably will include false positive results. Therefore credibility requires managing false positive calls and reporting their frequency. In order to balance between false positives and false negatives, users select an FDR threshold, input to several peak calling tools including CisGenome [**Chapter 9**, (22)], QuEST [Chapter 10, (50)], SiSSRs (52), FindPeaks (51), USEq (53), PeakSeq (54), and many others. These tools calculate the lowest value of the ranking statistics (e.g., density or signal-to-noise ratio) still not exceeding the selected FDR threshold. Not knowing all positive binding sites, FDR has to be estimated by either from the control experiments or from negative binomial or Poisson models of background noise distribution (**Chapter 9, section "Background Correction"**). Certain tools like QuEST (50) explicitly demand a control library but CisGenome [**Chapter 9** and ref. (22)], FindPeaks (51), and MACS (49) work with or without control experiments. FindPeaks (51) performs Monte Carlo simulations. Most recent tools improve peak calling by estimating the shift between the peaks on opposing strands (*see* **Chapter 10** and ref. (50)).

Having performed these analyses, researchers may find that immunoprecipitation is not selective enough or the first run of sequencing does not provide sufficient contrast between signal and background. Then a new antibody may be added and/or further sequencing is performed. It is prudent to have a contingency of resources for such experiments and analyses even at the cost of studying fewer regulatory proteins.

Since the called peaks are considerably wider than the actual TFBS, binding site motifs are further analyzed by pattern recognition methods (**Section 2.2**).

2.1.3. Measuring In Vitro Affinities of TFs to DNA

Transcriptional regulation is a temporal phenomenon, a conditional, short-term response to changing environmental and cellular conditions. Ideally, activator TFs bind only when the target gene needs to be upregulated and inhibitor TFs bind when the gene product is not needed at a given point of time. ChIP and subsequent motif discovery may also miss previously uncharacterized binding sites/motifs due to partial occupancy or low resolution. In 2009, Zhu et al. (55) estimated that almost half of the in vivo TFBS in yeast remained unknown. This is plausible since performing ChIP experiments for all possible conditions to find all biological sites in bound state remains an elusive proposition. Therefore, it is necessary to complement in vivo assays by in vitro assessments of the TF's affinity to double-stranded DNA k-mers. In vitro proteins, if in native conformation, bind to DNA probes by and large regardless of the conditions. It is also feasible to map the affinities of all k-mers for $k < 10$ for about 100 regulatory proteins. This can be achieved by Systematic Evolution of Ligands by EXponential enrichment (SELEX) and protein-binding microarrays as discussed below.

2.1.3.1. Systematic Evolution of Ligands by EXponential Enrichment (SELEX)

Rapid selection of nucleic acids (single- or double-stranded RNA or DNA) which have high affinity to a molecular target like a TF can be achieved by Systematic Evolution of Ligands by EXponential enrichment [SELEX, **Chapter 12** and refs. (56, 57)]. SELEX has been highly productive in the discovery of nucleic acid bound small molecule drug candidates (58) and 55 *Escherichia coli* TFs (14) among many other applications. The experiments are performed in multiple rounds. From an initial library of 10^{15}–10^{16} sequences, ligand-bound DNA is separated from free DNA and amplified. In the subsequent rounds, the library pre-selected in the previous round is reacted with the ligand again, separated, and amplified. The average ligand binding affinity of the selected DNA sequences increases exponentially with the number of rounds (16). While nonspecific binding occurs at every round, at the final round, the large majority of DNA sequences will be high-affinity binders. These DNA segments are sequenced recently using next-generation technology (**Section 2.1.2.2**). Over-selection, however, should be avoided since TFs in vivo bind to biologically important medium- or low-affinity loci as well. Besides the biological significance, de novo computational motif discovery critically depends on these sites as

An Overview of the Computational Analyses and Discovery   9

well. Since a typical SELEX starting library can exceed the size of the genome, many of the selected binder sequences may be absent from the genome. SELEX-derived affinities, consensus sequences, and PWMs are available in the SELEX_DB (59) and TRANSFAC (12, 13) databases. Larger data sets obtained by next-generation sequencing can be retrieved from the HTPSELEX database (60).

2.1.3.2. Protein-Binding Microarrays (PBMs)

Martha Bulyk and colleagues at Harvard characterize in vitro DNA affinities using protein-binding microarrays (PBMs). Among others, 30 previously uncharted and 59 other yeast TFs were characterized (55). PBMs are custom-designed microarrays with *double-stranded* DNA probes that include all possible ungapped and many gapped k-mers (61). A typical analysis applies 8-mers as follows (62). The selected TF is cloned in fusion with glutathione S-transferase and hybridized with the DNA on the microarray. DNA–TF–GST complexes are detected using fluorophore-conjugated anti-GST antibodies (63). Microarrays are scanned, mapped, background-corrected, and normalized. Then the TF's in vitro affinity to each k-mer is reported as a normalized enrichment score. A major criticism of PBMs is that in vitro affinities may differ from in vivo binding dependent on the current state of the chromatin environment. Also, position relative to regulatory regions matters, since in far intergenic regions, many DNA-bound complexes have no detectable effects on transcription as indicated by ENCODE observations (25). PBMs may miss bona fide sites when the association requires post-translational modifications or cofactors. Also, the cloned protein may fold into non-native conformations (55).

Computers work efficiently with thousands of k-mer affinities but humans cannot comprehend such massive data sets. Therefore the CRACR algorithm (64) converts affinities into more perceivable positional weight matrices (PWMs) to be discussed in **Section 2.2.2.1**. PBM results are generally compatible with ChIP-derived PWMs (55). In vitro observations were also confirmed (55) by regulatory patterns derived from knockout experiments (65) and condition-specific expression results from a compendium of gene expression experiments with 1,693 conditions (66).

2.2. Computational Analyses and Predictions

2.2.1. Ab Initio Predictions of TFBS from 3D Structures

Ideally, a 3D structure of a TF–DNA complex (**Chapter 4**) could allow us to predict binding sites. Such structures can be obtained from X-ray/NMR determination or homology modeling. In theory, molecular dynamics simulation and thermodynamic integration could facilitate the predictions of protein–DNA affinity for diverse nucleic acid sequences. Successful simulations and predictions were reported for the yeast MAT-2 homeodomain and GCN4 bZIP proteins (67, 68), but the widespread applicability of the methods still needs to be demonstrated.

2.2.2. De Novo Motif Discovery, Representation, and Validation

Binding sites of regulators evolved into often amazingly diverse sequences that pose major challenges for computational biology. Ten to fifteen years ago, binding sites were identified as overrepresented motifs in promoters of co-regulated genes in a single organism. Some motifs were easier to discover like palindromes where a sequence is identical with its reverse complement (CACGTG). Spaced dyads are associated with dimeric TFs. Co-regulated genes were identified from compendia of gene expression experiments [*see* **Chapters 23** and **24** and ref. (69)], ChIP, SELEX, PBM, and other experimental techniques. These methods find DNA segments that typically span much wider than the actual binding sites. ChIP-seq currently has a resolution of 50–200 bp (**Chapters 9** and **10**), and ChIP-chip has even more coarse resolution. ChIP-PCR experiments produce minimal flanking regions with scarce if any false positives. However, even decades of work generate relatively few sequences that poorly represent TFBS diversity. At the other extreme, gene co-expression analyses (**Chapters 23** and **24**) produce long lists of genes but binding sites need to be found in the promoter regions, possibly spanning over 1000 bp.

The task is to obtain a statistically representative sample of the variation including low-affinity but biologically important binding sites. Input to motif discovery is a set of overly long DNA sequences which contain the binding site for the TF in question. It is important to reduce flanking regions and false positive sequences as much as possible.

2.2.2.1. Computational and Visual Motif Representations

Motifs are concise representations of a set of TFBS. The simplest representation is the consensus sequence, where variable positions are shown in the IUPAC ambiguity code for nucleotides, for example, purines (A and G) are displayed as R, weak binders (A and T) by W, and N stands for any nucleotide. Consensus sequences, being only qualitative representations, cannot express important quantitative nucleotide preferences at a position.

The power of the computational representation is a key to the performance of motif discovery tools. First, we calculate positional frequency matrices (PFMs), which indicate the $P(b,i)$ probability of (di)nucleotide b at position i of the motif alignment. The background probability of nucleotide b is denoted by $P(b,0)$. In order to score a DNA segment for the motif, positional weight matrices (PWMs) are computed as follows. First, the (di)nucleotides b are counted at each position i. To avoid taking the logarithm of 0, some constant (typically 1) is added to each count. These values are divided by the number of sequences

An Overview of the Computational Analyses and Discovery   11

plus four times the constant. The base 2 logarithms of these ratios form the PWM (70).

2.2.2.2. k-mer (Word) Searches

The first group of motif discovery tools searches for k-mers (words) with c or fewer mismatches overrepresented as compared to the background. These alignment-free, deterministic searches are typically implemented as suffix tree algorithms. Suffix trees have been proven efficient for finding short k-mers with few mismatches in Weeder (71), the most sensitive and selective tool in Tompa et al.'s performance evaluations (7), and in the mismatch tree algorithm (MITRA) (72). Overlapping k-mers can be merged by graph theoretical methods in the WINNOWER (73) and cWINNOWER (74) tools. van Helden and colleagues (75) extended k-mer searches to include spaced dyads in a method accurate in yeast but less effective in higher eukaryotes. In general, k-mer search methods have the advantage of being rigorous and exhaustive but are less effective for long words and several mismatches than the probabilistic algorithms discussed below.

2.2.2.3. Probabilistic Motif Finding Algorithms

The second group of methods is typically based on either expectation maximization (76) or Gibbs search (77) as reviewed in **Chapter 6**.

2.2.2.3.1. Expectation Maximization

Expectation maximization (EM) (78) is a general statistical procedure that allows maximum likelihood estimates of parameters in probabilistic models depending on latent variables. Importantly, EM can make estimates even from incomplete data sets. For motif discovery, EM stems from progressive multiple alignments where the information content is being maximized. EM works with positional frequency matrices (PFMs), where $P(b,i)$ is the probability of (di)nucleotide b at position i. First, EM makes an initial guess for each $P(b,i)$ and also calculates the $P(b,0)$ background probabilities of nucleotides. In several iterations, the underlying multiple alignment and the $P(b,i)$ probabilities are refined so as to maximize the information in $P(b,i)$ relative to the background $P(b,0)$ (**Chapter 6**). Initial alignment is not necessary but the basic assumption is that each of the training sequences contains at least one occurrence of the motif.

Note that since the initial choice of the PFM determines the final outcome, it is prudent to improve this choice by restricting the length of input sequences to promoter regions and by assigning higher weights to alignments closer to the transcription start sites. When the strand bias of a TF is known, the search space may be limited to the preferred strand. Such choices are implemented in the motif elicitation by maximizing expectation (MEME) tool (76, 79–82). MEME can be instructed to remove the assumption that each input sequences contains the motif. Also, multiple occurrences of the same motifs within an input sequence can be handled. MEME also reports several different motifs.

2.2.2.3.2. Gibbs Sampling

To relieve from the bias of the initial PFM choice in EM, Charles Lawrence and colleagues (77) introduced random Gibbs sampling techniques. Starting from identical data with identical parameters, Gibbs sampling, unlike EM, typically ends up with different solutions, and the magnitude of these differences indicates the robustness of these solutions.

Gibbs sampling makes the assumption that each input sequence contains at least one occurrence of the motif and proceeds as follows. An input sequence is selected randomly and left out from the sample. From the remaining sequences, a random site is chosen and a PFM is calculated possibly by adding pseudocounts to avoid zero values. From this PFM and the background distributions, a positional weight matrix (PWM) is calculated (**Section 2.2.2.1** and **Chapter 6**). Each occurrence of the motif in the omitted sequence is scored using the PWM. Weighted by these scores, one of these sites is selected. Then some other sequence is left out, and a new PFM and the corresponding PWM are calculated. Iterations are performed until the score does not improve any more.

Note that Gibbs sampling cannot guarantee the global optimality of the solution. Therefore one has to perform several rounds of Gibbs sampling and analyze the convergence of solutions if any. Gibbs sampling was implemented, among others, in the BioProspector (83) and AlignACE (84) tools.

2.2.2.3.3. The Potential and the Limitations of EM and Gibbs Sampling

Both EM and Gibbs sampling have been proven useful for the de novo discovery of DNA motifs such as candidates for TFBS. It is important to note, however, that there is no guarantee that these methods find the motif because these methods depend on the presence of the motif in (almost) all input sequences, an overly strong constraint in noisy ChIP experiments for example (*see* **Chapter 11**). The lack of statistical significance in highly variable or short motifs may also lead to failure. PWMs are basically additive linear models of binding sites, while the free energy change during TF–DNA association maybe nonlinear (**Chapter 11**).

EM, Gibbs sampling, and other basic algorithms have been sophisticated in over 200 tools that apply a broad spectrum of models. For a comprehensive assessment of DNA motif finding algorithms, we recommend Das and Dai's review (85).

Until now, the installation and application of these tools had been a major burden for the users. In order to combine diverse statistical models and learning principles in a user-friendly, modular way, Ivo Grosse and colleagues (**Chapter 7**) developed *Jstacs*, an object-oriented Java framework for motif discovery. Importantly, the rapid and easy generation of diverse predictions allows the identification of those motifs that are reproduced by diverse models and learning principles and therefore more likely to represent biologically relevant TFBSs.

An Overview of the Computational Analyses and Discovery 13

2.2.2.4. Performance Estimates of the De Novo Motif Finding Algorithms

Over 200 tools have been published for the computational identification of DNA motifs including those of TFBS (*see* **Chapter 8** and http://biobase.ist.unomaha.edu/mediawiki/index.php/Main_Page). These methods perform the challenging statistical inference (generalization) from limited and noisy samples to a priori unknown sites. Evaluating their performance would require genomes where all binding sites were known but even in yeast, about half of the TFBS, particularly the weak binding sites, remains unknown (55). Such incomplete benchmark data sets unavoidably bias the evaluations of sensitivity and selectivity. While the numerical performance values remain low, some important lessons can be learnt. Quest and Ali (**Chapter 8** and ref. (86) introduced the motif tool assessment platform (MTAP) to assess the performance of over 20 motif discovery tools. Some tools excelled in a few motifs but over a diverse set of TFs and their binding sites, there was no single tool standing out in general performance. The balance between sensitivity and selectivity was compromised when using default parameter settings. This may indicate that certain binding sites may require specific learning principles, methods, and parameter settings that are not easily transferable to other motifs.

In a different study, Hu, Li, and Kihara (87) introduced an ensemble algorithm by combining prediction results from multiple runs of three heuristic motif discovery tools. This ensemble algorithm outperformed the popular MEME tool (82) by over 50% on the *E. coli* RegulonDB data set (88). Although prediction performance results obtained in bacterial or yeast genomes are difficult to scale up for much larger mammalian genomes, this finding supports the expert recommendation to analyze the same data set by using multiple tools and diverse parameter settings. It is also advisable to pursue not only the best hit but also the few top motifs (7). Most importantly, the highest Mathews correlation coefficient of 0.37 (**Chapter 8**) indicates low-to-medium prediction accuracy and calls for utilizing several additional lines of evidence as discussed below.

2.3. Supporting Evidence for TFBS Predictions

The most important evidence is the potential evolutionary conservation of regulatory sites in closely related genomes (**Section 2.3.1**). Binding sites also tend to cluster into *cis*-regulatory modules (**Section 2.3.2**). Further support comes from spatial correlation of transposable DNA elements and regulatory sites (**Section 2.3.3**).

2.3.1. Phylogenetic Footprinting

Potential evolutionary conservation of binding sites in orthologous promoter regions of closely related organisms increases the confidence in binding site predictions. This approach is termed phylogenetic footprinting (89). In contrast to ChIP, PBMs, and co-regulated genes, phylogenetic footprinting can work even on a single gene provided that sequences are available from multiple related species. The performance of footprinting is greatly improved by assigning weights to segment pairs in the function of evolutionary distance. The power of phylogenetic footprinting is remarkable in the case of TFBS discovery from ChIP-chip data in yeast. The original analyses using six motif discovery tools were limited to a single species, *Saccharomyces cerevisiae* (26). Later, phylogenetic footprinting over several yeast genomes using Phylocon and Converge allowed to create an improved map of the conserved regulatory sites (90). Binding sites for an additional 36 TFs, and in total, 636 novel regulatory interactions were identified. Phylogenetic footprinting is implemented in a number of state-of-the-art tools including PhyME (91), PhyloGibbs (92), and MITRA (72).

2.3.2. cis-Regulatory Modules

In multicellular organisms, several TFs regulate a typical gene and a TF may regulate a number of functionally related genes. The binding sites of such co-regulated genes are frequently organized into clusters termed *cis*-regulatory modules (CRMs) (93). These clusters are often conserved during evolution. The resulting spatial correlations among CRMs greatly increase the statistical power and confidence in TFBS discovery substantially as compared to predictions of individual binding sites (94). To discover new CRMs, researchers start with alignments of related genomes (e.g., those of mammalian, *Drosophila*, or yeast species) using, for example, the MULTIZ data sets (95). Either the alignments or the individual sequences are scored against known PWMs of TFBS, third-order Markov models are applied, and species-specific scores are calculated. Score significance is evaluated by the permutation test and subjected to multiple test corrections. In **Chapter 13**, Sebastian Schultheiss presents KIRMES, a support vector machine-based package for the large-scale predictions of CRMs. Gene sets that share CRMs may be compared with sets of genes co-expressed in large compendia of transcriptional profiling experiments like GNF Atlas II. Co-localization with DNase I hypersensitivity regions (18) further increases the confidence in the predicted CRMs.

2.3.3. Propagation of TFBS with Transposable Elements and Their Spatial Correlations

Transposable elements (TEs) are DNA segments that can frequently "transpose" from one genomic locus to another. TEs are present in all domains of life and account as much as for ~46% of the human genome (29). TEs are not only able to promote their own transcription but can provide alternative promoters to genes [**Chapter 14** and refs. (96, 97)]. Numerous TEs adapted to propagate new TFBS in the host genome (98). Most notably, TEs proliferated considerable parts of the *c-myc* (99) and p53 (100) regulatory networks.

An Overview of the Computational Analyses and Discovery   15

Chapter 14 provides an example of propagation of CCCTC-binding factor (CTCF) binding sites by TEs based on ChIP-seq data. One can predict potential binding sites as follows. Map sequencing reads to the genome by Bowtie (36). Then we rescue reads and perform the probabilistic assignments of multiple mapping reads using MuMRescueLite (101). Peaks can be called, among others, by the SISSRs package (52). Then the intersection of TE and TFBS locations is calculated using the University of California Santa Cruz Genome Browser tracks [*see* **Section 2.5** and ref. (102)]. Intersections can be found by a simple Structured Query Language query in a relational database or in any programming language, e.g., PERL.

2.4. Databases of TFS and Their Binding Sites

Computational analyses, systematic querying, and integration with diverse genic, genomic, and epigenomic observations require efficient (relational) databases. The scientific community attempts to annotate and organize comprehensive information about transcriptional regulation. The classic TRANSFAC (12, 13) and JASPAR (103) databases are focused on PFMs, PWMs, sequence logos, motifs, and their genomic coordinates. The latest release of JASPAR now also holds ChIP-chip and ChIP-seq data. Databases like the Open REGulatory ANNOtation [OregAnno, **Chapter 20** and ref. (104)] and PAZAR (105) systematically attempt to collect and organize quality information for high-throughput experiments, literature citations, text mining, expression, evolutionary conservation, and cellular reporter gene assays. **Chapters 2**, **20**, and **22** review regulatory databases with special emphasis on plant regulators (**Chapters 2** and **20**). *cis*-Lexicon and the Virtual Sea Urchin database tool is introduced by Sorin Istrail and colleagues in **Chapter 22**. SELEX_DB (59) stores high-quality Systematic Evolution of Ligands by EXponential enrichment (SELEX, **Chapter 12**) data and larger data sets obtained by next-generation sequencing can be retrieved from the HTPSELEX database (60).

2.5. Querying and Visualization Using Genome Browsers and Their Databases

Genome browsers (GBs) visually display a rich context of regulatory regions, DNase hypersensitive areas, genic and genomic landmarks, repetitive DNA, conserved regions, polymorphisms, and numerous other features. TFBS locations or even the locations of all individual ChIP-seq reads can be visualized in GBs. These displays facilitate the contextual analyses of TFBS with repetitive DNA elements, binding sites of other TFs, nucleosomes (106), or promoter, and other regulatory regions. GBs also help us to compare peak characteristics at likely sites (e.g., promoters) and unlikely regions (e.g., exons).

The most widely used GB was developed and is maintained at the University of California Santa Cruz (UCSC)

(107) (http://genome.ucsc.edu). GBrowse (108) is a BioPERL-based tool, the favorite choice of plant scientists. Hundreds of genomes are displayed in the ENSEMBL Browser at the European Bioinformatics Institute (109, 110) (http://www.ensembl.org/info/about/species.html). CisGenome [**Chapter 9** and ref. (22)] and Eagleview (111) help peak calling using coverage by visualizing sequencing reads separately at each strands. In **Chapter 22**, Sorin Istrail and colleagues demonstrate the *cis*-GRN-Browser specifically designed for the annotation and investigation of gene regulatory networks (GRNs). GRNs cannot be displayed in the usual one-dimensional browsers. Therefore the above authors also introduced the Virtual Sea Urchin system, a 4D interactive tool that visualizes the genomic regulatory network of the sea urchin embryo development in space and time.

All of the above GBs interface to underlying relational databases. Users can freely download the database tables in order to build their local MySQL or other database implementations for comprehensive statistical analyses. Researchers can also upload their own annotation tracks to the central UCSC server or a local implementation of the Browser directly on the "Custom Track" pages as described in **Chapter 10**. Using this opportunity, QuEST (50) and several other peak calling tools prepare input files for the UCSC Genome Browser (107).

2.6. Reconstruction of Transcriptional Regulatory Networks

Transcriptional regulation works in hierarchical, dynamic networks that change in response to environmental perturbations or internal stimuli (93, 112). TF binding can be mathematically represented as the edges of directed graphs connecting the vertices, TFs, and regulatory regions of genes. Mapping binding sites provides only qualitative information inadequate to predict the expression level of a gene. More quantitative information is provided by co-expression in large compendia of gene expression experiments (69). From co-expression under diverse conditions, we infer to sharing identical regulators. These agents may include TFs, microRNAs (8), DNA methylation (9), and specific histone modifications (10, 11). Lagged correlations in time series experiments may indicate regulatory relationships (113).

Simple linear network inference like clustering based on correlation coefficients is hindered by AND, OR, EXCLUSIVE OR and other nonlinear regulatory relationships. Such nonlinear dependencies can be captured using probabilistic methods including B-splines, clustering hidden Markov models, and most notably, dynamic Bayesian networks (**Chapters 23** and **24**). Probabilistic methods are more tolerant to errors so prominent in microarray experiments but require more sophisticated methods than correlation-based techniques. Bayesian networks can reveal causal relationships and allow the integration of heterogeneous data like ChIP-seq and protein–protein interactions. The first part of a Bayesian networks is a directed acyclic graph representing conditional independent relationships among nodes (TFs and genes). The second part is a set of parameters, which specify the conditional distribution for each TF and gene.

To reconstruct regulatory networks, Luo and Woolf (**Chapter 23**) propose three-way mutual information. 3MI measures the improvement in predictability when three variables are analyzed jointly versus considering them separately. Enumerating mutual information for each possible gene triplets, first local networks are built. These local structures are assembled into the global regulatory network.

3. Conclusions

The computational discovery of *cis*-regulatory modules typically produces acceptable results. For individual binding sites, motif finding results have to be combined with evolutionary conservation information, transposable elements, and experimental data. The latter includes ChIP-seq, protein-binding microarrays, SELEX, co-expression analyses, knockout mutants and knock-down by RNA interference, and protein–protein interactions. Note that many of these techniques generate very noisy observations and peaks derived from ChIP-seq and ChIP-chip observation which span much wider than the actual binding sites. Therefore, most experimental observations require filtering by computational motif discovery tools.

Although massive amounts of data are available in a wide array of databases, about half of the TFBS remains unknown even in such a primitive eukaryote as yeast. Major improvements are expected from the integration of all reliable observations and predictions. Even smaller current data sets allowed us to reconstruct sophisticated transcriptional regulatory networks in bacteria, and subnetworks in higher organisms.

Acknowledgments

The author is grateful to Yang Liu for the critical reading of the manuscript, and the NSF Grant EPS-0701892 for funding.

References

1. Maniatis, T., Ptashne, M., Backman, K. et al. (1975) Recognition sequences of repressor and polymerase in the operators of bacteriophage lambda. *Cell* 5, 109–113.
2. Stormo, G.D. (2000) DNA binding sites: representation and discovery. *Bioinformatics* 16, 16–23.
3. Cawley, S., Bekiranov, S., Ng, H.H. et al. (2004) Unbiased mapping of transcription factor binding sites along human chromosomes 21 and 22 points to widespread regulation of noncoding RNAs. *Cell* 116, 499–509.
4. Wei, C.L., Wu, Q., Vega, V.B. et al. (2006) A global map of p53 transcription-factor binding sites in the human genome. *Cell* 124, 207–219.
5. Nielsen, R., Pedersen, T.A., Hagenbeek, D. et al. (2008) Genome-wide profiling of PPARgamma:RXR and RNA polymerase II occupancy reveals temporal activation of distinct metabolic pathways and changes in RXR dimer composition during adipogenesis. *Genes Dev* 22, 2953–2967.
6. Hamza, M.S., Pott, S., Vega, V.B. et al. (2009) De-novo identification of PPARgamma/RXR binding sites and direct targets during adipogenesis. *PLoS One* 4, e4907.
7. Tompa, M., Li, N., Bailey, T.L. et al. (2005) Assessing computational tools for the discovery of transcription factor binding sites. *Nat Biotechnol* 23, 137–144.
8. Khan, A.A., Betel, D., Miller, M.L. et al. (2009) Transfection of small RNAs globally perturbs gene regulation by endogenous microRNAs. *Nat Biotechnol* 27, 549–555.
9. Jaenisch, R., and Bird, A. (2003) Epigenetic regulation of gene expression: how the genome integrates intrinsic and environmental signals. *Nat Genet* 33(Suppl.), 245–254.
10. Ito, T. (2007) Role of histone modification in chromatin dynamics. *J Biochem* 141, 609–614.
11. Barski, A., and Zhao, K. (2009) Genomic location analysis by ChIP-Seq. *J Cell Biochem* 107, 11–18.
12. Matys, V., Fricke, E., Geffers, R. et al. (2003) TRANSFAC: transcriptional regulation, from patterns to profiles. *Nucleic Acids Res* 31, 374–378.
13. Matys, V., Kel-Margoulis, O.V., Fricke, E. et al. (2006) TRANSFAC(R) and its module TRANSCompel(R): transcriptional gene regulation in eukaryotes. *Nucleic Acids Res* 34, D108–D110.
14. Robison, K., McGuire, A.M., and Church, G.M. (1998) A comprehensive library of DNA-binding site matrices for 55 proteins applied to the complete Escherichia coli K-12 genome. *J Mol Biol* 284, 241–254.
15. Liu, J., and Stormo, G.D. (2005) Combining SELEX with quantitative assays to rapidly obtain accurate models of protein-DNA interactions. *Nucleic Acids Res* 33, e141.
16. Djordjevic, M., and Sengupta, A.M. (2006) Quantitative modeling and data analysis of SELEX experiments. *Phys Biol* 3, 13–28.
17. Berger, M.F., and Bulyk, M.L. (2009) Universal protein-binding microarrays for the comprehensive characterization of the DNA-binding specificities of transcription factors. *Nat Protoc* 4, 393–411.
18. Hesselberth, J.R., Chen, X., Zhang, Z. et al. (2009) Global mapping of protein-DNA interactions in vivo by digital genomic footprinting. *Nat Methods* 6, 283–289.
19. Sabo, P.J., Humbert, R., Hawrylycz, M. et al. (2004) Genome-wide identification of DNaseI hypersensitive sites using active chromatin sequence libraries. *Proc Natl Acad Sci* 101, 4537–4542.
20. Workman, C.T., Mak, H.C., McCuine, S. et al. (2006) A systems approach to mapping DNA damage response pathways. *Science* 312, 1054–1059.
21. Elbashir, S.M., Harborth, J., Weber, K. et al. (2002) Analysis of gene function in somatic mammalian cells using small interfering RNAs. *Methods* 26, 199–213.
22. Ji, H., Jiang, H., Ma, W. et al. (2008) An integrated software system for analyzing ChIP-chip and ChIP-seq data. *Nat Biotechnol* 26, 1293–1300.
23. MacIsaac, K.D., and Fraenkel, E. (2006) Practical strategies for discovering regulatory DNA sequence motifs. *PLoS Comput Biol* 2, e36.
24. Viggiani, C.J., Aparicio, J.G., and Aparicio, O.M. (2009) ChIP-chip to analyze the binding of replication proteins to chromatin using oligonucleotide DNA microarrays. *Methods Mol Biol* 521, 255–278.
25. ENCODE Consortium. (2007) Identification and analysis of functional elements in 1% of the human genome by the ENCODE pilot project. *Nature* 447, 799–816.
26. Harbison, C.T., Gordon, D.B., Lee, T.I. et al. (2004) Transcriptional regulatory code of a eukaryotic genome. *Nature* 431, 99–104.

27. Ren, B., Robert, F., Wyrick, J.J. et al. (2000) Genome-Wide Location and Function of DNA Binding Proteins. *Science* 290, 2306–2309.

28. Johnson, D.S., Li, W., Gordon, D.B. et al. (2008) Systematic evaluation of variability in ChIP-chip experiments using pre-defined DNA targets. *Genome Res* 18, 393–403.

29. Lander, E.S., Linton, L.M., Birren, B. et al. (2001) Initial sequencing and analysis of the human genome. *Nature* 409, 860–921.

30. Quail, M.A., Kozarewa, I., Smith, F. et al. (2008) A large genome center's improvements to the Illumina sequencing system. *Nat Methods* 5, 1005–1010.

31. Margulies, M., Egholm, M., Altman, W.E. et al. (2005) Genome sequencing in micro-fabricated high-density picolitre reactors. *Nature* 437, 376–380.

32. Rougemont, J., Amzallag, A., Iseli, C. et al. (2008) Probabilistic base calling of Solexa sequencing data. *BMC Bioinformatics* 9, 431.

33. Erlich, Y., Mitra, P.P., delaBastide, M. et al. (2008) Alta-Cyclic: a self-optimizing base caller for next-generation sequencing. *Nat Methods* 5, 679–682.

34. Altschul, S.F., Madden, T.L., Schaffer, A.A. et al. (1997) Gapped BLAST and PSI-BLAST: a new generation of protein database search programs. *Nucleic Acids Res* 25, 3389–3402.

35. Kent, W.J. (2002) BLAT – the BLAST-like alignment tool. *Genome Res* 12, 656–664.

36. Langmead, B., Trapnell, C., Pop, M. et al. (2009) Ultrafast and memory-efficient alignment of short DNA sequences to the human genome. *Genome Biol* 10, R25.

37. Trapnell, C., and Salzberg, S.L. (2009) How to map billions of short reads onto genomes. *Nat Biotechnol* 27, 455–457.

38. Ozsolak, F., Platt, A.R., Jones, D.R. et al. (2009) Direct RNA sequencing. *Nature* 461, 814–818.

39. Dohm, J.C., Lottaz, C., Borodina, T. et al. (2008) Substantial biases in ultra-short read data sets from high-throughput DNA sequencing. *Nucleic Acids Res* 36, e105.

40. Hillier, L.W., Marth, G.T., Quinlan, A.R. et al. (2008) Whole-genome sequencing and variant discovery in C. elegans. *Nat Methods* 5, 183–188.

41. Vega, V.B., Cheung, E., Palanisamy, N. et al. (2009) Inherent signals in sequencing-based Chromatin-ImmunoPrecipitation control libraries. *PLoS One* 4, e5241.

42. Albert, I., Mavrich, T.N., Tomsho, L.P. et al. (2007) Translational and rotational settings of H2A.Z nucleosomes across the *Sac-charomyces cerevisiae* genome. *Nature* 446, 572–576.

43. Pepke, S., Wold, B., and Mortazavi, A. (2009) Computation for ChIP-seq and RNA-seq studies. *Nat Methods* 6, S22–S32.

44. Miller, M. (2009) The importance of being flexible: the case of basic region leucine zipper transcriptional regulators. *Curr Protein Pept Sci* 10, 244–269.

45. Yamada, K., and Miyamoto, K. (2005) Basic helix-loop-helix transcription factors, BHLHB2 and BHLHB3; their gene expressions are regulated by multiple extracellular stimuli. *Front Biosci* 10, 3151–3171.

46. Ladomery, M., and Dellaire, G. (2002) Multifunctional zinc finger proteins in development and disease. *Ann Hum Genet* 66, 331–342.

47. Klinck, R., Serup, P., Madsen, O.D. et al. (2008) Specificity of four monoclonal anti-NKx6-1 antibodies. *J Histochem Cytochem* 56, 415–424.

48. Benjamini, Y., and Hochberg, Y. (1995) Controlling the false discovery rate: a practical and powerful approach to multiple hypothesis testing. *J R Stat Soc B* 57, 289–300.

49. Zhang, Y., Liu, T., Meyer, C.A. et al. (2008) Model-based analysis of ChIP-Seq (MACS). *Genome Biol* 9, R137.

50. Valouev, A., Johnson, D.S., Sundquist, A. et al. (2008) Genome-wide analysis of transcription factor binding sites based on ChIP-Seq data. *Nat Methods* 5, 829–834.

51. Fejes, A.P., Robertson, G., Bilenky, M. et al. (2008) FindPeaks 3.1: a tool for identifying areas of enrichment from massively parallel short-read sequencing technology. *Bioinformatics* 24, 1729–1730.

52. Jothi, R., Cuddapah, S., Barski, A. et al. (2008) Genome-wide identification of in vivo protein-DNA binding sites from ChIP-Seq data. *Nucleic Acids Res* 36, 5221–5231.

53. Nix, D.A., Courdy, S.J., and Boucher, K.M. (2008) Empirical methods for controlling false positives and estimating confidence in ChIP-Seq peaks. *BMC Bioinformatics* 9, 523.

54. Rozowsky, J., Euskirchen, G., Auerbach, R.K. et al. (2009) PeakSeq enables systematic scoring of ChIP-seq experiments relative to controls. *Nat Biotechnol* 27, 66–75.

55. Zhu, C., Byers, K.J., McCord, R.P. et al. (2009) High-resolution DNA-binding specificity analysis of yeast transcription factors. *Genome Res* 19, 556–566.

56. Oliphant, A.R., Brandl, C.J., and Struhl, K. (1989) Defining the sequence specificity of DNA-binding proteins by selecting binding sites from random-sequence

oligonucleotides: analysis of yeast GCN4 protein. *Mol Cell Biol* 9, 2944–2949.

57. Tuerk, C., and Gold, L. (1990) Systematic evolution of ligands by exponential enrichment: RNA ligands to bacteriophage T4 DNA polymerase. *Science* 249, 505–510.

58. Djordjevic, M. (2007) SELEX experiments: new prospects, applications and data analysis in inferring regulatory pathways. *Biomol Eng* 24, 179–189.

59. Ponomarenko, J.V., Orlova, G.V., Frolov, A.S. et al. (2002) SELEX_DB: a database on in vitro selected oligomers adapted for recognizing natural sites and for analyzing both SNPs and site-directed mutagenesis data. *Nucleic Acids Res* 30, 195–199.

60. Jagannathan, V., Roulet, E., Delorenzi, M. et al. (2006) HTPSELEX – a database of high-throughput SELEX libraries for transcription factor binding sites. *Nucleic Acids Res* 34, D90–D94.

61. Bulyk, M.L., Huang, X., Choo, Y. et al. (2001) Exploring the DNA-binding specificities of zinc fingers with DNA microarrays. *Proc Natl Acad Sci USA* 98, 7158–7163.

62. Philippakis, A.A., Qureshi, A.M., Berger, M.F. et al. (2008) Design of compact, universal DNA microarrays for protein binding microarray experiments. *J Comput Biol* 15, 655–665.

63. Forde, G.M. (2008) Preparation, analysis and use of an affinity adsorbent for the purification of GST fusion protein. *Methods Mol Biol* 421, 125–136.

64. McCord, R.P., Berger, M.F., Philippakis, A.A. et al. (2007) Inferring condition-specific transcription factor function from DNA binding and gene expression data. *Mol Syst Biol* 3, 100.

65. Choi, Y., Qin, Y., Berger, M.F. et al. (2007) Microarray analyses of newborn mouse ovaries lacking Nobox. *Biol Reprod* 77, 312–319.

66. Hughes, J.D., Estep, P.W., Tavazoie, S. et al. (2000) Computational identification of cis-regulatory elements associated with groups of functionally related genes in *Saccharomyces cerevisiae*. *J Mol Biol* 296, 1205–1214.

67. Liu, L.A., and Bader, J.S. (2009) Structure-based ab initio prediction of transcription factor-binding sites. *Methods Mol Biol* 541, 23–41.

68. Liu, L.A., and Bader, J.S. (2007) Ab initio prediction of transcription factor binding sites. *Pac Symp Biocomput* 12, 484–495.

69. Hughes, T.R., Marton, M.J., Jones, A.R. et al. (2000) Functional discovery via a compendium of expression profiles. *Cell* 102, 109–126.

70. Staden, R. (1984) Computer methods to locate signals in nucleic acid sequences. *Nucleic Acids Res* 12, 505–519.

71. Pavesi, G., Mauri, G., and Pesole, G. (2001) An algorithm for finding signals of unknown length in DNA sequences. *Bioinformatics* 17(Suppl. 1), S207–S214.

72. Eskin, E., and Pevzner, P.A. (2002) Finding composite regulatory patterns in DNA sequences. *Bioinformatics* 18(Suppl 1), S354–S363.

73. Pevzner, P.A., and Sze, S.H. (2000) Combinatorial approaches to finding subtle signals in DNA sequences. *Proc Int Conf Intell Syst Mol Biol* 8, 269–278.

74. Liang, S. (2003) cWINNOWER algorithm for finding fuzzy DNA motifs. *Proc IEEE Comput Soc Bioinformatics Conf* 2, 260–265.

75. van Helden, J., Andre, B., and Collado-Vides, J. (1998) Extracting regulatory sites from the upstream region of yeast genes by computational analysis of oligonucleotide frequencies. *J Mol Biol* 281, 827–842.

76. Bailey, T.L., and Elkan, C. (1994) Fitting a mixture model by expectation maximization to discover motifs in biopolymers. *Proc Int Conf Intell Syst Mol Biol* 2, 28–36.

77. Lawrence, C.E., Altschul, S.F., Boguski, M.S. et al. (1993) Detecting subtle sequence signals: a Gibbs sampling strategy for multiple alignment. *Science* 262, 208–214.

78. Dempster, A.P., Laird, N.M., and Rubin, D.B. (1977) Maximum Likelihood from Incomplete Data via the EM Algorithm. *J R Soc Ser B* 39, 1–38.

79. Grundy, W.N., Bailey, T.L., and Elkan, C.P. (1996) ParaMEME: a parallel implementation and a web interface for a DNA and protein motif discovery tool. *Comput Appl Biosci* 12, 303–310.

80. Grundy, W.N., Bailey, T.L., Elkan, C.P. et al. (1997) Meta-MEME: motif-based hidden Markov models of protein families. *Comput Appl Biosci* 13, 397–406.

81. Price, A., Ramabhadran, S., and Pevzner, P.A. (2003) Finding subtle motifs by branching from sample strings. *Bioinformatics* 19(Suppl. 2), ii149–ii155.

82. Bailey, T.L., Boden, M., Buske, F.A. et al. (2009) MEME SUITE: tools for motif discovery and searching. *Nucleic Acids Res* 37, W202–W208.

83. Liu, X.S., Brutlag, D.L., and Liu, J.S. (2002) An algorithm for finding protein-DNA binding sites with applications to chromatin-immunoprecipitation microarray experiments. *Nat Biotechnol* 20, 835–839.

84. Roth, F.P., Hughes, J.D., Estep, P.W. et al. (1998) Finding DNA regulatory motifs

within unaligned noncoding sequences clustered by whole-genome mRNA quantitation. *Nat Biotechnol* 16, 939–945.

85. Das, M.K., and Dai, H.K. (2007) A survey of DNA motif finding algorithms. *BMC Bioinformatics* 8(Suppl. 7), S21.

86. Quest, D., Dempsey, K., Shafiullah, M. et al. (2008) MTAP: the motif tool assessment platform. *BMC Bioinformatics* 9(Suppl. 9), S6.

87. Hu, J., Li, B., and Kihara, D. (2005) Limitations and potentials of current motif discovery algorithms. *Nucleic Acids Res* 33, 4899–4913.

88. Gama-Castro, S., Jimenez-Jacinto, V., Peralta-Gil, M. et al. (2008) RegulonDB (version 6.0): gene regulation model of Escherichia coli K-12 beyond transcription, active (experimental) annotated promoters and Textpresso navigation. *Nucleic Acids Res* 36, D120–D124.

89. Wasserman, W.W., and Fickett, J.W. (1998) Identification of regulatory regions which confer muscle-specific gene expression. *J Mol Biol* 278, 167–181.

90. MacIsaac, K.D., Wang, T., Gordon, D.B. et al. (2006) An improved map of conserved regulatory sites for *Saccharomyces cerevisiae*. *BMC Bioinformatics* 7, 113.

91. Sinha, S., Blanchette, M., and Tompa, M. (2004) PhyME: a probabilistic algorithm for finding motifs in sets of orthologous sequences. *BMC Bioinformatics* 5, 170.

92. Siddharthan, R., Siggia, E.D., and van Nimwegen, E. (2005) PhyloGibbs: a Gibbs sampling motif finder that incorporates phylogeny. *PLoS Comput Biol* 1, e67.

93. Davidson, E.H. (2001) Genomic regulatory systems: development and evolution. Academic Press, New York, NY.

94. Blanchette, M., Bataille, A.R., Chen, X. et al. (2006) Genome-wide computational prediction of transcriptional regulatory modules reveals new insights into human gene expression. *Genome Res* 16, 656–668.

95. Blanchette, M., Kent, W.J., Riemer, C. et al. (2004) Aligning multiple genomic sequences with the threaded blockset aligner. *Genome Res* 14, 708–715.

96. Cohen, C.J., Lock, W.M., and Mager, D.L. (2009) Endogenous retroviral LTRs as promoters for human genes: a critical assessment. *Gene* 448, 105–114.

97. Conley, A.B., Piriyapongsa, J., and Jordan, I.K. (2008) Retroviral promoters in the human genome. *Bioinformatics* 24, 1563–1567.

98. Feschotte, C. (2008) Transposable elements and the evolution of regulatory networks. *Nat Rev Genet* 9, 397–405.

99. Wang, J., Bowen, N.J., Chang, L. et al. (2009) A c-Myc regulatory subnetwork from human transposable element sequences. *Mol Biosyst* 5, 1831–1839.

100. Wang, T., Zeng, J., Lowe, C.B. et al. (2007) Species-specific endogenous retroviruses shape the transcriptional network of the human tumor suppressor protein p53. *Proc Natl Acad Sci USA* 104, 18613–18618.

101. Hashimoto, T., de Hoon, M.J., Grimmond, S.M. et al. (2009) Probabilistic resolution of multi-mapping reads in massively parallel sequencing data using MuMRescueLite. *Bioinformatics* 25, 2613–2614.

102. Rhead, B., Karolchik, D., Kuhn, R.M. et al. (2009) The UCSC genome browser database: update 2010. *Nucleic Acids Res*, doi:10.1093/nar/gkp1939.

103. Portales-Casamar, E., Thongjuea, S., Kwon, A.T. et al. JASPAR 2010: the greatly expanded open-access database of transcription factor binding profiles. *Nucleic Acids Res* 38, D105–D110.

104. Griffith, O.L., Montgomery, S.B., Bernier, B. et al. (2008) ORegAnno: an open-access community-driven resource for regulatory annotation. *Nucleic Acids Res* 36, D107–D113.

105. Portales-Casamar, E., Arenillas, D., Lim, J. et al. (2009) The PAZAR database of gene regulatory information coupled to the ORCA toolkit for the study of regulatory sequences. *Nucleic Acids Res* 37, D54–D60.

106. Wang, J., and Morigen. (2009) BayesPI – a new model to study protein-DNA interactions: a case study of condition-specific protein binding parameters for Yeast transcription factors. *BMC Bioinformatics* 10, 345.

107. Kuhn, R.M., Karolchik, D., Zweig, A.S. et al. (2009) The UCSC Genome Browser Database: update 2009. *Nucleic Acids Res* 37, D755–D761.

108. Stein, L.D., Mungall, C., Shu, S. et al. (2002) The generic genome browser: a building block for a model organism system database. *Genome Res* 12, 1599–1610.

109. Spudich, G., Fernandez-Suarez, X.M., and Birney, E. (2007) Genome browsing with Ensembl: a practical overview. *Brief Funct Genomic Proteomic* 6, 202–219.

110. James, N., Graham, N., Clements, D. et al. (2007) AtEnsEMBL. *Methods Mol Biol* 406, 213–227.

111. Huang, W., and Marth, G. (2008) EagleView: a genome assembly viewer for next-

generation sequencing technologies. *Genome Res* 18, 1538–1543.

112. Balazsi, G., Barabasi, A.L., and Oltvai, Z.N. (2005) Topological units of environmental signal processing in the transcriptional regulatory network of Escherichia coli. *Proc Natl Acad Sci USA* 102, 7841–7846.

113. Qian, J., Dolled-Filhart, M., Lin, J. et al. (2001) Beyond synexpression relationships: local clustering of time-shifted and inverted gene expression profiles identifies new, biologically relevant interactions. *J Mol Biol* 314, 1053–1066.

Chapter 2

Components and Mechanisms of Regulation of Gene Expression

Alper Yilmaz and Erich Grotewold

Abstract

The control of gene expression is a biological process essential to all organisms. This is accomplished through the interaction of regulatory proteins with specific DNA motifs in the control regions of the genes that they regulate. Upon binding to DNA, and through specific protein–protein interactions, these regulatory proteins convey signals to the basal transcriptional machinery, containing the respective RNA polymerases, resulting in particular rates of gene expression. In eukaryotes, in addition and complementary to the binding of regulatory proteins to DNA, chromatin structure plays a role in modulating gene expression. Small RNAs are emerging as key components in this process. This chapter provides an introduction to some of the basic players participating in these processes, the transcription factors and co-regulators, the *cis*-regulatory elements that often function as transcription factor docking sites, and the emerging role of small RNAs in the regulation of gene expression.

Key words: Promoter, DNA-binding, operon, *cis*-regulatory element, microRNA, small interfering RNA.

---

1. Introduction

Cells can be considered as membrane-enclosed environments in which many different proteins undertake one or several specific functions. Thus, the proper development and the functional integration of cells within an organism depend on controlling the accumulation of these proteins within some defined concentration restrictions, which are space and time dependent. Consistent with the central dogma of biology, which states that the genetic information flow is, in general terms, from DNA to RNA and then to the proteins, the instructions on how much and when a

I. Ladunga (ed.), *Computational Biology of Transcription Factor Binding*, Methods in Molecular Biology 674, DOI 10.1007/978-1-60761-854-6_2, © Springer Science+Business Media, LLC 2010

protein needs to be made are encoded in the DNA. The process of transcription transfers the code responsible for making proteins, the cell workhorses, from the DNA to RNA and translation converts a messenger RNA (mRNA) sequence into a sequence of amino acids in a protein. Thus, protein levels can be controlled at multiple stages, including transcription, translation as well as mRNA and protein transport and stability. This chapter will primarily focus on the control mechanisms associated with transcription and responsible for how much mRNA is being made for each of the thousands (or tens of thousands) protein-encoding genes in a cell.

2. Description

2.1. Mechanisms of Transcription

In simple terms, the process of transcription involves the unwinding double stranded DNA and the chemical synthesis of RNA, using one of the two genomic DNA strands as the template for the RNA sequence. This is achieved by DNA-dependent RNA polymerases (RNAP). In prokaryotes, there is a single type of RNAP, which is responsible for the generation of various types of RNA, such as messenger RNA (mRNA), transfer RNA (tRNA), and ribosomal RNA (rRNA). In eukaryotes, however, there are multiple RNAPs, each specialized in the production of particular types of RNA species. For example, RNAP I synthesizes rRNAs, RNAP II synthesizes mRNAs, and RNAP III synthesizes tRNAs. In addition, there are other RNAP with functions more restricted to particular kingdoms. For example, in plants, RNAP IV synthesizes small interfering RNA (siRNAs) (1, 2) and RNAP V transcribes intergenic and non-coding sequences, participating in the small interfering RNA(siRNA)-mediated transcriptional gene silencing (TGS) (3, 4).

To ensure proper gene expression levels, the activity of prokaryotic RNAP and eukaryotic RNAP II, in particular, are subjected to tight control. One of the best-studied mechanisms involved in regulating RNAP II activity is through the effect of transcription factors (TFs), which specify when and where RNAP II (and associated factors) is tethered to DNA, how RNAP II initiates (and re-initiates once a round of mRNA formation has been completed) transcription, and elongates nascent mRNAs. We define here TFs as proteins that bind DNA in a sequence-specific fashion to particular DNA sequences (*cis*-regulatory elements) located in the regulatory regions of the genes that they control. This definition excludes the large number of proteins that can affect gene expression without binding to specific DNA sequences. As these proteins often function by modulating the action of specific DNA-binding TFs, there are few common characteristics that permit their easy identification.

TFs are usually classified into families, based on the presence of specific structures in their DNA-binding or protein–protein interaction domains. In vitro, TFs usually recognize DNA sequences 6–8-bp long, length that is clearly insufficient for the exquisite regulatory specificity that they display in vivo, suggesting that large number of TFs form the active regulatory complexes and providing the bases for the principle of combinatorial gene regulation (5).

In prokaryotes, binding of RNAP to specific regions is achieved by a particular protein factor, the sigma (σ) subunit. This prokaryotic TF increases the affinity of RNAP to certain promoter regions while decreasing its affinity to non-specific DNA. The σ factor responsible for the regulation of most "housekeeping" genes in *Escherichia coli* is σ^{70} and σ^A in *Bacillus subtilis*, which are responsible for initiating transcription from most promoters. Other σ factors are usually stress induced, to allow organisms to become virulent or adapt to any number of environmental changes such as hyperosmolarity, heat shock, oxidative stress, nutrient deprivation, and variations in pH (6, 7).

2.2. Organization of Gene Regulatory Sequences

2.2.1. Operons and Other Gene Clusters

One strategy by which prokaryotic organisms control the expression of genes that participate in a common process is to group the genes into operons, which are usually transcribed from a unique promoter resulting in a single (poly-cistronic) mRNA that is translated into multiple proteins, allowing the cell to streamline the control of transcription. Here, we describe the *lac* operon as an archetypical bacterial operon, as an example of how prokaryotes negotiate the control of gene expression (**Fig. 2.1**).

The *lac* operon encodes for three enzymes (*lacZ* encoding β-galactosidase, *lacY* encoding a lactose permease, and *lacA* encoding a trans-acetylase) necessary for the uptake and metabolism of lactose. Only when lactose but no glucose, a more favorable carbon source, is present in the environment, the *lac* operon is expressed. When grown in glucose, for example, regardless of whether lactose is present or not, the *lacZYA* genes are not expressed, a consequence of a repressor protein (*lac* repressor) recognizing the operator sequence of the operon regulatory region, preventing the recruitment of RNAP to the DNA. When lactose is present, this small molecule recognizes the *lac* repressor, preventing it from binding the operator sequence.

In eukaryotes, operon-like structures have been described, although they clearly differ from bacterial operons, since they

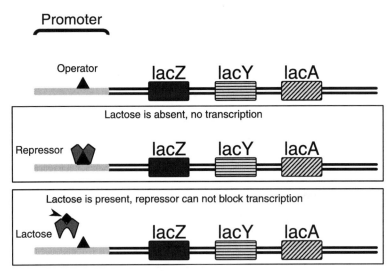

Fig. 2.1. Single RNAP transcribes multiple genes in an inducible lac operon. The repressor protein can bind to the operator region and hinder RNAP binding to the promoter region in the absence of lactose (lac). When lac is present, this small molecule binds to the repressor and dissociates it from operator, allowing RNAP to transcribe the *lacZYA* genes.

do not appear to produce poly-cistronic RNAs. Most of these gene clusters encode enzymes that participate in a common pathway. Plants have the best described examples. These gene clusters encode enzymes for multiple catalytic steps that synthesize compounds defending the host against pathogens (8–11). So far, the mechanisms involved in the coordinate regulation of these complex gene clusters have not been established.

2.2.2. The Organization of the Regulatory Regions of RNAP II-Transcribed Genes

The region of a gene, usually proximal to the transcription start site (TSS), to which RNAP II and associated factors are initially recruited, consists of the core or basal promoter. It assembles as a complex formed by the basal transcription factors (BTF). The precise boundaries of the core promoter must be empirically determined for each gene, but as a rule of thumb, it is considered to comprise ~50 bp to each site of the TSS. Note that the convention is to number the first nucleotide represented in the mRNA as +1, thus this interval can be represented as [−50; +50]. Core promoters contain a number of *cis*-regulatory elements, which include the TATA box and an Initiator (Inr) element (12–14). However, there is no *cis*-regulatory element that is universally present in all core promoters. Even the broadly distributed TATA motif involved in the recruitment of the TATA-binding protein (TBP), a central BTF involved in the assembly of the transcriptional pre-initiation complex (PIC), is present in just ~30% of all eukaryotic promoters (5). BTFs receive signals from other regulatory factors, the TFs, most likely mediated by mediator proteins (15). Textbooks indicate that the regulatory regions of genes are usually located upstream of the TSS. However, notable recent evidence in large part provided by the Encyclopedia of DNA Elements (ENCODE) consortium suggest that regulatory sequences can be found in 5′- and 3′-untranslated regions (5′- and 3′-UTRs), introns, and even coding regions (16). Thus, it is clear that the definition of what the typical regulatory region of a gene includes needs to be broadened.

2.3. Transcription Factors as Key Regulators of Transcription

TFs are responsible for providing signals necessary for the correct assembly of the PIC and are therefore primarily responsible for controlling the time, amplitude, and duration of gene transcription. About 5–7% of the genome of an eukaryotic organism encodes for TFs (17), which can be grouped into 50–60 distinct groups of families. Some families have dramatically expanded while others might be absent altogether from particular organisms or kingdoms. For example, the MYB family, named after the avian *myelo*b*lastocys* virus from where the first protein harboring this domain was first identified (18, 19), is very large in plants (>180 members in *Arabidopsis*), while animal genomes contain just a handful of genes encoding proteins with this domain.

TFs can activate or repress transcription. If they function as transcriptional activators, they often harbor a transcriptional activation domain (TAD), responsible for interacting with mediator or other BTFs. The structure of TADs is significantly less conserved than the folds that characterize DNA-binding domains, and they are classified into various types (acidic, proline-rich, glutamine-rich, etc.) (20). The structure of the acidic TAD of the herpes simplex virus VP16 was determined and key residues identified for function (21).

2.3.1. De Novo Identification of TFs and Target Sites

Important questions that the biologist often encounters include (1) how to determine if a protein functions as TF or not and (2) what are the direct targets (defined as the genes directly regulated) of a TF.

2.3.1.1. De Novo Identification of TFs

For the identification of TFs from genome sequence or Expressed Sequence Tag (EST) information, specific signatures characteristic of TFs can be followed. As described earlier, TFs can be classified into families based on particular folds of the respective DNA-binding domains. These structures can often share little sequence identity, resulting in the need to investigate relatedness by using profiles that capture weak similarities or even information on neighbor amino acids. The PFAM database (http://pfam.sanger.ac.uk/) is a large collection of protein families, each represented by multiple sequence alignments and Hidden Markov Model (HMM) profiles (22). Within a protein family, multiple alignments reveal similarity in particular regions due to conserved amino acid sequences. These protein fragments correspond to one or more functional regions termed domains. PFAM contains profiles of domains that carry DNA-binding protein–protein interaction functions and this information is used to predict if an unknown protein corresponds to a TF with a previously described DNA-binding domain or not.

2.3.1.2. Identification of Gene Directly Regulated by a TF

The second problem that the experimentalist often encounters is how to identify the genes that a TF directly regulates. In studying TF function, it is important to establish which DNA sequences they can bind to. This can be accomplished through in vitro protein–DNA interaction techniques that include electrophoretic mobility shift assays (EMSA) in combination with footprinting approaches or by the systematic evolution of ligands by exponential enrichment (SELEX). Using information derived from such experiments to predict TF targets in silico, however, is not trivial, as in vitro DNA-binding specificities established, for example, by SELEX are often not correlated with the sequences that a TF binds in vivo – a good example being provided by E2F factors (23). Thus, the alternative is to experimentally identify the in vivo targets of a TF. The participation of a TF in a given regulatory process can be inferred from mutant analyses or from gene expression profile clusters. However, determining the ultimate function of a TF depends on identifying which genes it can directly activate. Two main approaches are currently available to identify direct targets of TFs: (a) by expressing a fusion of the TF to the hormone-binding domain of the glucocorticoid receptor and identifying the mRNAs induced/repressed in the presence of the GR ligand (dexamethasone, DEX), in the presence of an inhibitor of translation (e.g., cycloheximide, CHX), or (b) by identifying the DNA sequences that a TF binds in vivo, using chromatin immunoprecipitation (ChIP) assays, which can be coupled with next generation sequencing methods (ChIP-Seq) (24) or by using the immunoprecipitated DNA to hybridize a tiling or promoter array representing all the genes in an organism (ChIP-chip) (25, 26). Information on TFs and their binding sequences for a number of species is available at several databases (**Table 2.1**).

2.4. Transcriptional Networks

TFs function in networks, in which a regulatory protein controls the expression of another, which in turn may modulate the expression of other regulatory proteins or control genes encoding structural proteins or enzymes. These hierarchical arrangements allow specific signals to be amplified, providing the information necessary for given sets of genes to be deployed with particular spatial and temporal patterns motifs (27). Gene regulatory networks (GRNs) are formed by motifs, and the dynamic properties of these motifs significantly contribute to the overall behavior of the network (28). MicroRNAs and other small RNAs (briefly described in the next section) are also emerging as key

Table 2.1
Online TF databases for various species. Online resources related to TFs are listed and marked for information provided on TF sequence (TFs), TF binding sequences (TF binding), promoter sequences, and TF binding locations in target gene promoters (Promoters) and regulatory networks. Circuitry of regulatory networks combines individual TF–target gene relationships into single comprehensive view. A list of plant *cis*-element resources and detailed discussion is available in (33)

Name	URL	TFs	TF binding	Promoters	Regulatory networks	Reference
AGRIS	arabidopsis.med.ohio-state.edu	✓	✓	✓	✓	(34)
DBD	www.transcription factor.org	✓				(35)
GRASSIUS	grassius.org	✓	✓	✓	a	(17)
JASPAR	jaspar.cgb.ki.se		✓			(36)
PAZAR	www.pazar.info	✓	✓	✓		(37)
PLANTTFDB	planttfdb.cbi.pku.edu.cn	✓				(38)
PLNTFDB	plntfdb.bio.uni-potsdam.de	✓				(39)
TFCONES	tfcones.fugu-sg.org	✓	✓			(40)
TFdb	genome.gsc.riken.jp/TFdb	✓				(41)
TRANSFAC[b]	www.gene-regulation.com	✓	✓	✓	✓	(42)

[a]Planned feature.
[b]Some features are available in commercial package.

components of GRNs [e.g., (29)], often participating in mixed network motifs (27).

2.5. Small RNAs and Gene Expression

One of the most significant discoveries of the past few years is the realization that most of the DNA that lies between genes is not really "junk," but that it participates in the formation and is the subject of regulation of a large number of non-coding RNAs, often groups under the term small RNA (to distinguish them from the longer mRNA, tRNA, or rRNA populations). Small RNAs have indeed been called the "Guardians of the Genome" (30), and one of their main functions appears to be to keep transposons (pieces of DNA that can move around the genome) at bay, preventing major genome damage. Small RNAs can be of different types and usually have lengths 20–30 nucleotides long. They appear to be broadly distributed in all eukaryotes, and even prokaryotes express small RNAs with unique regulatory activities (31). One class of small RNAs, the microRNAs (miRNAs) participate in the post-transcriptional regulation of mRNA translation and stability. In contrast, small interfering RNAs (siRNAs) control gene expression by specifically targeting particular sequences for silencing in the process of TGS that involves histone modifications and DNA methylation (32).

Acknowledgments

Support in the Grotewold lab for projects involving regulation of gene expression is provided by NRI Grant 2007-35318-17805 from the USDA CSREES, DOE Grant DE-FG02-07ER15881, and NSF grant DBI-0701405. A.Y. is supported by NIH Ruth L. Kirschstein National Research Service Award 5 T32 CA106196-05 from NCI.

References

1. Herr, A.J., Jensen, M.B., Dalmay, T., and Baulcombe, D.C. (2005) RNA polymerase IV directs silencing of endogenous DNA. *Science* 308, 118–120.
2. Pikaard, C.S., Haag, J.R., Ream, T., and Wierzbicki, A.T. (2008) Roles of RNA polymerase IV in gene silencing. *Trends Plant Sci* 13, 390–397.
3. Wierzbicki, A.T., Haag, J.R., and Pikaard, C.S. (2008) Noncoding transcription by RNA polymerase Pol IVb/Pol V mediates transcriptional silencing of overlapping and adjacent genes. *Cell* 135, 635–648.
4. Wierzbicki, A.T., Ream, T.S., Haag, J.R., and Pikaard, C.S. (2009) RNA polymerase V transcription guides ARGONAUTE4 to chromatin. *Nat Genet* 41, 630–634.
5. Grotewold, E., and Springer, N. (2009) Decoding the transcriptional hardwiring of the plant genome. In: *Plant systems biology* (Coruzzi, G., and R.A. Gutierrez, Eds.) pp. 196–228, Wiley-Blackwell, Chichester.
6. Gruber, T.M., and Gross, C.A. (2003) Multiple sigma subunits and the partitioning of bacterial transcription space. *Annu Rev Microbiol* 57, 441–466.
7. Kazmierczak, M.J., Wiedmann, M., and Boor, K.J. (2005) Alternative sigma factors and their roles in bacterial virulence. *Microbiol Mol Biol Rev* 69, 527–543.
8. Field, B., and Osbourn, A.E. (2008) Metabolic diversification–independent assembly of operon-like gene clusters in different plants. *Science* 320, 543–547.
9. Jonczyk, R., Schmidt, H., Osterrieder, A., Fiesselmann, A., Schullehner, K., Haslbeck, M. et al. (2008) Elucidation of the final reactions of DIMBOA-glucoside biosynthesis in maize: characterization of Bx6 and Bx7. *Plant Physiol* 146, 1053–1063.
10. Osbourn, A.E., Field, B. (2009) Operons. *Cell Mol Life Sci* 66, 3755–3775.
11. Qi, X., Bakht, S., Leggett, M., Maxwell, C., Melton, R., and Osbourn, A. (2004) A gene cluster for secondary metabolism in oat: implications for the evolution of metabolic diversity in plants. *Proc Natl Acad Sci USA* 101, 8233–8238.
12. Gurley, W.B., O'Grady, K., Czarnecka-Verner, E., and Lawit, S.J. (2006) General transcription factors and the core promoter: ancient roots. In: *Regulation of transcription in plants* (Grasser, K., Eds.) pp. 1–27. Blackwell Pub, Oxford.
13. Smale, S.T. (2001) Core promoters: active contributors to combinatorial gene regulation. *Genes Dev* 15, 2503–2508.
14. Smale, S.T., and Kadonaga, J.T. (2003) The RNA polymerase II core promoter. *Annu Rev Biochem* 72, 449–479.

15. Gustafsson, C.M., and Samuelsson, T. (2001) Mediator – a universal complex in transcriptional regulation. *Mol Microbiol* 41, 1–8.

16. Birney, E., Stamatoyannopoulos, J.A., Dutta, A., Guigo, R., Gingeras, T.R., Margulies, E.H. et al. (2007) Identification and analysis of functional elements in 1% of the human genome by the ENCODE pilot project. *Nature* 447, 799–816.

17. Yilmaz, A., Nishiyama, M.Y., Jr., Fuentes, B.G., Souza, G.M., Janies, D., Gray, J. et al. (2009) GRASSIUS: a platform for comparative regulatory genomics across the grasses. *Plant Physiol* 149, 171–180.

18. Klempnauer, K.H., Gonda, T.J., and Bishop, J.M. (1982) Nucleotide sequence of the retroviral leukemia gene v-myb and its cellular progenitor c-myb: the architecture of a transduced oncogene. *Cell* 31, 453–463.

19. Klempnauer, K.H., Ramsay, G., Bishop, J.M., Moscovici, M.G., Moscovici, C., McGrath, J.P. et al. (1983) The product of the retroviral transforming gene v-myb is a truncated version of the protein encoded by the cellular oncogene c-myb. *Cell* 33, 345–355.

20. Roberts, S.G. (2000) Mechanisms of action of transcription activation and repression domains. *Cell Mol Life Sci* 57, 1149–1160.

21. Uesugi, M., Nyanguile, O., Lu, H., Levine, A.J., and Verdine, G.L. (1997) Induced alpha helix in the VP16 activation domain upon binding to a human TAF. *Science* 277, 1310–1313.

22. Sonnhammer, E.L., Eddy, S.R., Birney, E., Bateman, A., and Durbin, R. (1998) Pfam: multiple sequence alignments and HMM-profiles of protein domains. *Nucleic Acids Res* 26, 320–322.

23. Rabinovich, A., Jin, V.X., Rabinovich, R., Xu, X., and Farnham, P.J. (2008) E2F in vivo binding specificity: comparison of consensus versus nonconsensus binding sites. *Genome Res* 18, 1763–1777.

24. Mardis, E.R. (2007) ChIP-seq: welcome to the new frontier. *Nat Methods* 4, 613–614.

25. Buck, M.J., and Lieb, J.D. (2004) ChIP-chip: considerations for the design, analysis, and application of genome-wide chromatin immunoprecipitation experiments. *Genomics* 83, 349–360.

26. Herring, C.D., Raffaelle, M., Allen, T.E., Kanin, E.I., Landick, R., Ansari, A.Z. et al. (2005) Immobilization of Escherichia coli RNA polymerase and location of binding sites by use of chromatin immunoprecip-

itation and microarrays. *J Bacteriol* 187, 6166–6174.

27. Re, A., Cora, D., Taverna, D., and Caselle, M. (2009) Genome-wide survey of microRNA-transcription factor feed-forward regulatory circuits in human. *Mol Biosyst* 5, 854–867.

28. Alon, U. (2007) Network motifs: theory and experimental approaches. *Nat Rev Genet* 8, 450–461.

29. Card, D.A., Hebbar, P.B., Li, L., Trotter, K.W., Komatsu, Y., Mishina, Y. et al. (2008) Oct4/Sox2-regulated miR-302 targets cyclin D1 in human embryonic stem cells. *Mol Cell Biol* 28, 6426–6438.

30. Malone, C.D., and Hannon, G.J. (2009) Small RNAs as guardians of the genome. *Cell* 136, 656–668.

31. Waters, L.S., and Storz, G. (2009) Regulatory RNAs in bacteria. *Cell* 136, 615–628.

32. Matzke, M., Kanno, T., Huettel, B., Daxinger, L., and Matzke, A.J. (2007) Targets of RNA-directed DNA methylation. *Curr Opin Plant Biol* 10, 512–519.

33. Brady, S.M., and Provart, N.J. (2009) Web-queryable large-scale data sets for hypothesis generation in plant biology. *Plant Cell* 21, 1034–1051.

34. Palaniswamy, S.K., James, S., Sun, H., Lamb, R.S., Davuluri, R.V., and Grotewold, E. (2006) AGRIS and AtRegNet. a platform to link cis-regulatory elements and transcription factors into regulatory networks. *Plant Physiol* 140, 818–829.

35. Kummerfeld, S.K., and Teichmann, S.A. (2006) DBD: a transcription factor prediction database. *Nucleic Acids Res* 34, D74–D81.

36. Sandelin, A., Alkema, W., Engstrom, P., Wasserman, W.W., and Lenhard, B. (2004) JASPAR: an open-access database for eukaryotic transcription factor binding profiles. *Nucleic Acids Res* 32, D91–D94.

37. Portales-Casamar, E., Kirov, S., Lim, J., Lithwick, S., Swanson, M.I., Ticoll, A. et al. (2007) PAZAR: a framework for collection and dissemination of cis-regulatory sequence annotation. *Genome Biol* 8, R207.

38. Guo, A.Y., Chen, X., Gao, G., Zhang, H., Zhu, Q.H., Liu, X.C. et al. (2008) PlantTFDB: a comprehensive plant transcription factor database. *Nucleic Acids Res* 36, D966–D969.

39. Riano-Pachon, D.M., Ruzicic, S., Dreyer, I., and Mueller-Roeber, B. (2007) PlnTFDB: an integrative plant transcription factor database. *BMC Bioinformatics* 8, 42.

40. Lee, A.P., Yang, Y., Brenner, S., and Venkatesh, B. (2007) TFCONES: a database of vertebrate transcription factor-encoding genes and their associated conserved noncoding elements. *BMC Genomics* 8, 441.

41. Kanamori, M., Konno, H., Osato, N., Kawai, J., Hayashizaki, Y., and Suzuki, H. (2004) A genome-wide and nonredundant mouse transcription factor database. *Biochem Biophys Res Commun* 322, 787–793.

42. Wingender, E., Dietze, P., Karas, H., and Knuppel, R. (1996) TRANSFAC: a database on transcription factors and their DNA binding sites. *Nucleic Acids Res* 24, 238–241.

Chapter 3

Regulatory Regions in DNA: Promoters, Enhancers, Silencers, and Insulators

Jean-Jack M. Riethoven

Abstract

One of the mechanisms through which protein levels in the cell are controlled is through transcriptional regulation. Certain regions, called *cis*-regulatory elements, on the DNA are footprints for the *trans*-acting proteins involved in transcription, either for the positioning of the basic transcriptional machinery or for the regulation – in simple terms turn on or turn off – thereof. The basic transcriptional machinery is DNA-dependent RNA polymerase (RNAP) which synthesizes various types of RNA and core promoters on the DNA are used to position the RNAP. Other nearby regions will regulate the transcription: in prokaryotic organisms operators are involved; in eukaryotic organisms, proximal promoter regions, enhancers, silencers, and insulators are present. This chapter will describe the various DNA regions involved in transcription and transcriptional regulation.

Key words: *cis*-regulatory element, core promoter, silencer, enhancer, insulator.

1. Introduction

The complexity of transcriptional regulation greatly increases from prokaryotic to simple, single-cell, eukaryotic organisms and again increases in metazoan eukaryotes. It has been postulated that the increase in complexity in transcriptional regulation, together with alternative splicing, post-translational modification of proteins, and chromatin modification and reordering, is a mechanism through which a relative small number of genes are used to produce ever-increasing complexity both physiological and behavioral (1).

I. Ladunga (ed.), *Computational Biology of Transcription Factor Binding*, Methods in Molecular Biology 674,
DOI 10.1007/978-1-60761-854-6_3, © Springer Science+Business Media, LLC 2010

In prokaryotes, some co-regulated genes are organized into operons on neighboring loci, to be transcribed together via a single promoter region. Within the promoter region, two hexamers help position the RNA polymerase (RNAP) I adjacent to the transcription start site (TSS) – they are located approximately at 10 and 35 bases upstream of the TSS (+1) and are hence in literature often referred to as the −10 and −35 sequences. Specific σ factors bound to the RNAP I increase the affinity for these hexamers.

Operators, other DNA motifs within the promoter, activate or repress transcription through the binding of gene regulatory proteins. Repression functions by binding of proteins to operators and thereby blocking the binding of RNAP I to the DNA. An example of such a mechanism is the *trp* operon in *Escherichia coli*. When tryptophan level is low in the cell, RNAP I can bind and transcribe the *trp* operon, but when tryptophan levels increase the activated tryptophan repressor protein occupies the operator, disabling further transcription (2).

Depending on the position near or within the promoter, the same regulatory protein may act as either a repressor or an activator, e.g., the bacteriophage *lambda* repressor (3). Operons can also be regulated by multiple signals, where both an activator and a repressor motif are present within or near the promoter. The *E. coli lac* operon is under dual control (4): inhibition via the *lac* repressor (within the promoter) and activation via the CAP-binding site (just upstream of the promoter). Yeast regulates its metabolism so as to utilize glucose as the primary carbon source and metabolize lactose only in the absence of glucose. Only during glucose depletion, cyclic AMP binds to CAP, and only when at the same time lactose is present, the *lac* repressor is not bound to the operator, together enabling transcription of the *lac* operon.

The two-signal regulation as exemplified in the *lac* operon is very simple; however, the same mechanism cannot be expanded to include many different signals as there is just not enough room in the promoter to accomplish this. In eukaryotes, several other agents and mechanisms have evolved to allow a more complex and combinatorial regulation of transcription. These include gene regulatory proteins that can influence transcription even when they are bound to DNA far away from the gene locus, basal transcription factors that are necessary for RNAP II binding, and the packaging of DNA into chromatin.

This chapter will mainly deal with the first two mechanisms and give an overview of the core promoter, enhancer, silencer, and insulator regions and proximal promoter and upstream activator elements (*see* **Fig. 3.1**).

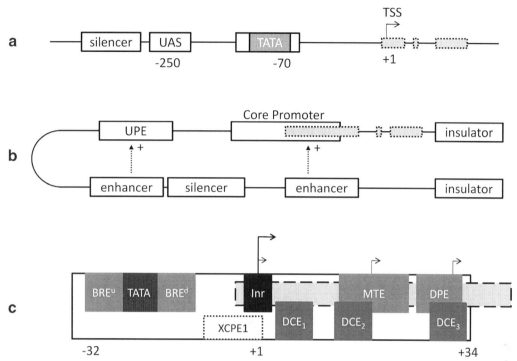

Fig. 3.1. Transcriptional regulatory units in eukaryotes. Schematic overview (**a** and **b** not drawn to scale) of the various elements in the transcriptional units in simple eukaryotes (**a**, yeast) and higher eukaryotes (**b**, mammalian) and a detailed overview (**c**) of the promoter region in mammals. Exons are shown as *gray boxes* with *dashes lines*. **a**, In yeast, a promoter region is shown with a TATA box at −70 bp and upstream activating sequences (UASs) around 250 bp upstream from the transcription start site (TSS). **b**, the mammalian transcriptional unit is more complex, with a large core promoter overlapping the first exon and upstream promoter elements (UPE, or proximal promoter elements) further upstream. The DNA loop shows that enhancers can be brought physically close to either the core promoter or the UPE. **c**, detailed architecture of the core promoter. Various elements, many of them optional, are shown roughly to scale. Darker shaded promoter elements are more frequent. Abbreviations: TATA box (TATA), initiator element (Inr), TFIIB recognition element (BRE, upstream and downstream), motif ten element (MTE), downstream core element (DCE, subunits 1, 2, and 3), X core promoter element 1 (XCPE1), and the downstream promoter element (DPE). Note that the core promoter can be focused or dispersed, here shown by one *bold* TSS and many smaller TSSs, respectively.

2. Regions Involved in Transcription and Transcriptional Regulation

2.1. The Core Promoter

The classical, textbook definition of the core promoter is a region around the TSS (+1) of a gene, which contains several DNA elements that facilitate the binding of regulatory proteins. Binding of these proteins is required for the step-wise sequestering and formation of the PIC (pre-initiation complex). In one promoter architecture, the TATA box, an AT-rich sequence acts as a binding site for the TATA-binding protein (TBP) (5). TBP together with TATA-associated factors (TAFs) forms the multi-subunit initiator complex TFIID. The binding of TFIID to the TATA box is the first step in the creation of a stable transcriptional complex. Other basal transcription factors (TFIIA-J) together with the RNAP II itself will bind forming the PIC.

However, it has been known for some time that the TATA box is not the only element within the core promoter: several other elements can recruit TFIID: the initiator element (Inr) and the downstream promoter element (DPE). The BRE (TFIIB recognition element) is a motif that specifically interacts with the TFIIB complex. Furthermore, it has been shown that the assembly of PIC via the TATA box, at least in mammals, is more the exception than the rule (10–20%) (6, 7).

Separately from this, the model that a core promoter regulates the initiation of a single or very narrow range of co-located transcription start sites is not fully correct. In recent years it has come to light that core promoters can be roughly divided into two classes: those that have a single TSS or a distinct cluster of TSSs over a very narrow, focused region of several nucleotides and those that have a very broad or dispersed range of potential transcription start sites over a 50–100 bp region (8–10). Focused core promoters often contain a TATA box (11) and are the most ancient type of promoter conserved from *Archaea* to vertebrates, while the dispersed core promoters have an overrepresentation of CpG islands. Most of the genes in higher eukaryotes are under transcriptional control of dispersed core promoters.

In *Metazoa*, especially vertebrates, core promoter elements have been best characterized while less is known about the organization of unicellular eukaryotes, for example, *Saccharomyces cerevisiae* (baker's yeast). The next section describes mainly mammalian, insect, and plant core promoter elements (*see* **Table 3.1, Fig. 3.1c**) and will indicate where significant differences with simpler eukaryotes exist.

2.1.1. Core Promoter Elements

The TATA box is the best known and most ancient promoter element (12–14). Although the exact position of the TATA box consensus sequence TATAWAAR varies from 28 to 34 bp upstream from the TSS, a strong preference to 30–31 bp is observed (15, 16). In yeast, the TATA box is located between –70 and –120 bp (17).

The BRE element is present in a subset of the TATA-containing core promoters and can be located immediately upstream (BRE[u], consensus sequence SSRCGCC) as well as downstream (BRE[d], consensus sequence RTDKKKK) of the TATA box and can act in both a negative and a positive manner (18–20).

Table 3.1

Core promoter elements. Consensus sequences, (approximate) location in relation to the transcription start site (TSS), and organisms from which consensus sequences are derived are listed for the core promoter elements that are most frequently used in eukaryotic organisms. Consensus sequences are listed in IUPAC nucleotide code

	Element	Consensus	Location	Organism	Reference
Inr	Initiator element	YYANWYY	−2 to +5	Human	(46)
		TCAKTY	−2 to +4	*Drosophila*	
TATA	TATA box	TATAWAAR	−31 to −24		(10)
BRE$_u$	TFIIB recognition element (upstream)	SSRCGCC			(18)
BRE$_d$	TFIIB recognition element (downstream)	RTDKKKK			(19)
DPE	Downstream promoter element	RGWYVT	+28 to +33	*Drosophila*[a]	(22)
MTE	Motif ten element	CSARCSSAAC	+18 to +27	*Drosophila*	(24)
DCE S1	Downstream core element (subunit 1)	CTTC	+6 to +11	Human	(25)
DCE S2	Downstream core element (subunit 2)	CTGT	+16 to +21	Human	(25)
DCE S3	Downstream core element (subunit 3)	AGC	+31 to +34	Human	(25)
XCPE1	X core promoter element 1	DSGYGGRASM	−8 to +2	Human	(26)

[a]Similar sequences conserved from *Drosophila* to human.

Recent studies have shown that the Inr motif is the element that is the most prevalent (approx. 40–60%) (17, 21) in focused core promoters, more so than the TATA box. The Inr straddles the TSS, and the consensus sequence is YY<u>A</u>NWYY in humans and TC<u>A</u>KTY in *Drosophila*, with the underlined <u>A</u> frequently being the +1 start site.

The DPE (downstream promoter element) is another motif (consensus RGWYVT in *Drosophila*) (22) that is important for transcriptional activity (23) and is under the same strict positional control as the TATA box: it is located downstream +28 to +33 bp from the TSS and operates cooperatively with the Inr.

More core promoter elements are currently known, but are underrepresented when compared with the TATA, Inr, BRE, and DPE elements: the motif ten element (MTE) (24), the downstream core element (DCE) (25), and the X core promoter element 1 (XCPE1) (26).

2.2. Proximal Promoter Elements

In *Metazoa*, several other promoter elements exist which are located upstream of the core promoter: the proximal promoter elements. They do not always act as traditional activators or repressors; instead, it is postulated that they serve as tethering elements for active distant enhancers, enabling these enhancers to interact with the core promoter (27, 28).

2.3. Enhancers

One of the characteristics of eukaryotic gene expression is the existence of groups of specific DNA motifs that often from a great distance can sequester transcription factors (TF) to upregulate the rate of formation and binding of the pre-initiation complex to the core promoter. These enhancer regions can be found up- and downstream of the TSS, within exons or introns, in the 5′ and 3′ untranslated (UTR) regions of genes, and even as far as 10,000 bp in *Drosophila* or 100,000 bp in human and mouse away from the gene boundaries (1, 29, 30).

The exact mechanisms through which enhancers influence transcriptional activity are still under debate, but it is clear that enhancer activation often needs the binding of several transcription factors to *cis*-regulatory motifs to the enhancer. Once active, the enhancer can bind to the PIC or to tethering elements in the proximal region of the promoter and influence (the rate of) transcription by itself. Looping in chromatin (*see* **Fig. 3.1b**) plays a role in bringing enhancers physically close to the proximal or core promoter region of a target gene – these interactions have been shown via Chromosome Conformation Capture (3C) technology and successors (4C, 5C, Hi-C) (**Chapters 16**) (31, 32). How the looping is effectuated is still unclear: direct-contact models postulate that the interaction between enhancers and promoter elements is more by chance due to free motion of the chromatin strand; the tracking model hypotheses that the active enhancer–protein complex somehow tracks the chromatin strand until it encounters the promoter region.

Another model postulated to increase the transcriptional rate is for the enhancers to be instrumental in changing the subnuclear position of the target genes and bring them closer to a ready source of RNAP II: the RNAP II loci or factories (33–35).

Many of these enhancers are non-coding sequences that are strongly conserved over hundreds of millions of years (fish to mouse) (36) and regulate gene expression in highly specific tissues, developmental stages, or combinations of these (37). The importance of enhancers is also illustrated in their role in disease, where chromosome rearrangements, deletion of, or point mutations in enhancers can cause abnormal phenotypes. For example, thalassemia is a blood disease in which nonstoichiometric quantities of α- or β-globin are produced (these subunits need to be produced in equimolar quantities). For some patients no mutations or deletions could be detected in the coding region. Further studies showed that deletion or rearrangement of the enhancer caused the globin imbalance (38). Another example is preaxial polydactyly caused by point mutations in the limb-specific enhancer *ZRS* that regulates the sonic hedgehog (*SHH*) gene, which codes for an important signaling molecule (39, 40).

In contrast to higher eukaryotes, the majority of yeast genes do not have distant enhancer sites (*see* **Fig. 3.1a**). Yeast does, however, have upstream activating sequences (UASs) ~250 bp upstream of the TSS. UASs facilitate the binding of activating transcription factors (41). Each UAS often contains one or two closely linked *cis*-acting binding sites; activating transcription factors that bind to those sites positively regulate the PIC via TAFs. A well-studied example is the *GAL4p* transcription factor that binds to a UAS with a specific motif of $5'$-CGGN$_{11}$CCG-$3'$ and is responsible for the regulation of expression of *GAL* genes when galactose is utilized as a carbon source (42).

2.4. Silencers

The role of silencers in the downregulation of gene expression has been recognized much later and much less is known about these *cis*-regulatory elements than their enhancer counterparts. Two distinct classes of silencers exist: short, position-independent motifs that via their bound TF (repressors) proteins actively interfere with the PIC assembly are called silencer elements and are normally found upstream of the TSS and position-dependent silencers or negative regulatory elements (NREs) that passively prevent the binding of TFs to their respective *cis*-regulatory motifs and can be found both up- and downstream of the TSS and within introns and exons (43).

2.5. Insulators

Enhancers and silencers can act on multiple genes but in certain cases these interactions might be unwanted. Special *cis*-acting regulatory DNA sequence regions called insulators can block such interactions. Two distinct types of insulators have been discovered: enhancer-blocking insulators and barrier insulators (35). The enhancer-blocking insulators protect against gene activation by enhancers and interfere with the enhancer–promoter interaction only if the insulator is located between the enhancer and the promoter. Barrier insulators safeguard against the spread of heterochromatin, and thus of chromatin-mediated silencing, and lie on the border of eu- and heterochromatin domains.

Positional requirements for enhancer-blocking insulators were first described in the *Drosophila* insulator element *gypsy* (44). This insulator consists of 12 repeats of the consensus sequence YRYTGCATAYYY and is a target for the suppressor of hairy-wing *Su(Hw)* protein. The *Su(Hw)* protein binds with other proteins (*CP190, modifier of mdg4, ubiquitin ligase, topoisomerase-I-interacting protein*) to form complexes that bind to the nuclear lamina and as a result bring insulators together to form insulator bodies (35). This is also one of the proposed mechanisms how insulators block enhancers: by topological separation, for example, by resulting loop domains, of the enhancers from promoter sites.

In vertebrates, all currently identified enhancer-blocking insulators contain *cis*-regulatory binding motifs for the CCCTC-binding factor (*CTCF*) (45) and similar mechanisms like *gypsy* for enhancer-blocking activity involving *CTCF*-containing insulators have been proposed.

3. Conclusion

In the last 20 years, understanding about transcriptional regulation has greatly increased. It is now clear that promoters with TATA boxes are not the rule but the exception, and that several other less-known promoter elements are important, too. Added to that the combinatorial complexity of enhancer and silencer interaction, together with recent discoveries with regard to 3D localization and epigenetic control, has made clear that the 'textbook' models of gene regulation are now severely outdated.

Acknowledgements

The author is partly supported through the National Science Foundation grant EPSCoR EPS-0701892.

References

1. Levine, M., and Tjian, R. (2003) Transcriptional regulation and animal diversity, *Nature* 424, 147–151.

2. Yanofsky, C. (2004) The different roles of tryptophan transfer RNA in regulating *trp* operon expression in *E. coli* versus *B. subtilis*. *Trends Genet* 20(8), 367–374.

3. Ptashne, M., Backmann, K., Humayun, M.Z. et al. (1976) Autoregulation and function of a repressor in bacteriophage Lambda. *Science* 194, 156–161.

4. Malan, T.P., and McClure, W.R. (1984) Dual promoter control of the *Escherichia coli* lactose operon. *Cell* 39(1), 173–180.

5. Struhl, K., Kadosh, D., Keaveney, M. et al. (1998) Activation and repression mechanisms in yeast. *Cold Spring Harb Symp Quant Biol* 63, 413–421.

6. Cooper, S.J., Trinklein, N.D., Anton, E.D. et al. (2006) Comprehensive analysis of transcriptional promoter structure and function in 1% of the human genome. *Genome Res* 16, 1–10.

7. Gershenzon, N.I., and Ioshikhes, I.P. (2005) Synergy of human Pol II core promoter elements revealed by statistical sequence analysis. *Bioinformatics* 21, 1295–1300.

8. Sandelin, A., Carninci, P., Lenhard, B. et al. (2007) Mammalian RNA polymerase II core promoters: insights from genome-wide studies. *Nature Genet* 8, 424–436.

9. Carninci, P., Kasukawa, T., Katayama, S. et al. (2005) The transcriptional landscape of the mammalian genome. *Science* 309, 1559–1563.

10. Carninci, P., Sandelin, A., Lenhard, B. et al. (2006) Genome-wide analysis of mammalian promoter architecture and evolution. *Nature Genet* 38(6), 626–635.

11. Juven-Gershon, T., Hsu, J., Theisen, J.W.M. et al. (2008) The RNA polymerase II core promoter – the gateway to transcription. *Curr Opin Cell Biol* 20, 253–259.

12. Reeve, J.N. (2003) Archaeal chromatin and transcription. *Mol Microbiol* 48, 587–598.

13. Molina, C., and Grotewold, E. (2005) Genome wide analysis of *Arabidopsis* core promoters. *BMC Genomics* 6, 25.

14. Yamamoto, Y.Y., Ichida, H., Abe, T. et al. (2007) Differentiation of core promoter architecture between plants and mammals revealed by LDSS analysis. *Nucleic Acids Res* 35, 6219–6226.

15. Ponjavic, J., Lenhard, B., Kai, C. et al. (2006) Transcriptional and structural impact of TATA-initiation site spacing in mammalian core promoters. *Genome Biol* 7, R78.

16. Hahn, S. (2004) Structure and mechanism of the RNA polymerase II transcription machinery. *Nat Struct Mol Biol* 11, 394–403.

17. Yang, C., Bolotin, E., Jiang, T. et al. (2007) Prevalence of the initiator over the TATA box in human and yeast genes and identification of DNA motifs enriched in human TATA-less core promoters. *Gene* 1;389(1), 52–65.

18. Lagrange, T., Kapanidis, A.N., Tang, H. et al. (1998) New core promoter element in RNA polymerase II-dependent transcription: sequence-specific DNA binding by transcription factor IIB. *Genes Dev* 12, 34–44.

19. Deng, W., and Roberts, S.G. (2005) A core promoter element downstream of the TATA box that is recognized by TFIIB. *Genes Dev* 19, 2418–2423.

20. Deng, W., and Roberts, S.G. (2007) TFIIB and the regulation of transcription by RNA polymerase II. *Chromosoma* 116, 417–429.

21. Gershenzon, N.I., Trifonov, E.N., and Ioshikhes, I.P. (2006) The features of *Drosophila* core promoters revealed by statistical analysis. *BMC Genomics* 7, 161.

22. Kutach, A.K., and Kadonaga, J.T. (2000) The downstream promoter element DPE appears to be as widely used as the TATA box in *Drosophila* core promoters. *Mol Cell Biol* 20, 4754–4764.

23. Burke, T.W., and Kadonaga, J.T. (1996) *Drosophila* TFIID binds to a conserved downstream basal promoter element that is present in many TATA-box-deficient promoters. *Genes Dev* 10, 711–724.

24. Lim, C.Y., Santoso, B., Boulay, T. et al. (2004) The MTE, a new core promoter element for transcription by RNA polymerase II. *Genes Dev* 18, 1606–1617.

25. Lewis, B.A., Kim, T.K., and Orkin, S.H. (2000) A downstream element in the human beta-globin promoter: evidence of extended sequence-specific transcription factor IID contacts. *Proc Natl Acad Sci USA* 97, 7172–7177.

26. Tokusumi, Y., Ma, Y., Song, X. et al. (2007) The new core promoter element XCPE1 (X Core Promoter Element 1) directs activator-, mediator-, and TATA-binding protein-dependent but TFIID-independent RNA polymerase II transcription from TATA-less promoters. *Mol Cell Biol* 27, 1844–1858.

27. Su, W., Jackson, S., Tjian, R. et al. (1991) DNA looping between sites for transcriptional activation: self-association of DNA-bound Sp1. *Genes Dev* 5, 820–826.

28. Calhoun, V.C., Stathopoulos, A., and Levine, M. (2002) Promoter-proximal tethering elements regulate enhancer-promoter specificity in the *Drosophila* Antennapedia complex. *Proc Natl Acad Sci USA* 99, 9243–9247.

29. Birney, E., Stamatoyannopoulos, J.A., Dutta, A. et al. (2007) Identification and analysis of functional elements in 1% of the human genome by the ENCODE pilot project. *Nature* 447, 799–816.

30. Latchman, D.S. (2008) *Eukaryotic transcription factors* (5th edn). Academic Press, London.

31. Fullwood, M.J., and Ruan, Y. (2009) ChIP-based methods for the identification of long-range chromatin interactions. *J Cell Biochem* 107, 30–39.

32. Lieberman-Aiden, E., van Berkum, N.L., Williams, L. et al. (2009) Comprehensive mapping of long-range interactions reveals folding principles of the human genome. *Science* 326, 289–293.

33. Marenduzzo, D., Faro-Trindade, I., and Cook, P.R. (2007) What are the molecular ties that maintain genomic loops? *Trends Genet* 23, 126–133.

34. Miele, A., and Dekker, J. (2008) Long-range chromosomal interactions and gene regulation. *Mol Biosyst* 4, 1046–1057.

35. Gazner, M., and Felsenfeld, G. (2006) Insulators: exploiting transcriptional and epigenetic mechanisms. *Nat Genet* 7, 703–713.

36. Nobrega, M.A., Ovcharenko, I., Afzal, V. et al. (2003) Scanning human gene deserts for long-range enhancers. *Science* 302, 413.

37. Siepel, A., Bejerano, G., Pedersen, J.S. et al. (2005) Evolutionarily conserved elements in vertebrate, insect, worm, and yeast genomes. *Genome Res* 15, 1034–1050.

38. Visel, A., Rubin, E.M., and Pennacchio, L.A. (2009) Genomic views of distant-acting enhancers. *Nature* 461, 199–205.

39. Lettice, L.A., Hill, A.E., Devenney, P.S. et al. (2008) Point mutations in a distant sonic hedgehog *cis*-regulator generate a variable regulatory output responsible for preaxial polydactyly. *Hum Mol Genet* 17, 978–985.

40. Lettice, L.A., Horikoshi, T., Heaneya, S.J.H. et al. (2002) Disruption of a long-range *cis*-acting regulator for *Shh* causes preaxial polydactyly. *Proc Natl Acad Sci USA* 99, 7548–7553.

41. de Bruin, D., Zaman, Z., Liberatore, R.A. et al. (2001) Telomere looping permits gene activation by a downstream UAS in yeast. *Nature* 409, 109–113.

42. Campbell, R.N., Leverentz, M.K., Ryan, L.A. et al. (2008) Metabolic control of transcription: paradigms and lessons from *Saccharomyces cerevisiae. Biochem J* 414, 177–187.

43. Ogbourne, S., and Antalis, T.M. (1998) Transcriptional control and the role of silencers in transcriptional regulation in eukaryotes. *Biochemistry* 331, 1–14.

44. Holdridge, C., and Dorsett, D. (1991) Repression of hsp70 heat shock gene transcription by the suppressor of hairy-wing protein of *Drosophila melanogaster. Mol Cell Biol* 11, 1894–1900.

45. Bell, A.C., West, A.G., and Felsenfeld, G. (1999) The protein CTCF is required for the enhancer blocking activity of vertebrate insulators. *Cell* 98, 387–396.

46. Smale, S.T., and Baltimore, D. (1989) The "initiator" as a transcription control element. *Cell* 57, 103–113.

Chapter 4

Three-Dimensional Structures of DNA-Bound Transcriptional Regulators

Tripti Shrivastava and Tahir H. Tahirov

Abstract

Our understanding of the detailed mechanisms of specific promoter/enhancer DNA-binding site recognition by transcriptional regulatory factors is primarily based on three-dimensional structural studies using the methods of X-ray crystallography and NMR. Vast amount of accumulated experimental data have revealed the basic principles of protein–DNA complex formation paving the way for better modeling and prediction of DNA-binding properties of transcription factors. In this review, our intent is to provide a general overview of the three-dimensional structures of DNA-bound transcriptional regulators starting from the basic principles of specific DNA recognition and ending with high-order multiprotein–DNA complexes.

Key words: Transcription factor, gene expression, crystal structure, DNA-binding domain, DNA-binding motif, protein–DNA interaction, cooperative DNA binding, DNA recognition, promoter, enhancer.

1. Introduction

DNA-binding proteins play central roles in biology. Among other activities, they are responsible for replicating the genome, transcribing active genes, and repairing damaged DNA. One of the largest and most diverse classes of DNA-binding proteins is the transcription factors that regulate gene expression. Transcription factors regulate cell development, differentiation, and cell growth by binding to a specific DNA site (or set of sites) on promoters and enhancers and regulating gene expression. Gene expression requires transcription: making mRNA (messenger RNA) copies from the DNA.

I. Ladunga (ed.), *Computational Biology of Transcription Factor Binding*, Methods in Molecular Biology 674,
DOI 10.1007/978-1-60761-854-6_4, © Springer Science+Business Media, LLC 2010

For a deeper understanding of the transcriptional regulatory processes, it is essential to determine the structure and analyze the three-dimensional organization of multiprotein–DNA regulatory complexes. This is a formidable task; progress made in this direction has evolved from understanding of the basic mechanisms of specific DNA recognition by smaller DNA-binding motifs. Nowadays our knowledge is expanding based on structural studies of complexes with several transcriptional regulatory proteins acting cooperatively on DNA binding. In this review, our intent is to provide a general overview of the three-dimensional structure of DNA-bound transcriptional regulators starting from the basic principles of specific DNA recognition and ending with more complicated multiprotein–DNA complexes. We do not provide a comprehensive review; instead, we present some of typical examples demonstrating how transcriptional regulatory factors adopt the simple DNA-recognition motifs to form more complex regulatory assemblies.

2. Discussion

2.1. The Principle of DNA Recognition

Inspection of protein–DNA complexes at an atomic level reveals that contacts between DNA and protein can be explained in terms of two broadly defined mechanisms: direct and indirect readout. Direct readout of a DNA sequence is the sensing of base pair identity by direct hydrogen bonding and van der Waals interactions (1, 2). Indirect readout senses base pair identity without direct base–protein contact. It utilizes the sequence-dependent deformability of DNA (3). Consequently, indirect readout allows the sensing of the DNA sequence at a distance.

2.1.1. Hydrogen Bonds

The discovery of hydrogen bond formation in macromolecules solved the mysteries associated with the formation of secondary structures, i.e., α-helices and β-sheets in proteins (4) and double-helix formation from single strands in DNA (5). The grooves of DNA are rich in functional groups that can form hydrogen bonds (1). AT base pairs provide N3 (H-bond acceptor) and O2 (acceptor) atoms in the minor groove, and N7 (acceptor), NH2 (6-amino donor), and O4 (acceptor) atoms/groups in the major groove. GC base pairs provide NH2 (2-amino donor), N3 (acceptor), and O2 (acceptor) in the minor groove and N7 (acceptor), O6 (acceptor), and NH2 (4-amino donor group) in the major groove. If hydrogen bond formation were the sole mechanism of recognition, this information should suffice to distinguish GC/CG and AT/TA base pairs from each other because of the order in which the acceptor and donor functional groups appear in the grooves (6).

Hydrogen bonds are further classified into subgroups based on the type of interaction between amino acids and base pairs. These are either single interactions where only one hydrogen bond exists between the amino acid and its corresponding base, bidentate interactions where amino acids interact with bases by two or more bonds, or complex interactions where amino acids interact with more than one base simultaneously (7).

2.1.2. Water-Mediated Interactions

Water molecules can participate in hydrogen bonding networks that link side-chain and main-chain atoms with the functional groups on bases, and the anionic oxygens of the phosphodiester backbone (1). With improved crystal structure resolution beyond 3 Å, macromolecular crystallography revealed that water molecules can contribute significantly to stability and specificity (8–10). These water molecules mediate interactions between the two molecules and fill the gap arising from imperfect matching of the protein and DNA surfaces (6).

2.1.3. van der Waals Interactions

van der Waals contacts comprise 64.9% of all protein–DNA interactions (7). Most of the contacts occur along the DNA backbone, which is consistent with its high surface accessibility. The preference of DNA bases for van der Waals contacts is in the following order: thymine interacts most readily, then adenine, guanine, and cytosine (7). The C5M methyl group of thymine often confers the specificity of this base both by providing favorable van der Waals contacts and repulsion of unnecessary side chains.

Not all protein–DNA complexes are highly hydrated at the interface. For example, TATA box-binding protein (TBP) bound to DNA exhibits a hydrophobic interface (11, 12). TBP interacts along the minor groove of DNA, which is splayed open and curves away from the protein. The driving force for such complex formation seems to be primarily entropic. When a protein and DNA form a complex, water molecules left at the interface between the protein and DNA decrease the entropy of the system. Consequently, the surfaces of the protein and DNA tend to be exactly complementary so that none of the unnecessary water molecules remain when the complex forms and the surface is rendered inaccessible to solvent molecules.

2.1.4. Ionic Interactions

Some theoretical and experimental studies underscore the importance of long-range electrostatic effects in protein–DNA complexes. The negatively charged DNA is packed with nuclear proteins. To form chromatin, huge repulsive forces must be overcome. The positively charged amino acids Lys and Arg of the histone-fold domain form a distinct charged surface that directs DNA wrapping around the histone core (13). In another example, the repressor in the met J repressor–operator complex is activated by binding two positively charged S-adenosyl-methionine (SAM) molecules. During cooperative oligomerization, the protein undergoes small conformational changes and its affinity for the operator increases 1,000-fold. In the ternary complex, the positively charged SAM lies on the protein surface opposite of DNA. The altered charge distribution resulting from SAM binding may make the electrostatic surface more favorable to intra-protein interactions and DNA binding (14). c-Myb onco-gene recognizes DNA with three tandem DNA-binding domain repeats: R1, R2, and R3. The structure of DNA-bound R1R2R3 revealed involvement of R2 and R3 in specific DNA base interac-tions; however, no direct DNA interaction was observed for R1 (1H88) (15). Instead, the large positively charged surface of R1 is positioned facing the DNA major groove and the phosphate back-bone, forming long-range electrostatic interaction between R1 and the DNA. This interaction stabilizes the c-Myb–DNA com-plex explaining the reason why R1 increases the binding affinity of c-Myb for DNA five- to six-fold (15).

2.2. DNA Recognition Elements

Beyond the atomic level of interaction between DNA and pro-tein, the basic mode of interactions is based on protein secondary structures, i.e., α-helices, β-sheets, and loops.

2.2.1. α-Helices

α-Helices are the most common secondary structure elements used by transcriptional regulators and other DNA-binding pro-teins for base recognition. They typically interact through the major grooves. Maximal DNA base contacts are achieved when an α-helix inserts into the major groove with its axis parallel to the flanking DNA backbone. Typically, base contacts are made between the main-chain atoms and the side chain of the helix. Interferon regulatory factor, or IRF, interacts to DNA through the helix at the major groove of its winged helix–turn–helix (HTH) motif (16) (**Fig. 4.1a**). Other examples show incidences of α-helix involvement with bases on the minor groove. For example, in the LacI family of proteins, the accommodation of an α-helix at the minor groove distorts the DNA. In the purine repressor dimer (PurR) DNA complex, each monomer contains a helix-turn-helix domain which interacts with the major groove bases and forms a two-turn 'hinge' helix contact with the minor groove bases (17) (**Fig. 4.1d**).

2.2.2. β-Strands

β-Strands are the next secondary structure elements most fre-quently involved in specific DNA binding. Normally two or more β-strands form a β-sheet to mediate interactions with DNA. The NikR nickel-induced transcriptional repressor of the nickel ABC-type transporter, NikABCDE, is one such example. NikR forms a tetramer, which is arranged as dimer of dimers and binds to

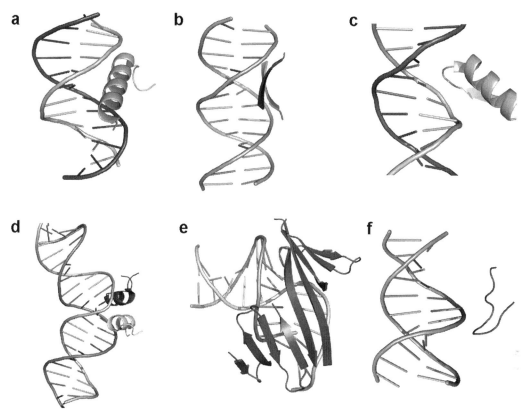

Fig. 4.1. Different modes of recognition. **a** α-Helix (pdb code 2IRF), **b** β-sheet (2HZV), and **c** loop (2GEQ) at major groove; **d** α-helix (2KEI), **e** β-sheet (1NVP), and **f** loop (1HJB) at minor groove. All figures were produced with PyMol.

palindromic DNA (18). In each half of the DNA palindrome, the subunits of the NikR dimer donate a β-strand to form a two-stranded antiparallel β-sheet that interacts with the DNA in the major groove (**Fig. 4.1b**). Insertion of three-stranded β-sheet into the major groove has been observed in the plant GCC box-binding domain (19). In contrast to the β-sheets recognition in the major grooves, the TATA-binding protein uses a surface of 10-stranded concave β-sheet to recognize the DNA at the minor groove (20). The insertion of 10 β-strands into the minor groove profoundly distorts the DNA (**Fig. 4.1e**).

2.2.3. Loops

Loops are the third type of structural element that characterizes protein–DNA interactions. Whereas α-helices and β-sheets provide rigid scaffolds for interaction with DNA, a superfamily of proteins having an immunoglobulin-like fold uses the flexible loops as their primary structural element for DNA recognition (21). The core domain of the p53 tumor suppressor (**Fig. 4.1c**), for example, forms a sandwich of antiparallel β-sheets with two helices (H1 and H2); the H2 helix along with the preceding loop interact with the DNA in the major groove (22). DNA recognition by a Runt domain of Runx1 via a minor groove dramatically increases its DNA-binding affinity and plays an important regulatory role (23) (**Fig. 4.1f**).

2.3. DNA-Binding Motifs

Proteins that bind DNA have common folding patterns known as *DNA-binding motifs*. The helix-turn-helix (HTH), Zn-finger (ZnF), basic leucine zipper (bZip), basic helix-loop-helix (bHLH), and β-ribbon-containing motifs are examples of frequently observed DNA-binding motifs.

2.3.1. The Helix-Turn-Helix Motifs

HTH is the best-characterized member of the DNA-binding motif (24, 25). Its simplest form is traditionally defined as a 20-amino acid segment of two perpendicular α-helices connected by short linker. The second helix, known as the 'recognition helix,' inserts into the DNA major groove and forms contacts with both DNA bases and the sugar–phosphate backbone. The first helix, while not embedded in the major groove, may make additional DNA contacts (26–28). In the simplest form of HTH, the α-helices are connected together by a short linker of three amino acids (29) (**Fig. 4.2a**). With the variation in the turn region of the HTH motifs, several topologies have been determined where β-sheets comprised of two or more strands interrupt, precede, or follow the helices involved in DNA binding (**Fig. 4.2b**). These β-sheets, which can participate both in DNA base and in backbone interactions, are packed against the helices of the motif.

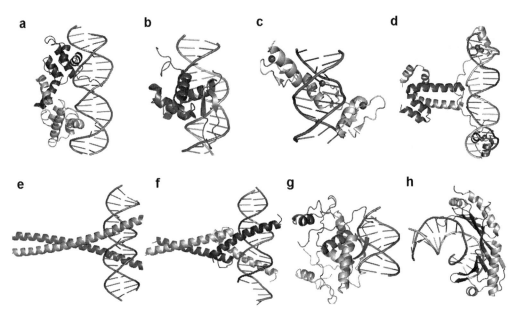

Fig. 4.2. DNA-binding motifs. **a** A HTH motif of λ repressor (1LMB) (56), **b** a wHTH motif of FoxO1 (3CO6) (57), **c** a Cys2His2-type ZnF of Zif268 (1AAY) (41), **d** a Zn_2Cys_6-type ZnF of GAL4 (3COQ) (58), **e** a bZip motif of Jun homodimer (2H7H), **f** a bHLH motif of sterol regulatory element binding protein (1AM9) (59), **g** a β-ribbon motif of Met J repressor of *E. coli* (1CMA) (14), and **h** a β-sheet motif of TATA box-binding protein (1NVP) (20).

Three-Dimensional Structures of DNA-Bound Transcriptional Regulators 49

This type of motif is known as a winged HTH (wHTH) domain (30–32).

2.3.2. Zn-Bearing Motifs

Proteins containing a domain with one or more coordinated zinc ions at their core form a superfamily of eukaryotic DNA-binding proteins. In each of the three classes of proteins within this superfamily, zinc plays a structural role in maintaining the protein fold and does not interact with the DNA. ZnF is the most prominent class, where an ~30 residue module with one Zn ion is coordinated by two cysteines and two histidines (33, 34). This class is also referred as Cys_2-His_2-type ZnF (**Fig. 4.2c**). The second class is an ~70 residue domain, found in steroid and related hormone receptors. Here each of the two Zn ions is liganded by four cysteine residues (35). The third class was discovered in GAL4 and other yeast activators (**Fig. 4.2d**), with two closely spaced Zn ions sharing six cysteines (36).

2.3.3. Zipper Group

The zipper group derives its name from its dimerization leucine zipper region. So far, this group has only been found in eukaryotic organisms (26, 37). The two known subfamilies in this group are bZip and bHLH proteins. The structure of bZip proteins is divided into two parts: the DNA-binding basic region interacting via the major groove and the dimerization leucine zipper region (26, 37) (**Fig. 4.2e**). The GCN_4, C/EBP, and Jun/Fos proteins are typical representatives of bZip family. The bHLH proteins differ from bZip proteins due to the presence of the loop separating the basic DNA binding and dimerization leucine zipper regions (**Fig. 4.2f**). This separation of the two segments by a loop provides more flexibility for DNA binding (37). The mouse Max protein makes a parallel, left-handed four-helix bundle that contributes a second dimerization interface (38).

2.3.4. β-Sheet Group

This group includes diverse DNA-binding proteins which use a β-sheet as their principal DNA recognition element. β-Ribbon/hairpin proteins use smaller two- or three-stranded β-sheets or hairpin motifs to bind to either the minor or the major grooves of the DNA (**Fig. 4.2g**). For example, the Met J repressor of *Escherichia coli* (14) and the *arc* repressor of *Salmonella* phage P22 (39) contain dimeric DNA-binding domains. Each dimer subunit consists of a helical bundle and a single β-strand; the strands from each subunit pack side by side forming an antiparallel β-sheet that binds DNA in the major groove. In contrast to β-ribbon/hairpin proteins, the TATA box-binding protein interacts with the DNA minor groove using its 10-stranded antiparallel β-sheet (20) (**Fig. 4.2h**).

2.4. Arrangement of Motifs

Transcription factors have evolved a variety of mechanisms to target a wider range of specific sites using fewer types of DNA-binding motifs. Well-known cases include DNA binding with a tandem repeat of similar motifs or formation of homo- and heterodimers.

2.4.1. Single Motif

Pancreatic and duodenal homeobox 1 (Pdx1) is a homeodomain transcription factor that binds to DNA as a monomer. This protein contains three α-helices and a flexible N-terminal arm. An α-helix termed as the recognition helix binds to DNA, whereas the N-terminal arm contacts the DNA bases through the minor groove (40).

2.4.2. Tandem Repeats of Motifs

Binding with tandem repeats often occurs among ZnF proteins. For example, three tandem Cys2-His2-type ZnF motifs of Zif268 bind at the DNA major groove (**Fig. 4.2c**) (41). HTH proteins also recognize DNA with tandem repeats. c-Myb contains three tandem repeats, R1, R2, and R3. R2 and R3 bind the DNA major groove and R1 enhances DNA-binding affinity by long-range electrostatic contacts (**Fig. 4.4e**) (15).

2.4.3. Homo- or Heterodimers

The specificity for more diverse regulatory DNA-binding sites is frequently achieved by homo- and heterodimer formation of highly related DNA-binding modules. Representative examples are the DNA complexes of NFκB proteins. These proteins contain two DNA-binding immunoglobulin domains connected by a flexible linker. NFκB p50 (**Fig. 4.3a**) and NFκB p65 (**Fig. 4.3b**) bind palindromic DNA sites as homodimers with participation of both domains in each subunit (42, 43). The carboxy-terminal domains of p50 and p65 form a dimerization interface between β-sheets using conserved residues. The conservation of dimerization interface allows the formation of a p50–p65 heterodimer which can bind to a nonpalindromic DNA with 5-base-pair 5′ subsite for p50 and a 4-base-pair 3′ subsite for p65 (**Fig. 4.3c**) (44). Some transcriptional factors, including Zn_2Cys_6 binuclear cluster proteins, contain a dimerization domain and a DNA-binding motif that are held together by a flexible linker

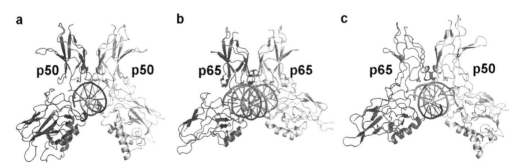

Fig. 4.3. Examples of homo- and heterodimers bound to DNA. **a** NFκB p50 homodimer (1NFK) (42), **b** NFκB p65 homodimer (1RAM) (43), and **c** NFκB p50–p66 heterodimer (1VKX) (44) complexes with DNA.

Three-Dimensional Structures of DNA-Bound Transcriptional Regulators 51

(45). Such flexibility allows variations in polarity and inter-half-site separation of common half-site DNA sequences (45).

2.4.4. Multiprotein–DNA Complexes

Transcription of eukaryotic genes is influenced by various regulatory elements within promoters, enhancers, and silencers (**Chapter 3**). These regulatory elements constitute the sites for the highly ordered cooperative assembly of multiprotein complexes for the activation or repression of transcription (46). The large molecular weight as well as transient and flexible nature of such complexes makes them difficult to study by conventional methods of X-ray crystallography and NMR. However, some progress has been achieved in understanding the protein–protein interactions leading to cooperative DNA binding. In general these interactions can be classified into three modes: first, those between a DNA-binding factor and a non-DNA-binding factor; second, those between DNA-binding factors recognizing adjacent cites on the promoter; and third, those between DNA-binding factor recognizing widely separated cites on the promoter (47). Among the representative structures of the first mode are the GABPα–GABPβ–DNA (48) and Runx1–CBFβ–DNA (23, 49) complex structures. In the latter complex (**Fig. 4.4a**), the non-DNA-binding CBFβ enhances the DNA-binding activity of Runx1 by allosteric mechanism (23, 50). The structures of Ets1–Pax5–DNA (51), MATα2–MCM1–DNA (52), and NFAT–Fos–Jun–DNA (53) are few of representative examples exhibiting the second mode of cooperative DNA binding (**Fig. 4.4b, d**). And finally, the structure of c-Myb-C/EBPβ–DNA complex represents the third mode of cooperation (**Fig. 4.4e**). Within this structure the protein–protein interaction is not between c-Myb and C/EBPβ bound to the same DNA fragment, but between these molecules bound to different fragments. The structure mimics a case in which c-Myb and C/EBPβ bind to widely separated sites on *mim-1* promoter and interact by mediating DNA loop formation (15).

Modeling of the obtained structures of individual parts of enhanceosome allows better description of its cooperative assembly. For example, the activation of the interferon-β (IFN-β) gene requires assembly of an enhanceosome containing several transcriptional regulatory factors. Cooperative binding of these factors to the IFN-β enhancer results in recruitment of coactivators and chromatin-remodeling proteins to the IFN-β promoter (54). The atomic model of the IFN-β enhanceosome was build using the structures of ATF-2/c-Jun/IRF-3/DNA (55) and IRF-3/IRF-7/NFκB p65/NFκB p50/DNA (54). The model shows that an association of eight proteins with the enhancer creates a continuous surface for cooperative recognition of a composite DNA-binding element (54).

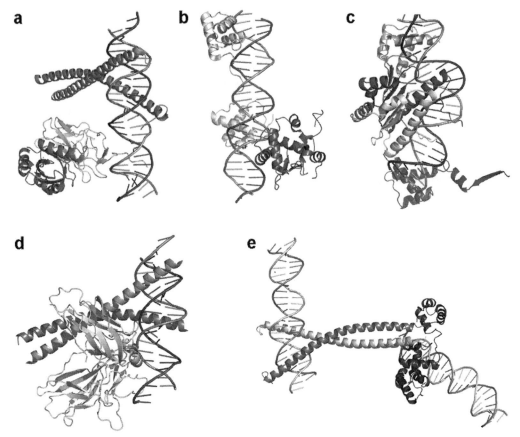

Fig. 4.4. High-order multiprotein–DNA complexes. **a** Runx1–CBFβ-C/EBPβ–DNA (1IO4), **b** Ets1–Pax5–DNA (1MDM) (53), **c** MATα2–MCM1–DNA (1MNM) (52), **d** NFAT–Fos–Jun–DNA (1A02) (51), **e** C/EBPβ–DNA/c-Myb–DNA (1H88) (15).

3. Conclusions

In spite of tremendous progress in structural studies of transcriptional regulatory factors, many barriers remain in the complete understanding of how these factors work. Among the frequently facing hurdles is the missing structure of residues located beyond the DNA-binding domains. These residues may fulfill a variety of regulatory functions, including autoinhibition, protein–protein interactions. However, the flexibility of these protein fragments also interferes with crystallization of macromolecules which is the most important step in structure determination. That is why, alongside with biophysical methods, development of alternative approaches, including computational biology methods capable of building three-dimensional models of transcription factor complexes, is necessary for full understanding the mechanisms of transcriptional regulation.

References

1. Seeman, N.C., Rosenberg, J.M., and Rich, A. (1976) Sequence-specific recognition of double helical nucleic acids by proteins. *Proc Nat Acad Sci USA* 73, 804–808.
2. Watkins, D., Hsiao, C., Woods, K.K. et al. (2008) P22 c2 repressor-operator complex: mechanisms of direct and indirect readout. *Biochemistry* 47, 2325–2338.
3. Zhang, Y., Xi, Z., Hegde, R.S. et al. (2004) Predicting indirect readout effects in protein-DNA interactions. *Proc Natl Acad Sci USA* 101, 8337–8341.
4. Pauling, L., and Corey, R.B. (1951) Atomic coordinates and structure factors for two helical configurations of polypeptide chains. *Proc Natl Acad Sci USA* 37, 235–240.
5. Watson, J.D., and Crick, F.H. (1953) Molecular structure of nucleic acids; a structure for deoxyribose nucleic acid. *Nature* 171, 737–738.
6. Jayaram, B., and Jain, T. (2004) The role of water in protein-DNA recognition. *Annu Rev Biophys Biomol Struct* 33, 343–361.
7. Luscombe, N.M., Laskowski, R.A., and Thornton, J.M. (2001) Amino acid-base interactions: a three-dimensional analysis of protein-DNA interactions at an atomic level. *Nucleic Acids Res* 29, 2860–2874.
8. Schwabe, J.W. (1997) The role of water in protein-DNA interactions. *Curr Opin Struct Biol* 7, 126–134.
9. Janin, J. (1991) Wet and dry interfaces: the role of solvent in protein-protein and protein-DNA recognition. *Structure* 7, R277–R279.
10. Woda, J., Schneider, B., Patel, K. et al. (1998) An analysis of the relationship between hydration and protein-DNA interactions. *Biophys J* 75, 2170–2177.
11. Nikolov, D.B., Chen, H., Halay, E.D. et al. (1996) Crystal structure of a human TATA box-binding protein/TATA element complex. *Proc Natl Acad Sci USA* 93, 4862–4867.
12. Kim, Y., Geiger, J.H., Hahn, S. et al. (1993) Crystal structure of a yeast TBP/TATA-box complex. *Nature* 365, 512–520.
13. Korolev, N., Vorontsova, O.V., and Nordenskiold, L. (2007) Physicochemical analysis of electrostatic foundation for DNA-protein interactions in chromatin transformations. *Prog Biophys Mol Biol* 95, 23–49.
14. Somers, W.S., and Phillips, S.E. (1992) Crystal structure of the met repressor-operator complex at 2.8 Å resolution reveals DNA recognition by beta-strands. *Nature* 359, 387–393.
15. Tahirov, T.H., Sato, K., Ichikawa-Iwata, E. et al. (2002) Mechanism of c-Myb-C/EBP beta cooperation from separated sites on a promoter. *Cell* 108, 57–70.
16. Fujii, Y., Shimizu, T., Kusumoto, M. et al. (1999) Crystal structure of an IRF-DNA complex reveals novel DNA recognition and cooperative binding to a tandem repeat of core sequences. *EMBO J* 18, 5028–5041.
17. Romanuka, J., Folkers, G.E., Biris, N. et al. (2009) Specificity and affinity of Lac repressor for the auxiliary operators O2 and O3 are explained by the structures of their protein-DNA complexes. *J Mol Biol* 390, 478–489.
18. Schreiter, E.R., Wang, S.C., Zamble, D.B. et al. (2006) NikR-operator complex structure and the mechanism of repressor activation by metal ions. *Proc Natl Acad Sci USA* 103, 13676–13681.
19. Allen, M.D., Yamasaki, K., Ohme-Takagi, M. et al. (1998) A novel mode of DNA recognition by a beta-sheet revealed by the solution structure of the GCC-box binding domain in complex with DNA. *EMBO J* 17, 5484–5496.
20. Bleichenbacher, M., Tan, S., and Richmond, T.J. (2003) Novel interactions between the components of human and yeast TFIIA/TBP/DNA complexes. *J Mol Biol* 332, 783–793.
21. Garvie, C.W., and Wolberger, C. (2001) Recognition of specific DNA sequences. *Mol Cell* 8, 937–946.
22. Ho, W.C., Fitzgerald, M.X., and Marmorstein, R. (2006) Structure of the p53 core domain dimer bound to DNA. *J Biol Chem* 281, 20494–20502.
23. Tahirov, T.H., Inoue-Bungo, T., Morii, H. et al. (2001) Structural analyses of DNA recognition by the AML1/Runx-1 Runt domain and its allosteric control by CBFb. *Cell* 104, 755–767.
24. Harrison, S.C., and Aggarwal, A.K. (1990) DNA recognition by proteins with the helix-turn-helix motif. *Annu Rev Biochem* 59, 933–969.
25. Pabo, C.O., and Sauer, R.T. (1984) Protein-DNA recognition. *Annu Rev Biochem* 53, 293–321.
26. Harrison, S.C. (1991) A structural taxonomy of DNA-binding domains. *Nature* 353, 715–719.
27. Struhl, K. (1989) Helix-turn-helix, zinc-finger, and leucine-zipper motifs for eukaryotic transcriptional regulatory proteins. *Trends Biochem Sci* 14, 137–140.

28. Wintjens, R., and Rooman, M. (1996) Structural classification of HTH DNA-binding domains and protein-DNA interaction modes. *J Mol Biol* 262, 294–313.

29. Brennan, R.G., and Matthews, B.W. (1989) The helix-turn-helix DNA binding motif. *J Biol Chem* 264, 1903–1906.

30. Brennan, R.G. (1993) The winged-helix DNA-binding motif: another helix-turn-helix takeoff. *Cell* 74, 773–776.

31. Lai, E., Clark, K.L., Burley, S.K. et al. (1993) Hepatocyte nuclear factor 3/fork head or "winged helix" proteins: a family of transcription factors of diverse biologic function. *Proc Natl Acad Sci USA* 90, 10421–10423.

32. Clubb, R.T., Omichinski, J.G., Savilahti, H. et al. (1994) A novel class of winged helix-turn-helix protein: the DNA-binding domain of Mu transposase. *Structure* 2, 1041–1048.

33. Miller, J., McLachlan, A.D., and Klug, A. (1985) Repetitive zinc-binding domains in the protein transcription factor IIIA from *Xenopus* oocytes. *EMBO J* 4, 1609–1614.

34. Brown, R.S., Sander, C., and Argos, P. (1985) The primary structure of transcription factor TFIIIA has 12 consecutive repeats. *FEBS Lett* 186, 271–274.

35. Freedman, L.P., Luisi, B.F., Korszun, Z.R. et al. (1988) The function and structure of the metal coordination sites within the glucocorticoid receptor DNA binding domain. *Nature* 334, 543–546.

36. Pan, T., and Coleman, J.E. (1990) GAL4 transcription factor is not a "zinc finger" but forms a Zn(II)2Cys6 binuclear cluster. *Proc Natl Acad Sci USA* 87, 2077–2081.

37. Luscombe, N.M., Austin, S.E., Berman, H.M. et al. (2000) An overview of the structures of protein-DNA complexes. *Genome Biol* 1, REVIEWS001.

38. Ferre-D'Amare, A.R., Prendergast, G.C., Ziff, E.B. et al. (1993) Recognition by Max of its cognate DNA through a dimeric b/HLH/Z domain. *Nature* 363, 38–45.

39. Raumann, B.E., Rould, M.A., Pabo, C.O. et al. (1994) DNA recognition by beta-sheets in the Arc repressor-operator crystal structure. *Nature* 367, 754–757.

40. Longo, A., Guanga, G.P., and Rose, R.B. (2007) Structural basis for induced fit mechanisms in DNA recognition by the Pdx1 homeodomain. *Biochemistry* 46, 2948–2957.

41. Elrod-Erickson, M., Rould, M.A., Nekludova, L. et al. (1996) Zif268 protein-DNA complex refined at 1.6 Å: a model system for understanding zinc finger-DNA interactions. *Structure* 4, 1171–1180.

42. Ghosh, G., van Duyne, G., Ghosh, S. et al. (1995) Structure of NF-kappa B p50 homodimer bound to a kappa B site. *Nature* 373, 303–310.

43. Chen, Y.Q., Ghosh, S., and Ghosh, G. (1998) A novel DNA recognition mode by the NF-kappa B p65 homodimer. *Nat Struct Biol* 5, 67–73.

44. Chen, F.E., Huang, D.B., Chen, Y.Q. et al. (1998) Crystal structure of p50/p65 heterodimer of transcription factor NF-kappaB bound to DNA. *Nature* 391, 410–413.

45. Marmorstein, R., and Fitzgerald, M.X. (2003) Modulation of DNA-binding domains for sequence-specific DNA recognition. *Gene* 304, 1–12.

46. Tjian, R., and Maniatis, T. (1994) Transcriptional activation: a complex puzzle with few easy pieces. *Cell* 77, 5–8.

47. Ogata, K., Sato, K., and Tahirov, T.H. (2003) Eukaryotic transcriptional regulatory complexes: cooperativity from near and afar. *Curr Opin Struct Biol* 13, 40–48.

48. Batchelor, A.H., Piper, D.E., de la Brousse, F.C. et al. (1998) The structure of GABPalpha/beta: an ETS domain- ankyrin repeat heterodimer bound to DNA. *Science* 279, 1037–1041.

49. Warren, A.J., Bravo, J., Williams, R.L. et al. (2000) Structural basis for the heterodimeric interaction between the acute leukaemia-associated transcription factors AML1 and CBFbeta. *Embo J* 19, 3004–3015.

50. Bartfeld, D., Shimon, L., Couture, G.C. et al. (2002) DNA recognition by the RUNX1 transcription factor is mediated by an allosteric transition in the RUNT domain and by DNA bending. *Structure* 10, 1395–1407.

51. Garvie, C.W., Hagman, J., and Wolberger, C. (2001) Structural studies of Ets-1/Pax5 complex formation on DNA. *Mol Cell* 8, 1267–1276.

52. Tan, S., and Richmond, T.J. (1998) Crystal structure of the yeast MATalpha2/MCM1/DNA ternary complex. *Nature* 391, 660–666.

53. Chen, L., Glover, J.N., Hogan, P.G. et al. (1998) Structure of the DNA-binding domains from NFAT, Fos and Jun bound specifically to DNA. *Nature* 392, 42–48.

54. Panne, D., Maniatis, T., and Harrison, S.C. (2007) An atomic model of the interferon-beta enhanceosome. *Cell* 129, 1111–1123.

55. Panne, D., Maniatis, T., and Harrison, S.C. (2004) Crystal structure of ATF-2/c-Jun and IRF-3 bound to the interferon-beta enhancer. *Embo J* 23, 4384–4393.

56. Beamer, L.J., and Pabo, C.O. (1992) Refined 1.8 Å crystal structure of the lambda repressor-operator complex. *J Mol Biol* 227, 177–196.

57. Brent, M.M., Anand, R., and Marmorstein, R. (2008) Structural basis for DNA recognition by FoxO1 and its regulation by posttranslational modification. *Structure* 16, 1407–1416.

58. Hong, M., Fitzgerald, M.X., Harper, S. et al. (2008) Structural basis for dimerization in DNA recognition by Gal4. *Structure* 16, 1019–1026.

59. Parraga, A., Bellsolell, L., Ferre-D'Amare, A.R. et al. (1998) Co-crystal structure of sterol regulatory element binding protein 1a at 2.3 Å resolution. *Structure* 6, 661–672.

Chapter 5

Identification of Promoter Regions and Regulatory Sites

Victor V. Solovyev, Ilham A. Shahmuradov, and Asaf A. Salamov

Abstract

Promoter sequences are the main regulatory elements of gene expression. Their recognition by computer algorithms is fundamental for understanding gene expression patterns, cell specificity and development. This chapter describes the advanced approaches to identify promoters in animal, plant and bacterial sequences. Also, we discuss an approach to identify statistically significant regulatory motifs in genomic sequences.

Key words: Promoter prediction, animal and plant promoters, bacterial promoters, regulatory motifs and homology inference.

1. Introduction

RNA polymerase II (Pol II) promoter is a key region that is involved in differential transcription regulation of eukaryotic protein-coding genes and some RNA genes. The gene-specific architecture of promoter sequences makes it extremely difficult to devise the general strategy for predicting promoters. Promoter 5′-flanking regions may contain dozens of short (5–10 bases long) motifs that serve as recognition sites for proteins providing initiation of transcription as well as specific regulation of gene expression.

The minimal promoter region called the core promoter is capable of initiating basal transcription. It contains a transcription start site (TSS) located in the initiator region (Inr), typically spanning from −60 to +40 bp relative to the TSS. About 30–50% of all known promoters contain a TATA-box at a position about 30 bp

---

I. Ladunga (ed.), *Computational Biology of Transcription Factor Binding*, Methods in Molecular Biology 674,
DOI 10.1007/978-1-60761-854-6_5, © Springer Science+Business Media, LLC 2010

upstream from the transcription start site. The TATA-box is the most general functional signal in eukaryotic promoters. In some cases, it can direct accurate transcription initiation by Pol II even in the absence of other control elements. Many highly expressed genes contain a strong TATA box in their core promoter. At the same time, large groups of genes including housekeeping genes, some oncogenes and growth factor genes possess TATA-less promoters. In these promoters, Inr or the recently found downstream promoter element (DPE), usually located ~25–30 bp downstream of TSS, may control the exact position of the transcription start. Many human genes are transcribed from several promoters (having multiple TSS) producing alternative first exons. Moreover, transcription initiation appears to be much less precise than initially assumed. In the human genome, it is not uncommon that the 5′-ends of mRNAs transcribed from the same promoter region are spread over hundreds of nucleotides (1–3).

The core promoter recruits the general transcriptional apparatus and supports basal transcription, while the proximal promoter (the region immediately upstream of the core promoter) engages various transcriptional factors, which are necessary for appropriate transcription activation or repression. Further upstream is located the distal part of promoter that may also contain transcription factor-binding sites and enhancer elements. A typical organization of Pol II promoters is shown in **Fig. 5.1**. The distal promoter part is usually the most variable region of promoters and

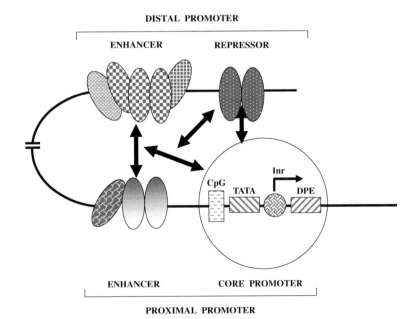

Fig. 5.1. Structural organization of RNA polymerase II promoter. Inr is the initiator region, usually containing TSS; DPE is the downstream promoter element, often appearing in TATA-less promoters; CpG is a CpG island.

generally poorly described; therefore, most of computational promoter recognition tools use the characteristics of only the core and/or proximal regions. A few reviews of eukaryotic promoters structure and computational identification have been published (2, 4–6).

To date, a number of databases provide information on known promoter sequences. Bucher and Trifonov (7) have initiated a first collection of experimentally mapped transcription start sites and surrounding sequences called Eukaryotic Promoter Database (EPD). Up to Release 72 (October 2002), EPD was a manually compiled database, relying exclusively on published experimental evidences. With Release 73, the curators started to exploit knowledge of 5′-ESTs from full-length cDNA clones as a new resource for defining promoters, and about a half of the EPD entries are based on 5′-EST sequences (8). There are few other databases that provide information about experimentally mapped TSSs. DBTSS (9) and PromoSer (10) are the large collections of mammalian promoters created using clustering of expressed sequence tag (EST) and full-length cDNA sequences. Promoters of genes of the hematopoietic system have been collected in HemoPDB, a specialized resource that provides also information on transcription factor binding sites (11). Orthologous promoters from various human/animal and plant species presented in the OMGProm (12), DoOP (13) and CORG (14) databases. There is a plant-specific PlantProm (15) database providing annotated, non-redundant collection of proximal promoter sequences for RNA polymerase II with experimentally determined transcription start sites. The second release of the PlantProm database contains 561 experimentally verified promoters and about 8,000 putative promoters with TSS predicted by using mapping full-length cDNAs on corresponding genomic sequences.

Regulatory sequences in promoter regions are composed of transcription factor binding sites called regulatory motifs. Often, a motif is fairly short (5–20 bp) sequence pattern and is observed in different genes or several times within a gene (16–18). A relational Transcription Factor Database (TFD) including large collection of regulatory factors and their binding sites was created by Gosh (19, 20). Over 7,900 sequences of transcriptional elements have been described in TRANSFAC database (21, 22). The other collections of functional motifs are TRRD (23), PlantCARE (24), PLACE (25), RegSite (http://softberry. com), PlantTFDB (*see* Chapter 2 and (26)), Osiris for rice (27), and Athena for *Arabidopsis* (28). RegsiteDB (Plant) contains about 1,942 experimentally discovered regulatory motifs of plant genes and detail descriptions of their functional properties. Annotating long genomic sequences using these motifs is not practical due to their short length and degenerate nature. For example, even if we will search for the well-described TATA-box motif using its weight matrix representation, there will be predicted one false-positive at every 120–130 bp (29). Nevertheless, the above-mentioned collections are invaluable for the detailed analysis of gene regulation and interpretation of experimental data, including microarray and gene networks studies.

The modular organization of transcription factors and regulatory sequences facilitates regulatory diversity and high level of specificity using relatively small number of different components (30). Therefore, to understand gene regulation, we need to study patterns of regulatory sequences rather than single elements. Searching for such patterns should produce much less false-positive predictions in new sequences compared to the recognition of single motifs. The simplest examples of regulatory patterns are composite elements (CE). They consist of modular arrangements of contiguous or overlapping binding sites for various factors, providing the possibility that the bound regulatory factors may interact directly. For example, the composite element of proliferin promoter comprises glucocorticoid receptor (GR) and AP-1 factor-binding sites. Both GR and AP-1 are expressed in most cell types, but the composite element demonstrated remarkable cell specificity: The hormone–receptor complex repressed the reporter gene expression in CV-1 cells, but enhanced its expression in HeLa cells and had no effect in the F9 cell (31). The database of composite elements (COMPEL) contains information about several hundred experimentally identified composite elements, where each element consists of two functionally linked sites (32, 33). A computational technique that provides possibility to identify statistically significant occurrences of motifs (or composite motifs) in a query sequence is described in **Section 2.5**.

2. Methods

2.1. Identification of Promoter Regions in Human DNA

Fickett and Hatzigeorgiou (18) presented one of the first reviews of promoter prediction programs. Among these were oligonucleotide content-based (34, 35), neural network (36–38) and the linear discriminant approaches (39). Although their relatively small test set of 18 sequences had several problems (40), the results demonstrated that the programs can recognize ~ 50% of true promoters with false-positive rate about 1 per 700–1,000 bp. To reduce false-positive predictions located within protein-coding genomic regions, some promoter-finding software include special modules for recognition of coding parts of gene inside promoter prediction programs (41, 42). However, modern gene prediction software, such as Genscan (43) or Fgenesh (44, 45), provides much better accuracy in the identification of coding exons and introns than any such procedures. Therefore, it was suggested (46, 47) to run gene prediction software first, and then execute promoter prediction on sequences upstream of the coding exons of predicted genes.

In this chapter we will describe successful algorithms implemented by us in a set of promoter prediction programs (TSSW, TSSG, Fprom, TSSP, TSSP-TCM and PromH). These tools use similar promoter features, but trained on different learning sets, or for different classes of organisms. While the initial version of the promoter recognition program TSSW (Transcription Start Site, W stands for using functional motifs from the Wingender et al. (21) database) (39)) has been developed more than 10 years ago, the above-mentioned programs are still among the most accurate ones (18, 46, 48, 49).

The approach implemented in TSSW will be described in detail and its modifications in other programs will be noted. Different features of a promoter region may have different power for promoter identification and might not be independent. Classical linear discriminant analysis provides a good method to combine such type of features in a discriminant function, which applied to a pattern yields its class membership. The discriminant analysis technique minimizes the error rate of classification (50). Let us assume that each sequence can be described by vector $\mathbf{X}$ of p characteristics $(x_1, x_2, \ldots, x_p)$, which could be measured (computed) for a given sequence fragment. The procedure of linear discriminant analysis is to find a linear combination of the measures (linear discriminant function or LDF) that provides maximum discrimination between sequences from class 1 and class 2.

The LDF

$$Z = \sum_{i=1}^{p} a_i x_i$$

classifies (X) into class 1 if $Z > c$ and into class 2 if $Z < c$. The vector of coefficients and threshold constant c are derived from the training set by maximizing the ratio of the between-class variation z to within-class variation and are equal to (50):

$$c = \vec{a}\,(\vec{m_1} + \vec{m_2})/2,$$

and

$$\vec{a} = S^{-1}(\vec{m_1} - \vec{m_2})$$

where $\overrightarrow{m_i}$ are the sample mean vectors of characteristics for class i; S is the pooled covariance matrix of characteristics,

$$S = \frac{1}{n_1 + n_2 - 2}(S_1 + S_2),$$

S_i is covariance matrix and n_i is the sample size of class i. Using these formulae, we can analytically calculate the coefficients of LDF and the threshold constant c using the values of characteristics computed on the training sets and then test the accuracy of LDF on the test set data. The significance of a given characteristic or a set of characteristics can be estimated by the generalized distance between two classes (the D^2 Mahalonobis distance):

$$D^2 = (\overrightarrow{m_1} - \overrightarrow{m_2})S^{-1}(\overrightarrow{m_1} - \overrightarrow{m_2}),$$

which is computed from values of the characteristics in the training sequences of classes 1 and 2. To find the most discriminative sequence features, a lot of possible characteristics can be generated and checked, such as score of weigh matrices, distances, oligonucleotide preferences within different sub-regions. Selection of the subset of significant characteristics **q** (among the tested **p**) is performed by step-wise discriminant procedure including only those characteristics that significantly increase the Mahalonobis distance. The procedure to test this significance of characteristics uses the fact that the quantity

$$F = \frac{n_1 + n_2 - p - 1}{p - q} \frac{n_1 n_2 (D_p^2 - D_q^2)}{(n_1 + n_2)(n_1 + n_2 - 2) + n_1 n_2 D_q^2},$$

has an $F(p - q, n_1 + n_2 - p - 1)$ distribution when testing hypothesis H_0: $\Delta_p^2 = \Delta_q^2$, where Δ_m^2 is the population Mahalonobis distance based on m variables. If the observations come from multivariate normal populations, the posterior probability that the example belongs to class 1 may be computed as

$$\Pr(class1/\overrightarrow{X}) = \frac{1}{1 + \frac{n_2}{n_1}\exp\{-Z + c\}}.$$

Potential TATA+ promoter sequences can be selected by the value of score computed using the TATA box weight matrix (51). Significant characteristics of promoter sequences from both groups found by discriminant analysis are presented in **Table 5.1**. This analysis shows that TATA+ and TATA– promoters should be analysed separately as they have different sequence characteristics. TATA– promoters have much weaker general features comparing with TATA+ promoters and they will be extremely difficult to predict by any general-purpose method.

Table 5.1
Significance of selected characteristics of TATA+ and TATA− human promoters

Characteristics of sequences	D^2 (TATA+ promoters)	D^2 (TATA− promoters)
•Hexaplets −200 to −45	2.6	1.4 (−100 to −1)
•TATA box score	3.4	0.9
•Triplets around TSS	4.1	0.7
•Hexaplets +1 to +40		0.9
•Sp1-motif content		0.9
•TATA fixed location	0.7	
•CpG content	1.4	0.7
•Similarity −200 to −100	0.3	0.7
•Motif density(MD) −200 to +1	4.5	3.2
•Direct/Inverted MD −100 to +1	4.0	3.3
Total Mahalonobis distance	11.2	4.3
No. of promoters/non-promoters	203/4,000	193/74,000

The TSSW program classifies each position of a given sequence as TSS or non-TSS based on two linear discriminant functions (for TATA+ and TATA− promoters) where the sequence characteristics are calculated within the (−200, +50) region around an analysed position. If the TATA-box weight matrix in the region ~30 bp upstream of the potential TSS gives a score higher than some threshold, then the position is classified based on LDF for TATA+ promoters, otherwise the LDF for TATA-less promoters is applied. Only one prediction with the highest LDF score is retained within any 300 bp region. If we observe a lower scoring promoter predicted by the TATA-less LDF near a higher scoring promoter predicted by TATA+ LDF, then the first prediction is also retained as a potential enhancer region.

Using the same approach but the TFD database of functional motifs (19) to calculate the density of functional sites in potential promoter region, we have developed the TSSG (39) program and later its variant Fprom (47) that used different learning set of promoter sequences. Examples of performance of TSSW, TSSG and Fprom programs on sequences upstream CDS regions of 10 genes with experimentally verified positions of transcriptional stat sites are presented in **Table 5.2**. In many cases the predicted TSS is located within a few bases of the experimental site. The programs produce one false-positive prediction per each 2,000–4,000 bp. TSSW outputs all potential TFBS around the predicted promoters or enhancers that includes the position, the strand (±), the TRANSFAC database identifier and the sequences of functional motifs found.

Table 5.2
Comparisons of promoter predictions to experimentally verified TSS annotated in GenBank

Gene	GenBank accession number	Length upstream CDS	True TSS position	TSSW	No. of false	TSSG	No. of false	Fprom	No. of false
CXCR4	AJ224869	2,720	2,632	2,631	2	2,631	1	2,632	1
HOX3D	X61755	2,354	2,280	2,278	2	2,278	2	2,223	2
DAF	M64356	815	733	744	0	760	0	760	0
GJB1	L47127	900	404	428	0	418	0	0	0
ID4	AF030295	1,470	1,090–1,096	1,081	1	1,085	1	1,095	1
C inhibitor	M68516	7,948	2,200	2,004	4	1,955	1	2,370	1
MBD1	AJ132338	2,951	1,964	1,876	1	1,891	0	1,891	0
Id –3	X73428	738	665	663	0	663	0	637	0
SFRS7	L41887	519	410–415	417	0	280	0	415	0
RDP22	EU338455	1,237	1,076	1,077	1	929	0	1,079	0

Identification of Promoter Regions and Regulatory Sites 65

A critical assessment of the promoter prediction accuracy has been done relative to the manual Havana gene annotation (48). As few as four programs participated in 'blind' predictions: two variants of McPromoter program (40, 52), N-scan (53) and Fprom (47). McPromoter and Fprom derived its predictions from a genomic sequence under analysis; N-scan used corresponding sequences of several vertebrate genomes. When the maximum allowed mismatch of the prediction from the reference TSS for counting true positive predictions on test sequences was 1,000 bp, N-scan achieved ~3% higher accuracy than Fprom, the next most accurate predictor. When the true positives predictions required be closer than 250 bp to the experimental TSS, Fprom demonstrated the best performance on most prediction accuracy measures (48). In these experiments, the sensitivity of computational promoter predictions was only 30–50% (relative to the 5′-gene ends of Havana annotation), but we should note that the TSS annotations from two experimentally derived databases also overlapping in only 48–58%. **Table 5.3** presents the relative accuracy of several popular promoter finding programs on genes with known full-length mRNAs investigated by Liu and States (46).

Table 5.3
Performance of promoter finding programs for genes with known 5′-ESTs

Program	Set 1 (133 promoters)		Set 2 (120 promoters)	
	True predictions	False predictions	True predictions	False predictions
PROSCAN1.7	32 (24%)	18 (36%)	30 (25%)	22 (42%)
NNPP2.0	56 (42%)	41 (42%)	26 (22%)	50 (66%)
PromFD1.0	88 (66%)	43 (33%)	69 (58%)	57 (45%)
Promoter2.0	8 (6%)	100 (93%)	14 (12%)	92 (88%)
TSSG	75 (56%)	10 (12%)	62 (52%)	18 (23%)
TSSW	57 (43%)	29 (34%)	58 (48%)	20 (26%)

2.2. Improving Promoter Identification by Using Homologous Sequences

The analysis of human–mouse conserved blocks in orthologous genes (those which are each other's closest homologues in the two organisms) specifically upregulated in skeletal muscle reported by Wasserman et al. (54) shows that 98% of experimentally defined transcriptional factors binding sites are confined to the 19% of human sequences most conserved during evolution. We have suggested using several types of conserved blocks to enhance sensitivity and specificity of promoter prediction programs by analysing alignment of orthologous genomic sequences (55). Since the sequences of a dozen of eukaryotic genomes are available, this strategy can be applied for improving promoter prediction in many organisms.

In most studies, researchers investigated conserved promoter elements in particular pairs of orthologous genes (54; *see also references therein*). However, we are interested in such conserved features that can be observed in many different pairs of orthologous gene promoters. Analysing pairwise sequence alignments of upstream regions of a set of mammalian genes, we have noticed that general similarity of upstream regions of related genes is relatively weak: for four pairs – about 30%, for five pairs – 40–50% and only for one pair (human and rat *MYL3* genes) – 61%. But at the same time we have observed many short blocks with very high level of conservation. We identified four classes of such blocks making meaningful contribution for predicting 'true' promoters.

TATA-box conserved region: Seventeen of the twenty 'true' TATA-promoters have interspecies conservation level in this region over 70% (six of them have 100%).

TSS conserved region: Thirteen of twenty-one genes have ≥77% (five have 100%) level of similarity, six genes have 66%, one gene has 41% and only one gene has 25%.

An average conservation level of regulatory motifs upstream of the TSS region: Sixteen of twenty-one genes have such similarity >70%, five genes have 45–56%.

Conservative region downstream of TSS. Thirteen of twenty-one genes have similarity in this region more than 70% and seven have >50%.

2.2.1. Promoter Prediction by the PromH Program

To take advantage of knowledge of conserved elements in 5′-regions of homologues genes, we have developed the PromH program that included four new features in the discriminant function, in addition to the features described in **Table 5.1**, conservation levels: around TSS and TATA-box (for TATA+ promoters), in area downstream of the potential TSS (40 bp) and in regulatory motifs observed upstream of the TSS.

The performance of the PromoterH program is shown in **Table 5.4**. The program identified 20 of 21 tested TATA-promoters. At the same time, TSSW program predicted only 15 true promoters and 3 false ones. The most TSSs predicted by PromH differs from the annotated pre-mRNA start positions by only 1–5 bp and the average distance between predicted and annotated TSSs is 2 bp. Examples of promoter predictions and their conservative blocks are shown in **Fig. 5.2**. The regulatory motifs, TATA-box and TSS of predicted TATA-promoters are highly conserved in orthologous genes, and these predictions correspond closely to the promoter annotations.

Some discrepancy is found in a few genes including *H-GLUT4*, *M-GLUT4* and *H-NPPA* between the predicted and the annotated TSS localization. There are several reasons for

Table 5.4
Test results of PromoterH on human, *Otolemur*, mouse and rat sequences

Gene[a]	GenBank accession number	Position of predicted TSS[b]	Conservation of TSS,[c] %	Conservation of TATA, %	Conservation of REs,[d] %	Total conservation level, %
H-HBB	U01317.1	-4^{mRNA}	66	100	82	47
OL-HBB	U60902	-10^{EST}	77	100	85	47
H-HBD	U01317.1	-4^{mRNA}	77	81	80	30
OL-HBD	U60902	-9^{EST}	88	81	51	30
H-HBE	U01317.1	$+6^{mRNA}$	88	75	88	48
OL-HBE	U60902	-27^{EST}	66	71	88	48
H-HBGA	U01317.1	$+1^{mRNA}$	66	71	81	48
H-HBGG	U01317.1	$+1^{mRNA}$	66	71	81	50
OL-HBGG	U60902	-53^{CDS}	66	71	75	50
H-MYL3	M76408	-5^{UTR}	100	87	89	61
R-MLC1V	X16325	$+4^{mRNA}$	41	75	79	61
H-MLC1emb	X58851, X55000	-1^{mRNA}	66	25	56	28
M-MLC1F	X12973	$+1^{mRNA}$	25	62	45	28
H-MYF4	AF050501	-27^{CDS}	100	100	90	43
M-MYOG	M95800	-2^{mRNA}	100	100	90	43
H-PGAM-M	J05073	$+1^{mRNA}$	88	81	71	40
R-PGAM2	Z17319	-1^{CDS}	77	81	82	40
H-NPPA	AL021155	-220^{CDS}	77	47	51	31
R-NPPA	J03267	$+1^{mRNA}$	100	100	87	31
H-GLUT4	M91463	-105^{mRNA}	88	25	52	30
M-GLUT4	M29660	-46^{mRNA}	88	e	86	30

[a] *H-HBB*: human beta-hemoglobin, *OL-HBB*: otolemur beta-hemoglobin, *H-HBD*: human delta-hemoglobin, *OL-HBD*: otolemur delta-hemoglobin, *H-HBE*: human epsilon-hemoglobin, *OL-HBE*: otolemur epsilon-hemoglobin, H-HBGA: human hemoglobin gamma A, *H-HBGG*: human hemoglobin gamma-G, *OL-HBGG*: otolemur hemoglobin gamma-G, *H-MYL3*: human ventricular myosin light chain, *R-MLC1V*: rat gene encoding alkali myosin ventricle light chain, *H-MLC1emb*: human embryonic myosin alkaline light chain, *M-MLC1F*: mouse myosin alkali light chain, *H-MYF4*: human myogenin (MYF4) gene, *M-MYOG*: mouse myogenin, *H-PGAM-M*: human phosphoglycerate mutase, *R-PGAM2*: rat phosphoglycerate mutase, H-NPPA: human atrial natriuretic factor ANF precursor (atrial natriuretic peptide ANP/prepronatriodilatin/isoform 2), *R-NPPA*: rat atrial natriuretic factor (ANF), *H-GLUT4*: human glucose transporter (GLUT4), *M-GLUT4*: mouse glucose transporter.

[b] Localizations of the predicted TSS are given in relation to mRNA or 5'-end of EST mapped on the promoter region, CDS or 5'-UTR.

[c] Interspecies conservation level around TSS (−3 . . . TSS . . . +5).

[d] Average interspecies conservation level of regulatory motifs left to TSS.

[e] Predicted promoter is TATA-less.

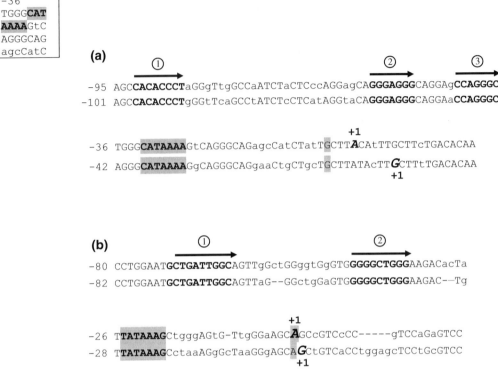

Fig. 5.2. Location of predicted TSSs and TATA boxes (*highlighted*) in aligned sequences of *H-HBB* and *OL-HBB* (**a**) and *H-PGAM*-M and *R-PGAM2* (**b**) orthologous gene pairs. Annotated start positions of pre-mRNAs are **boldfaced** and *italicized*. Some of the found conservative regulatory motifs are shown: in *H-HBB* and *OL-HBB* genes, 1 – HSSB I (rat; RSA01074), 2 – PERE (rat; RSA00900), 3 – P3-D (human; RSA00057); in *H-PGAM*-M and *R-PGAM2* genes, 1 – inverted CCAAT-box (human; RSA00526), 2 – Sp1 binding site (rat; RSA00253).

possible discrepancy between predicted and annotated promoters. The GenBank annotation for the *M-GLUT4* gene includes a putative weak TATA-box, which has not been supported by experiments. Our analysis of this region did not reveal any motif resembling the consensus of TATA box. The comparative analysis of human and mouse orthologous *GLUT4* gene pairs revealed that the upstream regions of both genes contain two high-scoring and well-conserved non-TATA promoters.

These results indicate that the information from alignment of orthologous genomic sequences can substantially improve the quality of promoter identification. The found conservation characteristics independent of gene type can be extracted from alignments of orthologous genes by using SCAN2-like (http://www.softberry.com/berry.phtml?topic=scan2&group=programs&subgroup=scanh) alignment programs parametrized for weak but significant similarities. This program was specifically designed to compare genomic sequences and aligning about 10,000 bp of a pair of 5′-regions for a second. This work also

Identification of Promoter Regions and Regulatory Sites 69

demonstrates that using orthologous sequences one can identify the start of transcription within about 1–5 bases for TATA-promoters, while a similar prediction of TATA-less promoters remains an open problem.

2.3. Plant Promoter Identification

To identify plant promoter regions, we have developed the TSSP program that uses the sequence features described in **Section 2.1** and has been trained on a set of plant promoter sequences (56). Promoter characteristic, including functional motifs density, was derived from our RegSite DB of plant regulatory elements (http://softberry. com) that contains ~1,800 known plant regulatory sequences. Recent tests demonstrated a high accuracy of plant promoter identification by TSSP: $Sn = 0.88$ and $Sp = 0.90$ (*see* Table **5** in ref. 49). Here we describe a variant of this program called TSSP_TCM developed by using a new learning and discriminative technique called Transductive Confidence Machine (57). Beyond making predictions, it also provides valid measures of confidence *in the predictions for each individual example* in the test set. Validity in our method means that if we set up a confidence level, say, 95%, then we are not expected to have more than 5 errors out of 100 examples.

Table 5.5
Statistics of testing procedure for 40 TATA and 25 TATA-less promoter sequences of 351 bp[a]

Promoter type	Mean prediction error in	Negative samples from CDSs, %	Negative samples from introns, %
TATA	Positive samples	7.4	3.5
	Negative samples	6.0	8.7
TATA-less	Positive samples	18.6	14.0
	Negative samples	16.9	29.5

[a]Forty various sets of 1,000 negative samples of the same length (351 bp), randomly chosen from CDSs (20 sets, totally 20,000 sequences) and introns (20 sets, totally 20,000 samples) of known plant genes. Confidence and credibility levels were =0.9 (90%) and = 0.06 (6%), respectively.

Learning machines such as the support vector machine (SVM; 58) perform well in a wide range of applications without requiring any parametric statistical assumptions about the source of data; the only assumption made is that the examples are generated from the same probability distribution independently of each other. However, a typical drawback of techniques such as the SVM is that they usually do not provide any useful measure of confidence in the predicted examples. Transductive confidence machine (TCM; 58–60) allows us to compute prediction of promoters and estimate its confidence.

2.3.1. Estimating the Confidence of Prediction

Here we outline the application of TCM and SVM techniques as implemented in TSSP-TCM program, following closely Shahmuradov et al. (56).

Suppose we are given a training set of examples $(x_1, y_1), \ldots, (x_l, y_l)$, where x_i is a vector of attributes and y_i is a label, and our goal is to predict the classifications $y_{l+1}, \ldots, y_{l+k}$ for a test set $x_{l+1}, \ldots, x_{l+k}$. We make only i.i.d. (identically and independently distributed) assumption about the data generating mechanism. When predicting y_{l+1}, we can estimate the 'randomness' (or 'conformity') of the sequence $(x_1, y_1), \ldots, (x_l, y_l), (x_{l+1}, \Upsilon)$ with respect to the i.i.d. model for every possible value Υ of y_{l+1}. The prediction can be confident if and only if exactly one of these two (in the case of binary classifications) sequences is typical.

If the randomness level can be computed, we can provide an algorithm for making predictions complemented by some measures of confidence and credibility. Let assume that we have training set $(x_1, y_1), \ldots, (x_l, y_l)$ and test set $x_{l+1}, \ldots, x_{l+k}$ (usually $k = 1$) and that our goal is to predict the classifications $y_{l+1}, \ldots, y_{l+k}$ for $x_{l+1}, \ldots, x_{l+k}$, then we can do the following:

1. Consider all possible values $\Upsilon_1, \ldots, \Upsilon_k$ for labels $y_{l+1}, \ldots, y_{l+k}$ and compute (in practice, approximate) the randomness level of every possible completion

$$(x_1, y_1), \ldots, (x_1, y_1), (x_{l+1}, \Upsilon_1), \ldots, (x_{l+k}, \Upsilon_k)$$

2. Predict the set $\Upsilon_1, \ldots, \Upsilon_k$ corresponding to the completion with the highest randomness level.

3. Output as the *confidence* in this prediction one minus the second largest randomness level.

4. Output as the *credibility* of this prediction the randomness level of the output prediction $\Upsilon_1, \ldots, \Upsilon_k$ (i.e. the largest randomness level for all possible predictions).

To illustrate the intuition behind confidence, let us choose a conventional 'significance level' such as 1%. If the confidence in our prediction exceeds 99% and the prediction is wrong, the actual data sequence belongs to an a priori chosen set of probability less than 1% (namely, the set of all data sequences with randomness level less than 1%). Low credibility means that either the training set is non-random or the test examples are not representative of the training set.

The randomness level can be approximated using the SVM technique. Let us consider the problem of binary classification with one test example. The basic idea of a support vector machine is to map the original set of vectors into a higher dimensional feature space, and then to construct a linear separating hyperplane in this feature space. In SVM approach, we should select a separating hyperplane with a small number of errors and a large margin by finding the minimum of the objective function

$$\frac{1}{2}(w \cdot w) + C\left(\sum_{i=1}^{l} \xi_i\right) \to \min \tag{1}$$

subject to the constraints

$$y_i((x_i \cdot w) + b) \geq 1 - \xi, \, i = 1, \ldots, l.$$

Here C is a fixed positive constant (maybe 8), w denotes the weights, b is the intercept, and ξ_i stands for the non-negative 'slack variables'.

As the mapping of the original set of vectors often leads to a problem in dealing with a very large number of parameters, Vapnik (57) suggested reformulating the problem using Lagrangian multipliers and replacing the original setting of the problem by the dual setting: maximize a quadratic form

$$\sum_{i=1}^{l} \alpha_i - \frac{1}{2}\sum_{i,j=1}^{l} y_i y_j \alpha_i \alpha_j K(x_i, x_j) \to \max$$

subject to the constraints

$$0 \leq \alpha_i \leq C, \, i = 1, 2, \ldots, l.$$

Here, K is the kernel and the values α_i, $i = 1, \ldots, l$ are the Lagrangian multipliers corresponding to the training vectors. For each non-zero α_i the corresponding vector x_i is called a *support vector*. The number of support vectors is typically a small fraction of the training set. If x is a new vector, the prediction $\hat{y}$ is

$$\hat{y} = \text{sign}\left(\sum_{i=1}^{l} \alpha_i y_i K(x_i, x) + b\right).$$

With every possible label $\Upsilon \in \{-1, 1\}$ for x_{l+1}, we associate the SVM optimization problem for the $l+1$ examples (the training examples plus the test example labelled with Υ). The solutions (Lagrangian multipliers) $\alpha_1, \alpha_2, \ldots, \alpha_{l+1}$, to this problem reflect the 'strangeness' of the examples (α_i being the strangeness of (x_i, y_i), $i = 1, \ldots, l$, and α_{l+1}, being the strangeness of the (x_{l+1}, Υ)). By using Lagrangian multipliers α_i, we can approximate from below randomness deficiency. Gammerman et al. (58) did that by introducing a function:

$$p(z_1, \ldots, z_{l+1}) = \frac{f(\alpha_1) + \cdots + f(\alpha_{l+1})}{f(\alpha_{l+1})(l+1)} \qquad (2)$$

Here f is some monotonic non-decreasing function with $f(0) = 0$ as an upper bound for the randomness level. The suggested specific function $f(\alpha)$ was $f(\alpha) = $ sign α (that is, $f(0) = 0$ and $f(\alpha) = 1$ when $\alpha > 0$). Gammerman et al.'s method (58) corresponds to using the SVM method for prediction and using function [2] for estimating confidence and credibility. The α_i variables are non-negative and, in practice, only few of them are different from zero (the support vectors). An easily computable approximation of the randomness level is given by the *p-values* associated with every completion $(x_1, y_1), \ldots, (x_l, y_l), (x_{l+1}, \Upsilon)$:

$$\frac{\#\{i : \alpha_i \geq \alpha_{l+1}\}}{l+1}.$$

So, the p-value is the proportion of α's, which are at least as large as the last α. It is possible to show that these p-values are valid in the sense that they define a randomness test.

2.3.2. Predicting Plant Promoters by TSSP-TCM Program

To characterize promoter sequences, we use sequence content and signal features that were found in our previous works as being significantly different in promoter and non-promoter sequences (*see* **Table 5.1**). For training, we used 132 TATA and 104 TATA-less promoters. Forty genes with the TATA promoter annotated and 25 genes with the TATA-less promoter annotated were selected for the testing. All promoter sequences and other information were taken from the PlantProm DB (15). As negative samples (non-promoter sequences), 50,000 sequences of CDS and 50,000 sequences of introns of plant genes annotated in GenBank were extracted. The accuracy of recognition is shown in **Table 5.5**.

While testing on sets of promoter and non-promoter sequences demonstrated a good classification accuracy of the suggested approach, in practice we need to identify the most probable promoter location in long genomic sequences. For testing our recognition function on such sequences, we used the genomic sequences of the same 65 test genes. The performance of the TSSP_TCM program is presented in **Table 5.6**. For 35 of 40 TATA promoters (87.5%) and 21 of 25 TATA-less promoters (84%), TSS very close to the known one was predicted. For 29 TATA promoter genes (72.5%) and 14 TATA-less promoter genes (67%), the distances between known and nearest predicted TSS were 0–5 bp (**Fig. 5.3**). As upstream regions of plant genes usually ~2 kb, the developed approach having the rate of true predictions ~ 85% and one false-positive prediction

Table 5.6
Prediction accuracy of TSSP-TCM on plant genomic sequences[a]

	Forty TATA promoters	Twenty-five TATA-less promoters
False negatives	5	4
False positives	14	9
False positives' density	1 per 5,375 bp	1 per ~4,720 bp

[a]The confidence level for prediction of both promoter classes was 95% or higher. The credibility level was ≥35% for TATA promoters and ≥60% for TATA-less promoters. For every class of promoters only one predicted TSS with highest credibility level in an interval of 300 bp was taken.

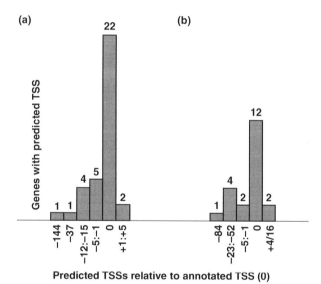

Fig. 5.3. The distance between the experimental and the closest predicted TSS in base pairs: **a** 35 (out of 40) genes with the annotated TATA promoters; **b** 21 (out of 25) genes with the annotated TATA-less promoters.

in ~4,000–5,000 bp can be successfully applied for identifying promoters in plant genomes.

2.4. Prediction of Bacterial Promoters

To identify discriminative features of bacterial promoter regions, we searched for conserved sequences in a set of known promoters from the *Escherichia coli* genome, which has the largest number of experimentally verified promoters. This set was used earlier in developing promoter prediction algorithm and described in Gordon et al. (61). Five relatively conserved sequence motifs represented by their weight matrices have been selected for a bacterial promoter model. Two most conserved motifs correspond to the well-characterized −10 and −35 sequence elements of promoters regulated by sigma70 factors. The third motif (upstream of the −35 box) with a length of 7 bp is searched in the area [−60 to −40]; the fourth motif (downstream of −10 block) with length 7 bp is searched in the area [−11 to +10] and the fifth motif (between −35 and −10 boxes) has length 5 bp and is searched in the area [−31 to −22] of potential promoter sequences. We applied linear discriminant analysis to derive the recognition function for discrimination between promoter and non-promoter sequences using as the 'negative' set of sequences from inner regions of protein-coding ORFs. We also considered the distance between −10 and −35 elements and the ratio of densities of octanucleotides overrepresented in known bacterial transcription factor binding sites relative to their occurrence in the coding regions. We used bacterial functional sites collected in the DPInteract database (62). The last feature was calculated similar to the one used in the eukaryotic promoter recognition programs such as PromoterScan (63) and TSSW (39). The linear discriminant function (LDF) implemented in Bprom demonstrated a sensitivity 83% and a specificity 84% in recognition of promoter and non-promoter sequences not included in the learning set. Bprom could be run at web servers of Royal Holloway (http://mendel. cs.rhul.ac.uk/mendel.php?topic=fgen) and Softberry, Inc. http://www.softberry.com/berry.phtml?topic=bprom&group= programs&subgroup=gfindb) or in combination with a hundred other bioinformatics software modules within the MolQuest package developed for Windows, MAC OS and Linux operation systems (http://www.molquest.com). Bprom has been used in numerous functional characterizations of bacterial sequences (64–67).

To reduce the rate of false-positive predictions, we recommend restricting prediction of promoters to the upstream regions of predicted ORFs in the annotation pipeline. The predicted promoters can help to refine the boundaries of operons as well.

2.5. Finding Statistically Significant Regulatory Motifs

Depending on cell/tissue type, developmental stage and extracellular signals (hormonal induction, stress, etc.) transcription factors (TFs) interact with their DNA-binding sites (regulatory elements, REs) and control gene expression. A gene expression pattern is primarily determined by the architecture of the promoter region including cooperativity and binding sites for alternatively functioning multiple TFs (*for review, see*: 68–70). The identification of REs is one of the critical steps in deciphering mechanisms of transcription regulation. Although large collections of various REs and corresponding TFs documented in several databases (20, 23–25, 33, 71) have been experimentally identified, we are extremely far from understanding a complex regulatory content of promoter regions. Moreover, as the experimental identification of TFs and their binding sites require enormous time and material resources, computer methods for predicting REs have particular significance.

There are two major approaches to this problem. The first one includes methods that identify REs based on available biological knowledge. The second approach relies on comparative analysis of homologous sequences (54, 72–74). To search for REs, most methods of the first type use real site consensus sequences expressed in terms of the IUPAC ambiguous nucleotide code or weight matrices (75–81). Pattern identification programs: SIGNAL SCAN (76), weight matrix-based approaches: ConsInspector (79), MatInspector (82) and MATRIX SEARCH (77) belong to this group. There are more complex approaches also have been applied to REs identification such as neural networks (83, 84), hidden Markov models (85, 86) and machine-learning methods (87). Benham (88) suggested detecting putative REs based on the prediction of possible sites of DNA duplex destabilization.

To account for involving multiple TFs/REs in the transcription regulation network, Kel et al. (89) created a *COMPELL* database of composite REs affecting gene transcription in vertebrates. Quandt et al. (80) have developed a software package GenomeInspector to detect potentially synergistic signals in genomes. A number of other computer-assisted promoter recognition methods devoted to the problem of combinatorial regulation of transcription have been published (63, 90). Thakurta and Stormo (91) reported a Co-Bind algorithm (Cooperative BINDing) for discovering DNA target sites of cooperatively acting TFs. At the same time for many promoter regions, information on their REs is not available yet and more complex composite REs are still remain to be discovered. In this connection, we should mention a group of methods for computational discovery of novel motifs (92–99; for a recent review, *see*: 100).

When we search for occurrence of functional motif in a query sequence, we consider the best alignments of the motif sequence with some of the sequence fragments. To assess whether a given alignment constitutes evidence for potential function of the aligned sequence, it helps to know how often such alignment can be expected from chance alone. We have suggested a probabilistic model to assess the statistical significance of the motif similarity (101).

2.5.1. Estimating the Statistical Significance of RE Sequences

Let us search for a site in a random nucleotide sequence of length N where the nucleotide frequencies are denoted by P_A, $P_{T(U)}$, P_G and P_C, respectively. If $P_1 = P_A$, $P_2 = P_G$, $P_3 = P_T$, $P_4 = P_C$, then the frequencies of the nucleotides of the other classes $P_j(j = 5, \ldots, 15)$ are determined as sums of frequencies of nucleotides of all the types included to the j-th class.

Simple (one − block) site. Let us consider a site of length L characterized by the values $N_l(l = 1, \ldots 15)$, where N_l is the number of nucleotides of the lth class belonging to the site and $N_1 + N_2 + \ldots N_{15} = L$.

Let assume that the site has M conserved positions characterized by the values $M_l(l = 1, \ldots, 14)$, where M_l is the number of conserved nucleotides of the lth class $(M_1 + M_2 + \ldots + M_{14} = M)$. Then $k(k = 0, 1, \ldots)$ mismatches between the site and the segment of length L belonging to the sequence under consideration are allowed only at $L-M$ variable positions. The number of mismatches between the consensus and the DNA segment of the lth class, $R_l(l = 1, \ldots, 14)$, should meet the following conditions: $0 \leq R_1 \leq \min(k, N_1 - M_1)$, $0 \leq R_2 < \min(k - R_1, N_2 - M_2), \ldots, 0 \leq R_{14} < \min K - R_1 - R_2 - \ldots - R_{13}, N_{14} - M_{14})$.

Assuming binomial distribution for matches and mismatches, the probability $P(L-k)$ of detecting the segment (L, k) of length L with mismatches in k variable positions between it and the site is

$$P(L, k) = \sum_{R_l=0}^{\min(k, N_1 - M_1)} \cdots \sum_{R_{15}=0}^{\min(k - R_1 - R_2 \ldots R_{14}, N_{14} - M_{15})}$$

$$C_{N_1 - M_1}^{R_1} P_1^{N_1 - P_1}(1 - P_1)^{R_1} \cdots C_{N_{15} - M_{15}}^{R_{15}} P_{15}^{N_{15} - R_{15}}(1 - P_{15})^{R_{15}} \qquad [3]$$

In this case the expected number $\bar{T}(L, k)$ of structures (L, k) in a random sequence of length N is

$$\bar{T}(L, k) = P(L, k) \times F_L$$

Here F_L is the number of possible site positions in the sequence: $F_L = N - L + 1$.

Let us assume that the mean number of motifs (L, k) in the random sequence is less than 1 or close to it. Then the probability of having precisely T structures (L, k) in the sequence may be estimated using the binomial distribution:

$$P(T) = C_{F_L}^T P^T(L, k)[1 - P(L, k)]^{F_L - T}. \qquad [4]$$

The probability of detecting in the sequence T structures with k or less mismatches is

$$P(T) = \sum_{z=0}^{k} C_{F_L}^T P^T(L, z)[1 - P(L, z)]^{F_L - T}. \qquad [5]$$

Now we can derive the upper boundary of the confidence interval T_0 (with the significance level q) for the expected number of structures in the random sequence:

$$\sum_{t=0}^{T_0 - 1} P(t) < q \quad \text{and} \quad \sum_{t=0}^{T_0} P(t) \geq q \qquad [6]$$

If the number of (L, k) structures detected in the sequence meets the condition $T \geq T_0$, they can be considered as potential functional sites with significance level q.

Composite (two – block) site. Let us consider a composite site containing two blocks of lengths L_1 and L_2 at a distance $D_t(D_1 < D < D_2)$, i.e. D_1 and D_2 are, respectively, the minimum and maximum allowed distances between the blocks). Let N_{1l} and N_{2l} be the number of nucleotides of the lth class in the first and second blocks, respectively ($l = 1, \ldots, 15$). It is clear that $N_{j1} + N_{j2} + \ldots + N_{j15} = L_j(j = 1, 2)$.

Let the first and second blocks have M_{1l} and M_{2l} conserved positions of the nucleotides of lth class. Then the probability $P(L_j, k_j)$ of finding in random sequence the segment (L_j, k_j) of size L_j differing in k_j non-conserved positions from the jth block of the site is calculated using equation [3] with the substitutions of L, k, N_l and M_l by L_j, k_j, N_{jl} and $M_{jl}(l = 1, \ldots, 14; j = 1, 2)$, respectively. The probability of simultaneous and independent occurrence of the segments (L_1, k_1) and (L_2, k_2) in the random sequence is

$$P(L_1, k_1, L_2, k_2) = P(L_1, k_1) \times P(L_2, k_2) \qquad [7]$$

The number of possible ways of arranging the segments (L_1, k_1) and (L_2, k_2) in the random sequence of length N is

$$F(L_1, L_2, D_1, D_2) = (D_2 - D_1 + 1)[N - L_1 - L_2 - \frac{D_1 + D_2}{2} + 1]$$
$$[8]$$

Thus, the expected number of structures $(L_1, k_1, L_2, k_2, D_1, D_2)$ is

$$\bar{T}(L_1, k_1, L_2, k_2, D_1, D_2) = F(L_1, L_2, D_1, D_2) \times P(L_1, k_1, L_2, k_2)$$

The probability $P(T)$ of detecting T structures $(L_1, k_1, L_2, k_2, D_1, D_2)$ in the random sequence is computed using equations [4] and [5] with the substitutions of F and $P(L, k)$ by $F(L_1, L_2, D_1, D_2)$ and $P(L_1, k_1, L_2, k_2)$ given in equations [7] and [8]. At last, using the obtained values $P(T)$, the upper boundary of the confidence interval T_0 can be computed from the conditions [6].

2.5.2. The Nsite, NsiteM and NsiteH Programs for Identification of Functional Promoter Elements

To search for regulatory motifs in human or animal genomic sequences, we can use available collections of functional elements such as TRANSFAC (102), TRANSCompel (33), TFD (20). Analysis of plant genomic sequences can be done with Regsite DB (http://linux1.softberry.com/berry.phtml?topic=regsite). Applying the statistical model described above, we have developed a group of computer programs for identification of statistically significant regulatory motifs including the *Nsite,*

Table 5.7
Web software for gene, promoters and functional signals prediction

Program & task	WWW address
Fgenesh HMM-based gene prediction (human, *Drosophila*, dicots, monocots, *C. elegans*, *S. pombe*, etc.)	http://sun1.softberry.com/berry.phtml?topic=fgenesh&group=programs&subgroup=gfind
Genscan HMM-based gene prediction (human, *Arabidopsis*, maize)	http://genes.mit.edu/GENSCAN.html
HMM-gene HMM-based gene prediction (human, *C.elegans*)	http://www.cbs.dtu.dk/services/HMMgene/
Fgenes Discriminative gene prediction (human)	http://sun1.softberry.com/berry.phtml?topic=fgenes&group=programs&subgroup=gfind
Fgenesh-M Prediction of alternative gene structures (human)	http://sun1.softberry.com/berry.phtml?topic=fgenesh-m&group=programs&subgroup=gfind
Fgenesh+/Fgenesh_c Gene prediction with the help of similar protein/EST	http://sun1.softberry.com/berry.phtml?topic=index&group=programs&subgroup=gfind
Fgenesh-2 Gene prediction using two sequences of close species	http://sun1.softberry.com/berry.phtml?topic=fgenes_c&group=programs&subgroup=gfs
BESTORF Finding best CDS/ORF in EST (human, plants, *Drosophila*)	http://sun1.softberry.com/berry.phtml?topic=bestorf&group=programs&subgroup=gfind
FgenesB Gene, operon, promoter and terminator prediction in bacterial sequences	http://sun1.softberry.com/berry.phtml?topic=index&group=programs&subgroup=gfindb
Mzef Internal exon prediction (human, mouse, *Arabidopsis*, yeast)	http://rulai.cshl.org/tools/genefinder/
FPROM/TSSP Promoter prediction (human/animals, plants) *NSITE* Search for functional motifs	http://sun1.softberry.com/berry.phtml?topic=index&group=programs&subgroup=promoter
Promoter 2.0 Promoter prediction	http://www.cbs.dtu.dk/services/Promoter/
CorePromoter Promoter prediction	http://rulai.cshl.org/tools/genefinder/CPROMOTER/index.htm
SPL/SPLM Splice site prediction (human, *Drosophila*, plants, etc.)	http://www.softberry.com/berry.phtml?topic=spl&group=programs&subgroup=gfind
NetGene2/NetPGene Splice site prediction (human, *C. elegans*, plants)	http://www.cbs.dtu.dk/services/NetPGene/
Scan2 Searching for similarity in genomic sequences and its visualization	http://sun1.softberry.com/berry.phtml?topic=scan2&group=programs&subgroup=scanh
RNAhybrid Prediction of microRNA target duplexes	http://bibiserv.techfak.uni-bielefeld.de/rnahybrid/

Identification of Promoter Regions and Regulatory Sites

NsiteM and *NsiteH* programs. Each of these programs performs searches for motifs of human/animal or plant REs, depending on the user's choice.

The Nsite program searches for one- or two-boxes statistically non-random REs using their sequences or consensuses in a single or a set of query sequences.

The NsiteM program searches for statistically significant REs motifs observed in a user defined percentage (default 50%) of a set of homologous sequences. The last condition serves as an additional criterion for selecting putative REs. As input data, it requires two or more sequences in FASTA format.

The NsiteH program discovers RE motifs with a given conservation level in a pair of aligned orthologous (homologous) sequences. Sequences are aligned beforehand by the SCAN2 program (http://softberry.com/scan.html).

3. Conclusions

For the prediction of promoters and the analysis of regulatory motifs, a wide array of programs are available through web servers (**Table 5.7**). The current accuracy is still not enough for their successful implementation as independent sub-modules to predict promoters on the whole genome sequences. It would be wise to use them in known or predicted upstream gene regions in combination with gene-recognition software tools. Many promoter prediction algorithms that use propensities of particular TF binding do not take into account the mutual orientation and positioning of these motifs. It would limit their performance, as the transcriptional regulation is a highly cooperative process involving simultaneous binding of many transcription factors. To make future progress in promoter identification, we need to study specific patterns of regulatory sequences, where definite mutual orientation and location of individual regulatory elements are necessary requirements for successful transcription initiation or its particular regulation.

References

1. Suzuki, Y., Taira, H., Tsunoda, T. et al. (2001) Diverse transcriptional initiation revealed by fine, large-scale mapping of mRNA start sites. *EMBO Rep* 2, 388–393.
2. Cooper, S., Trinklein, N., Anton, E. et al. (2006) Comprehensive analysis of transcriptional promoter structure and function in 1% of the human genome. *Genome Res* 16, 1–10.
3. Schmid, C.D., Perier, R., Praz, V., and Bucher, P. (2006) EPD in its twentieth year: towards complete promoter coverage of selected model organisms. *Nucleic Acids Res* 34, D82–D85.

4. Werner, T. (1999) Models for prediction and recognition of eukaryotic promoters. *Mamm Genome* 10, 168–175.

5. Pedersen, A.G., Baldi, P., Chauvin, Y., and Brunak, S. (1999) The biology of eukaryotic promoter prediction – a review. *Comput Chem* 23, 191–207.

6. Abnizova, I., Subhankulova, T., and Gilks, W. (2007) Recent computational approaches to understand gene regulation: mining gene regulation *in silico*. *Curr Genomics* 8, 79–91.

7. Bucher, P., and Trifonov, E. (1986) Compilation and analysis of eukaryotic POLII promoter sequences. *Nucleic Acids Res* 14, 10009–10026.

8. Schmid, C.D., Praz, V., Delorenzi, M. et al. (2004) The Eukaryotic Promoter Database EPD: the impact of *in silico* primer extension. *Nucleic Acids Res* 32, D82–D85.

9. Suzuki, Y., Yamashita, R., Sugano, S., and Nakai, K. (2004) DBTSS, DataBase of transcriptional start sites: progress report 2004. *Nucleic Acids Res* 32, D78–D81.

10. Halees, A.S., Leyfer, D., and Weng, Z. (2003) PromoSer: a large-scale mammalian promoter and transcription start site identification service. *Nucleic Acids Res* 31, 3554–3559.

11. Pohar, T.T., Sun, H., and Davuluri, R.V. (2004) HemoPDB: hematopoiesis promoter database, an information resource of transcriptional regulation in blood cell development. *Nucleic Acids Res* 32, D86–D90.

12. Palaniswamy, S.K., Jin, V.X., Sun, H., and Davuluri, R.V. (2005) OMGProm: a database of orthologous mammalian gene promoters. *Bioinformatics*, 21, 835–836.

13. Barta, E., Sebestyen, E., Palfy, T.B. et al. (2005) DoOP: databases of orthologous promoters, collections of clusters of orthologous upstream sequences from chordates and plants. *Nucleic Acids Res* 33, D86–D90.

14. Dieterich, C., Wang, H., Rateitschak, K. et al. (2003) CORG: a database for Comparative Regulatory Genomics. *Nucleic Acids Res* 31, 55–57.

15. Shahmuradov, I.A., Gammerman, A.J., Hancock, J.M. et al. (2003) PlantProm: a database of plant promoter sequences. *Nucleic Acids Res* 31, 114–117.

16. Wingender, E. (1988) Compilation of transcription regulating proteins. *Nucleic Acids Res* 16, 1879–1902.

17. Tjian, R. (1995) Molecular machines that control genes. *Sci Am* 272, 54–61.

18. Fickett, J., and Hatzigeorgiou, A. (1997) Eukaryotic promoter recognition. *Genome Res*, 7, 861–878.

19. Ghosh, D. (1990) A relational database of transcription factors. *Nucleic Acids Res* 18, 1749–1756.

20. Ghosh, D. (2000) Object-oriented Transcription Factors Database (ooTFD). *Nucleic Acids* Res 28, 308–310.

21. Wingender, E., Dietze, P., Karas, H., and Knuppel, R. (1996) TRANSFAC: a database of transcription factors and their binding sites. *Nucleic Acids Res* 24, 238–241.

22. Matys, V., Kel-Margoulis, O.V., Fricke, E. et al. (2006) TRANSFAC and its module TRANSCompel: transcriptional gene regulation in eukaryotes. *Nucleic Acids Res* 34, D108–D110.

23. Kolchanov, N.A., Ignatieva, E.V., Ananko, E.A. et al. (2002) Transcription regulatory Regions Database (TRRD): its status in 2002. *Nucleic Acids Res* 30, 312–317.

24. Lescot, M., Déhais, P., Thijs, G. et al. (2002) PlantCARE, a database of plant *cis*-acting regulatory elements and a portal to tools for *in silico* analysis of promoter sequences. *Nucleic Acids Res* 30, 325–327.

25. Higo, K., Ugawa, Y., Iwamoto, M., and Korenaga, T. (1999) Plant cis-acting regulatory DNA elements (PLACE) database: 1999. *Nucleic Acids Res* 27, 297–300.

26. Guo, A-Y., Chen, X., Gao, G. et al. (2008) PlantTFDB: a comprehensive plant transcription factor database. *Nucleic Acids Res* 36, D966–D969.

27. Morris, R.T., O'Connor, T.R., and Wyrick, J.J. (2008) Osiris: an integrated promoter database for *Oryza sativa L*. *Bioinformatics* 24, 2915–2917.

28. O'Connor, T.R., Dyreson, C., and Wyrick, J.J. (2005) Athena: a resource for rapid visualization and systematic analysis of *Arabidopsis* promoter sequences. *Bioinformatics* 21, 4411–4413.

29. Prestridge, D., and Burks, C. (1993) The density of transcriptional elements in promoter and non-promoter sequences. *Hum Mol Genet* 2, 1449–1453.

30. Tjian, R., and Maniatis, T. (1994) Transcriptional activation: a complex puzzle with few easy pieces. *Cell* 77, 5–8.

31. Diamond, M., Miner, J., Yoshinaga, S., and Yamamoto, K. (1990) Transcription factor interactions: selectors of positive or negative regulation from a single DNA element. *Science* 249, 1266–1272.

32. Kel, O., Romaschenko, A., Kel, A. et al. (1995) A compilation of composite regulatory elements affecting gene transcription in vertebrates. *Nucleic Acids Res* 23, 4097–4103.

33. Kel-Margoulis, O.V., Kel, A.E., Reuter, I. et al. (2002) TRANSCompel: a database on composite regulatory elements in eukaryotic genes. *Nucleic Acids Res* 30, 332–334.

34. Hutchinson, G. (1996) The prediction of vertebrate promoter regions using differential hexamer frequency analysis. *Comput Appl Biosci* 12, 391–398.

35. Audic, S., and Claverie, J. (1997) Detection of eukaryotic promoters using Markov transition matrices. *Comput Chem* 21, 223–227.

36. Guigo, R., Knudsen, S., Drake, N., and Smith, T. (1992) Prediction of gene structure. *J Mol Biol* 226, 141–157.

37. Reese, M.G., Harris, N.L., and Eeckman, F.H. (1996) Large scale sequencing specific neural networks for promoter and splice site recognition. In: *Biocomputing: Proceedings of the 1996 pacific symposium* (Hunter, L., and T. Klein, Eds.), World Scientific Publishing Co, Singapore.

38. Knudsen, S. (1999) Promoter2.0: for the recognition of PolII promoter sequences. *Bioinformatics* 15, 356–361.

39. Solovyev, V.V., and Salamov, A.A. (1997) The Gene-Finder computer tools for analysis of human and model organisms' genome sequences. In: *Proceedings of the 5th international conference on intelligent systems for molecular biology* (Rawlings, C., D. Clark, R. Altman, L. Hunter, T. Lengauer, and S. Wodak, Eds.) pp. 294–302, AAAI Press, Halkidiki, Greece.

40. Ohler, U., Harbeck, S., Niemann, H. et al. (1999) Interpolated Markov chains for eukaryotic promoter recognition. *Bioinformatics* 15, 362–369.

41. Scherf, M., Klingenhoff, A., Frech, K. et al. (2001) First pass annotation of promoters of human chromosome 22. *Genome Res* 11, 333–340.

42. Bajic, V.B., Seah, S.H., Chong, A. et al. (2002) Dragon promoter finder: recognition of vertebrate RNA polymerase II promoters. *Bioinformatics* 18, 198–199.

43. Burge, C., and Karlin, S. (1997) Prediction of complete gene structures in human genomic DNA. *J Mol Biol* 268, 78–94.

44. Salamov, A.A., and Solovyev, V.V. (2000) *Ab initio* gene finding in *Drosophila* genomic DNA. *Genome Res* 10, 516–522.

45. Solovyev, V.V. (2002) Finding genes by computer: probabilistic and discriminative approaches. In: *Current Topics in Computational Biology* (Jiang, T., T. Smith, Y. Xu, and M. Zhang, Eds.) pp. 365–401, The MIT Press, Cambridge, MA.

46. Liu, R., and States, D.J. (2002) Consensus promoter identification in the human genome utilizing expressed gene markers and gene modeling. *Genome Res* 12, 462–469.

47. Solovyev, V., Kosarev, P., Seledsov, I., and Vorobyev, D. (2006) Automatic annotation of eukaryotic genes, pseudogenes and promoters. *Genome Biol* 7(Suppl. 1), S10.1–S10.12.

48. Bajic, V., Brent, M., Brown, R. et al. (2006) Performance assessment of promoter predictions on ENCODE regions in the EGASP experiment. *Genome Biol* 7(Suppl. 1), S3.1–S3.13.

49. Anwar, F., Baker, S.M., Jabid, T. et al. (2008) Pol II promoter prediction using characteristic 4-mer motifs: a machine learning approach. *BMC Bioinformatics* 9, 4.

50. Afifi, A.A., and Azen, S.P. (1979) *Statistical analysis. A computer oriented approach.* Academic Press, New York, NY.

51. Bucher, P. (1990) Weight matrix descriptions of four eukaryotic RNA polymerase II promoter elements derived from 502 unrelated promoter sequences. *J Mol Biol* 212, 563–578.

52. Ohler, U., Liao, G.C., Niemann, H. et al. (2002) Computational analysis of core promoters in the *Drosophila* genome. *Genome Biol*, 3:1–12. RESEARCH0087.

53. Arumugam, M., Wei, C., Brown, R.H., and Brent, M.R. (2006) Pairagon+N-SCAN_EST: a model-based gene annotation pipeline. *Genome Biol* 7(Suppl. 1), S5.1–S5.10.

54. Wasserman, W.W., Palumbo, M., Thompson, W. et al. (2000) Human-mouse genome comparisons to locate regulatory sites. *Nat Genet* 26, 225–228.

55. Solovyev, V.V., and Shahmuradov, I.A. (2003) PromH: promoters identification using orthologous genomic sequences. *Nucleic Acids Res* 31, 3540–3545.

56. Shahmuradov, I.A., Solovyev, V.V., and Gammerman, A.J. (2005) Plant promoter prediction with confidence estimation. *Nucleic Acids Res* 33, 1069–1076.

57. Vapnik, V.N. (1998) *Statistical learning theory.* Wiley, New York, NY.

58. Gammerman, A., Vapnik, V.N., and Vovk, V. (1998) Learning by transduction. In: *Proceedings of the 14th conference on uncertainty in artificial intelligence,* 24–27 July, Madison, WI (Cooper, G.F., and S. Moral, Eds) pp. 148–156, Morgan Kaufmann, San Francisco, CA.

59. Vovk, V., Gammerman, A., and Saunders, C. (1999) Machine-learning applications of algorithmic randomness. In: *Proceedings of the 16th international conference on machine learning,* 27–30 June, Bled,

Slovenia (Bratko, I., and S. Dzeroski, Eds.) pp. 444–453, Morgan Kaufmann, San Francisco, CA.

60. Saunders, C., Gammerman, A., and Vovk, V. (2000) Computationally efficient transductive machines. In: *Proceedings of the 11th International Conference on Algorithmic Learning Theory,* 11–13 December, Sydney, Australia, Lecture Notes in Artificial Intelligence, Springer-Verlag, Berlin, pp. 325–333.

61. Gordon, L., Chervonenkis, A., Gammerman, A. et al. (2003) Sequence alignment kernel for recognition of promoter regions. *Bioinformatics* 19, 1964–1971.

62. Robison, K., McGuire, A.M., and Church, G.M. (1998) A comprehensive library of DNA-binding site matrices for 55 proteins applied to the complete Escherichia coli K-12 genome. *J. Mol. Biol* 284, 241–254.

63. Prestridge, D.S. (1995) Predicting Pol II promoter sequences using transcription factor binding sites. *J Mol Biol* 249, 923–932.

64. Pope, W.H., Weigele, P.R., Chang, J. et al. (2007) Genome sequence, structural proteins, and capsid organization of the cyanophage Syn5: a 'horned' bacteriophage of marine *Synechococcus. J Mol Biol* 368, 966–981.

65. Sriramulu, D.D., Liang, M., Hernandez-Romero, D. et al. (2008) *Lactobacillus reuteri* DSM 20016 produces cobalamin-dependent diol dehydratase in metabolosomes and metabolizes 1,2-propanediol by disproportionation. *J Bacteriol* 190, 4559–4567.

66. Singh, J., Banerjee, N. (2008) Transcriptional analysis and functional characterization of a gene pair encoding iron-regulated xenocin and immunity proteins of *Xenorhabdus* nematophila. *J Bacteriol* 190, 3877–3885.

67. Mariscotti, J.F., and García-Del Portillo, F. (2008) Instability of the *Salmonella* RcsCDB signalling system in the absence of the attenuator IgaA. *Microbiology* 154, 1372–1383.

68. Ptashne, M., and Gann, A. (1997) Transcriptional activation by recruitment. *Nature* 386, 569–577.

69. Lemon, B., and Tjian, R. (2000) Orchestrated response: a symphony of transcription factors for gene control. *Genes Dev* 14, 2551–2569.

70. Bonifer, C. (2000) Developmental regulation of eukaryotic gene loci. *Trends Genet* 16, 310–314.

71. Wingender, E., Karas, H., and Knüppel, R. (1997) TRANSFAC database as a bridge between sequence data libraries and biological function. *Pac Symp Biocomput* 1997, 477–485.

72. Duret, L., and Bucher, P. (1997) Searching for regulatory elements in human noncoding sequences. *Curr Opin Struct Biol* 7, 399–406.

73. Eisen, M.B., Spellman, P.T., Patrick, O. et al. (1998) Cluster analysis and display of genome-wide expression patterns. *Proc Natl Acad Sci USA* 95, 14863–14868.

74. Stormo, G.D. (2000) DNA binding sites: representation and discovery. *Bioinformatics* 16, 16–23.

75. Hertz, G.Z., Hartzell, G.W., and Stormo, G.D. (1990) Identification of consensus in unaligned DNA sequences known to be functionally related. *Comput Appl Biosci* 6, 81–92.

76. Prestidge, D.S., and Stormo, G. (1993) SIGNAL SCAN 3.0: new database and program features. *Comput Appl Biosci* 9, 113–115.

77. Chen, Q.K., Hertz, G.Z., and Stormo, G.D. (1995) MATRIX SEARCH 1.0: a computer program that scans DNA sequences for transcriptional elements using a database of weight matrices. *Comput Appl Biosci* 11, 563–566.

78. Eddy, S.R. (1996) Hidden Markov models. *Curr Opin in Struct Biol* 6, 361–365.

79. Frech, K., Herrmann, G., and Werner, T. (1993) Computer-assisted prediction, classification, and delimitation of protein binding sites in nucleic acids. *Nucleic Acids Res* 21, 1655–1664.

80. Quandt, K., Grote, K., and Werner, T. (1996) GenomeInspector: a new approach to detect correlation patterns of elements on genomic sequences. *Genomics* 33, 301–304.

81. Crowley, E.M., Roeder, K., and Bina, M. (1997) A statistical model for locating regulatory elements in genomic DNA. *J Mol Biol* 268, 8–14.

82. Quandt, K., Frech, K., Karas, H. et al. (1995) MatInd and MatInspector: new fast and versatile tools for detection of consensus matches in nucleotide sequence data. *Nucleic Acids Res* 23, 4878–4884.

83. Larsen, N.I., Engelbrecht, J., and Brunak, S. (1995) Analysis of eukaryotic promoter sequences reveals a systematically occurring CT-signal. *Nucleic Acids Res* 23, 1223–1230.

84. Milanesi, L., Muselli, M., and Arrigo, P. (1996) Hamming-clustering method for signals prediction in 5′ and 3′ regions of eukaryotic genes. *Comput Appl Biosci* 12, 399–404.

85. Frith, C.M., Hansen, U., and Weng, Z. (2001) Detection of cis-elements in

higher eukaryotic DNA. *Bioinformatics* 17, 878–889.

86. Pedersen, A.G., Baldi, P., Brunak, S., and Chauvin, Y. (1996) Characterization of prokaryotic and eukaryotic promoters using hidden Markov models. *Intel Sys Mol Biol* 4, 182–191.

87. Seledtsov, I.A., Solovyev, V.V., and Merkulova, T.I. (1991) New elements of glucocorticoid-receptor binding sites of hormone-regulated genes. *Biochim Biophys Acta* 1089, 367–376.

88. Benham, C.J. (1996) Computation of DNA structural variability – a new predictor of DNA regulatory regions. *Comput Appl Biosci* 12, 375–381.

89. Kel, A., Kel-Margoulis, O., Babenko, V., and Wingender, E. (1999) Recognition of NFATp/AP-1 composite elements within genes induced upon the activation of immune cells. *J Mol Biol* 288, 353–376.

90. Kondrakhin, Y.V., Kel, A.E., Kolchanov, N.A. et al. (1995) Eukaryotic promoter recognition by binding sites for transcription factors. *Comput Appl Biosci* 11, 477–488.

91. Thakurta, D.G., and Stormo, G.D. (2001) Identifying target sites for cooperatively binding factors. *Bioinformatics* 17, 608–621.

92. Staden, R. (1989) Methods for discovering novel motifs in nucleic acid sequences. *Curr Opin Struct Biol* 5, 293–298.

93. Bailey, T.L., and Elkan, C. (1995) Unsupervised learning of multiple motifs in biopolymers using expectation maximization. *Mach Learn* 21, 51–80.

94. Brazma, A., Jonassen, I., Vilo, J., and Ukkonen, E. (1998) Predicting gene regulatory elements in silico on a genomic scale. *Genome Res* 8, 1202–1215.

95. Mironov, A.A., Koonin, E.V., Roytberg, M.A., and Gelfand, M.S. (1999) Computer analysis of transcription regulatory patterns in completely sequenced bacterial genomes. *Nucleic Acids Res* 27, 2981–2989.

96. Geraghty, M.T., Bassett, D., Morrell, J.C. et al. (1999) Detecting patterns of protein distribution and gene expression *in silico. Proc Natl Acad Sci USA* 96, 2937–2942.

97. McGuire, A.M., and Church, G.M. (2000) Predicting regulons and their *cis*-regulatory motifs by comparative analysis. *Nucleic Acids Res* 28, 4523–4530.

98. Fujibuchi, W., Anderson, J.S., and Landsman, D. (2001) PROSPECT improves *cis*-acting regulatory element prediction by integrating expression profile data with consensus pattern searches. *Nucleic Acids Res* 29, 3988–3996.

99. Birnbaum, K., Benfey, P.N., and Shasha, D.E. (2001) *cis* element/transcription factor analysis (cis/TF): a method for discovering transcription factor/cis element relationships. *Genome Res* 11, 1567–1573.

100. Das, M.K., and Dai, H.K. (2007) A survey of DNA motif finding algorithms. *BMC Bioinformatics* 8(Suppl. 7), S21.

101. Shahmuradov, I.A., Kolchanov, N.A., Solovyev, V.V., and Ratner, V.A. (1986) Enhancer-like structures in middle repetitive DNA elements of eukaryotic genomes. *Genetika (Russ.)* 22, 357–367.

102. Wingender, E., Chen, X., Fricke, E. et al. (2001) The TRANSFAC system on gene expression regulation. *Nucleic Acids Res* 29, 281–283.

Chapter 6

Motif Discovery Using Expectation Maximization and Gibbs' Sampling

Gary D. Stormo

Abstract

Expectation maximization and Gibbs' sampling are two statistical approaches used to identify transcription factor binding sites and the motif that represents them. Both take as input unaligned sequences and search for a statistically significant alignment of putative binding sites. Expectation maximization is deterministic so that starting with the same initial parameters will always converge to the same solution, making it wise to start it multiple times from different initial parameters. Gibbs' sampling is stochastic so that it may arrive at different solutions from the same initial parameters. In both cases multiple runs are advised because comparisons of the solutions after each run can indicate whether a global, optimum solution is likely to have been achieved.

Key words: Expectation maximization, Gibbs' sampling, transcription factor binding sites, motif discovery, position weight matrices, position frequency matrices, regulatory sites, motif modeling.

1. Introduction

Frequently one can identify DNA segments that contain binding sites for specific transcription factors (TFs) but the resolution is not sufficient to identify the exact binding site positions. For example, one may have a set of genes that are controlled by a common TF and therefore expect to find binding sites for that TF in the regulatory regions of those genes. Depending on the species this may localize the binding site to a region of about 100 base pairs (bp) in bacteria, to several thousand base pairs (kbp) in higher eukaryotes. Another type of data is the physical evidence of binding of the TF to a specific region of DNA. For

I. Ladunga (ed.), *Computational Biology of Transcription Factor Binding*, Methods in Molecular Biology 674,
DOI 10.1007/978-1-60761-854-6_6, © Springer Science+Business Media, LLC 2010

example, chromatin-immunoprecipitation (ChIP) of DNA that is cross-linked to a TF can be hybridized to an array (ChIP-chip) or sequenced (ChIP-seq) to obtain regions that contain binding sites (*see* **Chapters 9, 10,** and **11**), and the resolution is often a few hundred base pairs. One might also perform selections of binding sites in vitro from random pools containing a large number of potential binding sites. Often the randomized regions are on the order of 10–30 bp, large enough that the exact location of the binding site may not be immediately obvious and some alignment procedure is necessary to find them. In each of these data sets the binding sites themselves are not precisely given, but one only has regions of sequences that can be inferred to contain them. In order to find the actual binding sites one employs a model that describes the features of the binding sites and an algorithm that attempts to find sites that conform to the model and are statistically significant. The motif should represent the specificity of the TF, and its high scoring occurrences in the set of sequences are predicted to be the individual binding sites. **Figure 6.1** is an abstract view of this type of data. Each line represents a sequence, such as one of the genomic regions identified as containing a binding site, and the thick segments within each line represent the unknown positions of the binding sites. The goal of motif discovery algorithms is to find the binding sites within the DNA segments and, in the process, determine the parameters of the motif that represent the specificity of the TF.

This type of problem has existed for many years, ever since one could sequence DNA and wished to determine the important features of regulatory sites. In the earliest days the segments containing the sites were generally quite short and there was extensive experimental data available that allowed one to

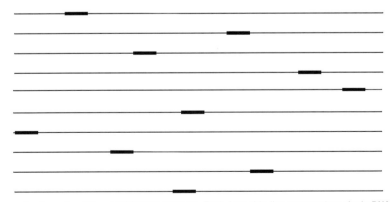

Fig. 6.1. A general schematic of the motif finding problem. Each *long thin line* represents a single DNA sequence. The *dark segments* within each line represent the binding sites whose positions are unknown in advance and we are trying to discover.

find the main features, such as the consensus sequence, "by eye" (1, 2). Within a few years an algorithm was published to automate the process of discovering consensus sequences from collections of unaligned DNA segments known to contain binding sites for a common factor (3). In the ensuing years many other approaches for discovering consensus sequences for regulatory sites have been published [for example (4, 5)]. The problem is non-trivial because individual binding sites are all similar to a consensus sequence but often have variations, which require that the algorithms for finding them must tolerate mismatches. However, it was also discovered in those early days that consensus sequences were limited in their ability to accurately represent the features of regulatory sites. Not only are mismatches to the consensus sequence common, but different positions within the binding sites have different degrees of variability, with some being highly conserved and others much less conserved (6). A weight matrix model (or position weight matrix, PWM) solves this problem by allowing different mismatches from the consensus sequence to have different scores. First developed using a discriminative learning procedure to find a PWM that would score known regulatory sites higher than similar, but non-functional sites (7), it was later used in a probabilistic model where the base distributions of the known sites were used directly in the PWM (8). **Figure 6.2** shows an example of this approach. **Figure 6.2a** is an aligned set of 10 binding sites for some TF. **Figure 6.2b** is just the count of each base at each position in the aligned set of sites, and **Fig. 6.2c** converts those directly to frequencies, which are sometimes referred to as a position frequency matrix (PFM). To convert to a PWM, in which the scores at each position are added to give the score for the entire site, the logarithms of the PFM elements are used in the PWM (8). **Figure 6.2d** shows a modified version of the PFM in which +1 has been added to each of the counts of the matrix of **Fig. 6.2b** before computing the frequencies because a small sample size may not represent the true distribution [and to avoid assigning a value to $\log(0)$]. The use of PWMs for motif discovery was initially developed using a progressive multiple alignment approach where an alignment with maximum information content was sought (9). Over the next few years new statistical approaches were introduced and in the intervening years many different algorithms for identifying TF binding motifs from unaligned DNA sequences have been developed (10, 11). The purpose of this chapter is to provide a primer on two important statistical methods, expectation maximization and Gibbs' sampling, that were initially adapted for the purpose of motif finding in biological sequences (both DNA and protein) by Charles E. Lawrence and colleagues (12, 13).

88   Stormo

A.

```
ATTCCGA
ATCACAA
TTTACGA
ATTGCGG
ACTTCGA
ATTAGGA
GTTACGA
ACTACCA
GTCTCGA
ATTTTGA
```

B.

Pos:	1	2	3	4	5	6	7
A	7	0	0	5	0	1	9
C	0	2	2	1	8	1	0
G	2	0	0	1	1	8	1
T	1	8	8	3	1	0	0

C.

Pos:	1	2	3	4	5	6	7
A	0.7	0.0	0.0	0.5	0.0	0.1	0.9
C	0.0	0.2	0.2	0.1	0.8	0.1	0.0
G	0.2	0.0	0.0	0.1	0.1	0.8	0.1
T	0.1	0.8	0.8	0.3	0.1	0.0	0.0

D.

Pos:	1	2	3	4	5	6	7
A	0.57	0.07	0.07	0.43	0.07	0.14	0.71
C	0.07	0.21	0.21	0.14	0.64	0.14	0.07
G	0.21	0.07	0.07	0.14	0.14	0.64	0.14
T	0.14	0.64	0.64	0.29	0.14	0.07	0.07

Fig. 6.2. **a** Alignment of binding sites. **b** The count matrix that shows how many each base (A, C, G, T) occurs at each position in the aligned sites. **c** The position frequency matrix (PFM) that converts the count matrix to a probability matrix by dividing by the total in each column. **d** An alternative PFM for the same data in which +1 has been added to each of the elements of the matrix of part **b**. This prevents any of the elements of the PFM from being 0 and may be important when the PFM is based on a small sample of binding sites.

2. Methods

2.1. Expectation Maximization

Expectation maximization (EM) is a general statistical procedure that allows for inferences when working with incomplete or missing data (14, 15). Its use in motif discovery is easily described in reference to **Fig. 6.1**. We assume that each of the sequences is composed of two parts, a background genomic sequence and the embedded binding site, and that those two parts have different statistical properties. The binding sites are modeled as a PFM, as in the PFM of **Fig. 6.2c**, where the probability of observing a specific base depends on the position within the binding site. We refer to the specific probability of a base at a position within a binding site as $P(b,i)$, where b is the base (A,C,G or T) and i is the position (between 1 and l, the length of the binding site). We also assume that the background can be described by an overall probability for each of the four bases. We will refer to this probability at $P(b,0)$ where 0 refers to the base coming from the background sequence rather than any of the positions in the binding site. In the simplest case this would just be 25% each of A, C, G, and T, but may be different in different species. It can probably be well estimated just from the overall composition of all the sequences, especially if the embedded binding sites make up only a small fraction of the total sequence. Consider each of the following scenarios where different information is available.

Scenario 1: If we had complete information, we knew $P(b,i)$ for the binding sites and $P(b,0)$ for the background, and we were also told where the binding site was located in each sequence, then we could calculate the probability of the observed collection of sequences. For example, suppose we know that there is a binding site at position J in the first sequence. Then the probability for the l positions starting at J (J to $J–l+1$) would be product of the $P(b,i)$ values from the PFM for the bases at each position in the binding site. All other positions in the sequence would be assigned the probabilities from the $P(b,0)$ distribution, and the product of the probabilities of every base in the sequence would be the product of those individual probabilities. And the probability of the entire set of sequences would be the product of the probabilities from each separate sequence. That value, by itself, will not be very useful. Since all of the probabilities are <1, and there are a very large number of bases in the entire data set, the total probability of any particular data set will be exceedingly small. But comparing different probabilities can be very useful, especially if there is some uncertainty in the information we are given.

Scenario 2: Suppose we are given the PFM for the binding sites, $P(b,i)$, and $P(b,0)$ for the background, but we are not told where the binding site is in each sequence. We could consider each possible position of the binding site, from $J=1$ to $J = L - l + 1$ (if the entire sequence is L-long) and for each possible choice of J calculate the probability of the entire sequence. A comparison of those probabilities will tell us which is the most likely (highest probability) choice for the binding site position J. And since each sequence is independent of the other sequences, we could do that for each sequence separately to get the most likely position of the binding site in each sequence and the probability of all the sequences assuming those binding sites.

Scenario 3: Suppose instead we are given the binding site positions in each sequence, but we are given neither the PFM, $P(b,i)$, for the TF nor $P(b,0)$. Now we could simply line up the binding sites, as we do in **Fig. 6.2**, to determine the PFM for the TF. Then we would know everything, the same as in *Scenario 1* above, and we can calculate the probability of all the sequences with their designated binding sites. In fact, it is easy to show that if we are given the binding site positions, the most likely values of the PFM (the values that create the highest probability for the entire set of sequences) are obtained by the procedure of **Fig. 6.2a–c**.

Scenario 4: We are given only the sequences but neither the PFM for the binding site (assume we are given its length l, although this is not necessary) nor the binding site locations. We now have the task of determining both the binding site locations, J, in each sequence and the PFM for the TF, $P(b,I)$, and the background probability $P(b,0)$. This is the classic motif finding problem and the EM algorithm we now describe comes from (12).

Step 1: Make a guess for an initial PFM. In (12), the initial PFM is derived by assuming all of the possible binding sites, every position in every sequence, are equally likely to be the true binding site. The PFM is then obtained from an alignment of all possible binding sites, each weighted by $1/(L - l + 1)$ so that the sum of the probabilities on each sequence equal 1. $P(b,0)$ comes from the overall probability of each base in all of the sequences. It is possible that doing this would end up with equal probability of each base at each position, so that $P(b,i) = P(b,0)$ for all i, in which case there would be no information to use in the following steps to distinguish between possible binding sites. But this is highly unlikely and even a small divergence from equal probability can be used in the subsequent steps to increase the overall probability.

Step 2: Given the $P(b,i)$ and $P(b,0)$ values from the previous step, one calculates the probability of each sequence for all possible choices of the binding site, for $J = 1$ to $L - l + 1$. In general some possible binding sites will now have higher probability than the average and some will have lower.

Step 3: A new PFM is derived from the alignment of all possible binding sites, but each one weighted by its probability as determined in *Step* 2. Because the individual sequences are no longer equally weighted, the PFM, and the values of $P(b,i)$, will change after this step. The background probability is taken over all of the bases in each sequence, but now weighted by the probability that they are not part of the binding site, which will lead to new values for $P(b,0)$.

Step 4: Repeat *Steps* 2 *and* 3 until convergence, when the values of $P(b,i)$, $P(b,0)$, and the predicted binding site probabilities no longer change. The total probability of the sequences is guaranteed by this procedure to increase after every step, until it finally converges. Regardless of the initial choice of the PFM, this method converges to some answer, which includes the predicted binding sites and their associated PFM. However, there is no guarantee that it will converge to the correct solution, or the solution with the highest probability over all possible choices. It is generally suggested that one starts the procedure, at Step 1, with different choices of the initial PFM. If it always converges to the same solution it is more likely to be correct than if it converges to many different ones.

2.1.1. Further Considerations

We started with the assumption that we knew the length of the binding site, but usually this is not the case. We can rerun the whole procedure with different choices and pick the one with the highest probability. It is also true that we may get the correct locations of the binding sites even if we do not know the correct length exactly. In that case we can often determine the correct length by aligning the sites after convergence and see if there is significant non-randomness in the adjacent positions. If so, use that length and rerun again to get the best model.

Since the choice of the initial PFM determines the final answer (the method is deterministic), it not only makes sense to start it with more than one initial guess, but any prior information that can be used to improve that guess is useful. For example, in bacteria the binding site is most likely to be within about 100 bp of the start of the gene, and so that region is used as the starting sequence, but it is more likely to be closer than that so one might weight the closer regions somewhat higher in the initial estimate of the PFM. One may have reason to expect the protein binds as a homodimer, a common occurrence in bacterial TFs, and therefore the PFM is likely to be symmetric. If the binding site is not symmetric, it might occur in different orientations in different examples, so one may need to consider both DNA strands in the search for a common motif. If one knows the type of transcription factor that is binding to the sites, for example, a zinc finger protein or a homeodomain protein, that provides information about the type of motif being sought and an initial bias toward that motif can help to find it (16). Other types of useful information can be applied to increase the likelihood that the EM procedure will converge to the correct solution (17).

The MEME program suite (17–20) implements the EM algorithm with many options that may be useful. For instance, one can specify that there may be more than one site per sequence. Or perhaps one is concerned that some of the sequences have no binding sites, so the program can be instructed to not require that every sequence contribute a site to the estimate of the PFM. One can run it to identify multiple motifs within the set of sequences, since in many cases TFs act coordinately to control gene expression so it would not be surprising to find more than one significant motif.

And once motifs are found they can be used directly in the search of genomic sequences to identify more predicted binding sites.

2.2. Gibbs' Sampling

The Gibbs' sampling (13, 21) procedure has some similarities to EM, but also some important differences. The most important difference is that it is not deterministic but rather uses a random sampling step. This means that multiple runs, starting with the same initial parameters may end up in different solutions. The practical consequence is that Gibbs' sampling is less likely than EM to get stuck in local optima and more likely to find a global optimum if run long enough. It can still benefit from multiple independent initializations, but it is better at exploring the "search space" of possible solutions than is EM and therefore more likely to find the best solution for the set of sequences. The following steps, in parallel with the steps of EM, describe the basic Gibbs' sampling algorithm and its differences from EM.

Step 1: From the set of sequences, leave one out and from the others choose a single site at random. If there are a total of N sequences, this gives an alignment of N–1 sites, as in **Fig. 6.2a**. From this set of sites one determines a PFM, but because N may not be a large number it is important to add pseudocounts, as in **Fig. 6.2d**, to avoid any of the $P(b,i)$ values being 0. If N is large the addition of a pseudocount of +1 has a very small effect, but it can be critical if N is small. The pseudocount may be something other than +1 and it may be different for different bases, for example, it may be proportional to the background probabilities for each base, $P(b,0)$, but the choice of +1 is fairly typical.

Step 2: Using the current PFM, calculate the binding probability of each potential site in the sequence that was left out. Since this sequence did not contribute to the PFM the probabilities of all its sites are independent estimates of binding probability, given the current values of $P(b,i)$. Rather than using the probability based on $P(b,i)$ alone, one usually computes the probability ratio for each potential site coming from the PFM versus coming from the background model, $P(b,0)$ (13). For computational efficiency this can be done directly by converting the PFM to a PWM by taking the logarithm of the ratio of the sites model, $P(b,i)$, to the background model, $P(b,0)$ to get the PWM:

$$W(b, i) = \ln \frac{P(b, i)}{P(b, 0)}$$

and then summing $W(b,i)$ over the positions of each potential binding site, rather than multiplying the $P(b,i)$ values of the PFM (6).

Step 3: From the probability ratios for each potential binding site (from the log probability ratios if one uses the PWM) one chooses a single site from the sequence where the choice is weighted by that probability ratio. That is, the higher the probability ratio the more likely a specific site will be chosen, but any site within the sequence has some probability of being selected. One of the sequences that was used in the previous step is now left out, and its site, which contributed to the previous PFM, is replaced by this new site. A new PFM is made from the alignment of this current set of sites as in **Fig. 6.2** (and a PWM if desired).

Step 4: *Steps* 2 *and* 3 are repeated as many times as desired. Unlike EM this procedure does not converge, although it tends to increase in score [generally a log-likelihood ratio score such as "information content" (6, 13)] until reaching a plateau that it fluctuates around. How long this takes is unknown, and independent runs may take quite different times [*see* **Fig. 3** of ref. (13)]. Usually the program is run multiple times for a fixed number of steps each time and then the final results after each run are compared. If they are all the same, or very similar, it has probably found the global optimum, but if they are each different it could be that none is really the optimum solution.

2.2.1. Further Considerations

Variations of the basic Gibbs' sampling approach have been developed by different groups, sometimes customized for specific types of data [for example (22, 23)]. The basic Gibbs' sampling procedure described can easily be extended to allow these multiple sites per sequence, as well as looking for multiple different motifs. Other constraints, as in the EM algorithm, can also be applied, such as requiring symmetric sites. And similar to the EM approach, prior information can be incorporated that biases the initial PFM toward the expected motif which can increase the likelihood of finding the correct solution.

3. Conclusions

EM and Gibbs' Sampling are both powerful statistical methods that are capable of identifying motifs and binding site de novo, without any prior information. There is no guarantee that they will succeed, either in finding the correct solution or in finding the highest probability of all possible solutions. Multiple runs that return the same solution are likely to be correct, but if many different solutions are found perhaps none are correct. Prior information of various types can be used to help find the correct solution. And it is easy to look for multiple motifs that may correspond to sets of factors that coordinately control gene expression.

Despite their usefulness, methods such as EM and Gibbs' sampling may fail to find the correct solution or any significant solution. This could be simply because the sequence set contains enough "incorrect data" (sequences that do not actually contain binding sites) that the motif occurrence is not statistically significant. It is also possible that the motif is present but does not conform to the PFM model. For example, the PFM model assumes that the positions of the binding sites contribute independently to the binding activity and that is an approximation that may or may not be true (24). Another potential problem is that the probability model of the PFM does not fit well and an energy model that takes into account the non-linear relationship between binding energy and binding probability is needed instead (25, 26). But even in such cases one may be able to get a good approximation to the true specificity of the factors being studied.

References

1. Pribnow, D. (1975) Nucleotide sequence of an RNA polymerase binding site at an early T7 promoter. *Proc Natl Acad Sci USA* 72, 784–788.
2. Rosenberg, M., and Court, D. (1979) Regulatory sequences involved in the promotion and termination of RNA transcription. *Annu Rev Genet* 13, 319–353.
3. Galas, D.J., Eggert, M., and Waterman, M.S. (1985) Rigorous pattern-recognition methods for DNA sequences. Analysis of promoter sequences from Escherichia coli. *J Mol Biol* 186, 117–128.
4. Pavesi, G., Mauri, G., and Pesole, G. (2001) An algorithm for finding signals of unknown length in DNA sequences. *Bioinformatics* 17(Suppl. 1), S207–S214.
5. Marschall, T., and Rahmann, S. (2009) Efficient exact motif discovery. *Bioinformatics* 25, i356–i364.
6. Stormo, G.D. (2000) DNA binding sites: representation and discovery. *Bioinformatics* 16, 16–23.
7. Stormo, G.D., Schneider, T.D., Gold, L., and Ehrenfeucht, A. (1982) Use of the 'Perceptron' algorithm to distinguish translational initiation sites in E. coli. *Nucleic Acids Res* 10, 2997–3011.
8. Staden, R. (1984) Computer methods to locate signals in nucleic acid sequences. *Nucleic Acids Res* 12, 505–519.
9. Stormo, G.D., and Hartzell, G.W., 3rd. (1989) Identifying protein-binding sites from unaligned DNA fragments. *Proc Natl Acad Sci USA* 86, 1183–1187.
10. Das, M.K., and Dai, H.K. (2007) A survey of DNA motif finding algorithms. *BMC Bioinformatics* 8(Suppl. 7), S21.
11. GuhaThakurta, D. (2006) Computational identification of transcriptional regulatory elements in DNA sequence. *Nucleic Acids Res* 34, 3585–3598.
12. Lawrence, C.E., and Reilly, A.A. (1990) An expectation maximization (EM) algorithm for the identification and characterization of common sites in unaligned biopolymer sequences. *Proteins* 7, 41–51.
13. Lawrence, C.E., Altschul, S.F., Boguski, M.S., Liu, J.S., Neuwald, A.F., and Wootton, J.C. (1993) Detecting subtle sequence signals: a Gibbs sampling strategy for multiple alignment. *Science* 262, 208–214.
14. Dempster, A.P., Laird, N.M., and Rubin, D.B. (1977). Maximum likelihood from incomplete data via the EM algorithm. *J R Stat Soc. Ser B (Methodol)* 39, 1–38.
15. Little, R.J.A., and Rubin, D.B. (2002). *Statistical analysis with missing data*, 2nd edn. Wiley, New York, NY.
16. Narlikar, L., Gordân, R., Ohler, U., and Hartemink, A.J. (2006) Informative priors based on transcription factor structural class improve de novo motif discovery. *Bioinformatics* 22, e384–e392.
17. Bailey, T.L., and Elkan, C. (1995) The value of prior knowledge in discovering motifs with MEME. *Proc Int Conf Intell Syst Mol Biol* 3, 21–29.
18. Bailey, T.L., and Elkan, C. (1994) Fitting a mixture model by expectation maximization to discover motifs in biopolymers. *Proc Int Conf Intell Syst Mol Biol* 2, 28–36.
19. Bailey, T.L., and Elkan, C.P. (1995) Unsupervised learning of multiple motifs in biopolymers using expectation maximization. *Mach Learn* 21, 51–80.
20. Bailey, T.L. (2002) Discovering novel sequence motifs with MEME. *Curr Protoc Bioinformatics* Chapter 2, Unit 2.4.

21. Liu, J.S., Neuwald, A.F., and Lawrence, C.E. (1995) Bayesian models for multiple local sequence alignment and Gibbs sampling strategies. *J Am Stat Assoc* 90, 1156–1170.

22. Roth, F.P., Hughes, J.D., Estep, P.W., and Church, G.M. (1998) Finding DNA regulatory motifs within unaligned noncoding sequences clustered by whole-genome mRNA quantitation. *Nat Biotechnol* 16, 939–945.

23. Liu, X., Brutlag, D.L., and Liu, J.S. (2001) BioProspector: discovering conserved DNA motifs in upstream regulatory regions of co-expressed genes. *Pac Symp Biocomput* 2001, 127–138.

24. Benos, P.V., Bulyk, M.L., and Stormo, G.D. (2002) Additivity in protein-DNA interactions: how good an approximation is it? *Nucleic Acids Res* 30, 4442–4451.

25. Djordjevic, M., Sengupta, A.M., and Shraiman, B.I. (2003) A biophysical approach to transcription factor binding site discovery. *Genome Res* 13, 2381–2390.

26. Zhao, Y., Granas, D., and Stormo, G.D. (2009) Inferring binding energies from selected binding sites. *PLoS Comp Bio*, 5, e1000590.

Chapter 7

Probabilistic Approaches to Transcription Factor Binding Site Prediction

Stefan Posch, Jan Grau, André Gohr, Jens Keilwagen, and Ivo Grosse

Abstract

Many different computer programs for the prediction of transcription factor binding sites have been developed over the last decades. These programs differ from each other by pursuing different objectives and by taking into account different sources of information. For methods based on statistical approaches, these programs differ at an elementary level from each other by the statistical models used for individual binding sites and flanking sequences and by the learning principles employed for estimating the model parameters. According to our experience, both the models and the learning principles should be chosen with great care, depending on the specific task at hand, but many existing programs do not allow the user to choose them freely. Hence, we developed Jstacs, an object-oriented Java framework for sequence analysis, which allows the user to combine different statistical models and different learning principles in a modular manner with little effort. In this chapter we explain how Jstacs can be used for the recognition of transcription factor binding sites.

Key words: Transcription factor binding sites, probabilistic models, generative learning, discriminative learning.

1. Introduction

Hundreds of different computer programs for the prediction of transcription factor binding sites have been developed over the last decades. However, many of them yield contradictory predictions, leading to long debates and a lot of frustration on the hallways of many biology departments all over the world. One of the reasons why there are so many different programs is that binding of transcription factors to their binding sites, and unbinding from

I. Ladunga (ed.), *Computational Biology of Transcription Factor Binding*, Methods in Molecular Biology 674,
DOI 10.1007/978-1-60761-854-6_7, © Springer Science+Business Media, LLC 2010

them, is an extremely complex process. Existing programs differ by taking into account different aspects of that complexity, by modeling the same aspects in a different manner, by taking into account different sources of additional information, and by pursuing different objectives.

Programs for *de novo motif discovery*, for example, (1–6) obtain as input a set of promoter sequences containing unaligned binding sites of unknown binding motifs. In contrast, programs for the *recognition* or *classification* of binding sites (7–13) are supplied with sets of known binding motifs. Orthologous promoters are used as additional information in approaches of phylogenetic footprinting (14, 15) or phylogenetic shadowing (16) and expression data (6, 17) and/or ChIP-Seq data (18) can be used as valuable additional information, too.

In addition to these differences, programs for the prediction of transcription factor binding sites often differ at an elementary level by the statistical models used for individual binding sites and flanking sequences and by the learning principles employed for estimating the model parameters. Many existing programs for the prediction of transcription factor binding sites do not allow the user to choose the statistical models or the employed learning principle. However, according to our experience, both should be chosen with great care, depending on the specific task at hand. Hence, we developed Jstacs (www.jstacs.de), an object-oriented Java framework for sequence analysis, which allows the user to combine different statistical models and different generative and discriminative learning principles in a modular manner with little effort.

In this chapter, we focus on the recognition of transcription factor binding sites and explain step by step how Jstacs can be used for this task. By choosing a simple example we illustrate how models can be learned based on different learning principles, how each of the model combinations can be evaluated based on independent test sets, and how the resulting classifier can finally be used for the prediction of binding sites of steroid hormone receptors in human promoter sequences. We provide the complete program combining all of the source code snippets used in this chapter as well as example data sets as supplementary material at www.jstacs.de/index.php/MiMB.

2. Software

All code examples are based on Jstacs, an open-source Java framework for statistical analysis and classification of biological sequences. Jstacs is easy to use and readily extensible due to its strictly object-oriented design.

Jstacs comprises an efficient representation and convenient handling of sequence data and provides ready-to-use implementations of many statistical models for sequence data (**Section 3.2**). These models can be learned generatively (**Section 3.3**) or discriminatively (**Section 3.4**) and can be combined to constitute classifiers. Jstacs comes with assessment methods which are used for comparing different classifiers on test data sets or by hold-out experiments. For evaluating classifiers, the user may choose from several performance measures, e.g., sensitivity or specificity. Jstacs also provides classes for de novo motif discovery spanning from generative approaches using the EM algorithm (*See* **Chapter 6**, This volume) to more recent discriminative (19, 20) discovery algorithms.

Jstacs is capable of handling a great variety of data and is not restricted to DNA sequences. Data sets are called `Samples` in Jstacs and consist of a number of `Sequences`. For convenience, we implement the class `DNASample` that allows to easily load data sets comprising DNA sequences.

Jstacs comes not only with implementations of statistical models for sequence analysis, which help experimentalists to analyze their data, but it is also based on an object-oriented infrastructure, which assists the implementation and assessment of new models. To this end, Jstacs provides interfaces and abstract classes for statistical models, e.g., `AbstractModel`, and classifiers, e.g., `AbstractClassifier`. New statistical models that implement and extend the required interfaces and abstract classes may be combined for obtaining a classifier without further implementation overhead. If that classifier extends a predefined abstract class, it is ready to be trained and to be evaluated on given data and to be used for classification of new data.

To get started with Jstacs, a Java Runtime Environment (JRE)[1] of at least version 5 is required. The easiest way to run the example code is to download the Jstacs binaries, which are publicly available at www.jstacs.de, and extract them into a directory of your choice. Download the example Java code file and the data sets from www.jstacs.de/index.php/MiMB and follow the instructions given in *Getting started*.

We present and explain parts of the example code in the sequel. Line numbers in front of the code snippets allow to quickly identify these parts in the example-code file, where also detailed comments are supplied.

---

[1] www.sun.com/java

3. Methods

In this section we present some of the theoretical concepts for probabilistic prediction of binding sites and demonstrate these concepts using some of the basic functions of Jstacs. As a specific example we choose the prediction of binding sites of mammalian transcription factors – namely androgen receptors (AR), glucocorticoid receptors (GR), and progesterone receptors (PR) – from the family of steroid hormone receptors, which we refer to as AR/GR/PR, in a set of human promoter sequences. Specifically, we present how to load the training data of experimentally verified AR/GR/PR binding sites into Jstacs and how to obtain the binding motif of these sites using different learning principles. Subsequently we show how to evaluate the performance of the resulting classifiers and how to perform the final recognition of AR/GR/PR binding sites in human promoter sequences using the resulting classifier.

3.1. Classification

Taking a probabilistic approach to transcription factor binding site prediction requires the definition of probabilities for each possible sequence $x = (x_1, x_2, \ldots, x_L)$ corresponding to a putative binding site of fixed length L, where each x_ℓ is from the alphabet $\Sigma = \{A, C, G, T\}$ of the four nucleotides A, C, G, and T. Our goal is to distinguish binding sites – called foreground sequences and abbreviated by fg – from flanking regions – called background sequences and abbreviated by bg. Hence, we need likelihoods $P(x|c, \theta)$ with parameters θ for the occurrence of a sequence for both classes $c \in C = \{\text{fg}, \text{bg}\}$.

For classification it is common to use the *Bayes classifier* that decides for class c^* with

$$c^* = \underset{c \in C}{\operatorname{argmax}} P(c|x, \theta) = \underset{c \in C}{\operatorname{argmax}} P(c, x|\theta) \qquad [1]$$

where $P(c|x, \theta)$ denotes the posterior probability of class c given sequence x and parameters θ and where $P(c, x|\theta)$ denotes the joint likelihood of class c and sequence x given parameters θ. In case of two classes and a properly chosen threshold, this classifier is equivalent to the *likelihood ratio classifier*, which decides for the foreground if $P(x|\text{fg}, \theta)/P(x|\text{bg}, \theta)$ exceeds a given threshold and which decides for the background otherwise.

The main challenge for probabilistic approaches is to estimate the likelihood $P(x|c, \theta)$. In addition, an estimation of the class probability $P(c|\theta)$ is formally required, although is not critical in most applications. For classification, we are in the case of supervised learning and are given a data set of N labeled data points (x_n, c_n), which we denote by $D = (x_1, \ldots, x_N)$ and

$\boldsymbol{c} = (c_1, \ldots, c_N)$ in the following. The data points $(\boldsymbol{x}_n, c_n)$ are assumed to be independent and identically distributed (i.i.d.) according to the joint likelihood $P(c, \boldsymbol{x}|\boldsymbol{\theta})$.

Approaches for obtaining these probabilities from a set of training data differ mainly by the families of statistical models (**Section 3.2**) chosen for the likelihoods $P(\boldsymbol{x}|c, \boldsymbol{\theta})$ and by the learning principle (**Sections 3.3** and **3.4**) chosen for estimating the parameters $\boldsymbol{\theta}$ of these models.

To solve any classification problem we need to handle data sets. The package `de.jstacs.data` of Jstacs contains Java classes to represent data. Here we use the class `DNASample` for handling the foreground and background data, assuming that the training sequences are stored in FastA-files `foreground.fa` and `background.fa`:

```
53 Sample fgData = new DNASample( "foreground.fa" );
54 Sample bgData = new DNASample( "background.fa" );
```

The data sets `fgData` and `bgData` are subsequently used for training statistical models. Jstacs also supports plain text files and, via the `BioJavaAdapter`, all formats and data bases accessible from BioJava.

3.2. Statistical Models

To characterize the distribution of binding sites, the prevalent statistical model is currently still the position weight matrix (PWM) model (e.g. (7, 8), *See* **Chapter 6**, this volume). This model assumes statistical independence of the nucleotides observed at different positions. As a consequence, the likelihood of a sequence decomposes as

$$P(\boldsymbol{x}|\mathrm{fg}, \boldsymbol{\theta}_{\mathrm{fg}}) = \prod_{\ell=1}^{L} P_{\ell}(x_{\ell}|\mathrm{fg}, \boldsymbol{\theta}_{\mathrm{fg}}) \qquad [2]$$

where the parameter[2] $\boldsymbol{\theta}_{\mathrm{fg}}$ denotes the matrix of four rows and L columns called PWM. Here, the matrix element in row 1, 2, 3, or 4 and column ℓ contains the probability of finding nucleotide A, C, G, or T, respectively, at position ℓ in the binding site of length L (7, 8) (*See* **Chapter 6**, this volume). The index ℓ in $P_{\ell}(x_{\ell}|\mathrm{fg}, \boldsymbol{\theta}_{\mathrm{fg}})$ emphasizes that these probabilities may vary from position to position. **Figure 7.1** shows the PWM of binding sites of the AR/GR/PR-family and the corresponding consensus sequence and sequence logo (21). The first three entries of row 1

---

[2] Note that the parameters $\boldsymbol{\theta}$ contain the parameters for each class, e.g., $\boldsymbol{\theta}_{\mathrm{fg}}, \boldsymbol{\theta}_{\mathrm{bg}}$, and the class probabilities.

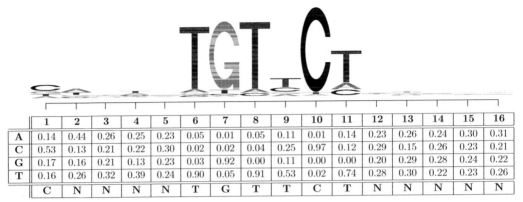

Fig. 7.1. Sequence logo, position weight matrix, and consensus sequence of the binding sites of the AR/GR/PR family.

of this PWM show that nucleotide *A* occurs with a probability of 0.14 at position 1, 0.44 at position 2, and 0.26 at position 3 of the binding sites of the AR/GR/PR family. The consensus sequence is composed of the consensus nucleotides, i.e., the nucleotides with the highest probability at each position. Here, we replaced consensus nucleotides with a probability of less than 0.5 by *N*.

As this strong assumption of independence is questionable in general (4), inhomogeneous Markov models (iMMs) of higher order have been used for modeling bindings sites (22, 23). In these models the probability of observing a nucleotide at a given position depends on nucleotides observed at previous positions. This results in the likelihood:

$$P(\boldsymbol{x}|\text{fg}, \boldsymbol{\theta}_{\text{fg}}) = \prod_{\ell=1}^{L} P_\ell(x_\ell | x_{\ell-m_\ell}, \ldots, x_{\ell-1}, \text{fg}, \boldsymbol{\theta}_{\text{fg}}) \quad [3]$$

where $x_{\ell-m_\ell}, \ldots, x_{\ell-1}$ with $m_\ell = \min\{m, \ell - 1\}$ defines the context of at most m nucleotides on which the nucleotide at position ℓ depends on. The maximal length m of the context defines the model order of the Markov model. Clearly, the PWM is an iMM of order zero. **Note 1** provides an intuitive justification and further extensions of Markov models.

3.3. Generative Learning Principles

Generative learning principles aim at an accurate description of the probability distributions of binding sites and background sequences. This may seem the only sensible way of estimating model parameters, but we see in the next sections that other learning principles are conceivable and potentially superior in many applications of binding site recognition.

3.3.1. Maximum Likelihood Principle

The maximum likelihood (ML) principle is probably the most popular learning principle (**Chapter 5**). It suggests to choose

Probabilistic Approaches to Transcription Factor Binding Site Prediction   103

those parameters θ that maximize the joint likelihood $P(\boldsymbol{D}, \boldsymbol{c}|\theta)$ of the labeled data set $(\boldsymbol{D}, \boldsymbol{c})$:

$$\hat{\theta}^{\text{ML}} = \underset{\theta}{\operatorname{argmax}} P(\boldsymbol{D}, \boldsymbol{c}|\theta) \qquad [4]$$

The estimate $\hat{\theta}^{\text{ML}}$ is used in rule (1) for the decision of the classifier. Note that the parameters θ contain the parameters for each class – according to the chosen family of distributions – and the class probabilities.

If the parameters of different classes are assumed to be independent, which is usually appropriate, maximization can be performed for each class separately. In case of PWM models, ML estimation amounts to counting frequencies of nucleotides in the data set, i.e., the ℓth column of the ML estimate $\hat{\theta}^{\text{ML}}_{\text{fg}}$ contains the relative frequencies of the four nucleotides A, C, G, and T at position ℓ. For example, the AR/GR/PR data set contains 104 binding sites, out of which 15 binding sites start with an A, 46 binding sites have an A at position 2, and 27 binding sites have an A at position 3, resulting in the relative frequencies $\frac{15}{104} = 0.14$, $\frac{46}{104} = 0.44$, and $\frac{27}{104} = 0.26$. **Figure 7.1** contains the full $4 \times L$ matrix $\hat{\theta}^{\text{ML}}_{\text{fg}}$.

As with all estimation methods, care must be taken with regard to *overfitting*, which is the effect of over-adaptation of the estimated parameters to noise and/or randomness in the training data. Overfitting results in a weak ability of generalizing to new data, i.e., of predicting transcription factor binding sites in promoter sequences not used for training. In general, the risk of overfitting increases with decreasing sample size and with increasing model complexity. For a PWM model overfitting can easily be understood for a case where at some position ℓ some nucleotide was not observed by chance in the training data set, resulting in an estimated probability of zero for this nucleotide at this position. This essentially "forbids" such sites, although they may not occur in the training data just by chance. To alleviate this problem, often *pseudo-counts* are added to the data.

3.3.2. Maximum A Posteriori Principle

The maximum a posteriori (MAP) principle takes a Bayesian view on parameter estimation. This learning principle employs a prior density $P(\theta|\alpha)$ for the parameters θ, which is used for representing prior knowledge or assumptions. The prior is chosen from a family of distributions and α denotes the hyperparameters of the prior. For the MAP principle, the objective is to choose those parameters θ that maximize the posterior. Decomposing the posterior yields

$$\hat{\theta}^{\text{MAP}} = \underset{\theta}{\operatorname{argmax}} P(\theta|\boldsymbol{D}, \boldsymbol{c}, \alpha) = \underset{\theta}{\operatorname{argmax}} P(\boldsymbol{D}, \boldsymbol{c}|\theta) P(\theta|\alpha) \qquad [5]$$

This shows that maximizing the posterior can be viewed as maximizing the data likelihood multiplied by the corresponding density of the parameters. Furthermore, the posterior can be interpreted as the knowledge we have about the parameters updating the prior with the observed data. **Note 2** presents more information on Bayesian approaches.

If we choose the prior from the family of Dirichlet distributions (**Note 3**) using consistent hyperparameters α (**Note 4**), the resulting MAP estimator corresponds to using pseudo-counts derived from a set of virtually observed pseudo-data. The amount of pseudo-data used is called the *equivalent sample size* (ESS) and determines the influence of the prior on the parameter estimate. In analogy to the ML estimate of a PWM, the MAP estimate of a PWM can be easily obtained from absolute frequencies of the data plus pseudo-counts stemming from the prior distribution. For more complex models the product-Dirichlet prior can be used and allows alleviating the problem of overfitting (**Note 4**).

3.3.3. ML and MAP Learning in Jstacs

The Jstacs package `de.jstacs.models` and its sub-packages contain classes for models that can be trained generatively. The class `BayesianNetworkModel` of the sub-package `de.jstacs.models.discrete.inhomogeneous` is a ready-to-use implementation of a Bayesian network. Inhomogeneous Markov models (**Section 3.2**), which we use for representing foreground and background sequences, are special cases of Bayesian networks.

In our example, we choose order 0 for the foreground model and the background model, i.e., we decide for a PWM model for both classes. We decide for MAP parameter estimation with ESS = 4 for the foreground model and ESS = 1,024 for the background model. For the foreground model, we create a `BayesianNetworkModelParameterSet`, which is a container of external model parameters, and which can be used for instantiating a `BayesianNetworkModel`:

```
59 BayesianNetworkModelParameterSet pars =
60   new BayesianNetworkModelParameterSet(
61     fgData.getAlphabetContainer(),
62     fgData.getElementLength(), 4, "fg model",
63     ModelType.IMM, (byte)0, LearningType.ML_OR_MAP );
```

External parameters are the alphabet, the length of the sequences, the ESS, a description of the model, the type of the model, and the order of the model. The last external parameter determines how the parameters are estimated, where `LearningType.ML_OR_MAP` indicates that the ML or the MAP learning principle is used. If we set t the ESS to 0 instead of 4, we obtain ML instead of MAP parameter estimation.

We instantiate the foreground model by calling

```
69 Model fgModel = new BayesianNetworkModel( pars );
```

We construct the background model `bgModel` in analogy to the foreground model. Details are given in the example code file.

As described above, we can estimate the parameters of the foreground model, the background model, and the a priori probabilities of the classes independently of each other. While we could perform these steps by hand, Jstacs provides a convenient implementation in the class `ModelBasedClassifier` of package `de.jstacs.classifier.modelBased`.

```
85 ModelBasedClassifier cl =
86   new ModelBasedClassifier( fgModel, bgModel );
```

A `ModelBasedClassifier` is a subclass of `Abstract Classifier`. As such it has a method `train(Sample...)`, which can be used for training the models as well as the class probabilities.

```
87 cl.train( fgData, bgData );
```

The variable `cl` now holds a classifier that comprises a generatively trained PWM model as foreground model and a generatively trained PWM model as background model. We can next assess the classification performance of that classifier (**Section 3.5.3**) or we can use it for recognizing binding sites (**Section 3.6**).

3.4. Discriminative Learning Principles

Discriminative learning principles (**Note 5**) have been introduced to bioinformatics in the last decade as a promising alternative to generative learning principles. While the latter aim at an accurate representation of the probability distributions of the data in each of the classes, discriminative learning principles focus on an accurate discrimination of the data.

3.4.1. Maximum Conditional Likelihood Principle

The maximum conditional likelihood (MCL) principle (19, 24) is the discriminative analog of the ML principle. It suggests to choose those parameters $\boldsymbol{\theta}$ that maximize the conditional likelihood $P(\boldsymbol{c}|\boldsymbol{D},\boldsymbol{\theta})$:

$$\hat{\boldsymbol{\theta}}^{\mathrm{MCL}} = \underset{\boldsymbol{\theta}}{\mathrm{argmax}}\, P(\boldsymbol{c}|\boldsymbol{D},\boldsymbol{\theta}) \qquad [6]$$

The maximization of the conditional likelihood is motivated by the classification rule (1) and the MCL principle was successfully applied to the classification of biological sequences (25, 13).

Solving this maximization problem is more involved than for the ML and MAP principles. First, maximization cannot be performed for the classes independently. Second and more severely, optimization cannot be done analytically for many popular families of distributions including Markov models. For Markov models considered in this chapter, numerical methods converging to the global maximum are available for a properly chosen parameterization of the Markov model and its prior distribution (e.g., (19)). A sensible choice of the initial values of the parameters are the corresponding generative estimates, which are called plug-in parameters in Jstacs. In general, the resulting numerical methods are computationally more demanding than the analytical solution in the generative setting.

3.4.2. Maximum Supervised Posterior

The effects of overfitting due to limited data may be even more severe when using the discriminative MCL principle compared to the generative ML principle (24). To overcome this problem, the maximum supervised posterior (MSP) principle has been proposed as another discriminative learning principle (20). The MSP principle suggests to choose those parameters that maximize

$$\hat{\boldsymbol{\theta}}^{\mathrm{MSP}} = \operatorname*{argmax}_{\boldsymbol{\theta}} P(\boldsymbol{c}|\boldsymbol{D},\boldsymbol{\theta},\alpha)P(\boldsymbol{\theta}|\alpha) \qquad [7]$$

Comparing this equation [7] to equation [6], we see that the MSP principle is the Bayesian analog of the non-Bayesian MCL principle. Comparing equations [7] to [5], we see that it can also be interpreted as the discriminative analog of the generative MAP principle, because the supervised posterior is defined as the product of the conditional likelihood [6] and the prior $P(\boldsymbol{\theta}|\boldsymbol{\alpha})$. As for the MAP principle, prior knowledge on the parameters is introduced via the distribution $P(\boldsymbol{\theta}|\boldsymbol{\alpha})$ and again frequencies of zero are compensated for. The remarks made with regard to optimization for the MCL principle apply to the MSP principle as well.

Figure 7.2 summarizes the four learning principles described.

	non-Bayesian	Bayesian
Generative	ML	MAP
Discriminative	MCL	MSP

Fig. 7.2. The tableau distinguishes the four learning principles introduced with regard to the generative or discriminative objective and with regard to the use of prior knowledge.

Probabilistic Approaches to Transcription Factor Binding Site Prediction   107

3.4.3. MCL and MSP Learning in Jstacs

In **Section 3.3.3** we have seen how Jstacs can be used for creating a `BayesianNetworkModel`. The discriminative counterpart of that model is the `BayesianNetworkScoring Function` located at `de.jstacs.scoringFunctions. directedGraphicalModels`. In analogy to the instantiation of a `BayesianNetworkModel`, we first define the external parameters of the foreground model:

```
92 BayesianNetworkScoringFunctionParameterSet parsD =
93   new BayesianNetworkScoringFunctionParameterSet(
94     fgData.getAlphabetContainer(),
95     fgData.getElementLength(), 4, true,
96     new InhomogeneousMarkov( 0 ) );
```

where `true` in line 95 indicates that we use plug-in parameters for initializing the parameters. Using `new InhomogeneousMarkov(0)` results in an inhomogeneous Markov model of order 0, i.e., a PWM model, and the remaining parameters have the same meaning as in the generative case. We use these parameters for instantiating the foreground model by

```
13 BayesianNetworkScoringFunction fgFun =
14   new BayesianNetworkScoringFunction( parsD );
```

We instantiate the background model `bgFun` accordingly with $ESS = 1,024$ instead of 4. Details can be found in the example code file.

We combine these models in a `MSPClassifier` of package `de.jstacs.classifier.scoringFunctionBased.msp`, which learns the parameters of the models by the discriminative MSP principle. To instantiate this classifier, we specify its external parameters by

```
112 GenDisMixClassifierParameterSet clPars =
113   new GenDisMixClassifierParameterSet(
114     fgData.getAlphabetContainer(),
115     fgData.getElementLength(),
116     Optimizer.QUASI_NEWTON_BFGS, 1E-6, 1E-6, 1,
117     false, KindOfParameter.PLUGIN, true, 1 );
```

where `QUASI_NEWTON_BFGS, 1E-6, 1E-6, 1` define the method for numerical optimization and parameters thereof. `KindOfParameter.PLUGIN` indicates that we want to use plug-in parameters for the class probabilities as well.

We instantiate the classifier from these parameters by

```
125 MSPClassifier cll = new MSPClassifier(
126    clPars, new CompositeLogPrior(), fgFun, bgFun );
```

where `new CompositeLogPrior()` may be replaced by `null` for obtaining the MCL principle.

We can now train this classifier in analogy to the generative case by calling

```
133 cll.train( fgData, bgData );
```

which starts the numerical optimization and results in an discriminatively trained classifier. In **Section 3.5.3** we show how the classification performance of such a classifier can be assessed and in **Section 3.6** we show how it might be used for the recognition of binding sites.

3.5. Comparison of Models and Learning Principles

In the previous sections, we considered Markov models of different orders on the one hand and different learning principles on the other. However, we do not know in advance which combination of models and learning principle is best for a certain problem and a certain data set. Hence, we typically scrutinize the performance of different classifiers – using different pairs of models and different learning principles – on the specific data set using several performance measures and we strongly recommend this approach to everyone working on the recognition of transcription factor binding sites.

3.5.1. Performance Measures

All performance measures considered in this chapter can be derived from the *confusion matrix*. The general schema of a confusion matrix is depicted in **Fig. 7.3**. Given that the data are

		Actual		
		fg	bg	
Predicted	fg	$TP = 51$	$FP = 959$	$\bar{p} = 1{,}010$
	bg	$FN = 1$	$TN = 27{,}556$	$\bar{n} = 27{,}557$
		$p = 52$	$n = 28{,}515$	$N' = 28{,}567$

Fig. 7.3. Confusion matrix. The entries of the matrix are computed for PWM models on the AR/GR/PR data set using a classification threshold of 1, corresponding to a threshold of 0 on the log-likelihood ratios.

partitioned into a training and a test data set and that a classifier has been learned on the training data set, we use this classifier for predicting the class of each of the sequences in the test data set. Subsequently we determine the number of correctly classified sequences from the foreground class (true positives, TP) and background class (true negatives, TN) as well as the number of sequences classified incorrectly as belonging to the foreground class (false positives, FP) or background class (false negatives, FN). The sum $TP + FN$ is equal to the number of foreground sequences p, $FP + TN$ is equal to the number of background sequences n, $TP + FP$ is the number of sequences $\bar{p}$ classified into the foreground class, and $TN + FN$ is the number of sequences $\bar{n}$ classified into the background class. Finally, $p + n = \bar{p} + \bar{n} = N'$ is the size of the test data set.

It is important to note that the class labels of the data points in the training and the test data set do not reflect some absolute truth, but only some relative truth based on currently available experiments. For example, if a set of sequences is partitioned into those that are bound by a given transcription factor and those that are not, then this partitioning is based on some data set, for example, some set of ChIP-Seq data. However, these data are intrinsically noisy, containing both biological and technical variation. Hence, the foreground set of ChIP-Seq-positive sequences is typically contaminated by some sequences that are not bound by the immuno-precipitated transcription factor and vice versa. Hence, class labels do not correspond to biological reality, but strictly speaking they correspond only to currently available experimental observation. In case of ChIP-Seq data, the degree of cross-contamination is still quite high, which often leads to frustratingly low classification performance.

Based on the entries of the confusion matrix, several performance measures can be computed:

- Classification rate $\text{cr} = \frac{TP+TN}{N'}$ is the percentage of correct predictions ($\frac{27,607}{28,567} = 0.9664$ in the example).

- Sensitivity $\text{Sn} = \frac{TP}{p} = \frac{TP}{TP+FN}$ is the percentage of foreground sequences correctly predicted (0.9898).

- Positive predictive value $\text{ppv} = \frac{TP}{\bar{p}} = \frac{TP}{TP+FP}$ is the percentage of correct predictions among the sequences predicted as foreground (0.0505).

- Specificity $\text{Sp} = \frac{TN}{n} = \frac{TN}{FP+TN}$ is the percentage of correctly predicted background sequences (0.9664).

- False-positive rate $\text{fpr} = 1 - Sp$ is the percentage of erroneously predicted background sequences (0.0336).

From the confusion matrices for different thresholds (**Section 3.1**), we can compute pairs of Sn and fpr, which can

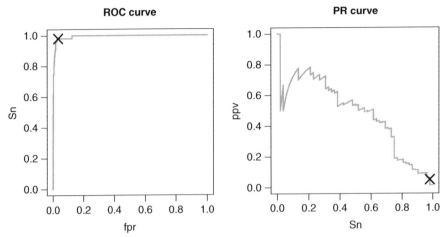

Fig. 7.4. ROC curve (*left*) and PR curve (*right*) for PWM models on the AR/GR/PR data set. The *cross* on the ROC curve illustrates the pair of Sn and fpr for a threshold of 0 on the log-likelihood ratios, corresponding to the confusion matrix presented in **Fig. 7.3**, and the *cross* on the PR curve illustrates the corresponding pair of ppv and Sn. The ROC curve is biased by the unbalanced sizes of the foreground and background data set, yielding a spuriously inflated area under the ROC curve close to 1, whereas the PR curve gives a more realistic view on the performance of the classifier.

be used for plotting a receiver operating characteristic (ROC) curve. Another view on the classification performance can be obtained by plotting ppv against Sn resulting in the precision–recall (PR) curve. Examples for both curves are given in **Fig. 7.4**. On first sight, the ROC curve indicates an almost perfect classification. However, the test data set contains approximately 600 times as many background sequences as foreground sequences, which strongly biases the ROC curve (*see* also **Note 6**). In contrast to the ROC curve, the PR curve reveals that ppv decreases by approximately the same amount as Sn can be increased. Hence, in cases of very unbalanced data sets, the PR curve is a more adequate measure of classification performance.

If a large number of classifiers need to be compared, the visual comparison of curves is not always manageable. In this case, it is helpful to aggregate the ROC and PR curves into scalar values by computing the areas under the curves, denoted by *AUC-ROC* and *AUC-PR*, respectively. **Note 6** contains additional recommendations regarding performance measures.

3.5.2. Cross-Validation and Holdout Sampling

Data are limited for many applications in bioinformatics. This is especially true for transcription factor binding sites, where typical data sets of verified binding sites comprise 20–250 sequences, although chromatin immunoprecipitation combined with next generation sequencing produces several thousand low-confidence sites. For such small data sets a simple approach of splitting all data available in training and test data does not yield reliable results.

Here we discuss two approaches for a reliable assessment of classifiers on small data sets. The first approach is a k-fold cross-validation: partition the data set into k non-overlapping

Probabilistic Approaches to Transcription Factor Binding Site Prediction — 111

parts of approximately the same size. Successively use each of the k data sets for testing and train the classifier on the remaining $k-1$ data sets. Finally, average the performance measures of the results over the k folds. The maximum possible number k of cross-validation folds is the number of sequences in the data set, which results in a *leave-one-out cross-validation*.

Another approach is holdout sampling: randomly partition the data set into a training and a test data set, comprising, for example, 90 and 10% of the original data set, respectively. Use the training data set to train the classifier and test its performance on the test data set. Repeat this procedure k times and average the performance measures over the k runs. Holdout sampling allows the possibility of choosing a large number k of repetitions even for small data sets. However, it cannot be assured that each sequence is used exactly once for testing, which is the case for cross-validation.

For cross-validation as well as holdout sampling, it is recommended to partition the data in a *stratified* manner, i.e., to assure that the proportion of foreground and background sequences remains approximately the same for the training and the test partition.

3.5.3. Assessment of Classifiers in Jstacs

First, we demonstrate how to assess an already trained classifier on a separate test data set. To this end, each subclass of `AbstractClassifier` including `ModelBasedClassifier` and `CLLClassifier` contains a method `evaluateAll`.

We first choose the desired performance measures by instantiating `MeasureParameters` of package `de.jstacs.classifier`:

```
175 MeasureParameters mp =
176   new MeasureParameters( true, 0.999, 0.95, 0.95 );
```

where 0.999 is the fixed Sp for computing Sn and the two values of 0.95 correspond to the fixed Sn for computing fpr and ppv, respectively (**Note 6**).

Next we call the `evaluateAll`-method on the trained classifier `cl` on the foreground and background test data sets `fgTest` and `bgTest`.

```
183 ResultSet rs =
184   cl.evaluateAll( mp, true, fgTest, bgTest );
185 System.out.println(rs);
```

We obtain a `ResultSet` as a container for the performance measures, which can be printed using standard methods. An example output is depicted in **Fig. 7.5**. In this case, the

```
0.9664 = Classification rate (...)
0.5577 = Sensitivity for fixed specificity (...)
3.9974 = Threshold for sensitivity (...)
...
0.9942 = Area under ROC curve (...)
0.4770 = Area under PR curve (...)
[table]  Receiver operating characteristic curve (...)
[table]  Precision recall curve (...)
```

Fig. 7.5. Output of the evaluation of a classifier. The last two entries indicate that the ROC curve and the PR curve have been computed in the evaluation.

`ResultSet` also contains the points of the ROC and PR curves, which can be directly plotted using R (26) from within Jstacs, resulting in the plots of **Fig. 7.4**. The example-code file contains helpful comments for setting up communication between R and Jstacs.

The package `de.jstacs.classifier.assessment` contains classes for cross-validation and holdout sampling. In the following we decide for a 1,000-fold stratified holdout sampling. Again, we must first define the external parameters

```
230 RepeatedHoldOutAssessParameterSet parsA =
231   new RepeatedHoldOutAssessParameterSet(
232     Sample.PartitionMethod.PARTITION_BY_NUMBER_OF
   _SYMBOLS,
233     fgData.getElementLength(), true, 1000,
234     new double[]{ 0.1, 0.1 } );
```

where `Sample.PartitionMethod.PARTITION_BY_NUMBER_OF_SYMBOLS` indicates that we want to measure the size of the partitions by the number of symbols, $1,000$ is the number of repetitions, and the array of `doubles` defines the relative size of the sampled foreground and background test data sets.

We assess the performance of the classifiers `cl` and `cll` by calling

```
240 RepeatedHoldOutExperiment exp =
241   new RepeatedHoldOutExperiment( cl, cll );
242 ListResult lr = exp.assess( mp, parsA, fgData, bgData );
243 System.out.println( lr );
```

where `mp` are `MeasureParameters` as before. By printing the `ListResult` to standard out, we obtain a table of the classifiers and corresponding values of the performance measures. We can use these results for comparing the performance of different classifiers.

As an example, we present the results of a 1,000-fold stratified holdout sampling on the AR/GR/PR data set in **Fig. 7.6**. We

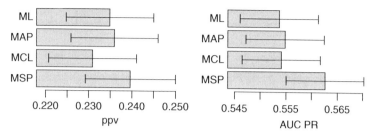

Fig. 7.6. Results of a 1,000-fold stratified holdout sampling for PWM models on the AR/GR/PR data set. The parameters of the PWM models have been trained by the ML, MAP, MCL, and MSP learning principle. *Whiskers* indicate the two-fold standard error. While we find no significant differences regarding ppv, we find a just significantly improved performance of the classifier learned by the MSP principle on this data set considering the area under the PR curve.

compare the performance of PWM models learned by the generative ML and MAP learning principles as well as the discriminative MCL and MSP learning principles considering ppv and *AUC-PR* as performance measures. We find no significant differences regarding ppv, but we find a significantly improved performance of the classifier learned by the MSP principle on this data set considering *AUC-PR* (**Note 7**). These results suggest we use such a classifier for recognizing new AR/GR/PR binding sites in the next section.

3.6. Recognition

The final goal is to predict transcription factor binding sites in some set of genomic regions such as promoters of differentially expressed genes, regions bound by one or several TFs obtained by ChIP-Chip or ChIP-Seq experiments, or conserved regions of orthologous promoters of evolutionarily related species. Here, we consider the specific example of predicting the putative binding sites of the AR/GR/PR-family of transcription factors in a set of human promoter sequences, each of length 500 bp, obtained from the human promoter database (http://zlab.bu.edu/mfrith/HPD.html). Since PWM models learned by the MSP principle achieved the best performance of the classifiers we studied (**Section 3.5**), we now use such a classifier for recognizing AR/GR/PR binding sites. The chosen classifier cll has a method getScore(sub,class), which returns a score for sequence sub belonging to class class. Here class=0 means foreground and class=1 means background.

We load the promoter sequences into Jstacs and compute the log-likelihood ratio for each sub-sequence of length 16 bp of each promoter sequence.

```
258 Sample promoters = new DNASample( "human_promoters.fa" );
259
260 for( Sequence seq : promoters ){
261     for(int l=0;l<seq.getLength()-cll.getLength()+1;l++){
```

```
262      Sequence sub = seq.getSubSequence( 1, cll.getLength() );
263      llr = cll.getScore( sub, 0 ) - cll.getScore( sub, 1 );
264      out.print( llr + "\t" );
265    }
266    out.println();
267  }
```

In this example, we consider only the forward strand of the promoters. The same analysis can be repeated for the backward strand, if we replace `getSubSequence` by `reverseComplement`. The log-likelihood ratios are printed for further analysis. In **Fig. 7.7**, we present a plot of the log-likelihood ratios for a sub-sequence of one of these promoter sequences. We apply a threshold of 2 in this example and predict one potential occurrence ("CATTTTGTCCTAAACA") of a putative AR/GR/PR binding site within this sub-sequence. Comparing this occurrence to the sequence logo of **Fig. 7.1** we find that this occurrence is in good accordance with the motif of the AR/GR/PR-family. Interestingly, despite its large log-likelihood ratio, this putative binding site cannot be found by searching for the consensus sequence, as it does not match the consensus "T" at position 9 of the motif.

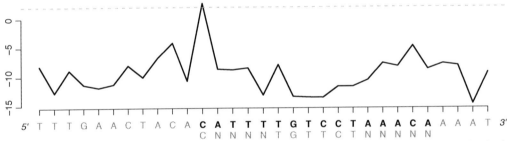

Fig. 7.7. Plot of the log-likelihood ratios for a sub-sequence of a promoter sequence, where the log-likelihood ratios for each sub-sequence of length 16 bp, which is the length of AR/GR/PR binding sites, are plotted above its first nucleotide. If we apply a classification threshold of 2 (*horizontal line*), we recognize one putative AR/GR/PR binding site (*boldface letters*). The position of this binding site is determined by the log-likelihood ratio at its first nucleotide (from 5′ to 3′ end) and the length of AR/GR/PR binding sites.

4. Notes

1. One way of understanding inhomogeneous Markov models (**Section 3.2**) is to start with the standard factorization of an arbitrary distribution $P(\mathbf{x})$, where we omit the class and parameters for brevity:

$$P(\mathbf{x}) = P_1(x_1) \prod_{\ell=2}^{L} P_\ell(x_\ell | x_1, \ldots, x_{\ell-1}) \qquad [8]$$

This factorization holds for arbitrary distributions, whereas an iMM(m) restricts the context to a maximal length of m nucleotides. This may tempt us to use large model orders m for capturing all potential dependencies. However, the number of parameters increases exponentially with the model order m, resulting in difficulties in estimating the parameters from data due to overfitting (**Section 3.3.1**). To overcome this problem, variable order Markov models have been introduced in (27) and applied to DNA and protein sequence analysis, e.g., in (28, 29). The idea is to shorten the context in those cases where the training data suggest that a longer context does not contain "strong additional" dependencies.

Depending on the problem at hand a shortcoming of iMMs is the strict sequential order imposed on the dependencies. Generally this is appropriate for time series, but not obviously for binding sites (4, 9). Bayesian networks (BNs) (30) do not suffer from this limitation. BNs allow, for each position $\ell = 1, \ldots, L$, statistical dependencies on an arbitrary set of other positions as long as no cycles of statistical dependencies are induced. Alternatively this can be understood by first imposing a suitable permutation on the L positions of the sequence, applying the standard factorization, and choosing for each position appropriate predecessors to which statistical dependencies are allowed. Examples of applications to sequence data are (4, 12, 31). The web server VOMBAT available at https://www2.informatik.uni-halle.de:8443/VOMBAT/ allows the recognition of transcription factor binding sites based on variable order Markov models and variable order Bayesian trees (32, 33).

2. The MAP principle introduced in **Section 3.3.2** is sometimes called the first level of Bayesian analysis and sometimes not considered truly Bayesian. Loosely speaking, this controversy stems from the fact that the MAP principle uses only the location of the maximum of the posterior and ignores all other information of the posterior. To exploit all information of the posterior, the classification rule of [1] can be adapted as follows:

$$c^* = \underset{c \in \mathcal{C}}{\operatorname{argmax}} \, P(c | \mathbf{x}, \mathbf{D}, \mathbf{c}, \boldsymbol{\alpha})$$

$$= \underset{c \in \mathcal{C}}{\operatorname{argmax}} \int_{\boldsymbol{\theta}} P(c, \mathbf{x} | \boldsymbol{\theta}) P(\boldsymbol{\theta} | \mathbf{D}, \mathbf{c}, \boldsymbol{\alpha}) \, d\boldsymbol{\theta} \qquad [9]$$

Here, the decision takes not only one, but all of the possible parameter values into account, and weighs the class probabilities accordingly for making a decision.

3. For the MAP and MSP learning principle (**Sections 3.3.2** and **3.4.2**), a prior density $P(\theta|\alpha)$ is needed that represents the prior knowledge or assumptions about the parameters θ. In case of inhomogeneous Markov models including the PWM model (**Section 3.2**), a popular prior is the product-Dirichlet prior defined as a product of Dirichlet densities.

   For a PWM model it is a product of L four-dimensional Dirichlet densities with hyperparameters α equal to the pseudo-counts mentioned in **Section 3.3.2.** The extension of the product-Dirichlet prior to inhomogeneous Markov models of higher order and to more complex models such as variable order Markov models, Bayesian networks, or variable order Bayesian networks (**Note 1**) is straightforward.

4. The values of the hyperparameters α for the product-Dirichlet prior should be chosen with care, as they can strongly influence the recognition and thus all subsequent results. Fortunately, there is an intuitive interpretation of the hyperparameters of a product-Dirichlet prior.

   The sum of all $\alpha_{\ell,a}$ of position ℓ is called equivalent sample size and denoted by ESS. Often, it is beneficial to use hyperparameters α that satisfy the *consistency* condition (34, 35), resulting in an identical ESS at each position. Under this condition, each hyperparameter $\alpha_{\ell,a}$ can be interpreted as the – possibly real valued – amount of pseudo-data observed. The product-Dirichlet prior cannot be used for different Markov models without further premises, since these models differ in the number of parameters. Building on the consistency condition, it is advisable to use hyperparameters that represent uniform pseudo-data in order to avoid artificial biases (34) that favor certain models over others. The general assumption of uniform pseudo-data does not prevent different ESS in different classes and this freedom can and should be used for representing different a priori class probabilities.

5. Discriminative learning approaches have a long tradition in bioinformatics. For example, the first application of weight matrices in bioinformatics (7) employs a discriminative learning algorithm called *perceptron algorithm*. The weight matrix of (7) contains integer values instead of probabilities as it is the case for discriminatively trained PWMs. Another very popular example for discriminatively learned classifiers are support vector machines (SVMs) (36). SVMs aim at finding a number of *support vectors*, i.e., examples of the training data, which define a *hyperplane* separating the foreground and the background class. Although SVMs achieve good performance for many sequence classification tasks (e.g., (37)), their parameters are less easy to interpret than those of the

probabilistic approaches presented in this chapter. However, the interpretability of SVMs has been improved lately using so-called positional oligomer importance matrices (POIMs) (38).

6. The entries of the confusion matrix and, consequently, the point measures cr, Sn, ppv, Sp, and fpr, depend on the threshold (**Section 3.1**) used for classification. By varying the threshold, it is trivial to yield, e.g., a sensitivity of one, where all sequences are classified as binding sites, obviously at the price of a specificity of zero. Hence, one typically chooses the threshold in such a way that one of the performance measures is fixed to a predefined value and then reports the resulting value of a second performance measure. For example, we may choose the threshold such that the specificity is fixed to 0.999 and then use the sensitivity as performance measures, which quantifies the sensitivity if one false prediction per 1,000 negative sequences is allowed. Another common example is to use the false-positive rate for a fixed sensitivity of 0.95, which quantifies the amount of false positives if 95% of the binding sites are predicted correctly.

Not all measures are suited for unbalanced test data sets. For example, the test data set may comprise 9,900 background and 100 foreground sequences. We can easily achieve a cr of 0.99 if we classify all sequences into the background class without considering sequence information. A similar problem can be encountered for the ROC curve, which is also dominated by a large number of background sequences. Consider the case of **Section 3.5**, where the test data set comprises 600 times as many background sequences as foreground sequences. Assume that we achieve $Sn = 1$ for some threshold, i.e., all foreground sequences in the test data set are classified correctly. Further assume that for the same threshold we observe for each correct positive prediction on average 10 additional, however, incorrect, positive predictions. This would result in an fpr of approximately 0.0167 for an Sn of 1, although we would consider the classification result as far from perfect.

In such cases, measuring the ppv or the PR curve is more adequate for quantifying differences in the performance of classifiers.

7. Irrespective of the value of k chosen for cross-validation or holdout sampling, the obtained results depend on the chosen data sets, and typically the results vary substantially from data set to data set. Hence, we recommend to not rely on the error bars obtained from cross-validation or holdout sampling of only one data set, but to repeat all studies on several

different data sets. The choice of appropriate data sets, however, is a highly non-trivial task and due to the condition that the final results strongly depend on the chosen data sets we recommend this choice to be made with great care and in a problem-specific manner. This choice is typically influenced by a priori knowledge on both the expected binding sites and the targeted genome regions. Examples of features that are often considered when choosing appropriate data sets are the GC content of the target region, their association with CpG islands, or their size and proximity to transcription start sites.

Carefully choosing appropriate training and test data sets is of additional advantage if the set of targeted genome regions is not homogeneous, e.g., comprising both GC-rich and GC-poor regions, CpG islands and CpG deserts, TATA-containing and TATA-less promoters, upstream regions with and without binding sites of another transcription factor. In this case, one often finds that different combinations of models and/or different learning principles work well for different subgroups, providing the possibility of choosing subgroup-specific prediction approaches. These considerations are vital for a successful prediction of transcription factor binding sites, but beyond the scope of this chapter, so we choose only one foreground data set and only one background data set in the presented example. Specifically, we choose the data set of second exons used for training the PWM models of Transfac, implying that the specific results obtained in **Section 3.5** are probably optimistic.

References

1. Lawrence, C.E., Altschul, S.F., Boguski, M.S. et al. (1993) Detecting subtle sequence signals: a Gibbs sampling strategy for multiple alignment. *Science* 262, 208–214.
2. Bailey, T.L., and Elkan, C. (1994) Fitting a mixture model by expectation maximization to discover motifs in biopolymers. In *Proceedings of the 2nd International Conference on Intelligent Systems for Molecular Biology.*
3. Pavesi, G., Mauri, G., and Pesole, G. (2001) An algorithm for finding signals of unknown length in dna sequences. *Bioinformatics* 17, S207–S214.
4. Barash, Y., Elidan, G., Friedman, N. et al. (2003) Modeling dependencies in protein-DNA binding sites. In *Proceedings of the Annual International Conference on Research in Computational Molecular Biology (RECOMB).* pp.28–37.
5. Smith, A. D., Sumazin, P., and Zhang, M. Q. (2005) Identifying tissue-selective transcription factor binding sites in vertebrate promoters. *Proc Natl Acad Sci U S A* 102, 1560–1565.
6. Elemento, O., Slonim, N., and Tavazoie, S. (2007) A universal framework for regulatory element discovery across all genomes and data types;. *Mol Cell* 28, 337–350.
7. Stormo, G.D., Schneider, T.D., Gold, L.M. *et al.* (1982) Use of the 'perceptron' algorithm to distinguish translational initiation sites. *Nucleic Acids Res* 10, 2997–3010.
8. Staden, R. (1984) Computer methods to locate signals in nucleic acid sequences. *Nucleic Acids Res* 12, 505–519.
9. Zhao, X., Huang, H., and Speed, T. P. (2004) Finding short dna motifs using permuted markov models. In *Proceedings of*

the *8th Annual International Conference on Computational Molecular Biology pp.*, 68–75. ACM, San Diego, CA.

10. Kel, A.E., Güssling, E., Reuter, I. et al. (2003) Match: a tool for searching transcription factor binding sites in dna sequences. *Nucleic Acids Res* 31, 3576–3579.

11. Sinha, S., van Nimwegen, E., and Siggia, E.D. (2003) A probabilistic method to detect regulatory modules. *Bioinformatics* 19, 292–301.

12. Ben-Gal, I., Shani, A., Gohr, A. et al. (2005) Identification of transcription factor binding sites with variable-order Bayesian networks. *Bioinformatics* 21, 2657–2666.

13. Grau, J., Keilwagen, J., Kel, A. et al. (2007) Supervised posteriors for DNA-motif classification. In *German Conference on Bioinformtics.* pp. 123–134.

14. Blanchette, M., and Tompa, M. (2002) Discovery of regulatory elements by a computational method for phylogenetic footprinting. *Genome Res* 12, 739–748.

15. Zhang, Z., and Gerstein, M. (2003) Of mice and men: phylogenetic footprinting aids the discovery of regulatory elements. *J Bio* 2, 11.

16. Boffelli, D., McAuliffe, J., Ovcharenko, D. et al. (2003) Phylogenetic shadowing of primate sequences to find functional regions of the human genome. *Science* 299, 1391–1394.

17. Halperin, Y., Linhart, C., Ulitsky, I. et al. (2009) Allegro: analyzing expression and sequence in concert to discover regulatory programs. *Nucleic Acids Res* 37, 1566–1579.

18. Ji, H., Jiang, H., Ma, W. et al. (2008) An integrated software system for analyzing chip-chip and chip-seq data. *Nat Biotech* 26, 1293–1300.

19. Roos, T., Wettig, H., Grünwald, P. et al. (2005) On discriminative Bayesian network classifiers and logistic regression. *Mach Learn* 59, 267–296.

20. Cerquides, J., and De Mántaras, R. (2005) Robust Bayesian linear classifier ensembles. In *Proceedings of the 16th European Conference Machine Learning, Lecture Notes in Computer Science.* Citeseer, pp. 70–81.

21. Schneider, T.D., and Stephens, R.M. (1990) Sequence logos: a new way to display consensus sequences. *Nucleic Acids Res* 18, 6097–6100.

22. Zhang, M., and Marr, T. (1993) A weight array method for splicing signal analysis. *Comput Appl Biosci* 9, 499–509.

23. Salzberg, S.L. (1997) A method for identifying splice sites and translational start sites in eukaryotic mRNA. *Comput Appl Biosci* 13, 365–376.

24. Ng, A., and Jordan, M. (2002) On discriminative vs. generative classifiers: a comparison of logistic regression and naive bayes. In Dietterich, T. S. Becker, and Z. Ghahramani (Eds.) *Advance in neural information processing systems* volume 14, pp.605–610. MIT Press, Cambridge, MA.

25. Yakhnenko, O., Silvescu, A., and Honavar, V. (2005) Discriminatively trained Markov model for sequence classification. In *ICDM '05: Proceedings of the 5th IEEE International Conference on Data Mining.* IEEE Computer Society, Washington, DC, pp. 498–505.

26. R Development Core Team. (2009) *R: a language and environment for statistical Computing.* R Foundation for Statistical Computing, Vienna. ISBN 3-900051-07-0.

27. Rissanen, J. (1983) A universal data compression system. *IEEE Trans Inform Theory* 29, 656–664.

28. Bejerano, G., and Yona, G. (2001) Variations on probabilistic suffix trees: statistical modeling and prediction of protein families. *Bioinformatics* 17, 23–43.

29. Orlov, Y.L., Filippov, V.P., Potapov, V.N. et al. (2002) Construction of stochastic context trees for genetic texts. *In Silico Bio* 2, 233–247.

30. Pearl, J. (1988) *Probabilistic reasoning in intelligent systems: networks of plausible inference.* Morgan Kaufmann, San Francisco, CA.

31. Castelo, R., and Guigo, R. (2004) Splice site identification by idlbns. *Bioinformatics* 20, i69–i76.

32. Grau, J., Ben-Gal, I., Posch, S. et al. (2006) VOMBAT: prediction of transcription factor binding sites using variable order Bayesian trees. *Nucleic Acids Res* 34, W529–W533.

33. Posch, S., Grau, J., Gohr, A. et al. (2007) Recognition of cis-regulatory elements with VOMBAT. *J Bioinfor Comput Bio* 5, 561–577.

34. Buntine, W.L. (1991) Theory refinement of Bayesian networks. In *Uncertainty in artificial intelligence.* Morgan Kaufmann, San Francisco, CA, pp. 52–62.

35. Heckerman, D., Geiger, D., and Chickering, D.M. (1995) Learning Bayesian networks: the combination of knowledge and statistical data. *Mach Learn* 20, 197–243.

36. Cortes, C., and Vapnik, V. (1995) Support-vector networks. *Mach Learn* 20, 273–297.

37. Schweikert, G., Sonnenburg, S., Philips, P. et al. (2007) Accurate splice site prediction using support vector machines. *BMC Bioinformatics* 8, S7.

38. Sonnenburg, S., Zien, A., Philips, P. et al. (2008) POIMs: positional oligomer importance matrices – understanding support vector machine-based signal detectors. *Bioinformatics* 24, 6–14.

Chapter 8

The Motif Tool Assessment Platform (MTAP) for Sequence-Based Transcription Factor Binding Site Prediction Tools

Daniel Quest and Hesham Ali

Abstract

Predicting transcription factor binding sites (TFBS) from sequence is one of the most challenging problems in computational biology. The development of (semi-)automated computer-assisted prediction methods is needed to find TFBS over an entire genome, which is a first step in reconstructing mechanisms that control gene activity. Bioinformatics journals continue to publish diverse methods for predicting TFBS on a monthly basis. To help practitioners in deciding which method to use to predict for a particular TFBS, we provide a platform to assess the quality and applicability of the available methods. Assessment tools allow researchers to determine how methods can be expected to perform on specific organisms or on specific transcription factor families. This chapter introduces the TFBS detection problem and reviews current strategies for evaluating algorithm effectiveness. In this chapter, a novel and robust assessment tool, the Motif Tool Assessment Platform (MTAP), is introduced and discussed.

Key words: Transcription Factor Binding Sites (TFBS), prediction algorithms, assessment tools, Motif Tool Assessment Platform (MTAP).

1. Introduction

Transcription factors and other regulatory proteins bind to DNA primarily around the transcription start site, interact with RNA polymerase, and then facilitate or inhibit transcription of the gene. Most transcription factors bind to DNA at sequence-specific positions along the chromosome, called transcription factor binding sites (TFBS). The (partially) conserved sequence pattern found at several sites bound by the same transcription factor is called

I. Ladunga (ed.), *Computational Biology of Transcription Factor Binding*, Methods in Molecular Biology 674,
DOI 10.1007/978-1-60761-854-6_8, © Springer Science+Business Media, LLC 2010

a motif. Motifs co-occur near transcription start sites for genes that are regulated by the same transcription factor. Many computational approaches have been developed to find conserved motifs in the regulatory regions upstream of genes that have similar expression patterns. Computational approaches complement experimental approaches because they are less labor intensive and costly. In addition, a predictive computational model is very useful when experimental data are limited.

1.1. TFBS Detection Problem

In prokaryotes, given a set of genes that are differentially expressed, i.e., partially controlled by the same set of transcription factors, the TFBS identification problem is to mark conserved patterns in the regulatory regions of the differentially expressed genes. The patterns can be represented as a set of k-mers (words of length k) or as a Position Specific Scoring Matrix (PSSM) among others. When the pattern is represented as a set of k-mers, the objective function to be minimized is the number of mismatches in the set of words such that there exists a binding site in close proximity to the transcription start site for each differentially expressed gene. When the pattern is represented as a PSSM, the objective is to maximize the probability that a PSSM of a given length co-occurs in the promoters of each of the differentially expressed genes (1).

Regardless of the approach taken to represent motifs at the binding sites, practitioners must balance a set of complex trade-offs when building tools to solve the TFBS detection problem. Hence, in the detection process, motif representation is the first step. After motifs are represented, all possible motif instances in the differentially expressed promoters are indexed. Then, a distance function is used to discriminate motif instances that exist in the promoters of the differentially expressed genes but do not exist in background sequence. Finally, likely matches are extracted, ranked, and reported.

Currently, there are almost 200 tools to find TFBS motifs given a set of differentially expressed genes. For the current list, refer to http://biobase.ist.unomaha.edu/mediawiki/index.php/Main_Page. For many practitioners, the most pressing question is 'what prediction tool should I use?' Experts in the field commonly recommend running a set of tools and manually comparing the outputs. This has some merit, but a more formal methodology is needed to rank tools for different problem characteristics.

1.2. Algorithm Evaluation

One way of choosing the most appropriate tool for a specific problem is to run each possible tool on a related problem where there is experimental evidence. The experimental evidence can then be used as a standard to measure tool predictive performance. Ideally, one would just run each tool as a black box to

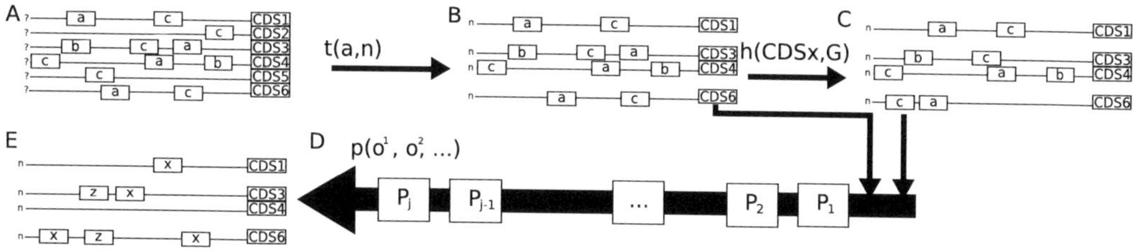

The MTAP for Sequence-Based Transcription Factor Binding Site Prediction Tools

Fig. 8.1. Evaluating TFBS discovery algorithms. **a** First, all known regulatory regions from a genome are assembled into a database. We then apply a reduction function, t, over all regulatory elements to determine a set of co-bound regulatory sequences (**a**). Function t uses evidence from ChIP-chip, ChIP-seq, or a TFBS database to include only regions bound by transcription factor i. The result of this pruning is shown in B. This results in n subsets $B_1, B_2, \ldots, B_n$ one for each transcription factor. For each regulatory subset (B_i) we apply additional functions, $h_1, h_2, \ldots, h_n$, to collect background sequence data, to collect the orthologous regulatory regions in other genomes. These sequences are then fed into the prediction pipeline (D), which calculates the background probability of a pattern in the sequences in (**a**) and from any other sequences collected in (**c**). The pipeline then generates a set of predictions corresponding to possible binding sites. Prediction positions are marked in a standard format shown in (**e**).

mark TFBS and then compare the TFBS predictions with the known binding sites found in the database. The most appropriate tool for a problem is one that correctly predicts the largest percentage of known binding sites (true positive predictions) while at the same time marking the least amount of non-sites, regions that have similar sequence composition to known motifs but are not known to be bound by a transcription factor (false-positive predictions).

Each prediction algorithm requires multiple and different stages in order to make a prediction. Each stage corresponds to a unique added value implemented by the method. Some methods implement novel approaches for modeling background sequences, other methods implement cross-species conservation models, while others include data from other sources such as expression arrays. Thus, diverse TFBS prediction algorithms cannot be treated as black boxes with the same input and outputs. Evaluation of several different TFBS detection methods requires that we build pipelines for all methods. These pipelines allow access to all of the same data sources and standardize the outputs so that they can be compared (**Fig. 8.1**). Once the predictions are generated from all of the tools, statistics are collected that measure the number of overlaps found between predictions and known TFBS.

2. Materials

The software discussed in this chapter, Motif Tool Assessment Platform (MTAP), was implemented in Python, C/C++, Java, and Perl. A large assortment of languages was used because many effective algorithm techniques from other authors are included in the MTAP download and are implemented in several different languages. MTAP is open source, free, and community supported. Enhancements are welcome. A community supported list of known TFBS finding algorithms can be viewed at http://biobase.ist.unomaha.edu/mediawiki/index.php/Main_Page. MTAP and a User's Manual can also be downloaded from this site. The installation of MTAP, due to its complex dependencies, is far from being trivial as described in **Note 1**. For running MTAP, *see* **Note 2**.

3. Methods

3.1. Algorithms

The central challenge in evaluating how well tools predict TFBS is collecting data sets that in some way constitute a meaningful representation of a (small) part of the transcription regulatory networks. It is likely that the large number of prediction tools exists primarily because the problem is difficult to pose. Despite this, computational predictions have proven useful in narrowing the search space for many known TFBS. Recently, databases have been developed that contain binding site information for a large number of transcription factors. High-throughput sequencing technologies and high-density micro-array-based technologies enable the construction of such TFBS databases. One application of these databases is to use it as a source of comparison with prediction algorithms, which should enable refinement of the tools and models used for TFBS prediction. The ability to accurately evaluate how well TFBS prediction algorithms correspond to TFBS databases is critical to understanding the faults of current methods and possible avenues for improvement.

3.2. TFBS Databases

As experimental evidence mounts, TFBS locations have been collected and entered into regulatory databases. Significant progress has been made identifying regulatory genes, signaling pathways, and transcription factor binding sites. Pathways responsible for a wide variety of cellular processes have been identified in *Escherichia coli, Bacillus subtilis,* yeast, worms, fruit flies, sea urchins, zebra fish, frogs, chicken, mice, and humans, just to name a few. The most substantial progress in constructing multicellular organism regulatory maps has been with the sea urchin embryo (2), the characterization of the dorsal–ventral patterning of early *Drosophila* embryo (3), and the detailed map of *Ciona intestinalis* (4).

The *E. coli* regulatory map (5) has been built by combining the work of Shen-Orr et al. (6), the curators of RegulonDB (7) and the maintainers of EcoCyc (8). This unique annotation contains a large network topology representing current understanding of gene regulation in *E. coli K12*. The annotation includes binding positions derived from experimental evidence for *E. coli K12* regulatory proteins. Transfac (9), DBTBS (10), RegTransBase (11), and Prodoric (12) are all examples of TFBS databases that have been developed in recent years to annotate the regulatory network in other organisms.

A standard for annotating binding sites is still emerging. In most databases, a TFBS is annotated in a database with a start position, end position, and strand information. Some databases contain additional information such as the genes regulated, the protein family of the transcription factor, and the type of regulation (e.g., activation or repression). The information found in these databases has not been standardized. Consequently, many useful properties such as the strength of the interaction between the transcription factor and the binding site are not available. Some databases differentiate between DNA regions that are bound by transcription factor and those that lie between interacting sites. Despite the need for protein–DNA interaction information in protein structure data, few databases incorporate structure data in the annotation. The structure and representation of TFBS information in the database limit the accuracy of TFBS detection. Many researchers currently believe that even the most comprehensive databases miss many sites, especially those with weak interactions. Some researchers build synthetic data sets to circumvent these issues, but these approaches are limited by the level of fidelity of synthetic test representing the biology of transcription factor binding. Appropriate evaluation metrics are essential for determining the type and structure of data that should be cataloged to improve TFBS prediction.

3.3. Core Evaluation Statistics

Algorithm performance is evaluated by comparing the positions of predicted sites to the positions of known sites. For each position marked, seven core statistics are collected. The first four core statistics, shown in **Table 8.1**, are nTP – nucleotide true positives, nFN – nucleotide false negatives, nFP – nucleotide false positives, and nTN – nucleotide true negatives. They are collected by adding the number of each occurrence for each position in the regulatory regions.

The site-level statistics (sTP – site true positives, sFN – site false negatives, and sFP – site false positives) are the final three core statistics. A site-level statistic encompasses the idea that a group of adjacent nucleotides, marked as binding positions for a specific transcription factor, is representative of a binding site annotation (**Fig. 8.2**). A site is a true positive if the prediction

Table 8.1
Nucleotide-level statistics. $u_{i,j}$ **is the upstream regulatory sequence** j **at position** i

Statistic	Definition
nTP	$u_{i,j}$ is both annotated and predicted
nFN	$u_{i,j}$ is annotated but not predicted
nFP	$u_{i,j}$ is predicted but not annotated
nTN	$u_{i,j}$ is neither annotated nor predicted

Nucleotide Level Scoring

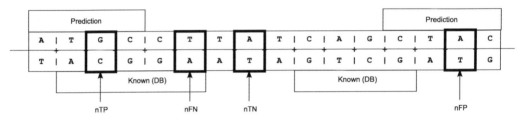

Site Level Scoring

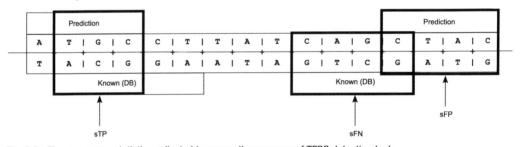

Fig. 8.2. The seven core statistics collected to assess the accuracy of TFBS detection tools.

overlaps the annotation by no less than τ percent (a threshold) of the site. Site true negatives (sTN) represent any collection of adjacent bases that are not predicted or annotated to be a site. The total number of such sites grows as a triangular number (13). However, once a site is annotated or predicted, all possible overlapping sites can no longer be marked a sTN. This makes sTN less meaningful because it can increase or decrease depending on the number of predictions and annotations in the data set. In practice, it is best to set this number sufficiently large so that it is always greater than sTP, sFN, and sFP and always positive and consistent regardless of the number of predictions and annotations in the regulatory regions. Our convention sets this value to the length of all sequences in the upstream set, $nTP + nFN + nFP + nTN$ divided by the number of sequences in the co-bound set. We then subtract the number of predictions and annotations

Table 8.2
Site-level statistics

Statistic	Definition
sTP	Number of known sites overlapped by predicted sites
sFN	Number of known sites not overlapped by predicted sites
sFP	Number of predicted sites not overlapped by known sites
sTN	$sTN = \frac{nTP+nFP+nFN+nTN}{Number\ Sequences} - sTP - sFN - sFP$

from this total when calculating statistics. This ensures that sTN will always be a strictly positive number that is independent of the number of predicted and annotated sites. The site-level statistics are shown in **Table 8.2**.

When evaluating site-level statistics, setting the value of the threshold τ is important. Tompa et al. (14) set τ to 25%. Assuming this overlap, if an experimentalist were to remove the site, a change in expression should be observed. In some organisms, such as bacteria, this threshold is too strict because the width of known binding sites is too large for some tools to ever achieve a sTP. Many motif discovery programs have fixed motif widths (e.g., 8 base pairs), a threshold of 25% would not be sufficient to mark $sTPs$ (e.g., an annotated site of width 60 and a site prediction of length 8). Site-level motifs could be ranked based on a percentage of the prediction width instead of the motif width in the annotated database, but this would give an unfair advantage to methods that predict larger sites. In the example benchmarks in this chapter, τ is set equal to the maximum annotated site width in the data set divided by the minimum expected motif width predicted by the suite of programs times 25%. A degree of overlap indicates that computational and biological refinement of site predictions can still find the site. **Table 8.3** illustrates the seven core statistics collected for algorithm evaluation.

Table 8.3
Statistics for evaluating motif prediction algorithm implementations

Sensitivity	$xSN = \frac{xTP}{xTP+xFN}$	[1]
Specificity	$nSP = \frac{nTN}{nTN+nFP}$	[2]
Positive predictive value	$xPPV = \frac{xTP}{xTP+xFP}$	[3]
Matthews correlation coefficient	$nCC = \frac{nTP^*nTN-nFN^*nFP}{\sqrt{(nTP+nFN)(nTN+nFP)(nTP+nFP)(nTN+nFN)}}$	[4]
Correlation coefficient	$xPC = \frac{xTP}{xTP+xFN+xFP}$	[5]
Site-level average site performance	$sASP = \frac{sSN+sPPV}{2}$	[6]

For each transcription factor, a set of regulatory regions n bases upstream of the controlled genes is collected, and each of the seven core statistics is collected for each of the upstream regions. Note that a given tool will be run for each set of co-bound regulatory regions separately but that the annotation is considered only once. For example, consider a set of regulatory regions bound by two transcription factors, A and B. A and B cooperate in the same regulon to control a set of genes X (those genes only controlled by A), Y (those genes controlled by both A and B), and Z (those genes only controlled by B). Consider the set of regulatory regions collected to calculate the TFBS for A (regulatory regions for X and Y). Predictions from regulatory regions regulating Y that overlap B's TFBS can be marked as false positives because they predict a TFBS other than the protein of interest.

The allowed prediction threshold indicates how many TFBS predictions are allowed by a tool. TFBS predictions come in sets and each set represents a highest scoring representative of binding sites for one transcription factor. In many cases, the highest scoring representative is just a sequence that happens to co-occur in the co-bound regulatory sequences and is not representative of a TFBS. For this reason, practitioners often accept more than one prediction. Allowing more predictions than one from a tool has the advantage that more true sites can be detected and tools can then better represent the combinatorial and co-operative regulatory cellular interactions that often occur. Varying the allowed prediction threshold has dramatic impacts on tool performance characteristics.

3.4. Derived Evaluation Statistics

More advanced metrics for performance evaluation can be calculated from the seven core statistics. Tompa et al. (14) recommended the six informative statistics shown in **Table 8.3**. Each of these statistics has its merit and is informative in different ways, depending on the objectives of the assessment. It is difficult to build tools that have high sensitivity and specificity. The sensitivity/specificity trade-off and the Matthews correlation coefficient, nCC (nCC takes values -1 to 1 with 0 representing not correlated), are often viewed as an overarching measurement for performance.

There are two central problems in the TFBS databases used for evaluation. First, most data sets are incomplete, since many TFBS are not annotated in the data set. The best way to avoid misleading scoring of a TFBS detection method is to construct a data set that is as complete as possible to diminish the possibility of false positives. The second problem shows up when a method is over-fit to the known data repositories. Over-fitting occurs when the method training set and testing set are too similar. Because so little data have been available on TFBS, many methods have been optimized to find binding sites that have already been discovered. It is impossible to say how well tools will detect unknown binding sites in the future.

In the machine learning community, cross-validation is often used in supervised learning problems. Leave one out cross-validation refers to training of an algorithm on a subset of the available data and testing on the subset of the data that is left out. Leave n out cross-validation refers to an iterative training and testing process where the data are partitioned into many sets. At a given stage in the cross-validation process a subset of the data is either used for training or testing. All sets are eventually used for both training and testing. The benchmark of algorithm performance is constructed from combining the values from multiple benchmarks on each partition of the data set dedicated to testing. In the context of new algorithm development, MTAP can be used in either of these ways. Historically, most algorithm developers did not view the TFBS detection problem as a supervised learning problem; instead it was viewed as an unsupervised learning problem. In other words, tool developers did not divide known TFBS instances into testing and training sets, actually, most often a training set did not exist. Instead, the goal was to build methods that could discover the first TFBS with the eventual goal of constructing large data sets for supervised learning. Most often, this was because of the lack of known TFBS. It is impossible to determine what TFBS influenced tool developers in the development process and should therefore be discarded in evaluation metrics. The most common goal of MTAP is to rank tools on a particular data set given a recommended runtime procedure recommended by the tool author. The data set used in this evaluation is assumed to be independent from the data used by the algorithm developer to construct the technique.

3.5. Combining, Viewing, and Evaluating Data Sets

Once each tool is run over data sets $D = \{d_1, d_2, \ldots\}$, the results need to be illustrated in a meaningful way. There is an ongoing discussion on the best approach. Tompa et al. (14) proposed three methods for combining the results into one graph. Sandve et al. (15) proposed a method for evaluating results based on how well instances of the motif conform to known binding models. We proposed a method based on ROC (receiver operating characteristic) curves for combining and evaluating data sets and relative performance graphs for viewing data sets relative performance over a suite of tools. This chapter covers five known statistics for each data set in D.:

(1) *Arithmetic mean.* The arithmetic mean of M scores is calculated after the derived statistics are calculated for each data set in D.

(2) *Normalized.* For each data set in D, normalize the score by subtracting the mean score over all tools then divide

by the standard deviation. Combine scores by calculating the arithmetic mean of M normalized scores. Note that this procedure is called standardization in statistics.

(3) *Combined*. For each data set in D add nTP, nFP, nFN, nTN, sTP, sFP, and nFN as if it were one data set. Calculate the derived statistical measures over the summed totals.

(4) *Relative Performance Graph*. Do not combine the M scores. Construct a graph with each data set along the X-axis and relative performance of the T tools along the Y-axis. Construct one graph for each derived statistics.

(5) *Receiver Operating Characteristic (ROC) Graph*. Determine an algorithm parameter P. Vary P so that algorithm sensitivity continues to increase while 1–specificity continues to decrease. The area under the curve (AUC) is an absolute measure of performance, comparable across methods.

Over most derived statistics, Mean, Average, Normalized and Combined summing methods correlate reasonably well on current data sets (14, 16). **Figure 8.3** shows the predictions of five different motif prediction methods. Therefore, none of those methods reported false positives on this region. When a tool fails to make predictions over a large number of regions like this, it appears (unfairly) that the tool is specific in locating binding sites because the number of false positives is small. On the other extreme, many tools tend to predict nearly the entire region instead of localizing to the TFBS. Thus, when reading the derived statistics it would appear that such methods are sensitive, when they are in fact predicting large contiguous regions of binding sites. For this reason, genome-wide comparisons of binding sites and predicted sites serve as an important sanity check when evaluating statistics.

ROC curves are excellent for comparing the sensitivity/specificity trade-offs of a single parameter (16). ROC curves track the performance of an algorithm over changes over a single parameter. Traditionally, ROC curves have been applied to changing internal algorithm parameters from tight thresholds to more lenient thresholds. This produces a ROC curve that travels straight up and then to the right when the algorithm corresponds exactly to the data set. Algorithms that poorly represent the problem or problems that are ill-conceived produce a curve that will travel straight to the right and then up. Random predictions produce a diagonal line.

We applied ROC analysis to (1) the width of the regulatory regions taken upstream of the gene, (2) the class (protein family) of transcription factor, (3) internal algorithm prediction parameters, and (4) cross-species regulatory region extraction techniques. The methods presented here provide insights into the overall trends but some information is lost. Generalization is dif-

The MTAP for Sequence-Based Transcription Factor Binding Site Prediction Tools 131

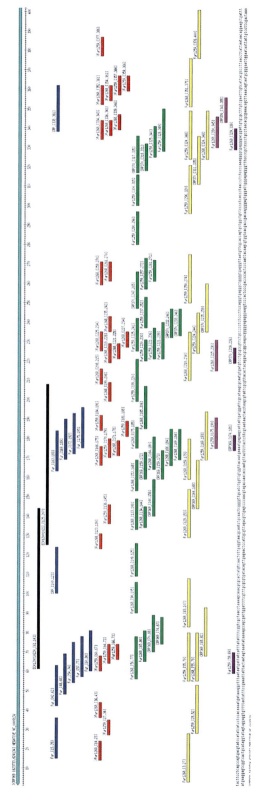

Fig. 8.3. An example of regulatory region 400-bp upstream of *fepB* and downstream *entC* in the *E. coli K12* genome. Features are mapped on the extracted upstream regulatory region, shown in *black*. From *top* to *bottom* features are mapped to the region as follows: (1) promoters in the database, (2) known TFBS from RegulonDB. Predictions from five TFBS prediction algorithms are marked below those from the experimentally determined TFBS from the database: Weeder (3), AlignACE (4), MEME (5), Glam (6), and Gibbs (7). Two promoters (sigma factor binding sites) lie within the regulatory region (shown by *dashed lines*) containing binding sites for transcription factors *CRP* and *Fur*. Experimentally determined *CRP* positions are furthest to the right and seventh from the left. *Fur* sites occupy the rest of the verified binding positions. Collectively, the five TFBS prediction methods cover nearly the entire region, a typical outcome in current prediction technology. Some tools like Weeder do not make predictions for both TFBS classes.

ficult since some data sets consist of a few examples, while others have motif instances in the hundreds; some motif instances are highly conserved while others contain a great variability.

3.6. Results

This section presents three illustrative example results of benchmarks that can be generated using a robust database of almost all transcription factor binding sites in the cell and several TFBS prediction methods. **Figure 8.4** illustrates derived statistics summed via the combined method introduced in the previous section. For each transcription factor in RegulonDB, all TFBS were used to create $D = \{d_1, d_2, \ldots\}$ where D represents the series of all

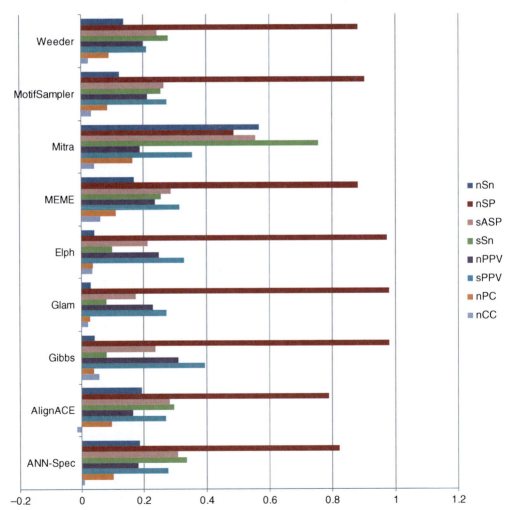

Fig. 8.4. Derived statistics for nine regulatory motif detection methods.

tests and $d_1, d_2, \ldots$ each represent a set of co-bound genes, each bound by the same transcription factor. Pipelines were developed for nine different TFBS detection methods and run over D and predictions compared to RegulonDB annotations to determine the core statistics. Each tool was allowed to make three predictions for a single transcription factor. All of the tools in **Fig. 8.4** illustrate performance profiles that could be improved. Mitra (17) is an example that, at the thresholds in this example, is sensitive but not specific. Mitra predicts sites over the entire regulatory region. For this reason, it discovers many of the annotated sites, but not because the algorithm is able to find patterns that correspond to sites. Elph, Glam (16), and Gibbs (18) are at the other extreme. These make very few predictions on this data set, resulting in perceived high specificity. Weeder (19), MotifSampler (20), MEME (21), AlignACE (22), and ANN-Spec (23) all appear to strike a better balance; however nCC remains between 0 and 0.37 for all approaches. There is some possibility that further refinement of tool parameters could yield better performance. Assessments like this provide an overview of where current methods stand and suggest ways of improvements. **Figure 8.5** shows seven-motif discovery methods evaluated in an ROC curve. Both single species and cross-species techniques are represented.

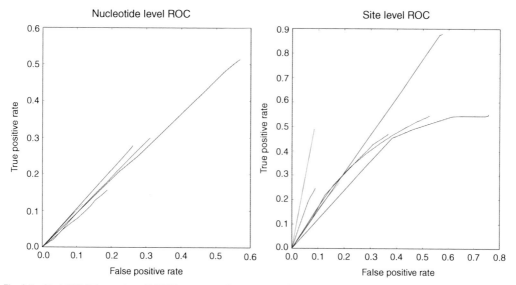

Fig. 8.5. PhyloMEME (a version of MEME run on regulatory regions from multiple species), PhyME, PhyloGibbs-MP, Phylo-Weeder, ANN-Spec, Motifsampler, and Mitra represent state-of-the-art motif detection algorithms. This composite ROC curve shows a side-by-side comparison of phylogenetic-assisted and purely sequence-based tools. At the nucleotide level, performance is virtually random. At the site level, phylogenetics-based tools such as PhyloGibbs-MP outperform single-genome methods. However, several methods perform hardly better than random. This result is expected given the performance of several methods shown in **Fig. 8.4**. Tools cover large portions of the regulatory region with predictions, many of them overlapping known binding sites.

The benchmark was originally generated using over 20 motif discovery methods but only the top seven tools were plotted for clarity. The ROC curves were generated by running each of the algorithms 10 times and increasing the amount of predictions each algorithm was allowed to produce (using algorithmic thresholds) after each run. This plot shows some advantages of incorporating cross-species information. Not every tool maintains good prediction accuracy as it is allowed to make additional predictions. On this data set, PhyloGibbs-MP appears to continue to make good predictions as the algorithmic thresholds are lowered.

An example using a relative performance graph is shown in **Fig. 8.6**. All relative performance graphs contain at least three axes, one for the tools in the study, and one for the transcription factor classes in the study, and one for the performance metric. Relative performance graphs are advantageous because they show an in-depth look at relative performance over parts of the data set for one statistic. The figure shows an in-depth look at nCC over the RegulonDB data set. Along the X-axis is each TFBS evaluated in the assessment. The top graph shows the relative total correlation for all tools in the assessment combined. On the bottom is the relative contribution for each tool for a specific TFBS. Note that columns in the graph have no relationship to one another. TFBS are sorted in this graph by conservation of the sequence at the binding site. This graph illustrates that sequence conservation at the binding site is not enough for accurate TFBS detection.

Figure 8.6 also demonstrates that no tool in this assessment is clearly dominant in detecting all binding sites of the same transcription factor. Some tools are very good at detecting TFBS for some sites, but not others. Some sites are more easily detected by all tools and some sites challenge all tools. These results indicate that one tool for detecting all TFBS may not be possible, instead multiple methods for different classes of problems may be more appropriate.

Fig. 8.6. (continued) content at the conserved site (*bottom*) to low-information content (*top*). *nCC* values do not increase as conservation at the site increases, most likely due to competing background signals in the upstream regulatory regions. In this table, *nCC* is negatively impacted for ELPH, Gibbs, Glam, PhyloMEME, PhyloGibbs, PhyloGibbs-MP, and PhyME because the number of predictions is low (easily overcome for some tools by considering more sites). Binding sites for some TFs such as *AgaR* are relatively accurately predicted by many tools. Other TFs such as *EvgA* pose a greater challenge. JAMM-b is a Bayesian filter for combining multiple methods. JAMM-i is the same filter, with a length-based constraint relative to the promoter. No tool is clearly dominant for every transcription factor.

The MTAP for Sequence-Based Transcription Factor Binding Site Prediction Tools

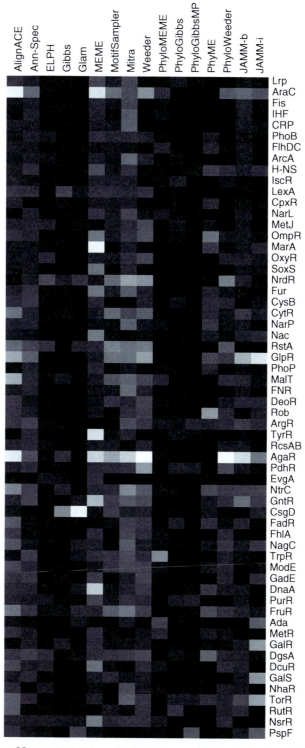

Fig. 8.6. *nCC* was calculated over RegulonDB by considering the top 3 predictions from each tool. *nCC* over 16 tools assessed in this study represented in a heatmap. *nCC* values range from 0 (*black*) to 0.37 (*white*). Binding sites are sorted from high-information

3.7. Conclusions

We introduced an assessment methodology for the performance of TFBS detection algorithms. Platforms such as MTAP make it easier to rank algorithms on multiple criteria and to find effective techniques that solve. We provided a new methodology and tool to compare methods and rank them based on how well they perform on certain subsets of the TFBS detection problem. The key is finding sub-problems of the overall TFBS detection problem that can be solved with reasonable expectation that the algorithm results correspond to real binding sites.

This new methodology is not without problems. First, high-quality data sets of TFBS locations need to be standardized and collected in order to use this technology effectively. Second, great care needs to be taken when looking at benchmarking outcomes, as numerical summaries cannot always convey intuition about why certain approaches fail.

We introduced four principal methods for understanding the TFBS detection problem: (1) tabulated results of derived statistics; (2) ROC graphs; (3) sedimentation graphs; and (4) genome-wide prediction visualization. We also introduced a platform, MTAP, for performing these comparisons. MTAP provides the raw data needed to perform comparisons shown in this section. These raw outputs can be customized by the users for diverse interpretations. A possible application is to compare the pipelines implemented in MTAP with new methods.

When performing assessment, it is important to consider the assumptions of the assessment and ensure that they are in line with the assumptions of the tools being assessed. MTAP was built to assess how well current tools work at automated annotation of genomes. Current tools are expected to perform much better with hand-picked motifs from TFBS databases, although this introduces a certain bias. Assessment must be viewed as a part of the overall system of discovery and verification. It is therefore important that any assessment has a scope consistent with prediction objectives.

---

4. Notes

1. *Installing MTAP*: MTAP was developed and tested on Ubuntu Linux. To set up an MTAP run, the user needs to (1) install motif tool dependencies – this will install tools such as BLAST and MLAGANS that many TFBS tools use as part of their pipelines, (2) install motif discovery tools, (3) format known motif databases, (4) configure MTAP, and (5) run the MTAP analysis. To install MTAP, download

the `mtap.tar.gz` file from `http://biobase.ist.unomaha.edu/mediawiki/index.php/Main_Page` and place it in a directory indicated by the `$MTAP_ HOME` environmental variable. Untar and unzip MTAP to `$MTAP_ HOME` with the command: `tar -xzvf mtap.tar.gz`. In this chapter, we will set `$MTAP_HOME=/home`. Then the tar command will create the following directory structure:

`/home/mtap/pipeline/bin`	`bin` contains useful scripts for running MTAP and installing motif tools
`/home/mtap/pipeline/conf`	`conf` contains configuration files for use by scripts in `bin`
`/home/mtap/pipeline/dumpdir`	`dumpdir` is the location MTAP will place all tool prediction and raw statistics
`/home/mtap/pipeline/lib`	`lib` contains libraries needed to run MTAP. Make sure to run `compile Java.py` before attempting to run MTAP
`/home/mtap/pipeline/motifTools`	`motifTools` contains TFBS prediction software from other institutions
`/home/mtap/pipeline/reqs`	`reqs` contains libraries and tools for motif prediction tools
`/home/mtap/pipeline/src`	`src` contains the MTAP source code
`/home/mtap/pipeline/tmp`	`tmp` is where TFBS databases are placed for MTAP runs and where intermediate results are stored in tool pipelines

1.1 *Install tool dependencies*: Scripts to install tool dependencies exist in `/home/mtap/pipeline/bin`. First install biopython, bioperl, and Java SDK and place them in your path. To install system-level dependencies run from bin: `./installPrereqs/home/matp/pipeline/motif Tools/Linux-i386/`. Cross-species regulatory region detection requires RSD. Install RSD with the following command: `./installRSDreqs/home/mtap/pipeline/reqs/RSD-bin/`.

Some motif tools can be distributed with MTAP. For these tools, we provide an automated script for installation. To install run `./installMotifTools/home/matp/pipeline/motifTools/Linux-i386/` More information on motif tools is in the next section. MTAP is made aware of the dependencies for your specific architecture through the MTAPglobals.py file found in `/home/mtap/pipeline/src/runManager/MTAPglobals.py`. Edit `MTAPglobals.py` to change RUNMANAGERPATH and PROGHOME to

- RUNMANAGERPATH = "/home/mtap/pipeline/src/runManager"
- PROGHOME = "/home/dquest/mtap/pipeline".

`MTAPglobals.py` also contains many variables to change the MTAP runtime characteristics such as number of predictions allowed, number of sequences required, threshold for site true positives, and other concepts discussed earlier in the chapter. Edit the following variable lines to make MTAP aware of the local installation:

```
motifToolBinHOME = "/home/mtap/pipeline/motifTools
/Linux-i386"
REQSPATH = "/home/mtap/pipeline/motifTools/"
PHYLOREQS = "/home/mtap/pipeline/motifTools/"
```

1.2. *Installing TFBS Databases*. To create an MTAP run, one needs to first create the setup files in `/tmp`. First make a directory for the name of the run. MTAP includes several examples (e.g., `/tmp/pito`) that can be copied and modified to create new MTAP runs. Following is the directory structure that is needed to create new runs (creating a new run called "Run1"):

/home/mtap/pipeline/tmp/Run1	The root directory ($RUN) for the new run
$RUN/accoc	Used to hold the accocs.txt association file for relating TFBS location data to Genbank files
$RUN/conf	Used to backup MTAP configuration files for this specific run
$RUN/gbks	Holds Genbank files containing genome sequences
$RUN/kmraws	Holds databases for known TFBS-binding locations
$RUN/motiflists	Holds the motif.list file containing a unique listing of every transcription factor annotated in the kmraw database
$RUN/phylo	Holds the Phylo.txt file for relating the .gbk files and for storing 16sRNA phylogenetic trees and multiple sequence alignments
$RUN/protein	Holds translated .faa files for each coding sequence in the .gbk file and blast databases for searching
$RUN/RSD	Holds ortholog tables for cross-species comparisons
$RUN/xmls	Location to store xml configuration files used by java components

Once the directory structure is made, the MTAP user needs only to copy Genbank files into the `gbks` directory, copy tab-delimited TFBS data into the `kmraw` directory, create the `phylo.txt` file, and create the `accocs.txt` file. MTAP automatically creates the rest of the needed information for the run. Then the MTAP user should change the settings for the MTAP run in `MTAPglobals.py`

The MTAP for Sequence-Based Transcription Factor Binding Site Prediction Tools   139

and `MTAPdbSetup.py` to ensure MTAP will run correctly. At this time, they are ready to run MTAP (`python /home/bin/MTAP.py`). Examples of known motif databases acceptable by MTAP and Genbank files are in the `MTAP.tar.gz` download for reference.

2. *Running Motif Detection Tools*: Motif tools are as variable as the people that develop them. Each takes a multi-Fasta file representing multiple regulatory regions. That is where the similarities end. Each tool produces a specifically formatted output file and takes a specific array of inputs. MTAP unifies all tools by converting each arbitrary output into a unified format (`.gff`) that represents predicted features from each tool in the data set. MTAP also produces each arbitrary input needed by the program. When MTAP is run, it creates a run database consisting of all run tests in the database. This database consists of run tuples of the form:

```
["runName","phylo/genic","#bpUpstream","cr/sr",
"real/markov","",
"MotifList","GenbankFile","knownmotifDatabase",
"fastaUpstreamFile",
"MotifName","MotifTool"]
```

These run tuples are constructed dynamically by permuting all options available in the src/runManager/ MTAPdbSetup.py configuration file and from the data found in the TFBS database in /tmp (*see* previous section). Each MTAP run consists of all possible permutations of the variables found in MTAPdbSetup.py file. These permutations are translated into jobs. The collection of all MotifTool pipeline jobs in a single MTAP run is logged in a file called RUNFILE. RUNFILE exists in /home/dumpdir/runName/RUNFILE. The directory structure under /home/dumpdir/ corresponds to the tuples in the RUNFILE. For example, consider a run using Weeder to find binding sites for the CRP transcription factor in *E. coli K12* (NC_000913.gbk). Assume RegulonDB is the data set we wish to use for evaluation and that we want to take 400 bp upstream of every gene regulated by CRP as annotated in RegulonDB. The MotifList file for all unique motifs in RegulonDB is called v2008_NC_000913. Assuming we call the run "Run1", the tuple for this job will look like

```
["Run1","genic","400","cr","real","v2008_NC_000913",
"knownmotifs.regulondb.v2008","NC_000913","CRP",
"Weeder"]
```

The job tuple indicates the location where the tool will be run on the local file structure. The above example will be run in $RUNDIR =/home/mtap/pipeline/dumpdir

/Run1/genic/400/cr/real/. Unified .gff files for plotting in tools such as gbrowse are available in $RUNDIR/gff. The raw statistics files for analysis are available in $RUNDIR/stats. Specific tool thresholds and pipelines can be modified by changing the tool driver scripts found in /home/mtap/pipeline/src/runManager/motifTools. Sophisticated data collection scripts for analyzing gene regulatory networks with graph theory and for plotting changes across run parameters are available in src/runManager.

References

1. Das, M., and Dai, H. (2007) A survey of DNA motif finding algorithms. *BMC Bioinformatics* 8, 1–13.
2. Davidson, E.H., Rast, J.P., Oliveri, P. et al. (2002) A genomic regulatory network for development. *Science* 295, 1669–1678.
3. Stathopoulos, A., and Levine, M. (2005) Genomic regulatory networks and animal development. *Dev Cell* 9, 449–462.
4. Imai, K., Levine, M., Satoh, N. et al. (2006) Regulatory blueprint for a chordate embryo. *Science* 312, 1183–1187.
5. Salgado, H., Santos-Zavaleta, A., Gama-Castro, S. et al. (2006) The comprehensive updated regulatory network of Escherichia coli K-12. *BMC Bioinformatics* 7, 1–5.
6. Shen-Orr, S.S., Milo, R., Mangan, S. et al. (2002) Network motifs in the transcriptional regulation network of Escherichia coli. *Nat Genet* 31, 64–68.
7. Salgado, H., Gama-Castro, S., Martaenez-Antonio, A. et al. (2004) RegulonDB (version 4.0): transcriptional regulation, operon organization and growth conditions in Escherichia coli K-12. *Nucleic Acids Res* 32, D303–D306.
8. Karp, P., Riley, M., Saier, M. et al. (2002) The EcoCyc database. *Nucleic Acids Res* 30, 56–58.
9. Wingender, E., Dietze, P., Karas, H. et al. (1996) TRANSFAC: a database on transcription factors and their DNA binding sites. *Nucleic Acids Res* 24, 238–241.
10. Ishii, T., Yoshida, K.-I., Terai, G. et al. (2001) DBTBS: a database of Bacillus subtilis promoters and transcription factors. *Nucleic Acids Res* 29, 278–280.
11. Kazakov, A.E., Cipriano, M.J., Novichkov, P.S. et al. (2006) RegTransBase – a database of regulatory sequences and interactions in a wide range of prokaryotic genomes. *Nucleic Acids Res* 35 (Database Issue), D407–D412.

12. Maench, R., Hiller, K., Barg, H. et al. (2003) PRODORIC: prokaryotic database of gene regulation. *Nucleic Acids Res* 31, 266–269.
13. Meng, H., Banerjee, A., and Zhou, L. (2006) BLISS: binding site level identification of shared signal-modules in DNA regulatory sequences. *BMC Bioinformatics* 7, 287.
14. Tompa, M., Li, N., Bailey, T. et al. (2005) Assessing computational tools for the discovery of transcription factor binding sites. *Nat Biotechnol* 23, 137–144.
15. Sandve, G., Abul, O., Walseng, V. et al. (2007) Improved benchmarks for computational motif discovery. *BMC Bioinformatics* 8, 193.
16. Frith, M.C., Hansen, U., Spouge, J.L. et al. (2004) Finding functional sequence elements by multiple local alignment. *Nucleic Acids Res* 32, 189–200.
17. Eskin, E., and Pevzner, P.A. (2002) Finding composite regulatory patterns in DNA sequences. *Bioinformatics* 18(Suppl. 1), S354–S363.
18. Lawrence, C.E., Altschul, S.F., Boguski, M.S. et al. (1993) Detecting subtle sequence signals: a Gibbs sampling strategy for multiple alignment. *Science* 262, 208–214.
19. Pavesi, G., Mereghetti, P., Mauri, G. et al. (2004) Weeder Web: discovery of transcription factor binding sites in a set of sequences from co-regulated genes. *Nucleic Acids Res* 32, W199–W203.
20. Thijs, G., Lescot, M., Marchal, K. et al. (2001) A higher-order background model improves the detection of promoter regulatory elements by Gibbs sampling. *Bioinformatics* 17, 1113–1122.
21. Bailey, T.L., and Elkan, C. (1994) Fitting a mixture model by expectation maximization to discover motifs in biopolymers. *Proc Int Conf Intell Syst Mol Biol ISMB Int Conf Intell Syst Mol Biol* 2, 28–36.

22. Hughes, J.D., Estep, P.W., Tavazoie, S. et al. (2000) Computational identification of cis-regulatory elements associated with groups of functionally related genes in Saccharomyces cerevisiae. *J Mol Biol* 296, 1205–1214.

23. Workman, C.T., and Stormo, G.D. (2000) ANN-Spec: a method for discovering transcription factor binding sites with improved specificity. *Pac Symp Biocomput* 5: 467–478.

Chapter 9

Computational Analysis of ChIP-seq Data

Hongkai Ji

Abstract

Chromatin immunoprecipitation followed by massively parallel sequencing (ChIP-seq) is a new technology to map protein–DNA interactions in a genome. The genome-wide transcription factor binding site and chromatin modification data produced by ChIP-seq provide invaluable information for studying gene regulation. This chapter reviews basic characteristics of ChIP-seq data and introduces a computational procedure to identify protein–DNA interactions from ChIP-seq experiments.

Key words: Transcription factor binding site, high-throughput sequencing, peak detection, false discovery rate.

1. Introduction

Chromatin immunoprecipitation (ChIP) followed by massively parallel sequencing (ChIP-seq) is a new technology to map protein–DNA interactions in genomes (1–4). In this technology, a protein of interest (POI) is cross-linked to chromatin. Chromatin is sheared into small fragments. The POI and its bound chromatin fragments are immunoprecipitated using an antibody specific to the protein. After reversing the cross-links, a DNA sample called "ChIP sample" is obtained. In many studies, a negative control sample is prepared in parallel using a similar protocol that bypasses the immunoprecipitation step. Compared to the control sample, the ChIP sample is enriched in DNA fragments bound by the protein of interest. After size selection and further processing, DNA fragments in the samples are sequenced from both ends using one of the recently developed high-throughput sequencing

I. Ladunga (ed.), *Computational Biology of Transcription Factor Binding*, Methods in Molecular Biology 674,
DOI 10.1007/978-1-60761-854-6_9, © Springer Science+Business Media, LLC 2010

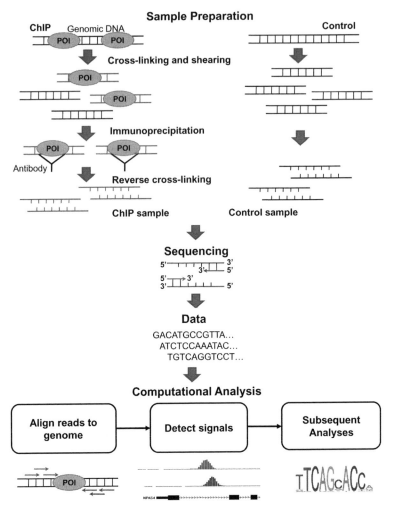

Fig. 9.1. Workflow for ChIP-seq.

platforms (5). This produces tens of millions of sequence tags, also known as sequence reads. By computationally mapping these reads to a reference genome and looking for genomic regions where ChIP reads are enriched, genomic loci with protein–DNA interactions can be identified (**Fig. 9.1**). Currently, this technology is widely used to study transcription factor binding sites (TFBS) (1, 2) and chromatin modifications (3, 4). The genome-wide transcription factor binding site and chromatin state data produced by ChIP-seq provide invaluable information for studying gene regulation.

An earlier technology to map protein–DNA interactions in genomes is ChIP-chip (6, 7), which uses chromatin immunoprecipitation to enrich protein-bound DNAs and hybridizes the enriched DNA fragments to genome tiling arrays. Compared to ChIP-chip, ChIP-seq has several advantages (8). First, ChIP-seq does not rely on array hybridization. As a result, it does not suffer from the biases and noise caused by cross-hybridization, the varying GC content of probe sequences and other issues related to hybridization chemistry, although ChIP-seq may have its own biases that are not well understood currently. Second, ChIP-chip measures enrichment by intensities of hybridization which may saturate at high signal, whereas ChIP-seq measures enrichment by tag counts which can handle signals in a much broader dynamic range. Third, protein–DNA interactions detected by ChIP-chip are restricted to genomic regions for which probes are available. Repetitive regions in the genome usually are excluded from the array design. In contrast, ChIP-seq can be used to study protein–DNA interactions in any part of the genome as long as reads can be unambiguously aligned to places where they are originally produced. For this reason, ChIP-seq is able to offer much less biased genome coverage. Fourth, for mapping TFBS, ChIP-seq is able to locate binding sites at 50–100 base pair (bp) resolution. This represents a significantly improved precision compared to the 300–1,000 bp resolution provided by ChIP-chip. Other advantages of ChIP-seq include requirement of less input materials and ability to provide extra information to study allele-specific protein binding. Thanks to these advantages, as the cost of high-throughput sequencing continues to decrease, ChIP-seq has the potential to become the dominant technology for creating genome-wide maps of protein–DNA interactions.

ChIP-seq creates unprecedented amounts of data. Extracting information from the data is not trivial. Typically, the analysis is a multiple step procedure (**Fig. 9.1**). First, raw sequence reads are mapped to the reference genome. Next, genomic regions in which ChIP reads are enriched are identified and the statistical significance of the predicted genomic regions is evaluated. Regions that satisfy certain significance criteria are reported. Subsequently, the reported regions are analyzed in various ways to help scientists understand their functional implications. These include adding gene annotations, finding or mapping transcription factor binding motifs, and correlating the protein–DNA interactions with gene expression information. The purpose of this chapter is to briefly review some basic characteristics of ChIP-seq data and introduce a computational procedure to analyze the data. We will mainly focus on describing a method to identify protein–DNA interactions and estimate the false discovery rates (FDR). Tools to perform subsequent analyses will be discussed briefly.

1.1. Types of ChIP-seq Experiments

We focus on two types of ChIP-seq experiments, namely the "one-sample experiment" and the "two-sample experiment." A two-sample experiment involves sequencing both a ChIP sample and a negative control sample. In contrast, a one-sample

146   Ji

experiment only involves sequencing a ChIP sample. Readers are referred to **Note 1** for a discussion on how to analyze experiments that have technical or biological replicates.

Compared to the two-sample experiment, the one-sample design is more cost effective. However, the negative control sample in the two-sample experiment allows one to build a better model to describe locus-dependent background noise, which can significantly reduce the number of false positives and false negatives in the subsequent data analyses (9, 10).

1.2. Models for Background Noise

In both one-sample and two-sample experiments, protein–DNA interactions can be identified by searching for enrichment of ChIP reads. A key component of ChIP-seq data analysis is to understand what level of enrichment is required to distinguish signals from noise.

1.2.1. Background Model for One-Sample Experiments

First consider a one-sample experiment. Assume that the length of the genome is L bps and the sample has N uniquely mapped reads in total. Consider a w bp window in the genome, and let n be the number of reads mapped to the window. Studies of negative control samples show that if the window does not contain any protein–DNA interaction of interest, n can be approximately modeled by a negative binomial distribution $NB(\alpha, \beta)$ (9). In other words, $\Pr(n = k) = \begin{pmatrix} k + \alpha - 1 \\ \alpha - 1 \end{pmatrix} \left(\dfrac{\beta}{\beta + 1} \right)^{\alpha} \left(\dfrac{\beta}{\beta + 1} \right)^{k}$.

Here all background windows in the genome have the same values of α and β. Based on this result, one approach to characterize the background noise is to find appropriate parameter values of α and β using the observed data. When estimating α and β, one should keep in mind that the data (i.e., the ChIP sample) usually consist of a mixture of background windows and windows that contain signals; however, α and β are parameters to describe background noise only. An algorithm that estimates the background parameters α and β from a mixture of signal and noise windows will be described in **Section 3.2.1**.

Another natural way to model the read count of a background window is to assume that n follows a Poisson distribution with a rate parameter λ (i.e., $\Pr(n = k) = \lambda^{k} e^{-\lambda}/k!$). Recent studies show that the Poisson distribution with a fixed rate λ does not perform well to characterize the background variability in real data (9–11). For example, in **Table 9.1**, a negative control sample from a ChIP-seq experiment in mouse embryonic stem cells (12) is analyzed by both the Poisson background model and the negative binomial model. The genome is divided into 100 bp long non-overlapping windows and the number of uniquely mapped reads in each window is counted. The negative control sample contains no protein–DNA interactions of interest;

Table 9.1
Comparison of the Poisson and negative binomial background model

Read count	Observed frequency	Expected by Poisson	Expected by NB
0	0.792664	0.792664	0.792230
1	0.164843	0.164843	0.164753
2	0.034140	0.017140	0.034122
3	0.006587	0.001188	0.007057
4	0.001320	0.000062	0.001459
5	0.000288	0.000003	0.000301
6	0.000075	0.000000	0.000062
7	0.000023	0.000000	0.000013
...	...	...	...

hence all windows represent background noise. The second column of the table shows the observed frequency that a window contains k reads. The third and fourth columns show frequencies expected by the Poisson and negative binomial models, respectively. This table clearly shows that the Poisson model is not able to describe the heavy tail of the empirical read count distribution and the negative binomial model performs much better.

Using a fixed rate Poisson model assumes that background reads are generated at the same rate for all loci in the genome or, in other words, background reads are distributed uniformly across the genome. **Table 9.1** illustrates that this assumption does not fit well with the real data. In the negative binomial model, it is implicitly assumed that the background reads are generated by Poisson distributions with different rates at different loci, and as a result, the background reads are not uniformly distributed across the genome. In order to see this, we note that a negative binomial distribution can be related to a Poisson distribution via a hierarchical model. Let us divide the genome into w bp long non-overlapping windows and assume that different windows generate reads independently. Let λ_i be the rate to generate reads in the ith window, n_i be the number of reads in window i, and assume that $n_i|\lambda_i \sim$ Poisson (λ_i). If we allow λ_i to vary across the genome but assume that λ_i's are random samples drawn independently from a locus-independent gamma distribution Gamma(α, β) (the probability density function for Gamma(α, β) is $f(x) = \frac{\beta^\alpha}{\Gamma(\alpha)} x^{\alpha-1} e^{-\beta x}$), then the marginal distribution of n_i of a background window, $\Pr(n_i = k|\alpha, \beta) = \int \Pr(n_i = k|\lambda_i) f(\lambda_i|\alpha, \beta) d\lambda_i$, has the same probability density function as that of the $NB(\alpha, \beta)$.

The hypothesis that read sampling rates vary across the genome is supported by analyses of independent samples from

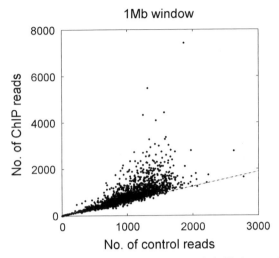

Fig. 9.2. Correlation of read numbers at the same genomic loci between a ChIP sample and a control sample. The samples are obtained from a ChIP-seq experiment that maps the NRSF TFBSs (1). The human genome is divided into non-overlapping windows, each window containing 1 million base pairs. For each window, ChIP and control reads are counted and plotted as a *dot*.

the same experiment (10). As an example, **Fig. 9.2** shows a scatter plot that compares the window read counts between a ChIP sample and a matching negative control sample in an experiment involving transcriptional repressor NRSF (1). The plot has a positive slope and the counts from the two samples in the same genomic window are clearly correlated. This indicates that the rate for generating reads is locus dependent and is not a constant across the genome. Unfortunately, in a one-sample experiment, background reads in a particular window cannot be separated from reads that represent biological signals in the same window. For this reason, the locus-dependent Poisson rate cannot be estimated without making additional assumptions. The negative binomial model makes the assumption that the background rates λ_is follow a common gamma distribution. By making this assumption, information from all windows can be used to infer the common parameters α and β, which are then used to describe the background for each individual window. This is the underlying rationale for using a negative binomial distribution as the background model (*see* **Note 2** for an alternative solution).

1.2.2. Background Model for Two-Sample Experiments

Now consider a two-sample experiment that involves a control sample in addition to a ChIP sample. Assume that the ChIP sample has N uniquely mapped reads in total and the control sample has M uniquely mapped reads. For a w bp window indexed by i, let n_i be the number of ChIP reads mapped to the window, and m_i be the number of control reads. In the previous section, it has been shown that the read counts in background windows can be viewed as Poisson random variables with varying rates across the genome (which results in negative binomial marginal distributions). In light of this observation, one can assume that $n_i \sim$ Poisson(μ_i) and $m_i \sim$ Poisson(λ_i), where μ_i and λ_i are rates at which reads are produced in window i in the ChIP and control samples, respectively, and we allow μ_i and λ_i to have different values at different loci in the genome. For each genomic window, μ_i can be decomposed into two parts $\mu_i = \mu_{i1} + \mu_{i0}$, where μ_{i0} is the rate at which background reads are generated and μ_{i1} is the rate to generate reads corresponding to signals. Often, it is reasonable to assume that the background rates in the ChIP and control samples, μ_{i0} and λ_i, are equal up to a proportionality constant, i.e., $\mu_{i0} = c\lambda_i$. The proportionality constant c reflects the observation that the total numbers of reads in the ChIP and control samples are usually not the same. Under the assumption that $\mu_{i0} = c\lambda_i$, information from the negative control sample can be used to describe the background read sampling rate in the ChIP sample. As a result, the assumption used in the one-sample analysis that background read sampling rates from different genomic windows follow a common probability distribution is no longer required.

For a window that does not contain any protein–DNA interactions, $\mu_i = \mu_{i0} = c\lambda_i$. It is known that the sum of two independent Poisson random variables $X \sim$ Poisson(λ_1) and $\Upsilon \sim$ Poisson(λ_2) follows a Poisson distribution, Poisson$(\lambda_1 + \lambda_2)$, and conditional on the sum, X, follows a binomial distribution. In other words, $X \mid X + \Upsilon = n \sim$ Bin(n, p), where $p = \lambda_1/(\lambda_1 + \lambda_2)$ (i.e., Pr$(X = k \mid X + \Upsilon = n) = \binom{n}{k} p^k (1-p)^{n-k}$).

Using these results, the number of ChIP reads in a background window conditional on the total number of reads in that window should follow a binomial distribution, i.e., $n_i \mid m_i + n_i \sim$ Bin$(m_i + n_i, p_0)$, where $p_0 = c/(1 + c)$ represents the expected proportion of ChIP reads in a background window. If p_0 is known, the enrichment of ChIP reads in any window can be evaluated. This evaluation does not require the knowledge of the actual values of the background sampling rates, λ_i.

In order to estimate p_0, one should keep in mind that the ratio $N/(M + N)$ based on the total read numbers in the two samples is a biased estimate. This is because the ChIP sample contains both background reads and reads that represent signals, whereas p_0 is related only to the background. If we divide the genome into w bp long non-overlapping windows (indexed by i) and assume that read numbers in different windows follow independent Poisson distributions, then $N \sim$ Poisson$(\sum_i \mu_{i0} + \sum_i \mu_{i1})$ and $M \sim$ Poisson$(\sum_i \lambda_i)$. As a result, $N|M+N \sim \text{Bin}(M+N, q)$, where $q = (c+d)/(1+c+d) \neq c/(1+c)$ and $d = \sum_i \mu_{i1} / \sum_i \lambda_i$. It can be shown that given λ_i, μ_{i1}, and c, the expectation of $N/(M+N)$ is q which is not equal to p_0. An algorithm that estimates p_0 and uses the binomial distribution to evaluate the enrichment of ChIP reads will be described in **Section 3.2.2**. An alternative approach to evaluate background variability for two-sample experiments is discussed in **Note 3**.

1.3. Normalization

The proportionality constant $c = p_0/(1-p_0)$ in the two-sample analysis can be viewed as a way to normalize the read counts of two different samples. This normalizing constant can be used to compute the fold enrichment of ChIP reads, which is defined by (9) as the ratio $(n_i + 1)/(cm_i + 1)$. Here m_i and n_i are read

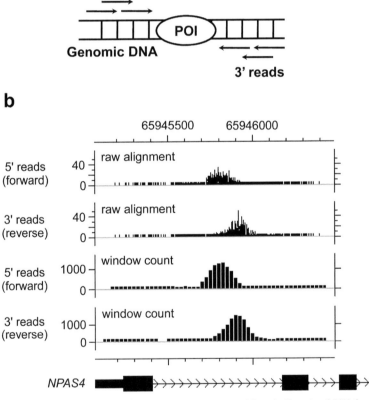

Fig. 9.3. Peak shape for a TFBS. **a** Reads are generated from both ends of DNA fragments. **b** 5′ reads are aligned to the forward strand of the reference genome, and 3′ reads are aligned to the reverse complement strand. These two types of reads form two separate peaks. The binding site is located between the modes of the peaks. From *top* to *bottom*, the four signal tracks are the number of 5′ reads aligned to each genomic position, number of 3′ reads aligned to each position, 5′ read count in a 100 bp sliding window, and 3′ read count in a 100 bp sliding window. The read counts in sliding windows form *smooth curves*. The modes of the curves define boundaries of binding sites.

numbers in the control and ChIP samples in a window indexed by i and a regularization constant one is added to both the numerator and the denominator to avoid dividing by zero.

1.4. Peak Shape

In most current high-throughput sequencing platforms, sequence reads are produced from both ends of DNA fragments. Surrounding a TFBS on the chromosomal map, reads that are aligned to the forward strand of the genome will form a peak upstream of the binding site, and reads that are aligned to the reverse complement strand will form a peak downstream of the binding site (13, 14) (**Fig. 9.3**). This forms a characteristic peak shape that contains useful information for distinguishing bona fide binding sites from false positives. Predicted TFBSs without this bimodal peak shape are often false positives and should be eliminated from the final results. The bimodal shape is also useful for making high-resolution binding site predictions. The bona fide binding site should fit in between the modes of the two peaks. Using this information, a TFBS can usually be mapped to a 50~100 bp long region (9, 11, 14–16).

2. Software

The methods described in this chapter for building background models and detecting protein–DNA interactions from mapped sequence reads are implemented in the open-source software CisGenome which is available at http://www.biostat.jhsph.edu/~hji/cisgenome (9). CisGenome provides a user-friendly graphic interface and it can also be used to perform various types of subsequent analyses. Sequence reads can be mapped to a reference genome using one of the following software tools: Eland provided by Illumina, Inc., Bowtie at http://bowtie.cbcb.umd.edu (17), MAQ at http://maq.sourceforge.net/ (18), SeqMap at http://biogibbs.stanford.edu/~jiangh/SeqMap/ (19), Corona Lite provided by the Life Technologies (http://solidsoftwaretools.com/gf/project/corona/), and SHRiMP at http://compbio.cs.toronto.edu/shrimp/ (20).

3. Methods

In this section, we describe a procedure to detect protein–DNA interactions from ChIP-seq data. Alternative methods are discussed in **Note 4**.

152   Ji

3.1. Align Sequence Reads

The first step of data analysis is to align sequence reads to a reference genome. A number of software tools have been developed to support fast mapping of millions of short-sequence tags to complex genomes. Examples include Eland (Cox, unpublished), Bowtie (17), MAQ (18), SeqMap (19), and SHRiMP (20). For data generated by the Life Technologies' SOLiD platform, alignment needs to be performed in color-space using tools such as Corona Lite (unpublished) and SHRiMP (20). From now on, we assume that all sequence reads are mapped, and reads that are uniquely aligned to the genome are retained for subsequent analyses.

3.2. Building Background Models

Using the mapped reads, build a background model using CisGenome (9).

3.2.1. Background Model for Analyzing One-Sample Experiments

Divide the genome into non-overlapping windows. The window size w should be chosen to roughly match the expected length of enrichment signals. For TFBS analysis, the window size w is typically set to 100 bp (*see* **Note 5** for more discussions). The entire set of windows can be viewed as a mixture of windows that represent background noise and windows that contain protein–DNA interactions of interest. Let π_0 denote the proportion of background windows. π_0 is unknown and needs to be estimated from the data.

For each window, count the number of reads that are uniquely aligned to the window. Let n_i be the number of reads within the ith window. It is assumed that for background and non-background windows, n_i follows two different probability distributions for which density functions are $f_0(n)$ and $f_1(n)$, respectively. Under this assumption, the data generating distribution for n_i can be described by a mixture distribution $g(n) = \pi_0 f_0(n) + (1 - \pi_0)f_1(n)$. Use the empirical distribution of n_i, i.e., the observed frequencies that $n_i = n(n = 0, 1, 2, \ldots)$, to estimate $g(n)$.

Based on the discussions in **Section 1.2.1**, the background distribution $f_0(n)$ can be modeled by a negative binomial distribution $NB(\alpha, \beta)$. In order to estimate α and β, we assume that windows with small number of reads are mostly background. Under this assumption, the background parameters α and β can be estimated using windows with no more than two reads. For a random variable n that follows negative binomial distribution $NB(\alpha, \beta)$, define $r_1 = \Pr(n = 1)/\Pr(n = 0)$ and $r_2 = \Pr(n = 2)/\Pr(n = 1)$. Since $r_1 = \alpha/(\beta + 1)$ and $r_2 = (\alpha + 1)/[2(\beta + 1)]$, we have $\alpha = r_1/(2r_2 - r_1)$ and $\beta = 1/(2r_2 - r_2) - 1$. Therefore, to estimate α and β count the number of windows that contain k reads and denote it as u_k. Use u_1/u_0 to estimate r_1 and use u_2/u_1 to estimate r_2. Plug the estimated values of r_1 and r_2 into $r_1/(2r_2 - r_1)$ and $1/(2r_2 - r_1) - 1$ to obtain the estimates of α and β.

In order to estimate π_0, we assume that most windows with no mapped read represent background noise. Under this assumption, $g(0) \approx \pi_0 f_0(0)$ and $\pi_0 \approx g(0)/f_0(0)$. Therefore, π_0 can be estimated by $u_0 / [(\sum_k u_k)\hat{f}_0(0)]$. Finally, using the estimated π_0, $f_0(.)$ and $g(.)$, one can estimate the local false discovery rate (local FDR) for any w bp window as follows: $lfdr$ (window i) $= \pi_0 f_0(n_i)/g(n_i)$. Here, n_i is the observed read count for window i.

3.2.2. Background Model for Analyzing Two-Sample Experiments

Divide the genome into w bp long non-overlapping windows. For each window, count the number of reads that are uniquely aligned to the window. For window i, let n_i and m_i denote the number of reads in the ChIP and control samples, respectively, and let $t_i = n_i + m_i$ be the total read count.

Using windows for which t_i is small (we usually use windows that contain only one mapped read, i.e., indices i for which $t_i = 1$), estimate the expected proportion of ChIP reads in background windows as $\hat{p}_0 = \sum_i n_i / \sum_i (n_i + m_i)$. This implicitly assumes that windows with small read counts mainly represent background. Estimate the normalizing constant $\hat{c} = \hat{p}_0/(1 - \hat{p}_0)$.

Next, group windows based on their total read counts t_i. For each group of windows for which $t_i = t(t = 0, 1, 2, \ldots)$, compute the observed frequency that $n_i = n(n = 0, 1, \ldots, t)$. Derive the function $g_{\text{obs}}(n|t) = \{$number of windows for which $t_i = t$ and $n_i = n\}$ / $\{$number of windows for which $t_i = t\}$. Define $f_{\text{Bin}}(n|t, p_0) = \Pr(X = n)$ where $X \sim \text{Bin}(t, p_0)$. For a window that contains t reads among which n are ChIP reads, estimate the local FDR as $f_{\text{Bin}}(n|t, \hat{p}_0)/g_{\text{obs}}(n|t)$. When t becomes big, there will be fewer windows available for estimating $g_{\text{obs}}(n|t)$. In order to get robust local FDR estimates, if there are fewer than 100 independent windows for a particular t, we suggest extrapolating the local FDR estimates from windows with smaller total read counts. In other words, find the biggest $t' < t$ that has more than 100 windows. For a window that contains t reads and n ChIP reads, the local FDR is estimated as $f_{\text{Bin}}(n'|t', \hat{p}_0)/g_{\text{obs}}(n'|t')$, where $n' = \lfloor t'n/t \rfloor$ and $\lfloor x \rfloor$ represents the maximal integer that is not bigger than x.

3.3. Detect Protein–DNA Interactions

Using CisGenome (9), scan the reference genome using a w bp long-sliding window. Compute the local FDR for each window. For analyzing a one-sample experiment, use the estimated background model described in **Section 3.2.1**. For analyzing a two-sample experiment, use the procedure described in **Section 3.2.2**. For the two-sample analysis, also compute a fold enrichment for each window: $(n_i + 1)/(\hat{c}m_i + 1)$. Here n_i is the number of ChIP reads in the window, m_i is the number of control reads, and $\hat{c}$ is the normalizing constant estimated using the method in **Section 3.2.2**.

Select all windows with local FDR smaller than a given cutoff (usually ≤ 10%). Merge overlapping windows into a single region. Report all regions obtained after merging. During the process in which windows are merged, use the smallest local FDR among the overlapping windows as the local FDR for the merged region. For the two-sample analysis, use the biggest fold enrichment among all the overlapping windows as the fold change of the merged region.

3.4. Improve Predictions of Transcription Factor Binding Sites

If the purpose of the ChIP-seq experiment is to locate TFBSs, the reported regions should be further processed using CisGenome as follows to improve the results.

3.4.1. Determine the Binding Site Boundary

Use a w bp sliding window to scan each reported region. For each window, count reads in the ChIP sample that are aligned to the forward strand of the genome and those that are aligned to the reverse complement strand. This creates two smooth curves of read counts (**Fig. 9.3**). Identify the locations where the two curves achieve their maxima (i.e., the modes of the curves) and use these locations to define boundaries of binding sites.

3.4.2. Adjust for DNA Fragment Length

For each reported region, compute the distance between the modes of the peaks on the forward and reverse complement strands. Compute the median of all distances and denote it as L. Shift all reads toward the center of the DNA fragments by $L/2$ base pairs. Reads aligned to the forward strand of the genome are shifted toward $3'$ of the reference genome and reads aligned to the reverse complement strand are shifted toward $5'$ of the reference genome. Using the shifted reads, perform the analyses described in **Sections 3.2** and **3.3** again. For the reported regions, determine the binding site boundaries using unshifted reads as described in **Section 3.4.1**.

3.5. Subsequent Analyses

Having identified protein-binding regions, they can be analyzed in different ways to study the biological implications. Here we suggest a few common analyses, most of which can be carried out using CisGenome (9). First, compute frequencies that reported regions occur in intragenic and intergenic regions, exons, introns, promoter regions, and other structural features of genes and compute the average level of conservation across species for each region. These two analyses may provide information on functional contexts and importance of the reported regions. Second, extract genes in the neighborhood of the reported regions as a gene set and perform Gene Set Enrichment analysis (http://www.broadinstitute.org/gsea/) (21) and Gene Ontology analysis (http://www.geneontology.org/GO.tools.shtml). These analyses may provide information on functional categories or pathways that are involved in the biological system in question. Third, perform de novo motif discovery or map the known motifs to the reported transcription factor binding regions and their flanking regions. Identify motifs that are enriched in the binding regions compared to control genomic regions using CisGenome. These analyses may identify motifs that are recognized by the transcription factor in question. They may also suggest collaborating factors. In addition, the motif analysis provides a way to verify that the reported TFBSs are bona fide signals. For example, if the ChIP-seq experiment studies a transcription factor and the binding motif of the transcription factor is known, then the motif is expected to be enriched in the reported binding regions. If this is not the case, it may indicate problems in the ChIP-seq experiment or data analyses. Last but not least, it is always a good idea to visualize the ChIP-seq data along with other structural and functional annotations of the genome. Both the CisGenome Browser and the Genome Browser at UCSC (http://genome.ucsc.edu/) (22) can be used to interactively visualize the data. Interesting patterns may emerge by simply eye balling the data. These patterns may create new hypotheses and suggest future research directions.

4. Notes

1. Analysis of experiments with replicate samples. The methods introduced in this chapter are developed for analyzing experiments that contain a single replicate. If an experiment contains more than one replicates, the analysis can be carried out in two steps. First, merge the replicate data into a combined ChIP sample and a combined control sample (there will be no control sample in a one-sample experiment). The combined sample can then be analyzed using the methods described in **Section 3**. Second, for the reported peaks, extract read counts from individual replicate samples. Normalize the read counts by multiplying the raw read numbers with the normalizing constants obtained using the approach described in **Section 3.2.2**. The normalized read counts can then be analyzed using existing methods developed for detecting differentially expressed genes in microarray experiments (e.g., limma (23)) to remove regions for which the observed ChIP enrichment over the controls can be explained by the random variability among replicates. Suppose that the normalized read counts are saved in a

tab-delimited text file named "data.txt," the R commands below show how limma can be used to perform the analysis in the second step.

```
> library(affy)
> library(limma)
> exprs <- as.matrix(read.table("data.txt",
  header =TRUE,
sep="\t", row.names=1, as.is=TRUE))
> exprs <- log2(exprs)
> eset<-new("ExpressionSet", exprs=exprs)
> design<-cbind(Base=1, ChIP=c(1,1,1,0,0,0)) ##
  3 ChIP vs.
3 controls
> fit<-lmFit(eset,design)
> fit<-eBayes(fit)
```

2. An alternative approach to estimate background in a one-sample experiment. Zhang et al. (15) proposed another approach to estimate the background Poisson rate. To estimate the rate λ_i for a genomic window (usually dozens of base pairs in length), this approach considers a few larger windows (usually 5 and 10 kb in a one-sample analysis) surrounding the window in question. λ_i is estimated using read occurrence rates derived from these larger windows. The underlying assumption of this method is that small windows (with a few dozens of base pairs) close to each other have similar background read sampling rate and reads in the larger surrounding windows are mostly background reads. This is usually a reasonable assumption for analyzing TFBSs. However, it may not hold true in data which contain broad signals or where signals occur at high frequency in the genome. When the assumption is true, this method may provide higher statistical power for detecting signals.

3. An alternative approach to estimate background in a two-sample experiment. Statistical significance of the observed enrichment in the ChIP-control comparison can also be assessed by swapping the sample labels (15). In other words, one treats the ChIP sample as the control and treats the control sample as the ChIP. One then applies the same peak detection procedure to detect "signals" in the label-swapped data. Any "signals" reported in this analysis should represent noise. The false discovery rate for a given enrichment level in the original analysis can be estimated by the ratio {number of regions reported in the label-swapped data}/ {number of regions reported in the original data}. This approach requires that the two samples have about the same number of background reads in order to produce correct FDR estimates. If two samples have different number of reads, a

random subset of reads is usually drawn from the larger sample to create a subsample that has roughly the same number of reads as the other sample. Because this procedure excludes some data from the analysis, it may sacrifice some statistical power. This procedure attempts to match the total number of reads between the two samples, which is not equivalent to matching the number of background reads. In light of discussions in **Section 1.2.2**, this may introduce bias into the FDR estimates. Compared to this approach, the approach described in **Section 3.2.2** does not require the two samples that have the same read numbers. However, since it depends on assumptions about the underlying data generating distribution, it may produce biased estimates as well if the model assumptions do not hold true in the data.

4. Alternative approaches to detect peaks from ChIP-seq data. Several other methods have been developed for detecting "enrichment peaks" from ChIP-seq data. QuEST (14) (*see* also **Chapter 10**) uses a kernel density estimation approach to build density profiles for forward and reverse reads separately. It then combines the two profiles to detect peaks. FDR is estimated by dividing the control sample into two halves and comparing the two subsets of the control. This requires one to have twice as many reads in the control sample as in the ChIP sample. SISSRs (16) detects points in the genome where the net difference between the forward and reverse read counts in a sliding window switches from positive to negative. It then detects statistically significant binding sites by using a constant rate Poisson model to evaluate the enrichment of the total read counts in the windows surrounding the detected switching points. *MACS* (15) uses a sliding window to scan the genome, and uses a locally estimated Poisson rate to detect enrichment peaks, as discussed in **Note 3**. Other methods include FindPeaks (24), USeq (25), PeakSeq (10), and a ChIP-seq processing pipeline developed by Kharchenko et al. (11). Currently, relative performance of various methods has not been benchmarked. However, for locating TFBSs, all these methods provide similar spatial resolution (a few dozens of base pairs) and the difference among them is subtle compared to the difference between ChIP-chip and ChIP-seq.

5. The choice of window size. The choice of window size w represents a trade-off between sensitivity and specificity. When independent information is available, it may be used to guide the choice of w. For example, in an experiment that locates TFBSs with known motif(s), one can map the motif to the reported binding regions and compute the motif occurrence rates (i.e., the number of motif sites per 1 kb).

The motif occurrence rate is a measure of signal-to-noise ratio. It decreases when the window size becomes too small or too big (9). Motif occurrence rates for regions reported using different window sizes can be compared and the window size that maximizes the rate can be selected to generate the final analysis results. If the transcription factor binding motif is not known before the study, one may first perform de novo motif discovery and use the method described in (26) to identify the motif. It has been shown that the approach described in (26) can correctly identify binding motifs for most genome-wide ChIP studies that involve transcription factors recognizing sequence-specific binding patterns. If one is not able to get the motif information but gene expression data are available, the window size may also be chosen based on what fractions of binding regions are associated with a particular gene expression pattern of interest for different choices of window sizes.

References

1. Johnson, D.S., Mortazavi, A., Myers, R.M., and Wold, B. (2007) Genome-wide mapping of in vivo protein-DNA interactions. *Science* 316, 1497–1502.
2. Robertson, G., Hirst, M., Bainbridge, M. et al. (2007) Genome-wide profiles of STAT1 DNA association using chromatin immunoprecipitation and massively parallel sequencing. *Nat Methods* 4, 651–657.
3. Mikkelsen, T.S., Ku, M., Jaffe, D.B. et al. (2007) Genome-wide maps of chromatin state in pluripotent and lineage-committed cells. *Nature* 448, 553–560.
4. Barski, A., Cuddapah, S., Cui, K. et al. (2007) High-resolution profiling of histone methylations in the human genome. *Cell* 129, 823–837.
5. Shendure, J., and Ji, H. (2008) Next-generation DNA sequencing. *Nat Biotechnol* 26, 1135–1145.
6. Ren, B., Robert, F., Wyrick, J.J. et al. (2000) Genome-wide location and function of DNA binding proteins. *Science* 290, 2306–2309.
7. Cawley, S., Bekiranov, S., Ng, H.H. et al. (2004) Unbiased mapping of transcription factor binding sites along human chromosomes 21 and 22 points to widespread regulation of noncoding RNAs. *Cell* 116, 499–509.
8. Park, P.J. (2009) ChIP-seq: advantages and challenges of a maturing technology. *Nat Rev Genet* 10, 669–680.
9. Ji, H., Jiang, H., Ma, W. et al. (2008) An integrated software system for analyzing ChIP-chip and ChIP-seq data. *Nat Biotechnol* 26, 1293–1300.
10. Rozowsky, J., Euskirchen, G., Auerbach, R.K. et al. (2009) PeakSeq enables systematic scoring of ChIP-seq experiments relative to controls. *Nat Biotechnol* 27, 66–75.
11. Kharchenko, P.V., Tolstorukov, M.Y., and Park, P.J. (2008) Design and analysis of ChIP-seq experiments for DNA-binding proteins. *Nat Biotechnol* 26, 1351–1359.
12. Marson, A., Levine, S.S., Cole, M.F. et al. (2008) Connecting microRNA genes to the core transcriptional regulatory circuitry of embryonic stem cells. *Cell* 134, 521–533.
13. Schmid, C.D., and Bucher, P. (2007) ChIP-Seq data reveal nucleosome architecture of human promoters. *Cell* 131, 831–832.
14. Valouev, A., Johnson, D.S., Sundquist, A. et al. (2008) Genome-wide analysis of transcription factor binding sites based on ChIP-seq data. *Nat Methods* 5, 829–834.
15. Zhang, Y., Liu, T., Meyer, C.A. et al. (2008) Model-based analysis of ChIP-seq (MACS). *Genome Biol* 9, R137.
16. Jothi, R., Cuddapah, S., Barski, A., Cui, K., and Zhao, K. (2008) Genome-wide identification of in vivo protein-DNA binding sites from ChIP-seq data. *Nucleic Acids Res* 36, 5221–5231.
17. Langmead, B., Trapnell, C., Pop, M., and Salzberg, S.L. (2009) Ultrafast and memory-efficient alignment of short DNA sequences to the human genome. *Genome Biol* 10, R25.

18. Li, H., Ruan, J., and Durbin, R. (2008) Mapping short DNA sequencing reads and calling variants using mapping quality scores. *Genome Res* 18, 1851–1858.

19. Jiang, H., and Wong, W.H. (2008) SeqMap : mapping massive amount of oligonucleotides to the genome. *Bioinformatics* 24, 2395–2396.

20. Rumble, S.M., Lacroute, P., Dalca, A.V. et al. (2009) SHRiMP: accurate mapping of short color-space reads. *PLoS Comput Biol* 5, e1000386.

21. Subramanian, A., Tamayo, P., Mootha, V.K. et al. (2005) Gene set enrichment analysis: a knowledge-based approach for interpreting genome-wide expression profiles. *Proc Natl Acad Sci USA* 102, 15545–15550.

22. Kent, W.J., Sugnet, C.W., Furey, T.S. et al. (2002) The human genome browser at UCSC. *Genome Res* 12, 996–1006.

23. Smyth, G.K. (2004) Linear models and empirical Bayes methods for assessing differential expression in microarray experiments. *Stat Appl Genet Mol Biol* 3, 1–9 (Article 3).

24. Fejes, A.P., Robertson, G., Bilenky, M. et al. (2008) FindPeaks 3.1: a tool for identifying areas of enrichment from massively parallel short-read sequencing technology. *Bioinformatics* 24, 1729–1730.

25. Nix, D.A., Courdy, S.J., and Boucher, K.M. (2008) Empirical methods for controlling false positives and estimating confidence in ChIP-seq peaks. *BMC Bioinformatics* 9, 523.

26. Ji, H., Vokes, S.A., and Wong, W.H. (2006) A comparative analysis of genome-wide chromatin immunoprecipitation data for mammalian transcription factors. *Nucleic Acids Res* 34, e146.

Chapter 10

Probabilistic Peak Calling and Controlling False Discovery Rate Estimations in Transcription Factor Binding Site Mapping from ChIP-seq

Shuo Jiao, Cheryl P. Bailey, Shunpu Zhang, and Istvan Ladunga

Abstract

Localizing the binding sites of regulatory proteins is becoming increasingly feasible and accurate. This is due to dramatic progress not only in chromatin immunoprecipitation combined by next-generation sequencing (ChIP-seq) but also in advanced statistical analyses. A fundamental issue, however, is the alarming number of false positive predictions. This problem can be remedied by improved peak calling methods of twin peaks, one at each strand of the DNA, kernel density estimators, and false discovery rate estimations based on control libraries. Predictions are filtered by de novo motif discovery in the peak environments. These methods have been implemented in, among others, Valouev et al.'s Quantitative Enrichment of Sequence Tags (QuEST) software tool. We demonstrate the prediction of the human growth-associated binding protein (GABPα) based on ChIP-seq observations.

Key words: Transcription factor, transcription factor binding site, regulatory protein binding, chromatin immunoprecipitation, next-generation sequencing, ChIP-seq, peak calling, false positive rate.

---

1. Introduction

Transcription from DNA to RNA in response to environmental stimuli or internal signals is regulated by complex networks of agents and mechanisms. These include transcription factors (TFs) and co-factors, nucleosomes and histone modifications, DNA methylation, microRNAs, and interactions of all of the above as reviewed in **Chapters 1, 2, 3,** and **4**. These interactions directly influence the recruitment and activation of the RNA polymerase complex. In the recent years, revolutionary progress in chromatin immunoprecipitation (ChIP) and ultra-high-throughput

I. Ladunga (ed.), *Computational Biology of Transcription Factor Binding*, Methods in Molecular Biology 674,
DOI 10.1007/978-1-60761-854-6_10, © Springer Science+Business Media, LLC 2010

sequencing allowed unprecedented high-throughput mapping of transcription factor binding sites (TFBS) to complex genomes. To reveal these complex networks, diverse results are integrated by sophisticated experimental and computational pipelines as discussed in **Chapter 1**. This chapter discusses conservative methods for the prediction of binding sites from ChIP-seq results. This task is not trivial since immunoprecipitation can pull down not only the DNA directly associated with the TF of interest, but also the DNA segments bound by a large array of other proteins (**Chapter 1**). Inherent challenges to mapping TF binding sites include mapping potential binding sites that may not be functional in the cell and missing some functional binding sites from signals below thresholds. ChIP depends on the sensitivity and selectivity of the antibody to the TF studied. Antibodies may frequently bind to other members of the TF family, causing a non-specific signal. In addition, TFs may be modified or bound by co-factors and not recognized by antibodies.

TFs, by definition, specifically bind to a limited range of DNA sequences primarily but not exclusively in the promoter region close to the transcription start site. TFs may also bind to distal promoter, enhancer, intronic regions, and even to exons (1). Since individual binding sites frequently have a footprint on DNA as short as 5 base pairs (bp), the computational pattern analysis of short TFBS *in isolation* is typically an infeasible problem. Fortunately though, these sites are frequently organized into *cis*-regulatory modules (CRMs) (2), for which computational prediction methods (*see* **Chapters 1, 6, 7, and 13**) are becoming increasingly accurate. As a rule, these computational predictions are based on very limited samples of experimentally verified binding sites, which considerably underrepresent the variability of the TFBS (**Chapter 1**). Based on such samples, generalizing the motifs, mathematical representations of the binding sites using expectation maximization (**Chapter 6**), Gibbs sampling (**Chapters 6 and 7**), or positional weight matrices (PWMs, **Chapter 6**) is an extremely challenging task. Considerably more representative samples are generated by recent in vivo high-throughput methods including chromatin immunoprecipitation (3) combined with either genomic tiling microarrays (ChIP-chip) (4, 5, 6) or next-generation sequencing (ChIP-seq (7)). This chapter focuses on ChIP-seq because it provides for finer resolution and higher accuracy than ChIP-chip [*see* **Section 1.4** and ref. (7)].

Even with these biological, experimental, and computational caveats, ChIP-seq, if and only if analyzed and interpreted by sophisticated computational methods, brings a leap in understanding transcriptional regulatory networks. To benefit both computational and experimental biologists, we briefly introduce ChIP-seq, discuss the theoretical and practical aspects of peak calling, validation by the identification of shared binding site motifs, and benchmarking the performance of the method, including false discovery rate estimations. The computational analyses are demonstrated on the biological example of a cell division regulator, the growth-associated binding protein α-chain (GABPα), and its binding sites in the human genome.

1.1. Chromatin Immunoprecipitation (ChIP)

In order to anchor the protein of interest to its in vivo DNA location, it is typically cross-linked to the DNA by formaldehyde (**Fig. 10.1**). Next, the DNA is either sonicated or sheared into few hundred base pair (bp) segments. The protein, still associated with the DNA, is incubated in the presence of a specific antibody, and immunoprecipitation is performed. Proteins are digested, then the ~150–200-bp long DNA segments are selected by

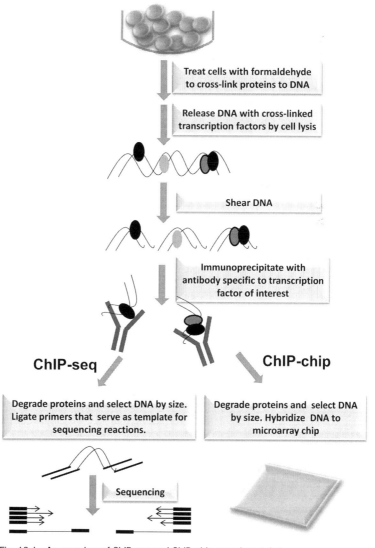

Fig. 10.1. An overview of ChIP-seq and ChIP-chip experimental steps.

gel electrophoresis. Note that this size considerably exceeds the 5–26 bp footprint of TFs on the DNA, compromising the resolution of ChIP-seq. In order to estimate background noise and false positive rate (**Section 3.6**), control libraries are created by either reversing the cross-links and ChIP, ChIP with a non-selective antibody like IgG or with no immunoprecipitation at all. Because in the absence of ChIP, little or no proteins are expected to be pulled down with DNA (8), the identified DNA segments are considered as background noise and used for estimating the false discovery rate (**Section 3.6**).

1.2. Identification of Chromatin-Bound DNA

ChIP-enriched DNA segments are identified either by hybridization to genomic tiling/promoter microarrays (5–7) (ChIP-chip) or by ultra-high-throughput sequencing (ChIP-seq). ChIP-chip works well on small genomes; in yeast, binding sites for over 100 TFs have been determined (9). Also, ChIP-chip conveniently limits the study to selected regions of the genome. Selected regions include promoter regions, which are primary loci for TF binding (10), chromosome 22 (11), and pilot-ENCODE regions. In the latter, carefully selected samples accounting for 1% the human genome (1) have been analyzed. Binding sites for CREB (12), the Polycomb group TFs (6), the mouse embryonic stem cell regulatory network (13), and the estrogen receptor (4) have been identified with ChIP-chip. While ChIP-chip is effective in yeast and other small genomes, its resolution is about 500 bp in higher eukaryotes (**Chapters 9** and **11**). ChIP-chip is a powerful tool, however, the high level of cross-hybridization and the need for a pre-designed chip are major drawbacks and its performance deteriorates in complex mammalian genomes (14).

One of the earliest methods proposed to overcome the limitations of ChIP-chip was Sequence Tag Analysis of Genomic Enrichment (STAGE) (15, 16). Revolutionary breakthrough in sequence coverage and affordability has brought by ultra-high-throughput sequencing, also called next-generation sequencing. Coupled with ChIP (ChIP-seq), immunoprecipitated DNA segments are sequenced by massively parallel technology including the Illumina (formerly Solexa) sequencing by synthesis (www.illumina.com) (17), Roche/454 pyrosequencing (www.454.com) (18), and Life Technologies's (formerly Applied Biosystems) SOLiD (http://solid.appliedbiosystems.com) platforms. ChIP-seq produces tens of millions of sequencing reads. From these massive but noisy data sets statistically significant peaks and binding sites can be found. Compared to ChIP-chip, ChIP-seq has a much finer resolution (25–200 bp), increased sensitivity, and selectivity by eliminating cross-hybridization effects. ChIP-seq is free from the hurdles of microarray design and manufacturing.

ChIP-seq revolutionizes the discovery of regulatory protein–DNA interactions, and binding sites for a number of TFs identified include p53 (19), NSRF (20), STAT-1 (21), c-Myc (22), and PPARγ (23, 24).

1.3. Computational Discovery of Binding Sites from ChIP-seq Observations

The density distribution of sequencing reads forms the basis for heuristic and algorithmic methods for calling peaks. From these peaks, the actual binding sites are inferred. The resolution, sensitivity, and selectivity of ChIP-seq critically depend on the choice of the heuristics or algorithms.

The development of algorithms and software tools of the ChIP-seq computational analyses are lagging far behind the progress of experimental technology. A diverse array of tools has been published as reviewed in **Chapter 1**. These tools implement fundamentally different methods for background correction, normalization, and analyzing bimodal (twin) peaks on opposing strands of the DNA. Certain tools like QuEST (25) explicitly demand a control library. CisGenome [**Chapter 9**, (26)], FindPeaks (27), and model-based analysis of ChIP-seq (MACS) (28) work without control libraries; CisGenome [**Chapter 9**, (26)] models background noise based on the negative binomial distribution, while SISSRs (29) and MACS (28) use the Poisson distribution for this purpose. Peaks are ranked by binomial p-values in USeq (30). Most recent tools improve peak calling by estimating the shift between the peaks on opposing strands (*see* **Section 3.3**). False discovery rate is calculated by QuEST (25) and MACS (28) on the basis of the control library, while FindPeaks (27) and spp (31) perform Monte Carlo simulations. ERANGE (20, 32) uses tag aggregation but calculates no p-values or FDR. F-Seq (33) also uses kernel density estimations, GLITR (34) and PeakSeq (35) evaluate peaks using FDR. The development of a more realistic FDR estimation would greatly benefit the discovery of TFBS (*see* **Note 2**).

Here we demonstrate the discovery of TFBS from ChIP-seq observations using the Quantitative Enrichment of Sequence Tags (QuEST) tool. QuEST was developed by Anton Valouev and colleagues at Stanford (25). QuEST takes the advantage of the directionality of the sequencing reads to find genomic regions enriched in TF-bound DNA fragments. It applies a nonparametric approach called kernel density estimation method to generate smoothed sequencing reads density, for which local maxima (regions with high density) are sought. With these approaches, QuEST can statistically analyze peak calls that indicate a higher likelihood of finding biologically relevant TFBS.

1.4. Example: Growth-Activated Binding Protein (GABPα)

We demonstrate below how to use QuEST to find putative binding sites for the growth-activated binding protein α-chain (GABPα, also known as E4TF1-60, nuclear respiratory factor 2 subunit α) (12) in human Jurkat cells. GABPα is a member of the Etf family of TFs, and it is both necessary and sufficient for restarting cell division (36, 37). GABPα has a 10–11 bp footprint on

166 Jiao et al.

the DNA, with low information content at positions 8–11 (38). ChIP was performed by using antibodies specific to this protein and sequencing was performed on the Illumina Genome Analyzer platform.

1.5. Overview

We describe how to obtain and install the QuEST software and how to format the input data in **Section 2** and in **Appendix 1**. In **Section 3**, we introduce the algorithms applied in QuEST. In **Section 4**, QuEST finds GABPα binding sites from the ChIP-seq reads mapped to the human genome and results are interpreted. In the **Notes** section, we discuss potential limitations and parameter settings.

2. Software and Data

Here we perform the prediction and analysis of TFBS on the basis of the ChIP-seq reads mapped to the genome using Quantitative Enrichment of Sequencing Tags (QuEST) tool developed by Valouev et al. (25). QuEST facilitated the discovery of thousands of binding sites for the human serum response factor (SRF), GABPα discussed above, and neuron-restrictive silencer factor (NRSF). The methods implemented in QuEST are discussed in **Section 3**. The installation is described in **Appendix 1**, and the computational protocol is detailed in **Appendix 2**.

3. Methods

TFBS discovery using QuEST is described in nine sub-sections. In **Section 3.1**, sequence reads are mapped to the genome. Then candidate peaks are called on both strands of the DNA based on the density distribution of the reads (**Section 3.2**). Next, the extent of the shift between the forward and reverse strand peaks on each side of a potential binding site is estimated (**Section 3.3**). This allows combining the density distributions on the two strands (**Section 3.4**). Then well-separated peaks with significant differences to the background library are called (**Section 3.5**). False discovery rate is estimated in order to reduce the number of potentially biologically irrelevant or statistically not significant peaks (**Section 3.6**). Running QuEST is discussed in **Section 3.7** and **Appendix 2**. The number of potentially missed sites is estimated by a saturation analysis in **Section 3.8**. Finally, in **Section 3.9** and **Appendix 3**, the called peaks are displayed in the University of California Santa Cruz Genome Browser.

3.1. Mapping ChIP-seq Reads to the Genome

ChIP-seq produces several million sequencing reads (tags) of varying length. These sequencing reads can be mapped onto the genome by a number of tools including Bowtie, the fastest tool at the time of this writing (39), MAQ (http://maq.sourceforge.net/), Eland (http://www.illumina.com), SHRiMP (40), SSAKE (41), SHARCGS (42), Exonerate (43), Corona Lite for the SoLiD platform (http://solidsoftwaretools.com/gf/project/corona/), and other packages. Inputs to QuEST are the genomic coordinates and strand of the sequencing reads. For every genomic position i, the number of high-quality forward reads $C_+(i)$ and reverse reads $C_-(i)$ is recorded.

QuEST Version 2.4 accepts aligned reads in QuEST, ELAND (http://illumina.com), Bowtie (39), and MAQ (http://maq.sourceforge.net/) formats.

3.2. Kernel Density Estimation

Loci significantly enriched in ChIP-seq reads may indicate biologically functional binding sites. QuEST computes enrichment by kernel density estimation (44), a nonparametric method of computing smooth estimates over noisy observations. First, we estimate the strand-specific smoothed density functions $H_+(i)$ from $C_+(i)$ and $H_-(i)$ from $C_-(i)$ at nucleotide position i in the genome for the forward strand:

$$H_+(i) = \frac{1}{h} \sum_{j=i-3h}^{i+3h} K\left(\frac{j-i}{h}\right) C_+(j) \qquad [1]$$

and analogously for the reverse sequencing reads:

$$H_-(i) = \frac{1}{h} \sum_{j=i-3h}^{i+3h} K\left(\frac{j-i}{h}\right) C_-(j), \qquad [2]$$

where $K(x) = \frac{1}{\sqrt{2\pi}} \exp\left(-\frac{x^2}{2}\right)$ is the normal kernel function and h is the kernel density bandwidth. The user-selectable kernel density bandwidth is the number of base pairs considered, used in the estimation formulae [1] and [2]. The kernel density estimator is a weighted moving average of the number of sequencing reads where $K\left(\frac{j-i}{h}\right)$ denotes the weight. The normal kernel is selected here for its computational efficiency. With increasing distance of j from i, the weight $K\left(\frac{j-i}{h}\right)$ decreases for $C_+(j)$ or $C_-(j)$. The bandwidth h is adjustable and the 30 bp default is recommended for the binding sites of GABPα, which has a 10–11 bp footprint on the DNA, with low information content in the last four positions (45). The optimal selection of bandwidth depends on the footprint width, the experimental characteristics, and the presence of co-binding proteins, and usually determined by trial and error. While the normal kernel function is widely used, alternative methods such as the Haar wavelets may perform better in certain applications (*see* **Note 3**).

3.3. Estimating the Peak Shift

The polymerase applied in amplification and sequencing attaches to the 5′ termini of the sample DNA segments. Moving toward the 3′ end, the polymerase dissociates from the DNA with a sharply increasing frequency. Therefore reads are overrepresented at the 5′ ends on both strands of ChIP-enriched DNA fragments as compared to their central and 3′ regions. Reads from the two strands form two peaks, one at each side of the binding site. QuEST estimates peak shift, the distance between the peaks on the forward and the reverse strands as follows (**Fig. 10.2**). For shift estimation, only twin peaks with high confidence are selected as follows. For each fixed length (default: 300 bp) sliding window r, the highest local maximum of forward reads is M_+^r and of the reverse reads is M_-^r; the second highest local maximum of forward reads is N_+^r and of reverse reads is N_-^r. Window r is selected for peak shift calculation if it satisfies the following three conditions:

1. Window r is covered by more than t reads (default: 600). This condition ensures robust estimates of local maxima.
2. $M_+^r > 20 N_+^r$ and $M_-^r > 20 N_-^r$. If the highest local maximum is much greater than the second highest local maximum, then the highest local maximum is more likely to be a real peak instead of some random spike.
3. $M_+^r > 20 c_+^r$ and $M_-^r > 20 c_-^r$, where c_+^r and c_-^r are the local maxima of the same window in the pseudo-ChIP library. This condition ensures that the peaks safely exceed the background level.

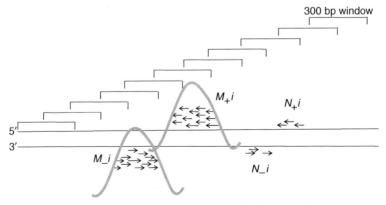

Fig. 10.2. Selection of peak calling windows. Users can configure the fixed length (default: 300 bp) of the sliding windows. A pair of peaks are formed by the M_+i and M_-i clusters of reads.

Let S denote the set of all selected windows. For every window $r \in S$, we compute the d_r distance between the highest peaks M_+^r and M_-^r. The peak shift λ is estimated by $\sum_{r \in S_1} d_r / 2M$, where M is the number of windows in S.

3.4. Combining Strand Densities

The above estimate for the peak shift allows us to calculate the combined densities of forward and reverse reads for both ChIP-seq and control library:

$$H(i) = H_+(i - \lambda) + H_-(i + \lambda).$$

The combined density is the basis for peak calling below.

3.5. Peak Calling

Peaks are defined as windows of high concentration of sequencing reads at a locus on the genome that may indicate TFBSs (**Figs. 10.2** and **10.3**). Candidate peaks are detected by scanning the genome using narrow sliding windows (default: 21 bp) for local maxima of the combined density. Let $p_1, \ldots p_B$ denote the positions of the candidate peaks; and let $c_1, \ldots, c_B$ denote the corresponding density in the control library. To facilitate conservative binding site predictions, a candidate at position p_i will be called if and only if it satisfies all of the following criteria:

1. $H(p_i) \geq t$, where t is a user-specified threshold (default: 30) to control the false discovery rate. By definition, the false discovery rate is the (estimated) frequency of false positives with a score equal to t or higher. Increasing t decreases the false discovery rate.

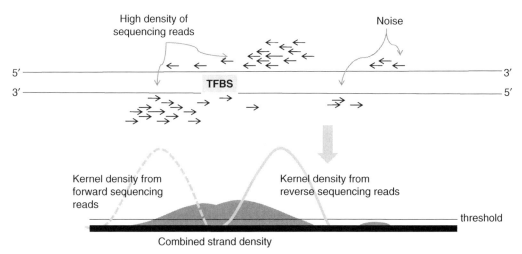

Fig. 10.3. Simplified depiction of peak calling by the kernel density estimator. A peak is called when the maximum value of combined read density exceeds a threshold. Peak pairs over the density threshold are called as a candidate TFBS. The density threshold is calculated so as the FDR would not to exceed the user-selected value.

2. Background test. Either $c_i \leq \tau$ or $H(p_i)/c_i > r$, where c_i is the background density at peak i and τ is the general background threshold, and r is a user-specified "rescue" ratio, which, by default, is set to 10.

3. To ensure clear separation ("valley") between neighboring peaks, a minimum of 10% drop in $H(j)$ read density is required.

$$0.9 \cdot \min\{H(p_{i-1}), H(p_i)\} \geq \max\{H(j)|p_{i-1} < j < p_i\}$$

and

$$0.9 \cdot \min\{H(p_i), H(p_{i+1})\} \geq \max\{H(j)|p_i < j < p_{i+1}\}.$$

The selection of the parameter values here are somewhat arbitrary. Values selected by the application of a systematic sensitivity analysis may increase performance (*see* **Note 4**).

3.6. False Discovery Rate

False discovery rate (46) is defined as the ratio of incorrect positive predictions. In the context of TFBS predictions, it is the proportion of erroneously called peaks that are either not binding sites or binding sites for proteins other than the TF of interest. These peaks are called since they score higher than the threshold and satisfy all the three conditions above. Conditions and thresholds are selected to strike a delicate balance of maximizing true positive and minimizing false positive predictions.

There is no reference set where each genomic position is reliably characterized as a binder or as a non-binder. Therefore false discovery rate is approximated by using control libraries created by reversing the cross-links and performing no ChIP (**Section 1.1**). The library with no IP is randomly split into a pseudo-ChIP-seq library and a background set. If a satisfactory number of pseudo-ChIP reads are available, splitting them into more than two sets could improve the accuracy of false positive rate estimations (*see* **Note 5**). For compatibility with the real IP experiment, the number of reads in the pseudo-ChIP-seq library must match the number of reads in the real ChIP-seq library. The peak calling procedure (**Sections 3.2, 3.3**, and **3.4**) is performed by comparing the pseudo-ChIP-seq library to the background set. Clearly, any pseudo-peak called in this comparison is false. Then peaks are called for real ChIP-seq library using the same background set. The approximated false discovery rate is the number of pseudo-peaks divided by the number of called peaks in the real ChIP-seq analysis. Valouev et al. (25) applies a threshold of 1%, but acceptable thresholds for false discovery rate are subject to the individual experimenter's discretion.

While the above described no-IP-based approximation is probably the best available choice, this procedure considerably underestimates the real false discovery rate in a ChIP environment. This procedure does not take into consideration antibody binding to untargeted proteins, which is a serious issue with large TF families with similar epitope structure, and major cause of false positives (3, 8). Another issue is that formaldehyde can cross-link DNA to close but unbound proteins (3). These concerns motivate further computational validation including the identification of shared motifs (**Section 3.10.2**) and correlation analyses with TFs that co-regulate certain genes with GABPα (**Results**).

3.7. Running QuEST on ChIP-Enriched Sequencing Reads

Sequencing reads obtained by the ChIP-seq experiments of the GABPα binding sites in human Jurkat lymphoblastoma cells were aligned to the human genome (Version hg18). Peaks are called and evaluated as described in **Appendix 2**.

3.8. Peak Saturation

How many peaks are missed when using one, two, or more sequencing lanes? This question can be answered by drawing a saturation curve where the number of peaks is a function of the number of reads. Saturation analysis can be performed by randomly selecting subsets of varying size from the original data and calculating the number of peaks for each subset as in **Sections 3.2, 3.3, 3.4, 3.5, 3.6,** and **3.7**.

3.9. Visualization in the Genome Browser

Peaks can be visualized in a rich context of diverse genomic information using the University of California Santa Cruz Genome Browser (http://genome.ucsc.edu) (47). QuEST prepares several custom tracks for this browser. Users can upload these tracks to either the UCSC server or a local implementation of the Genome Browser, as described in **Appendix 3**. As an example, **Fig. 10.4** shows the promoter region of the human G-protein

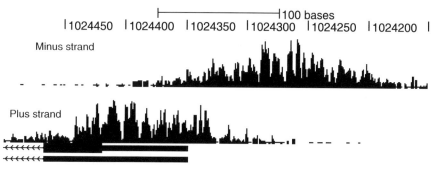

Fig. 10.4. Actual peaks and TFBS predictions in the promoter region of the gene encoding the human G-protein binding protein CRFG (GTPBP4) as displayed by the UCSC Genome Browser using custom tracks. Note that in the promoter region of this strand gene, sequencing reads in the forward orientation form (*upper subchart*) a peak upstream of the peak of the reverse strand reads (*lower subchart*). The two peaks combined span over 350 bp, much wider than the 10–12 bp footprint of the GABP TF.

binding protein CRFG (GTPBP4). Sequencing reads in the forward orientation cluster into a peak upstream of the reverse read peak. Note the coarse resolution of ChIP-seq: high read density extends to over 350 bp, much wider than the actual footprint of the GABPα protein on the DNA.

3.10. Results

3.10.1. Reproducibility

QuEST (25) reproduces several earlier identified target genes for GABPα including interleukin-16, cytochrome c oxidase subunits IV and Vb, and SRF-regulated FHI.2. The resulting binding sites are also in line with the co-occurrence of GABPα and SRF binding sites: 29% of the predicted SRF peaks were in the proximity of GABPα sites. The original publication (25) reported 6,442 peaks, however, using the parameters described in **Section 3.3**, we found 550 additional peaks. Saturation analysis performed as in **Section 3.5** indicates that the sequencing depth is sufficient to identify most peaks.

The reproducibility of peak positions and scores was estimated using an experiment targeting another TF, the neuron-restrictive silencer factor (NSRF). Potential binding sites of this TF were immunoprecipitated by both monoclonal and the polyclonal antibodies in different experiments. Peaks were called separately from both experiments. The standard deviation of the distances between the 2,320 comparable peaks was as low as 13.5 bp, and the scores were highly correlated ($r = 0.97$). These results demonstrate the high reproducibility of both the ChIP-seq technology and the QuEST methodology.

3.10.2. Shared Binding Site Motifs

As a rule, ChIP-enriched segments are considerably larger than the biological TFBS. This is due to sonicating the chromatin to ~500 bp segments, then size selection of the already chromatin-free DNA to ~150–200 bp by gel electrophoresis, experimental noise, and the presence of multiple proteins cross-linked by formaldehyde, including others than the TF being studied (3). Therefore a more accurate TFBS localization requires finding common motifs in the neighborhood of each called peak. This can be achieved by many of the ~200 motif prediction tools available to date, with varying performance for diverse TFs (reviewed in **Chapter 8**). In Valouev et al. (25), de novo motif finding in the 200 bp neighborhoods of called peaks was performed by MEME (Multiple Expectation Maximization for Motif Elicitation (48), discussed in **Chapters 6** and **11**. QuEST's reasonable specificity (the ability to reject false positives) is indicated by the observation that 71% of the peaks were significantly enriched in the canonical motif. The high accuracy of localization is shown by the 21.76 bp standard deviation of the distance between motif and peak centers.

4. Notes

1. The software developers recommend peak calling using parameter set Number 2. This, however, leads to an FDR of 12.7% for the test data. In contrast, with the more stringent parameter set Number 1, the FDR was 1.23%. This falls into the generally accepted FDR range of 1–10%.

2. We found that $-\log10(q\text{-value})$ is frequently in the order of 100 K even for some rejected peaks. This seems to be unrealistic therefore a more adequate method is required for estimating FDR.

3. In **Section 3.2**, QuEST uses a kernel smoother to estimate the density of both forward and reverse reads. An alternative for smoothing densities provided by Haar wavelets (49). The mother function of Haar wavelets can be written as

$$\psi(x) = \begin{cases} -1/\sqrt{2}, & -1 < x < 0 \\ 1/\sqrt{2}, & 0 < x < 1 \\ 0, & \text{otherwise} \end{cases}$$

Based on the mother wavelet, a family of child functions can be generated as $\psi_{j,t}(x) = 2^{j/2}\psi(2^j x - t)$, where j and t are indices for scale and location. Then the wavelet coefficients can be defined by

$$WT(x)\{j, t\} = \int \psi_{j,t}(x)f(x)\mathrm{d}x,$$

where $f(x)$ is the original density. For smoothing, some criteria are necessary to distinguish between wavelet coefficients that indicate true signals and those which reflect noise and should be eliminated. The Haar wavelet method was applied to denoise DNA copy number observations (50). Wavelet methods proved to be particularly well suited for handling the abrupt changes in (50), a situation similar to ChIP-seq results.

4. In **Section 3.3**, there are four criteria to call peaks and the values of several parameters are to be selected arbitrarily. A sensitivity analysis for the changes in these parameters could lead to more sensitive and selective predictions.

174   Jiao et al.

5. To estimate the false discovery rate in **Section 3.6**, the control library needs to be split randomly into two data sets. A more robust estimate of the false discovery rate could be obtained by averaging the results from multiple randomly split data sets.

Appendix 1: Installing QuEST

Source code for QuEST (including PERL and C++ modules) can be downloaded from http://www.stanford.edu/~valouev/QuEST/QuEST.html. It has been tested for the Linux/UNIX and Mac OS operation systems but no executables are available. The above web site provides installation instructions. System requirements include a local implementation of the Pattern Extraction and Regular Expression Language (PERL), a *gcc* compiler, 1 GB random access memory, and 30 MB disk space. Unpack and untar the archive using the command:

```
tar -zvxf QuEST_2.4.tar.gz
```

Replace the filename for the current version. In the source directory, configure QuEST by running

```
./configure.pl
```

Finally, compile and link QuEST by the

```
make
```

command to finish installation.

Appendix 2: Running QuEST on the GABPα-Enriched Sequencing Reads

Download the file http://mendel.stanford.edu/SidowLab/downloads/quest/GABP.align_25.hg18.gz containing input, total chromatin, sheared chromatin data and http://mendel.stanford.edu/SidowLab/downloads/quest/Jurkat_RX_noIP.align_25.hg18.gz for the control library. For the reference genome, use the Human hg18 genome table http://mendel.stanford.edu/SidowLab/downloads/quest/genome_table.gz.

These files may require 30 GB free disk space. Next, configure the parameters for the QuEST analysis (type on a single line):

```
<QuEST_Directory>/generate_QuEST_parameters.pl
-solexa_align_ChIP <Data_Directory>/GABP.align_25.hg18
```

Probabilistic Peak Calling and Controlling False Discovery Rate Estimations 175

```
-solexa_align_RX_noIP <Data_Directory>/Jurkat_RX_noIP.
align_25.hg18
-gt <Data_Directory>/genome_table -ap <Save_directory>
/QuEST_analysis
-ChIP_name GABP_Jurkat &,
```

where `<QuEST_Directory>` is the directory where QuEST is installed on the user's local system; `<Data_Directory>` is the directory to save unpacked data; the option "`solexa_align_ChIP`" specifies the sequencing platform's alignment. Other options are included for QuEST align file, Eland file, Eland extended file, Bowtie file, and MAQ file. The option "`-solexa_align_RX_noIP`" refers to the control data, where "noIP" is no immunoprecipitation, "`-gt`" specifies the input for reference genome, and "`-rp`" indicates that the input genome is in FASTA-files. Option "`-ap`" specifies the output directory.

When prompted to run QuEST with FDR analysis, choose "yes." For the ChIP experiment, select "1," for the peak calling parameters, choose "1" (*see* **Note 1**).

QuEST processes this experiment in ~1.5 h on a LINUX server with 2.33 GHz CPU.

Appendix 3: Displaying Peaks in the UCSC Genome Browser

Let us visualize a particular peak, e.g., *P-21-1* from the window (region) *R-21*.

Go to http://genome.ucsc.edu and select "Genomes" in the upper left corner. Then select "add custom tracks" and upload the following files:

```
tracks/wig_profiles/by_chr/ChIP_unnormalized/chr22.wig.gz
tracks/ChIP_calls.filtered.bed
tracks/data_bed_files/by_chr/ChIP/GABP_Jurkat.chr11.bed.gz
tracks/bed_graph/by_chr/ChIP/GABP_Jurkat.chr11.bedGraph.gz
```

Then click on "Go to Genome Browser" and "jump" to the region "chr22: 29160342-29162631".

Acknowledgments

IL thanks the NSF Grant EPS-0701892 for funding.

References

1. ENCODE Consortium. (2007) Identification and analysis of functional elements in 1% of the human genome by the ENCODE pilot project. *Nature* 447, 799–816.
2. Blanchette, M., Bataille, A.R., Chen, X. et al. (2006) Genome-wide computational prediction of transcriptional regulatory modules reveals new insights into human gene expression. *Genome Res* 16, 656–668.
3. Barski, A., and Zhao, K. (2009) Genomic location analysis by ChIP-Seq. *J Cell Biochem* 107, 11–18.
4. Carroll, J.S., Meyer, C.A., Song, J. et al. (2006) Genome-wide analysis of estrogen receptor binding sites. *Nat Genet* 38, 1289–1297.
5. Kim, T.H., Barrera, L.O., Zheng, M. et al. (2005) A high-resolution map of active promoters in the human genome. *Nature* 436, 876–880.
6. Lee, T.I., Jenner, R.G., Boyer, L.A. et al. (2006) Control of developmental regulators by Polycomb in human embryonic stem cells. *Cell* 125, 301–313.
7. Park, P.J. (2009) ChIP-seq: advantages and challenges of a maturing technology. *Nat Rev Genet* 10, 669–680.
8. Collas, P. (2009) The state-of-the-art of chromatin immunoprecipitation. *Methods Mol Biol* 567, 1–25.
9. Harbison, C.T., Gordon, D.B., Lee, T.I. et al. (2004) Transcriptional regulatory code of a eukaryotic genome. *Nature* 431, 99–104.
10. Ozsolak, F., Song, J.S., Liu, X.S. et al. (2007) High-throughput mapping of the chromatin structure of human promoters. *Nat Biotechnol* 25, 244–248.
11. Cawley, S., Bekiranov, S., Ng, H.H. et al. (2004) Unbiased mapping of transcription factor binding sites along human chromosomes 21 and 22 points to widespread regulation of noncoding RNAs. *Cell* 116, 499–509.
12. Euskirchen, G., Royce, T.E., Bertone, P. et al. (2004) CREB binds to multiple loci on human chromosome 22. *Mol Cell Biol* 24, 3804–3814.
13. Mathur, D., Danford, T.W., Boyer, L.A. et al. (2008) Analysis of the mouse embryonic stem cell regulatory networks obtained by ChIP-chip and ChIP-PET. *Genome Biol* 9, R126.
14. Johnson, D.S., Li, W., Gordon, D.B. et al. (2008) Systematic evaluation of variability in ChIP-chip experiments using predefined DNA targets. *Genome Res* 18, 393–403.
15. Kim, J., Bhinge, A.A., Morgan, X.C. et al. (2005) Mapping DNA-protein interactions in large genomes by sequence tag analysis of genomic enrichment. *Nat Methods* 2, 47–53.
16. Bhinge, A.A., Kim, J., Euskirchen, G.M. et al. (2007) Mapping the chromosomal targets of STAT1 by Sequence Tag Analysis of Genomic Enrichment (STAGE). *Genome Res* 17, 910–916.
17. Quail, M.A., Kozarewa, I., Smith, F. et al. (2008) A large genome center's improvements to the Illumina sequencing system. *Nat Methods* 5, 1005–1010.
18. Margulies, M., Egholm, M., Altman, W.E. et al. (2005) Genome sequencing in microfabricated high-density picolitre reactors. *Nature* 437, 376–380.
19. Wei, C.L., Wu, Q., Vega, V.B. et al. (2006) A global map of p53 transcription-factor binding sites in the human genome. *Cell* 124, 207–219.
20. Johnson, D.S., Mortazavi, A., Myers, R.M. et al. (2007) Genome-wide mapping of in vivo protein-DNA interactions. *Science* 316, 1497–1502.
21. Robertson, G., Hirst, M., Bainbridge, M. et al. (2007) Genome-wide profiles of STAT1 DNA association using chromatin immunoprecipitation and massively parallel sequencing. *Nat Methods* 4, 651–657.
22. Zeller, K.I., Zhao, X., Lee, C.W. et al. (2006) Global mapping of c-Myc binding sites and target gene networks in human B cells. *Proc Natl Acad Sci USA* 103, 17834–17839.
23. Hamza, M.S., Pott, S., Vega, V.B. et al. (2009) De-novo identification of PPARgamma/RXR binding sites and direct targets during adipogenesis. *PLoS One* 4, e4907.
24. Nielsen, R., Pedersen, T.A., Hagenbeek, D. et al. (2008) Genome-wide profiling of PPARgamma:RXR and RNA polymerase II occupancy reveals temporal activation of distinct metabolic pathways and changes in RXR dimer composition during adipogenesis. *Genes Dev* 22, 2953–2967.
25. Valouev, A., Johnson, D.S., Sundquist, A. et al. (2008) Genome-wide analysis of transcription factor binding sites based on ChIP-Seq data. *Nat Methods* 5, 829–834.
26. Ji, H., Jiang, H., Ma, W. et al. (2008) An integrated software system for analyzing ChIP-chip and ChIP-seq data. *Nat Biotechnol* 26, 1293–1300.
27. Fejes, A.P., Robertson, G., Bilenky, M. et al. (2008) FindPeaks 3.1: a tool for identifying

28. Zhang, Y., Liu, T., Meyer, C.A. et al. (2008) Model-based analysis of ChIP-Seq (MACS). *Genome Biol* 9, R137.

29. Jothi, R., Cuddapah, S., Barski, A. et al. (2008) Genome-wide identification of in vivo protein-DNA binding sites from ChIP-Seq data. *Nucleic Acids Res* 36, 5221–5231.

30. Nix, D.A., Courdy, S.J., and Boucher, K.M. (2008) Empirical methods for controlling false positives and estimating confidence in ChIP-Seq peaks. *BMC Bioinformatics* 9, 523.

31. Kharchenko, P.V., Tolstorukov, M.Y., and Park, P.J. (2008) Design and analysis of ChIP-seq experiments for DNA-binding proteins. *Nat Biotechnol* 26, 1351–1359.

32. Mortazavi, A., Williams, B.A., McCue, K. et al. (2008) Mapping and quantifying mammalian transcriptomes by RNA-Seq. *Nat Methods* 5, 621–628.

33. Boyle, A.P., Guinney, J., Crawford, G.E. et al. (2008) F-Seq: a feature density estimator for high-throughput sequence tags. *Bioinformatics* 24, 2537–2538.

34. Tuteja, G., White, P., Schug, J. et al. (2009) Extracting transcription factor targets from ChIP-Seq data. *Nucleic Acids Res* 37, e113.

35. Rozowsky, J., Euskirchen, G., Auerbach, R.K. et al. (2009) PeakSeq enables systematic scoring of ChIP-seq experiments relative to controls. *Nat Biotechnol* 27, 66–75.

36. Briguet, A., and Ruegg, M.A. (2000) The Ets transcription factor GABP is required for postsynaptic differentiation in vivo. *J Neurosci* 20, 5989–5996.

37. Rosmarin, A.G., Resendes, K.K., Yang, Z. et al. (2004) GA-binding protein transcription factor: a review of GABP as an integrator of intracellular signaling and protein-protein interactions. *Blood Cells Mol Dis* 32, 143–154.

38. Temple, M.D., and Murray, V. (2005) Footprinting the 'essential regulatory region' of the retinoblastoma gene promoter in intact human cells. *Int J Biochem Cell Biol* 37, 665–678.

39. Langmead, B., Trapnell, C., Pop, M. et al. (2009) Ultrafast and memory-efficient alignment of short DNA sequences to the human genome. *Genome Biol* 10, R25.

40. Rumble, S.M., Lacroute, P., Dalca, A.V. et al. (2009) SHRiMP: accurate mapping of short color-space reads. *PLoS Comput Biol* 5, e1000386.

41. Warren, R.L., Sutton, G.G., Jones, S.J. et al. (2007) Assembling millions of short DNA sequences using SSAKE. *Bioinformatics* 23, 500–501.

42. Dohm, J.C., Lottaz, C., Borodina, T. et al. (2007) SHARCGS, a fast and highly accurate short-read assembly algorithm for de novo genomic sequencing. *Genome Res* 17, 1697–1706.

43. Slater, G.S., and Birney, E. (2005) Automated generation of heuristics for biological sequence comparison. *BMC Bioinformatics* 6, 31.

44. Silverman, B. (1986) *Density estimation for statistics and data analysis.* Chapman and Hall, Boca Raton, FL.

45. Collins, P.J., Kobayashi, Y., Nguyen, L. et al. (2007) The ets-related transcription factor GABP directs bidirectional transcription. *PLoS Genet* 3, e208.

46. Benjamini, Y., Hochberg, Y. (1995) Controlling the false discovery rate: a practical and powerful approach to multiple hypothesis testing. *J R Statistic Soc B* 57, 289–300.

47. Rhead, B., Karolchik, D., Kuhn, R.M. et al. (2009) The UCSC genome browser database: update 2010. *Nucleic Acids Res*, doi:10.1093/nar/gkp1939.

48. Bailey, T.L., Williams, N., Misleh, C. et al. (2006) MEME: discovering and analyzing DNA and protein sequence motifs. *Nucleic Acids Res* 34, W369–W373.

49. Haar, A. (1910) Zur Theorie der orthogonalen Funktionensysteme. *Math Ann* 3, 331–371.

50. Hsu, L., Self, S.G., Grove, D. et al. (2005) Denoising array-based comparative genomic hybridization data using wavelets. *Biostatistics* 6, 211–226.

Chapter 11

Sequence Analysis of Chromatin Immunoprecipitation Data for Transcription Factors

Kenzie D. MacIsaac and Ernest Fraenkel

Abstract

Chromatin immunoprecipitation (ChIP) experiments allow the location of transcription factors to be determined across the genome. Subsequent analysis of the sequences of the identified regions allows binding to be localized at a higher resolution than can be achieved by current high-throughput experiments without sequence analysis and may provide important insight into the regulatory programs enacted by the protein of interest. In this chapter we review the tools, workflow, and common pitfalls of such analyses and recommend strategies for effective motif discovery from these data.

Key words: Motif discovery, sequence motifs, chromatin immunoprecipitation, ChIP-seq, ChIP-chip, transcriptional regulation.

---

1. Introduction

The regulatory programs enacted by transcription factors in response to developmental or environmental cues depend on specific interactions between these proteins and the genes whose expression they regulate. This specificity is provided, in large part, by short DNA sequences that are recognized and bound by transcription factors, thereby localizing them to their targets (1). In general, different transcription factors recognize different binding sites. The varying sequence specificities of these regulators and the genomic location of the sites they bind form a regulatory code whose decipherment has been an important area of research in molecular biology for over 40 years (2, 3).

I. Ladunga (ed.), *Computational Biology of Transcription Factor Binding*, Methods in Molecular Biology 674,
DOI 10.1007/978-1-60761-854-6_11, © Springer Science+Business Media, LLC 2010

There are a number of challenges that must be overcome in order to decipher this code. The interactions of transcription factors with DNA are transient, making detection difficult. In addition, it is clear that in vivo binding events vary extensively in their function; the same protein bound at different sites or at the same site under different conditions may activate, repress, or have no effect on transcription, depending on several factors including which proteins bind with it (4). For these reasons, a combination of condition-specific experimental data and computational analysis is critical for understanding transcriptional regulation.

One experimental technique that has provided significant insight into the regulatory code of eukaryotes is chromatin immunoprecipitation (ChIP). In a ChIP experiment, the transient interactions between proteins and DNA are stabilized by chemically cross-linking in vivo. After subsequent isolation and fragmentation of the cross-linked chromatin, protein-bound DNA fragments are immunoprecipitated using an antibody specific to the transcription factor of interest. Coupling this procedure to a high-throughput readout technique like microarrays (ChIP-chip) or massively parallel sequencing (ChIP-seq) allows the location of transcription factors to be experimentally profiled on a genome-wide basis (5, 6).

ChIP data provide a starting point for many types of analysis of transcription. In this chapter, we will focus on computational techniques that use these data to understand how a transcription factor is localized to its targets in a profiled tissue or cell type. This can involve identifying the sequences that are recognized and bound by the protein itself or sites bound by other proteins with which it cooperates to control gene expression. Since high-throughput ChIP experiments may have significant experimental noise, identifying sequences that have a strong statistical association with ChIP-enriched regions can provide additional confidence in the quality of the data and increase the resolution at which binding sites can be localized.

A variety of approaches have been proposed to represent the specificity of protein–DNA interactions, and the resulting models are commonly referred to as sequence motifs (7). The most intuitive representation of a sequence motif is the consensus sequence. A consensus sequence describes the binding site preference of a protein as a string of nucleotides. Sites where a range of nucleotides are accommodated are denoted using ambiguity codes. For example, the specificity of the Lrp regulatory protein from *Escherichia coli* can be described as YAGHAWATTWTDCTR (8). However, consensus sequences fail to capture the fact that transcription factors generally have a range of affinities for target sequences. An alternative model that conveys the range of affinities is the frequency matrix. Frequency matrices describe the binding site preference of a protein as a set of position-specific multinomial distributions over the four nucleotides A, C, G, and T. When an estimate of nucleotide frequencies is available for regions that are not bound by the transcription factor, frequency matrix motifs can be converted to "log-odds" matrices by taking the log of each entry and then subtracting the log of the background frequency for the appropriate nucleotide.

Log-odds matrix motif models have a link to underlying biophysical parameters like binding free energy (9, 10). For the purposes of analyzing ChIP data, biophysically based models often have the advantage of allowing more realistic modeling of transcription factor–DNA interaction. Because binding interactions are transient, a particular binding site is occupied in only a fraction of cells across the population. We refer to this fraction as the occupancy, θ. Consider a transcription factor present in the nucleus at a free concentration $[P]$. This protein can bind to a particular unbound site, U, of length N to form a bound complex, B.

$$P + U \leftrightarrow B$$

The association constant K_a, which is a measure of the protein's affinity for the site, is given by $K_a = \frac{[B]}{[U][P]}$. The occupancy of the site is related to this association constant and the transcription factor concentration according to $\theta = \frac{K_a[P]}{1+K_a[P]}$. Now assume that the free energy of protein binding to any site is given by a simple sum of nucleotide contributions at each position i. Because the association constant is related to the free energy by $K_a = \exp(-\Delta G/RT)$, we can re-write the expression for occupancy to take on a convenient logistic form:

$$\theta = \frac{1}{1 + \exp\left(-\log[P] + \sum_i \sum_j g_{i,j} n_{i,j}\right)} \qquad [1]$$

where the $g_{i,j}$ correspond to the position-specific free energy contributions (scaled by $1/RT$) of each nucleotide (indexed by $j = 1,\ldots,4$) and $n_{i,j}$ are binary variables in a 4 by N matrix indicating the presence or absence of nucleotide j at site i. We now derive a simple relationship between a standard sequence motif and the position-specific free energy contribution of each nucleotide. Let $p_{i,j}$ be the posterior probability of observing nucleotide j at position i in a genomic site, given that the site is bound in vivo. These probabilities correspond to the entries in the motif frequency matrix and are given by

$$p_{i,j} = \frac{P\left(\text{bound}|n_{i,j} = 1\right) P\left(n_{i,j} = 1\right)}{\sum_n P\left(\text{bound}|n_{i,j} = 1\right) P\left(n_{i,j} = 1\right)} \qquad [2]$$

where $P(n_{ij} = 1)$ denotes the prior probability of observing nucleotide j (i.e., its background frequency). If we now assume that the protein concentration is very low, then from equations [1] and [2], a site's occupancy is approximated by $\theta \approx \exp\left(\log[C] - \sum_i \sum_j g_{i,j} n_{i,j}\right)$. We define an occupancy estimate which ignores the contribution of nucleotide m at position k as $\theta^{\backslash k,m} = \exp\left(\log[C] - \sum_i \sum_j g_{i,j} n_{i,j} + g_{k,m} n_{k,m}\right)$. Then, equation [2] reduces to

$$p_{n,i} = \frac{\left[\sum_{S \backslash i} \theta_S^{\backslash i,j} P(S)\right] \exp(g_{i,j}) P(n_{i,j})}{\left[\sum_{S \backslash i} \theta_S^{\backslash i,j} P(S)\right] \sum_j \exp(g_{i,j}) P(n_{i,j})} \qquad [3]$$

$$= \frac{\exp(g_{i,j}) P(n_{i,j})}{\sum_j \exp(g_{i,j}) P(n_{i,j})}$$

In equation [3], the sum over $S \backslash i$ (which cancels out) denotes a summation over all possible binding site sequences holding position i constant. Taking the logarithm of equation [3] and rearranging gives

$$\log p_{i,j} - \log P\left(n_{i,j}\right) = g_{i,j} - \log \sum_j \exp\left(g_{i,j}\right) P\left(n_{i,j}\right)$$
$$= g_{i,j} - \log Z_i \qquad [4]$$

The entries of a log-odds matrix (the left-hand side of equation [4]), under some assumptions can be interpreted as the scaled relative free energy contributions of those nucleotides to the binding reaction.

In a high-throughput ChIP experiment where there is both positive and negative information about binding occupancy, it is possible to use this information to exploit the relationship of the biophysical and probabilistic approaches to learn accurate and interpretable binding specificity estimates. The simple logistic equation [1] relates occupancy, which has been measured in the ChIP experiment, to energies, and thus, as we showed above, sequence motifs. Given a set of bound and unbound example binding sites (*see* **Note 1**), fitting a motif model can be accomplished by simply training a logistic regression classifier to distinguish the two classes (11). Similar biophysically based approaches, making different simplifying assumptions, have been explored in the context of ChIP-chip data analysis and have been proven to be effective (*see* **Chapters 9** and **10** and refs. (12, 13)).

Sequence Analysis of ChIP Data 183

Fitting appropriate motif models to the ChIP data does not end the analysis. Most de novo motif discovery algorithms produce multiple motif hypotheses, and it is often advantageous to generate hypotheses using several different techniques. These motifs must be assessed for statistical significance, ranked, and clustered to reduce redundancy (*see* **Note 2**). Once a core set of non-redundant motifs has been identified, it is useful to be able to map them back to the genome to identify putative binding sites at high resolution (*see* **Note 3**). In this chapter we discuss tools and techniques for obtaining transcription factor binding specificity estimates from ChIP-chip and ChIP-seq data and for performing the downstream analyses that allows sequence information to be used to maximum effect alongside ChIP data.

2. Software

A wide variety of software packages exist to analyze ChIP data and perform motif discovery and an exhaustive overview of the various options is outside the scope of this chapter. Instead we offer several suggestions which may be used as starting points for these types of analyses. For identifying bound regions from a ChIP-chip experiment, the ChIPOTle (14), TiMAT (15), COCAS (16), and JBD (17) tools are all suitable for analysis of modern tiled microarray data. For ChIP-seq data, the MACS (18) and USeq (19) packages are publicly available, have demonstrated good performance, and allow for sophisticated statistical analyses of sequence reads. Dozens of motif discovery algorithms have been described in the literature and are publicly available for use including the Weeder (20), AlignACE (21), MEME (22), and MDScan (23) algorithms. These tools have been integrated into an online motif discovery package, WebMOTIFS (24). Suites of tools for motif analysis have also been developed including cisGenome (*see* also **Chapter 22**) (25), the MEME suite (26), and TAMO (27). The performance of select methods is discussed in **Chapter 8** using the Motif Tool Assessment Platform (MTAP).

3. Methods

The process of identifying biologically meaningful sequence motifs from a ChIP-chip or ChIP-seq experiment and mapping them back to the genome involves several steps. The overall workflow is summarized in the flow chart of **Fig. 11.1**.

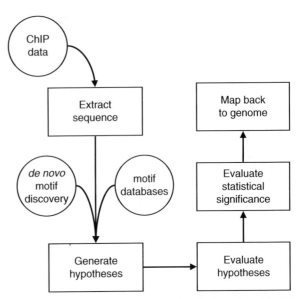

Fig. 11.1. Typical sequence motif analysis workflow for ChIP data. After identification of bound regions from the experiment, motif hypotheses are generated using de novo motif discovery algorithms or assembled from databases. Hypothesis quality is measured against the binding data using a quality score, and statistical significance testing is performed. Motifs may then be mapped back to the genome to improve the resolution of binding site identification.

3.1. Sequence Extraction

Motif analysis of high-throughput ChIP data begins by first identifying and extracting the DNA sequence of bound regions detected in the experiment. Most software packages used to analyze ChIP data will output the genomic coordinates of regions identified as bound in the experiment. For ChIP-chip, binding sites for the immunoprecipitated protein can be located several hundred base pairs away from the center of the peak identified by the ChIP analysis software. Data from ChIP-seq experiments is at significantly higher resolution, and the majority of bound regions identified by software packages have a putative binding site within a 300 bp window of the peak center as shown in **Fig. 11.2** for PPARγ ChIP-seq data (28). Even for ChIP-seq data, however, when the goals of a motif analysis include identification of binding sites corresponding to other transcription factors that may cooperate with the immunoprecipitated protein to enact a regulatory program, extending the sequences may allow these binding sites to be better captured. Of course, extending sequences also serves to decrease the signal-to-noise ratio in the data and makes motif discovery more challenging. **Figure 11.3** shows how the probability of observing a binding site in bound and randomly selected unbound regions changes as sequence length increases for the data set of **Fig. 11.2**. At relatively small sequence sizes, sensitivity is improved by increasing the size window since binding sites that are offset from the peak center are excluded when the length

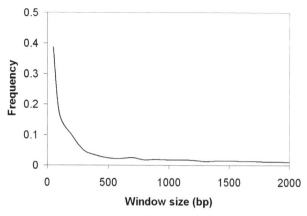

Fig. 11.2. Representative distribution of distances between ChIP-seq peaks and binding site matches. Genomic regions identified as bound by PPARγ in the study of Nielsen et al. were scanned for peroxisome proliferator response elements (PPREs) and the distribution of distances between the peak center and the closest PPRE is shown. The majority of bound regions have a PPRE within 250 bp of the peak center.

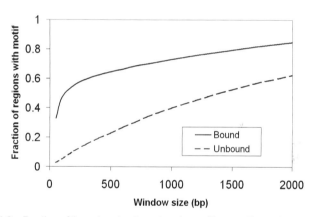

Fig. 11.3. Fraction of bound and unbound regions with a motif match as a function of region size. For the PPARγ ChIP data of Nielsen et al., increasing the size window around the ChIP peak centers increases the fraction of bound regions containing a PPRE. However, the fraction of randomly selected unbound regions that contain a PPRE also increases.

is too small. However, the probability of randomly observing a binding site in unbound sequence also increases. For the ChIP-seq data in **Fig. 11.3**, a sequence window size of approximately 250 bp adequately balances sensitivity and specificity considerations.

3.2. Hypothesis Generation

Once the sequences to be analyzed have been identified and extracted, the data can be mined for sequence motifs that may represent the binding specificity of the immunoprecipitated protein. This hypothesis generation step is often performed using one or more de novo motif discovery algorithms (29). These computer programs attempt to learn a representation of the protein's binding specificity directly from the sequence data in an unbiased manner and may be especially useful for immunoprecipitated proteins with unknown binding specificity and with no close homologs with known binding specificities. As the relative performance of particular algorithms has been shown to vary significantly from data set to data set (*see* ref. (30)), it is recommended that when de novo motif discovery is used as the primary hypothesis generation tool in an analysis two or more different programs be employed. It has been previously demonstrated that this can significantly improve the chances of identifying a motif consistent with the protein's true binding specificity (31). An alternate approach for generating hypotheses is to mine public or commercial databases for previously described DNA sequence motifs. When the DNA-binding domain family of the protein is known, hypotheses can be limited to motifs corresponding to transcription factors from that family (32, 33). A more comprehensive approach is to compile all motifs corresponding to transcription factors represented in a particular species or class and to treat these as motif hypotheses. For the PPARγ data set introduced in **Section 3.2**, we tested the large set of 101 DNA-binding domain-derived motifs reported in (33) to see which motif best represented the binding specificity of this transcription factor. To better make use of the sequence information at PPARγ-bound regions, we fit each motif to the binding data using an expectation maximization motif discovery approach (*see* **Chapter 6**). The resulting motifs were subsequently evaluated for quality (*see* **Section 3.3**).

3.3. Hypothesis Evaluation

The hypotheses that have been assembled, either by de novo motif discovery or by other methods, must be evaluated to determine which does the best job of representing the transcription factor's binding specificity. This involves calculating a score for each motif that measures its quality. Although most de novo motif discovery algorithms have built-in scoring methods for evaluating and ranking motifs, these scores are usually not directly comparable between different programs. Furthermore, for ChIP-chip and ChIP-seq data, it is natural, and generally desirable, to make use of the negative information in unbound regions from the experiment when evaluating different motif hypotheses (34); many de novo algorithms do not make use of this information. One particularly simple and useful scoring scheme is to calculate a p-value based on the hypergeometric distribution associated with each motif's occurrence in bound sequences relative to its occurrence in a pooled set of bound and unbound sequences from the experiment. Although the hypergeometric enrichment calculation produces a p-value, we will see in the next section that this statistic is not a reliable estimate of significance and should be treated like any other type of score. Alternatively, to avoid the difficulties associated with defining a match threshold for position weight matrix motif models, motif hypotheses can be ranked by evaluating the area under the receiver-operating characteristic curve for a motif-based classifier used to distinguish bound and unbound sequences. For the set of hypotheses generated from the PPARγ ChIP-seq data in **Section 3.3**, we used a similar approach, evaluating each motif's ability to correctly classify held out bound and unbound sequences. In **Fig. 11.4** we show the distribution of mean fivefold cross-validation errors for the 101 hypotheses. The motif with the lowest mean error, tgaCCTyTgNCCy, is an excellent match to the peroxisome proliferator response element bound by this transcription factor in vivo (35).

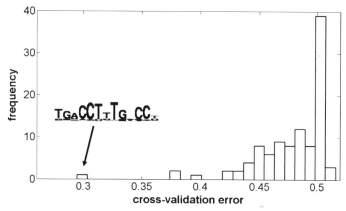

Fig. 11.4. Distribution of motif scores for PPARγ ChIP-seq data. A diverse set of 101 motif hypotheses were evaluated by assessing their ability to discriminate bound and unbound sequences. The resulting distribution of mean fivefold cross-validation errors is shown. The motif with the lowest cross-validation error matches the previously reported PPARγ binding specificity.

3.4. Evaluating the Statistical Significance of Motifs

High-scoring motifs may represent biologically meaningful transcription factor binding sites present in the immunoprecipitated regions identified by the experiment. To confidently link a particular motif to the binding data, however, it is necessary to estimate the level of statistical significance of the motif's score. Overfitting is a danger associated with any hypothesis generation scheme, like de novo motif discovery, that involves fitting a model to sequence data. Although even simple models can overfit the data, as model complexity increases (for example, by increasing the number of nucleotide positions in a position weight matrix model), overfitting becomes an increasingly serious problem.

p-Values are frequently used to evaluate the statistical significance of a motif. For motif analyses, the definition of a *p*-value is the probability of obtaining the same quality score or better for the motif when it is not bound by the transcription factor studied. A practical strategy for evaluating this probability is to estimate the null distribution using random sampling. The basic idea is as follows: the bound and unbound labels of the pooled set of sequences from the experiment are randomly permuted, the randomly sampled "bound" set is used to fit a motif model, and this model is then scored. Repeating this process many times allows the distribution of scores under the null hypothesis to be estimated. By comparing the scores of interesting motif hypotheses to this distribution, an empirical p-value for each motif can be obtained. When more than one hypothesis has been tested, it is important to account for this by performing a multiple test correction. There are several methods of performing multiple hypothesis correction including step-down False Discovery Rate (FDR) methods and Bonferroni correction (36). We tested the statistical significance of the top-ranked motif from our analysis in **Section 3.4** using this randomization strategy. After permuting bound and unbound labels in the PPARγ ChIP-seq data, we then fit the motif to this randomized data by EM. The ability of the resulting motif to classify "bound" and "unbound" sequences was assessed. Repeating this process 25 times we observed a mean cross-validation error of 0.49 with a standard deviation of 0.01. By comparison, on the actual data the mean cross-validation error of the top-ranked motif was 0.27, indicating that this motif is quite likely to have biological relevance.

3.5. Mapping Motifs Back to the Genome

The resolution of ChIP-chip and ChIP-seq experiments has improved tremendously, but unfortunately it still does not exceed a level of approximately 200 bp. For this reason, it is often of interest to identify putative in vivo transcription factor binding sites at higher resolution by mapping motifs back to the genome. Another important consideration is the noise in the data. Weakly bound regions with low ChIP enrichment may be biologically relevant. However, lowering the detection threshold may result in an unacceptably high level of false positives. Identification of motif matches allows sequence information to be used to adjust the confidence level associated with putatively bound regions detected in the experiment. The main challenge associated with identifying potential transcription factor binding sites using a sequence motif is in deciding what constitutes good enough agreement with the motif to be counted as a putative match. For matrix models, each genomic site can be scored using the matrix and a match threshold defined and used to identify putative binding sites. In the past, we have found that an empirically reasonable threshold to use is 0.6 times the maximum possible score of a log-odds matrix (29). However, more statistically principled methods for identifying motif matches can certainly be applied. It is always possible to associate an empirical p-value for a match score by evaluating the genome-wide distribution of scores for that motif. Alternatively, given a reasonable background model of nucleotide frequency in relevant genomic regions, a p-value can be obtained by calculating the log probability of the sequence under the motif and background models. We can then make the standard assumption that their ratio will be approximately chi-square distributed with degrees of freedom equal to the difference in degrees of freedom between the motif and background models. Of course, even when such p-values can be calculated, their relationship to the underlying biological reality is still unclear, and the motif match threshold selection problem has simply been converted into a p-value threshold selection problem. For this reason, we recommend a data-driven approach for picking a match threshold that takes advantage of the information the ChIP experiment has provided. By treating a motif as a feature that discriminates bound and unbound sequences in the experiment, reasonable criteria for selecting a match threshold naturally emerge. A threshold can be selected to keep FDR below some desired level, to minimize classification error, or to maximize sensitivity subject to a reasonable penalty on false positives. We define a true positive (TP) as a bound region in the ChIP experiment with a match to the motif, a false negative (FN) as a bound region with no match, a true negative (TN) as an unbound region with no match, and a false positive (FP) as an unbound region with a motif match. **Figure 11.5** shows how, on PPARγ-bound regions and an equally sized set of unbound regions, sensitivity (TP/(TP+FN)) and specificity (TN/(TN+FP)) change as the match threshold is increased for the PPARγ motif. For these data, a match threshold selected to maximize sensitivity while keeping the FDR below 20% recovers 56% of the bound sequences.

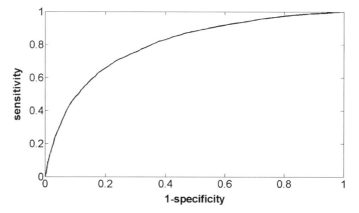

Fig. 11.5. Data-driven motif match threshold selection. Shown is sensitivity vs. 1 minus specificity curve for the PPARγ motif used as a classifier of bound and unbound sequences. A threshold selected to maximize sensitivity while keeping the false discovery rate below 20% recovers 56% of the bound sequences.

4. Notes

1. Selecting a background set of unbound regions: For ChIP-chip experiments it is often possible to use the entire set of unbound sequences represented on the array(s) as a background for either motif hypothesis generation or evaluation. For ChIP-seq experiments, this is infeasible and a representative set of unbound background regions often needs to be selected by the investigator. There are two important points to keep in mind when generating this background. First, the size distribution of bound and unbound sequences should be carefully matched in order to ensure that the predictive power of a particular hypothesis is accurately estimated. If the length of unbound sequences is too large, then the probability that a random unbound sequence will contain a motif match will be quite high, masking any true discriminative power that a particular motif may have. Second, different genomic regions have different nucleotide compositions. For example, promoter regions, which have the highest density of transcription factor binding sites, often contain GC-rich regions corresponding to CpG islands (37). A set of bound regions from a ChIP-seq experiment is likely to be enriched for high GC content even when the sequence recognized and bound by the transcription factor binding site itself is not GC rich. It is therefore often desirable to roughly match GC content between bound and unbound sets to avoid identifying uninformative GC-rich motifs during motif discovery and evaluation. For ChIP-chip data collected on promoter arrays, a simple and effective way of doing this, in our experience, is to match the distribution of distances to transcription start sites between the bound and unbound sets. Binding sites identified by the ChIP experiment that are enriched in promoter proximal regions will then be tested against a background that is also enriched in proximal regions, thereby controlling for the variations in nucleotide content between promoters and more distal sites. For ChIP-seq data, where many bound sites identified can be distal, a better strategy is to explicitly match the mean GC content of the bound regions and background.

2. Clustering motifs: In practice, motif discovery programs often produce several very similar, but not identical, motif hypotheses when run on a data set. If several programs are employed to analyze ChIP data, one is often faced with a pool of dozens or even hundreds of motifs with significant redundancy. Similarly, although the Jaspar database of transcription factor binding sites (38) has made an impres-

sive effort to eliminate redundancy, other databases do contain redundant motifs. When combining motif matrices or consensus sequences from multiple databases or literature sources this problem is exacerbated. One way of controlling redundancy is to employ a clustering algorithm to group similar motifs together. A representative motif can then be picked from each cluster to create a more manageable, non-redundant set of motifs to work with. In order to cluster motifs, one must first specify a motif similarity measure. For motifs that can be treated as frequency matrices, a very effective similarity score is the mean negative Kullback–Leibler (KL) divergence (39) between columns of the matrices. For the multinomial distributions given by two columns P and Q, the score is

$$ KL\left(P, Q\right) = -\sum_{j} P\left(n_j\right) \log \frac{P\left(n_j\right)}{Q\left(n_j\right)} \qquad [5] $$

Euclidean distance has also been used as a similarity score, and other specialized similarity scores have been suggested (40). No matter which measure is used, two additional issues must be addressed when calculating the similarity between a pair of motifs. First, both the forward and reverse complement orientations of the motifs must be considered. Second, because motifs often have different sizes the similarity measure should account for the different possible alignments of the matrices. One effective strategy is to evaluate the maximum similarity over all possible alignments (both forward and reverse complement) while enforcing a minimum overlap of six to eight nucleotides and to use this maximum as the similarity. Once a matrix of similarity scores has been calculated, an algorithm like affinity propagation (41) can then be used to perform the clustering itself.

3. Mapping motifs to the genome: Picking a single threshold to identify matches to a motif obscures a great deal of the complexity of transcription factor binding. The occupancy of a particular site in vivo will depend not only on the site's sequence but also on the protein's concentration in the nucleus. At low concentrations, most protein molecules will bind to very high-affinity sites, whereas at high concentrations, low-affinity sites may be bound and have biological function. It may therefore be more reasonable to predict an occupancy level between 0 and 100% on a site-by-site basis rather than to assign the sites binary labels indicating whether a site "matches" the motif. In practice, however, it is often more convenient to divide sites into matches and

non-matches. To this end, evaluation of sequence conservation across related species has been used to improve identification of functionally important transcription factor binding sites (42, 43). While it is reasonable to assume that conserved binding sites are likely to have functional importance, several studies have demonstrated that transcription factor binding can be surprisingly poorly conserved across species (44, 45). Enforcing stringent conservation thresholds on putative transcription factor binding sites is therefore likely to result in an underestimate of the true number of functional sites present in bound regions from the experiment.

References

1. Jacob, F., and Monod, J. (1961) Genetic regulatory mechanisms in the synthesis of proteins. *J Mol Biol* 3, 318–356.
2. Ptashne, M., and Hopkins, N. (1968) The operators controlled by the lambda phage repressor. *Proc Natl Acad Sci U S A* 60, 1282–1287.
3. Ippen, K., Miller, J.H., Scaife, J. et al. (1968) New controlling element in the Lac operon of *E. coli. Nature* 217, 825–827.
4. Liang, J., Yu, L., Yin, J. et al. (2007) Transcriptional repressor and activator activities of SMA-9 contribute differentially to BMP-related signaling outputs. *Dev Biol* 305, 714–725.
5. Robertson, G., Hirst, M., Bainbridge, M. et al. (2007) Genome-wide profiles of STAT1 DNA association using chromatin immuno-precipitation and massively parallel sequencing. *Nat Methods* 4, 651–657.
6. Ren, B., Robert, F., Wyrick, J.J. et al. (2000) Genome-wide location and function of DNA binding proteins. *Science* 290, 2306–2309.
7. Stormo, G.D. (2000) DNA binding sites: representation and discovery. *Bioinformatics* 16, 16–23.
8. Cui, Y., Wang, Q., Stormo, G.D. et al. (1995) A consensus sequence for binding of Lrp to DNA. *J Bacteriol* 177, 4872–4880.
9. Berg, O.G., and von Hippel, P.H. (1987) Selection of DNA binding sites by regulatory proteins. Statistical-mechanical theory and application to operators and promoters. *J Mol Biol* 193, 723–750.
10. Stormo, G.D., and Fields, D.S. (1998) Specificity, free energy and information content in protein-DNA interactions. *Trends Biochem Sci* 23, 109–113.
11. MacIsaac, K.D. (2009) Motifs, binding, and expression: computational investigations

of transcriptional regulation. *Department of Electrical Engineering and Computer Science.* Massachusetts Institute of Technology, Cambridge.
12. Djordjevic, M., Sengupta, A.M., and Shraiman, B.I. (2003) A biophysical approach to transcription factor binding site discovery. *Genome Res* 13, 2381–2390.
13. Foat, B.C., Morozov, A.V., and Bussemaker, H.J. (2006) Statistical mechanical modeling of genome-wide transcription factor occupancy data by MatrixREDUCE. *Bioinformatics* 22, e141–e149.
14. Buck, M.J., Nobel, A.B., and Lieb, J.D. (2005) ChIPOTle: a user-friendly tool for the analysis of ChIP-chip data. *Genome Biol* 6, R97.
15. Johnson, W.E., Li, W., Meyer, C.A. et al. (2006) Model-based analysis of tiling-arrays for ChIP-chip. *Proc Natl Acad Sci U S A* 103, 12457–12462.
16. Benoukraf, T., Cauchy, P., Fenouil, R. et al. (2009) CoCAS: a ChIP-on-chip analysis suite. *Bioinformatics* 25, 954–955.
17. Qi, Y., Rolfe, A., MacIsaac, K.D. et al. (2006) High-resolution computational models of genome binding events. *Nat Biotechnol* 24, 963–970.
18. Zhang, Y., Liu, T., Meyer, C.A. et al. (2008) Model-based analysis of ChIP-Seq (MACS). *Genome Biol* 9, R137.
19. Nix, D.A., Courdy, S.J., and Boucher, K.M. (2008) Empirical methods for controlling false positives and estimating confidence in ChIP-Seq peaks. *BMC Bioinformatics* 9, 523.
20. Pavesi, G., Mereghetti, P., Mauri, G. et al. (2004) Weeder Web: discovery of transcription factor binding sites in a set of sequences from co-regulated genes. *Nucleic Acids Res* 32, W199–W203.

21. Roth, F.P., Hughes, J.D., Estep, P.W. et al. (1998) Finding DNA regulatory motifs within unaligned noncoding sequences clustered by whole-genome mRNA quantitation. *Nat Biotechnol* 16, 939–945.
22. Bailey, T.L., and Elkan, C. (1994) Fitting a mixture model by expectation maximization to discover motifs in biopolymers. *Proc Int Conf Intell Syst Mol Biol* 2, 28–36.
23. Liu, X.S., Brutlag, D.L., and Liu, J.S. (2002) An algorithm for finding protein-DNA binding sites with applications to chromatin-immunoprecipitation microarray experiments. *Nat Biotechnol* 20, 835–839.
24. Romer, K.A., Kayombya, G.R., and Fraenkel, E. (2007) WebMOTIFS: automated discovery, filtering and scoring of DNA sequence motifs using multiple programs and Bayesian approaches. *Nucleic Acids Res* 35, W217–W220.
25. Ji, H., Jiang, H., Ma, W. et al. (2008) An integrated software system for analyzing ChIP-chip and ChIP-seq data. *Nat Biotechnol* 26, 1293–1300.
26. Bailey, T.L., Boden, M., Buske, F.A. et al. (2009) MEME SUITE: tools for motif discovery and searching. *Nucleic Acids Res* 37, W202–W208.
27. Gordon, D.B., Nekludova, L., McCallum, S. et al. (2005) TAMO: a flexible, object-oriented framework for analyzing transcriptional regulation using DNA-sequence motifs. *Bioinformatics* 21, 3164–3165.
28. Nielsen, R., Pedersen, T.A., Hagenbeek, D. et al. (2008) Genome-wide profiling of PPARgamma:RXR and RNA polymerase II occupancy reveals temporal activation of distinct metabolic pathways and changes in RXR dimer composition during adipogenesis. *Genes Dev* 22, 2953–2967.
29. Harbison, C.T., Gordon, D.B., Lee, T.I. et al. (2004) Transcriptional regulatory code of a eukaryotic genome. *Nature* 431, 99–104.
30. Tompa, M., Li, N., Bailey, T.L. et al. (2005) Assessing computational tools for the discovery of transcription factor binding sites. *Nat Biotechnol* 23, 137–144.
31. MacIsaac, K.D., Wang, T., Gordon, D.B. et al. (2006) An improved map of conserved regulatory sites for Saccharomyces cerevisiae. *BMC Bioinformatics* 7, 113.
32. Mahony, S., Auron, P.E., and Benos, P.V. (2007) DNA familial binding profiles made easy: comparison of various motif alignment and clustering strategies. *PLoS Comput Biol* 3, e61.
33. Macisaac, K.D., Gordon, D.B., Nekludova, L. et al. (2006) A hypothesis-based approach for identifying the binding specificity of regulatory proteins from chromatin immunoprecipitation data. *Bioinformatics* 22, 423–429.
34. Takusagawa, K.T., and Gifford, D.K. (2004) Negative information for motif discovery. *Pac Symp Biocomput* 9, 360–371.
35. Lemay, D.G., and Hwang, D.H. (2006) Genome-wide identification of peroxisome proliferator response elements using integrated computational genomics. *J Lipid Res* 47, 1583–1587.
36. Rice, T.K., Schork, N.J., and Rao, D.C. (2008) Methods for handling multiple testing. *Adv Genet* 60, 293–308.
37. Gardiner-Garden, M., and Frommer, M. (1987) CpG islands in vertebrate genomes. *J Mol Biol* 196, 261–282.
38. Sandelin, A., Alkema, W., Engstrom, P. et al. (2004) JASPAR: an open-access database for eukaryotic transcription factor binding profiles. *Nucleic Acids Res* 32, D91–D94.
39. Kullback, S., and Leibler, R.A. (1951) On information and sufficiency. *Ann Math Statist* 22, 79–86.
40. Habib, N., Kaplan, T., Margalit, H. et al. (2008) A novel Bayesian DNA motif comparison method for clustering and retrieval. *PLoS Comput Biol* 4, e1000010.
41. Frey, B.J., and Dueck, D. (2007) Clustering by passing messages between data points. *Science* 315, 972–976.
42. Wasserman, W.W., Palumbo, M., Thompson, W. et al. (2000) Human-mouse genome comparisons to locate regulatory sites. *Nature Genet* 26, 225–228.
43. Xie, X.H., Lu, J., Kulbokas, E.J. et al. (2005) Systematic discovery of regulatory motifs in human promoters and 3′ UTRs by comparison of several mammals. *Nature* 434, 338–345.
44. Borneman, A.R., Gianoulis, T.A., Zhang, Z.D.D. et al. (2007) Divergence of transcription factor binding sites across related yeast species. *Science* 317, 815–819.
45. Odom, D.T., Dowell, R.D., Jacobsen, E.S. et al. (2007) Tissue-specific transcriptional regulation has diverged significantly between human and mouse. *Nature Genet* 39, 730–732.

Chapter 12

Inferring Protein–DNA Interaction Parameters from SELEX Experiments

Marko Djordjevic

Abstract

Systematic Evolution of Ligands by EXponential enrichment (SELEX) is an experimental procedure that allows extraction, from an initially random pool of oligonucleotides, of the oligomers with a high binding affinity for a given molecular target. The highest affinity binding sequences isolated through SELEX can have numerous research, diagnostic, and therapeutic applications. Recently, important new modifications of the SELEX protocol have been proposed. In particular, a suitably modified SELEX experiment, together with an appropriate computational procedure, allows inference of protein–DNA interaction parameters with up to now unprecedented accuracy. Such inference is possible even when there is no a priori information on transcription factor binding specificity, which allows accurate predictions of binding sites for any transcription factor of interest. In this chapter we discuss how to accurately determine protein–DNA interaction parameters from SELEX experiments. The chapter addresses experimental and computational procedure needed to generate and analyze appropriate data.

Key words: In vitro selection, high-throughput SELEX, SELEX-SAGE, weight matrix, SELEX modeling, protein–DNA interactions, transcription factor binding sites.

1. Introduction

Systematic Evolution of Ligands by EXponential enrichment (SELEX) is a procedure that allows rapid selection of those oligonucleotides that have appropriate binding affinity to a given molecular target, starting from a large initial library of oligonucleotides (1). The oligonucleotide library can consist of either single-stranded oligonucleotides (RNA, ssDNA, modified RNA, or modified ssDNA) or double-stranded DNA (dsDNA). One most often starts with a large library of random oligonucleotides,

I. Ladunga (ed.), *Computational Biology of Transcription Factor Binding*, Methods in Molecular Biology 674,
DOI 10.1007/978-1-60761-854-6_12, © Springer Science+Business Media, LLC 2010

since little is known beforehand about the binding properties of a target molecule in many cases. While molecular targets can be either proteins or small molecules, we here concentrate on targets that are selective DNA-binding proteins (e.g., transcription factors). The basic steps of SELEX are protein binding, selection, and amplification. These steps are repeated successively, so that strong binders are finally selected from the initial library.

The first SELEX experiments were performed more than 19 years ago (1–3), and SELEX by now has numerous research, diagnostic, and therapeutic applications (4). Most of published SELEX experiments involve single-stranded oligonucleotides, while the experiments and applications involving dsDNA are comparably underrepresented. This bias is mostly due to that single-stranded oligos obtained through SELEX have important diagnostic and therapeutic applications. In particular, single-stranded oligos that bind with strong binding affinity can be identified for a large variety of molecular targets. Those strong binders can, for example, be used as alternatives to antibodies in many applications.

On the other hand, SELEX is also a very important tool to infer interactions of proteins with dsDNA. This is mainly because one often has a protein that interacts with dsDNA in vivo, but whose binding specificity is unknown. SELEX is then performed in order to identify dsDNA sequences that are the strongest (consensus) binders to the protein of interest (e.g., (2, 5, 6)). Furthermore, appropriate modifications of the standard SELEX protocol allow robust generation of a data set from which protein–DNA interaction parameters can be determined with high accuracy (7, 8). Such interaction parameters can consequently be used to accurately predict binding affinity of a transcription factor to any DNA segment of interest.

This chapter addresses how to use SELEX to enable accurate predictions of transcription factor binding sites, so we concentrate on SELEX experiments in which dsDNA is used. We will first discuss SELEX experimental protocol, and then consider how SELEX should be modified to generate a data set suitable for further analysis. We will then describe the computational analysis of the data set in order to accurately determine transcription factor binding parameters.

2. Methods

As indicated above, we here focus on SELEX experiments that are performed with dsDNA library and where the target is a DNA-binding protein (e.g., a transcription factor). The scheme of the

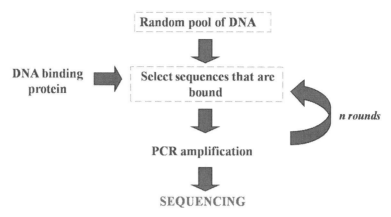

Fig. 12.1. The schema of the SELEX procedure. A certain number of experimental rounds (*n*) that consist of protein binding, selection, and amplification are performed. Some of the sequences from the last round of the experiment are extracted and sequenced.

SELEX procedure is shown in **Fig. 12.1**, and the experiments are performed as follows. One prepares a library of dsDNA segments that can be amplified, and the library is incubated with a DNA-binding protein of interest. Next, in the selection step, protein-bound DNA segments are separated from unbound ones (e.g., by gel shift or filtration through nitrocellulose). The selected segments are amplified by PCR. Binding, selection, and amplification steps are then repeated for certain number of rounds, and some of the segments that are selected in the final round of the experiment are extracted and sequenced.

The initial library of oligonucleotides typically consists of a large number (10^{15}–10^{16}) of random sequences. Larger libraries of up to 10^{20} oligonucleotides are technically feasible (9) but are rarely used in practice. Each oligonucleotide consists of a central region of random sequence, which is flanked by two regions of fixed sequence that enable amplification. The length of the random region is typically between 20 and 30 bps, while each flanking region is typically 15–25 bp long. One should note that the length of the random region is almost always larger than the length of transcription factor binding site, which has important implications for data analysis that will be discussed below.

2.1. Selection of Binding Sequences Through SELEX

We here discuss how binding sequences are selected through different rounds of SELEX. The selection process has to be understood in order to (i) determine how to modify standard SELEX procedure to generate a data set suitable for further analysis and (ii) understand how to analyze the assembled data so that parameters of transcription factor binding specificity are accurately determined.

A protein–DNA interaction may or may not be sequence specific. Sequence-specific interactions are due to hydrogen bonding and van der Waals interactions, while non-specific interactions are due to electrostatic interactions (10–12). When the sequence of a DNA segment is far from the consensus, interaction of the protein with DNA becomes sequence independent (11, 13). Therefore, most of the sequences in the starting (random) SELEX library will interact non-specifically with the target protein.

In addition to the selected sequence-specific binders, a number of non-specific binders will also be selected in each round of the experiment. This is a consequence of two conditions: first, a number of sequences are bound non-specifically by the protein. Second, during the selection step, only a partial separation is possible between bound and unbound sequences. The second effect is termed background partitioning (1, 14).

The selection of non-specific binders is an important effect in SELEX experiments. Non-specific interactions are typically characterized by several orders of magnitude smaller binding affinity compared to sequence-specific interactions (15). Also, background partitioning probability, the probability to select a sequence that is not bound by the protein, is likely low, e.g., 10^{-3} (1). However, these small numbers do *not* imply that the presence of non-specific binders can be neglected. For example, about 10^{12} *non-specific* binders will be selected in the first round of the experiment, just due to background partitioning, assuming the background partitioning probability of 10^{-3} and a starting library size of 10^{15} sequences. On the other hand, with protein–nucleic acid ratio of 10^{-3}, which is typical for SELEX experiments (e.g., (1)), less than 10^{12} specific binding sequences will be selected. Therefore, after the first round of the experiment, the number of non-specific binders is typically comparable or even larger than the number of specific binders. As a practical consequence, one must be sure to eliminate substantial noise due to non-specific binding, which would otherwise overrun useful signals. To achieve this, multiple rounds of SELEX experiment are performed. Also, selection of non-specific binders has to be taken into account when analyzing SELEX data, as will be discussed below.

It is important to understand how the affinity distribution of the selected sequence-specific binders changes through different experimental rounds. In a simple case of small protein to nucleic acid ratio, the average binding affinity of the selected sequence-specific binders increases exponentially with the number of performed SELEX rounds (8, 16, 14). Such exponential increase in binding affinity, during the first few rounds of the experiment, justifies the term "exponential" in Systematic Evolution of Ligands by EXponential enrichment. Finally, after a certain number of rounds, the maximum of the affinity distribution of the selected binders reaches an upper limit, which is determined by the affin-

Inferring Protein–DNA Interaction Parameters 199

ity of the strongest binder in the starting random library. At that point, most of the sequences in the selected pool will consist of the strongest binders. The observation that standard SELEX protocol rapidly selects only the strongest binders has important consequences which will be discussed in the next section.

2.2. The Standard SELEX Protocol Cannot Be Used to Accurately Determine Protein–DNA Interaction Parameters

As discussed above, the standard SELEX procedure can be used to efficiently converge to the strongest binders for a transcription factor of interest. However, the knowledge of only the strongest binders is usually not sufficient to determine transcription factor binding sites in the genome. This is because binding sites typically show considerable sequence variations (17). Moreover, a SELEX library is typically much larger than the size of a genome, so the consensus sequence may not be present in the genome at all. Therefore, a direct match with the consensus sequence cannot be used to identify binding sites in most cases. Alternatively, one may attempt to allow certain number of mismatches to the consensus sequence, in order to accommodate variability of binding sites. However, a general problem with this approach is that different positions in a binding site, as well as different mismatches at a given position, generally contribute very differently to protein–DNA interaction energy (15, 18). That is, while a certain mismatch can almost completely abolish sequence-specific binding, another mismatch may change the binding energy by only a small amount. Therefore, one needs to infer a more complete set of protein–DNA interaction parameters, in order to appropriately predict binding of transcription factors to DNA.

The interaction of proteins with dsDNA can be quantified by using the so-called independent nucleotide approximation (15). In this approximation, the binding energy of a protein to a dsDNA sequence is equal to the sum of contributions due to the presence of a given base at a given position in the binding site. The independent nucleotide approximation provides a very good parametrization of the binding energy in most cases (19–21), although there are some examples where binding at certain positions shows strong dependence on dinucleotide pairs (22–24).

Within the independent nucleotide approximation, one needs a total of $3L$ independent parameters (L is the binding site length),[1] in order to describe protein–DNA interaction. These parameters can be written in a form of a matrix with dimension $4 \times L$, which is called weight matrix (15, 17). Individual weight matrix elements are proportional to the contribution to the bind-

---

[1] One should observe that there is one parameter for each possible mismatch from a reference sequence.

ing energy due to the presence of a certain base at a certain position in a binding site (15, 25). Therefore, finding an accurate weight matrix is an important goal toward reliable predictions of transcription factor binding sites.

Weight matrices are typically determined from a set of aligned binding sites assembled in biological databases (26, 27). However, the majority of such weight matrices provide a low level of both specificity and sensitivity (28). In particular, there is a problem of a large number of false positives when most of these weight matrices are used to search for protein–DNA binding sites (29, 17). This problem is typically attributed to an inadequate data set from which most weight matrices are constructed (28) because (i) for most DNA-binding proteins, only a few binding sites are available in databases (26, 27), which is insufficient to accurately determine protein–DNA interaction parameters (24), and (ii) binding sites from databases are often assembled under diverse and ill-characterized conditions (25). Therefore, in order to improve the accuracy of weight matrices, it is highly desirable to be able to generate a data set consisting of a large number of binding sites assembled under well-controlled conditions.

Given that SELEX experiments are performed under controlled (uniform) conditions, it appears that SELEX may be used to generate such appropriate data set. However, can a standard SELEX protocol indeed be used to generate a suitable data set or the protocol has to be appropriately modified? To answer this, it is useful to note a comprehensive comparison between the weight matrices from eight available SELEX experiments with *Escherichia coli* transcription factors and the corresponding weight matrices constructed from natural binding sites (29). This comparison notes large discrepancies between the weight matrices derived from natural binding sites and from SELEX experiments, in seven out of those eight cases. Furthermore, in a SELEX experiment performed with a bacterial transcription factor LRP (30), it was noted that weight matrix scores inferred from SELEX experiments show a poor agreement with measured binding affinities. Similarly, it was noted (31) that a weight matrix constructed directly from sequences extracted in a standard SELEX procedure was not able to provide a good prediction of measured binding affinities. Therefore, it appears that the standard SELEX procedure is not appropriate to accurately determine protein–DNA interaction parameters.

Why does the standard SELEX procedure appear to fail in so many cases? To understand this, it is useful to consider what kind of data set is needed to construct an accurate weight matrix. First, a successful experiment has to eliminate non-specific binders from the data set, as discussed above. Second, overselection should be minimized, i.e., the selected sequences should not consist of only the strongest binding sites. To understand the second point, it is

Inferring Protein–DNA Interaction Parameters

useful to take the limit in which the data set consists from only the consensus binder, when it is evident that the weight matrix elements cannot be obtained from such information. This is also supported by a detailed statistical analysis, which shows that a significant fraction of medium affinity and weaker affinity binding sequences is needed to accurately determine weight matrix elements (7).

Actually, it turns out that the above two requirements, the elimination of non-specific binders and the absence of overselection, are very difficult to reconcile within the standard SELEX procedure. This conclusion directly follows from the theoretical modeling of the standard SELEX procedure (8). To intuitively understand this result, one should note the observation that the selected sequence-specific binders rapidly reach the highest affinity binding sites, and non-specific binders may not be eliminated from the pool of selected sequences by that time. Even if this does not happen, it is very difficult to reliably predict when to stop the experiment in practice, i.e., to determine in which SELEX round is the noise eliminated while the overselection has not happened yet. The reason for this is that the protein–DNA interaction parameters of the target protein are typically unknown a priori. Therefore, the appropriate number of rounds cannot be calculated. In the next section, we will discuss how the SELEX procedure can be appropriately modified in order to allow a robust generation of a data set from which accurate protein–DNA interaction parameters can be determined.

2.3. Fixed Stringency/High-Throughput SELEX Experiments

To understand how to appropriately modify SELEX, we first discuss the binding of proteins to DNA segments. The probability that a sequence S is bound by the protein, and consequently selected in the next round of the experiment, is given by the expression $[c]/(K_d(S) + [c])$ (25). Here $[c]$ and $K_d(S)$ are the concentration of free protein and the binding dissociation constant of the sequence S, respectively. Therefore, the selection stringency is determined by the concentration of *free* protein in solution. The formula for binding probability can be rewritten in terms of binding energy and chemical potential, so that $f(E - \mu) = 1/[\exp(E - \mu) + 1]$, where E is the interaction energy of the protein with the DNA sequence S and μ is a value of chemical potential, proportional to logarithm of free protein concentration. Both E and μ are measured in units of thermal energy ($k_B T$). Note that this binding probability is in statistical mechanics called Fermi–Dirac function.

In the standard SELEX protocol, most experiments are performed so that the *total* amount of protein and DNA is the same in each experimental round. Since the average binding affinity of the selected sequences increases with the number of performed rounds, the amount of free protein will decrease as a consequence of the increase in the amount of bound protein. Due to decrease in the free protein concentration selection stringency increases through the experiment. Such experimental design leads to a data set from which protein–DNA interaction parameters cannot be accurately determined, as we discussed above.

Let us now assume that the amount of *free* protein is constant in each round of SELEX. Since the selection stringency for any given sequence is then constant, we will further call this procedure *fixed stringency* SELEX. For fixed stringency SELEX, the change of the energy distribution of selected DNA sequences can be calculated from a theoretical model of SELEX experiments (*see* **Fig. 12.2** from (8)). We see that the maximum of the energy distribution for selected sequence-specific binders remains in the vicinity of the chemical potential, i.e., the maximum drifts very slowly toward the higher binding energies with the additional number of performed SELEX rounds. This is in a sharp contrast to the standard SELEX procedure, where the maximum of the energy distribution rapidly reaches the strongest affinity binders

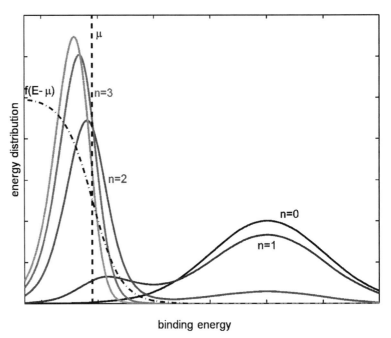

Fig. 12.2. The change of energy distribution through the SELEX procedure. *Solid curves* are energy distributions of selected DNA sequences for different number of performed SELEX rounds, in an experiment where the chemical potential μ is kept constant. Numbers above the curves indicate SELEX round, the position of the chemical potential is indicated by the *vertical dashed line*, the *dash–dotted line* indicates the binding probability $f(E - \mu)$. Note that once maximum of the energy distribution reaches μ, most of the selected sequences are in saturated regime, i.e., bound with probability close to one. This figure was adopted from (8).

(the strongest affinity binder corresponds to the leftmost point on the horizontal axis of **Fig. 12.2**). On the other hand, one can notice that the number of non-specific binders keeps decreasing with the increase in the number of performed SELEX rounds.

An important practical implication is that in the fixed stringency SELEX one can ensure that random binders are eliminated by performing larger number of SELEX rounds, without the risk that only the strongest sequences will be selected. One can theoretically show that the fixed stringency SELEX procedure leads to this desired behavior for all values of experimental parameters (8). Additionally, the procedure is robust, since it leads to a suitable data set for a large range of performed experimental rounds (in the example in **Fig. 12.2**, any round larger than two is suitable). Therefore, in conclusion, a fixed stringency SELEX experiment allows robust generation of a suitable data set for accurate determination of protein–DNA interaction parameters.

How can one experimentally implement the constraint of fixed free protein amount? An answer is a modification of the standard SELEX procedure by inclusion of the radiolabeled sequence (probe) S^* of moderate binding affinity, as described in an experiment by Roulet et al. (7). Additionally, the concentration of total DNA, added to the reaction mixture as a competitor to the radiolabeled probe, is adjusted in each round of the experiment, so that a fixed fraction of the probe is bound by protein in each SELEX round. Note that radiolabeling of the probe allows one to determine the fraction of the probe that is bound by the protein. Since the fraction of the bound probe is constant, the expression $[c]/([c] + K_d(S^*))$ has to be constant, where $K_d(S^*)$ is the dissociation constant of the probe. Therefore, the free protein amount ($[c]$) has to be constant as well, since of course $K_d(S^*)$ does not change.

Roulet et al. (32) introduced another important extension to combine the SELEX procedure with the SAGE (Serial Analysis of Gene Expression) protocol. This extension allows one to efficiently sequence up to several thousand binding sequences (7). The procedure was termed high-throughput SELEX or SELEX-SAGE protocol. As a recent development, a new generation of non-Sanger-based sequencing (33) may be used instead of SAGE procedure (34). In any case, the ability to generate a large data set provides an obvious advantage for a precise estimation of protein–DNA interaction parameters. Therefore, the combination of the fixed stringency procedure with ability to sequence a large number of DNA segments, which we call fixed stringency/high-throughput SELEX, allows both robust and accurate determination of protein–DNA interaction parameters. A database called HTPSELEX, specifically developed for storing large data sets obtained from high-throughput SELEX experiments, has recently become available (35). This complements SELEX_DB (36) and

204   Djordjevic

TRANSFAC (26) databases, which have been assembling the data obtained from standard SELEX experiments.

2.4. Computational Analysis of Fixed Stringency/High-Throughput SELEX Data

We here discuss how to accurately determine protein–DNA interaction parameters from sequences extracted in fixed stringency/high-throughput SELEX procedure. One should note that the length of the randomized part of DNA sequences is usually larger than the length of a transcription factor binding site. Therefore, one first needs to extract actual binding sites from these longer sequences. To do that, multiple local sequence alignment algorithms (MLSA) are used that allow identifying statistically overrepresented motifs in a set of DNA sequences. The algorithms for MLSA are typically based on either the Gibbs search (37) or expectation maximization (38) (*see* **Chapter 6**), and several computational implementations of these approaches exist.

In a typical data analysis, the set of aligned binding sites is used to construct an information theory based weight matrix (17). In the information theory based method, the weight matrix elements are equal to the logarithm of the ratio of probability to observe a given base at a given position in a collection of binding sites, compared to the base background probability. However, the information theory based weight matrix method has drawbacks, since it does not properly incorporate saturation in the binding probability (39, 25). That is, the information theory based method assumes that the probability that sequence S is bound by protein is given by $\exp(\mu - E)$, while the correct binding probability is given by Fermi–Dirac function with sigmoid form $f(E - \mu)$ as given above. This approximation is particularly inaccurate to use in analysis of fixed stringency SELEX experiments (8), since selected sequences rapidly reach saturated binding regime, where maximum of the binding energy distribution is in the vicinity of chemical potential (**Fig. 12.2**).

A procedure that correctly incorporates saturation effects is presented in (8). A key step in the procedure is using a maximum likelihood method: initially unknown parameters are inferred by maximizing the likelihood that the extracted set of DNA sequences is observed as the outcome of the experiment. The probability of extracting the given set of DNA sequences is calculated by taking into account the correct protein–DNA binding probability (see the formula for binding probability above). The set of equations resulting from varying the likelihood with respect to the unknown parameters is then numerically solved to compute the elements of the energy matrix. Computationally, we solve a set of $3L+1$ mutually coupled non-linear equations (L is length of the binding site, and one additional equation corresponds to solving for the unknown free protein concentration). For detailed implementation, please refer to (8).

While the above procedure leads to an accurate determination of protein–DNA interaction parameters, numerically solving a large number of coupled equations may be technically demanding. Therefore, it is useful to look at a limiting case of the above procedure, where sigmoid function is approximated by unit step function. In statistical physics, this is called "zero temperature approximation" and is appropriate to use in saturated binding regime (**Fig. 12.2**). It can be shown that this approximation leads to a quadratic programming procedure for determining protein–DNA interaction parameters, and this method was consequently termed QPMEME (25). Since the procedure involves finding the minimum of a convex function over a convex domain, finding a solution satisfying the Kuhn–Tucker condition (40), namely the condition for being a local minimum, is enough to find a global solution. There are standard numerical packages that can be used for solving quadratic programming problem (e.g., a simple to use but robust implementation is given in MATLAB's Optimization Toolbox). While the quadratic programming method is less accurate than the full procedure discussed above, it still leads to a significantly better false-positive/false-negative trade-off, as compared to the information theory weight matrix method (25). Due to its relative simplicity, a computational procedure for the quadratic programming method will be described below, while C and MATLAB codes for the method are available from the authors of (25).

We assume that after n rounds of SELEX, set A, which contains some number of sequences S, has been extracted and sequenced. Furthermore, we denote by $S^{(k)}$ the kth sequence in set A, and $S^{(k)}_{i,\alpha} = 1$ if base α is present at the position i in binding site, and $S^{(k)}_{i,\alpha} = 0$ otherwise. Furthermore, we denote by $\varepsilon_{i,\alpha}$ the energy matrix element that gives contribution to the binding energy due to presence of base α at position i in the binding site. As before, μ is the chemical potential. With this notation, the determination of energy matrix elements amounts to minimizing a quadratic form subject to linear inequality conditions:

$$\sum_{i,\alpha}\left(\frac{\varepsilon_{i,\alpha}}{\mu}\right)S^{(k)}_{i,\alpha} > 1 \qquad [1]$$

$$\sum_{i,\alpha}\left(\frac{\varepsilon_{i,\alpha}}{\mu}\right)^2 = \min \qquad [2]$$

The above equations can be solved for $\varepsilon_{i,\alpha}/\mu$ (i.e., the energy matrix elements are in the units of chemical potential) by standard numerical packages for quadratic programming.

The above equations have the following intuitive interpretation (**Fig. 12.3**). The figure shows such distribution of the

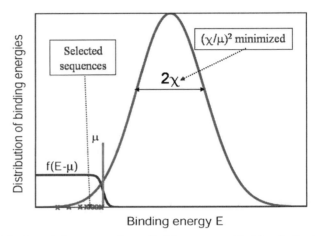

Fig. 12.3. A quadratic programming method for energy matrix determination. The distribution of binding energies for a set of random sequences is approximately Gaussian, as indicated in the figure. The binding probability $f(E - \mu)$ and the value of chemical potential μ is also indicated. The binding energies of the selected sequences in the final round are indicated by *crosses*. Width of the random energy distribution χ is also shown. The quadratic programming procedure minimizes $(\chi/\mu)^2$, while at the same time requiring that binding energies of all the selected sequences are below the chemical potential.

binding energy which corresponds to random DNA sequences. The first equation requires that binding energy of all sequences in set A is smaller than the chemical potential. Note that in the unit step function approximation, all sequences that have binding energy smaller than chemical potential are bound by the transcription factor with probability equal to 1. Therefore, the first equation requires that all sequences that are selected through SELEX procedure are bound by the transcription factor. The second equation corresponds to minimizing the ratio of the width of energy distribution to the value of chemical potential. Since all DNA sequences with energy below chemical potential are bound by transcription factor, it is straightforward to see that equation [**2**] corresponds to minimizing the number of random sequences that are bound by the transcription factor. Therefore, the quadratic programming procedure amounts to the requirement that all binding sites observed in the experiment (set A) are indeed bound by the transcription factor, while at the same time, the "noise" (the number of bound random sequences) is minimized.

3. Notes

An accurate energy matrix, which is obtained through an appropriate analysis of fixed stringency/high-throughput SELEX data, can be used to reliably detect putative protein binding sites in genomic DNA. Therefore, such methodology can be applied to a large number of different DNA-binding proteins, which would facilitate comprehensive understanding of gene regulation. The procedure described in this chapter shows how synergy of theoretical modeling, novel experimental developments, and data analysis based on physical understanding of the underlying process can significantly contribute to an important problem in computational biology. We below note some practical issues relevant for modeling and data analysis of SELEX experiments.

1. Terminology

We here note how the term "weight matrix" is used, since this term is often associated with different meanings. The most general definition is that weight matrix is any matrix of "weights"; "weights" are contributions of different bases to a score used to classify whether or not a sequence is a binding site (17). In a biophysical interpretation, which is also used in this chapter, weights in the matrix are defined as contributions of different bases at different positions to the binding energy (15). The term energy matrix is also often associated with this biophysical interpretation of the weight matrix (8, 25).

Other definitions of weight matrix have been frequently used, most notably the one coming from information theory (17). Such weight matrix is sometimes called information theory weight matrix (25), and weights in this matrix are equal to the logarithm of the ratio of probability to observe a given base at a given position in a collection of binding sites, compared to the probability of observing the base in the genome as a whole. One can show that the biophysical and the information theory definitions coincide in the limit of small transcription factor concentration (unsaturated limit) (25). That is, in this limit, the information theory weight matrix gives an accurate estimate of transcription factor binding energy. However, when the saturation effects become important (as in fixed stringency SELEX experiments), a different procedure has to be used for estimating protein–DNA interaction parameters, as described above.

2. Modeling SELEX experiments

We first note that stochastic effects can be generally neglected in SELEX experiments. This is a consequence of the fact that the size of the oligonucleotide library is so large that the relevant sequence space is in most cases completely saturated. For example, each possible sequence segment of length 20 is expected to appear about 10^4 times in the library of size 10^{15}. Therefore, since binding sites of transcription factors are typically less than 20 bp long, each possible sequence variant to which this protein can bind will be represented in a large number in the SELEX library. Accordingly, stochastic effects were not included in numerical simulations (41, 14) and theoretical models (8) of SELEX experiments.

Mutations can generally also be neglected in SELEX modeling. The term "evolution" in the name SELEX (Systematic Evolution of Ligands by EXponential enrichment) implies that both selection and mutation are important in the SELEX procedure, and some mutations are necessarily present due to errors in PCR amplification. However, it is not difficult to estimate that this effect can be neglected (4). To observe this, the following estimate is useful. High-fidelity DNA polymerase, which is typically used in SELEX, has a mutation rate of 10^{-4} per cycle per base (42). Furthermore, let us assume that a total of seven SELEX rounds are performed, that there are 10 PCR cycles per round, and that the length of DNA sequences is 25 bp. Under these (typical) SELEX conditions, a DNA sequence selected at the end of the experiment experiences, on average, a total of less than one mutation during the whole experiment (i.e., $10^{-4} \times 25 \times 10 \times 7 < 1$). Consequently, quantitative models of SELEX do not take mutations into account (41, 14, 8).

Finally, considerable mathematical simplifications can be achieved in modeling by noting that the amount of free protein in solution can be in most cases neglected. This is because the amount of DNA used in the experiments is almost always in a large excess over the amount of protein. Due to this most of the protein will end up bounded by DNA, and a very small amount of protein will remain free in solution (8).

3. Computational analysis

An issue to consider in the data analysis is how many sequences have to be extracted from SELEX in order to be able to extract sufficiently accurate protein–DNA interaction parameters. An estimate for this is provided by Roulet et al. (7), who obtained that few thousand sequences are needed to obtain an accurate weight matrix. Another estimate is provided by O'Flanagan et al. (24) who found that one to two sequences per weight matrix parameter are needed, e.g., for a transcription factor with 16 bp long binding site, one needs around 100 binding sequences. Accordingly, Nagaraj et al. (43) reported a reasonably accurate weight matrix for a bacterial transcription factor CRP (16 bp long binding region) with around 70 binding sites extracted in a SELEX procedure. Therefore, while a larger data set is an obvious advantage, it is likely that several hundred binding sites will lead to high-quality protein–DNA interaction parameters in most cases.

Furthermore, a highly non-trivial step of data analysis is to extract actual binding sites from longer sequences obtained through SELEX. As described in **Section 2**, MLSA (multiple local sequence alignment) algorithms are used for this task. However, two difficulties emerge when one does MLSA in analysis of SELEX data. First, due to non-specific binding and background partitioning, some of the selected sequences will not contain transcription factor binding site. This difficulty is not hard to overcome in practice, since most MLSA methods allow that some of the sequences do not contain the shared motif. One should, however, ensure that the noise is limited, which can be achieved by performing sufficient number of SELEX rounds (*see* **Section 2**), so that most of the non-specific binders are eliminated.

The second difficulty is due to a large number of sequences that are typically produced by high-throughput SELEX experiments. That is, one may obtain several thousand DNA sequences from a high-throughput SELEX procedure, and such a large data set is very demanding to align. Indeed, most MLSA implementations have difficulty in producing an accurate alignment for a large number of sequences. However, in the author's experience, an implementation of Gibbs search (The Gibbs Motif Sampler, see also **Chapter 6**) (44) consistently led to reliable results, even for very large data sets from high-throughput experiments.

Finally, the full procedure for determining protein–DNA interaction parameters involves using Fermi–Dirac binding probability and numerically solving a set of mutually coupled nonlinear equations (*see* **Section 2**). While this procedure is technically demanding, the following simplification can be used. One can first start by solving the zero temperature approximation, which leads to computationally much less demanding quadratic programming. The quadratic programming solution can then be improved, by using it as an initial guess for solving the set of equations in the full procedure. Such approach is equivalent to calculating finite temperature corrections to a zero temperature solution in statistical physics.

References

1. Tuerk, C., and Gold, L. (1990) Systematic evolution of ligands by exponential enrichment: RNA ligands to bacteriophage T4 DNA polymerase. *Science* 249, 505–510.
2. Oliphant, A.R., Brandl, C.J., and Struhl, K. (1989) Defining the sequence specificity of DNA-binding proteins by selecting binding sites from random-sequence oligonucleotides: analysis of yeast GCN4 protein. *Mol Cell Biol* 9, 2944–2949.
3. Ellington, A.D., and Szostak, J.W. (1990) In vitro selection of RNA molecules that bind specific ligands. *Nature* 346, 818–822.
4. Djordjevic, M. (2007) SELEX experiments: new prospects, applications and data analysis in inferring regulatory pathways. *Biomol Eng* 24, 179–189.
5. Blackwell, T.K., and Weintraub, H. (1990) Differences and similarities in DNA-binding preferences of MyoD and E2A protein complexes revealed by binding site selection. *Science* 250, 1104–1110.
6. Wright, W.E., Binder, M., and Funk, W. (1991) Cyclic amplification and selection of targets (CASTing) for the myogenin consensus binding site. *Mol Cell Biol* 11, 4104–4110.
7. Roulet, E., Busso, S., Camargo, A.A. et al. (2002) High-throughput SELEX SAGE method for quantitative modeling of transcription factor binding sites. *Nat Biotechnol* 20, 831–835.
8. Djordjevic, M., and Sengupta, A.M. (2006) Quantitative modeling and data analysis of SELEX experiments. *Phys Biol* 3, 13–28.
9. Gold, L. (1995) Oligonucleotides as research, diagnostic, and therapeutic agents. *J Biol Chem* 270, 13581–13584.

10. Jones, S., Daley, D.T., Luscombe, N.M. et al. (2001) Protein-RNA interactions: a structural analysis. *Nucleic Acids Res* 29, 943–954.

11. Gerland, U., Moroz, J.D., and Hwa, T. (2002) Physical constraints and functional characteristics of transcription factor-DNA interaction. *Proc Natl Acad Sci USA* 99, 12015–12020.

12. Magee, J., and Warwicker, J. (2005) Simulation of non-specific protein-mRNA interactions. *Nucleic Acids Res* 33, 6694–6699.

13. Winter, R.B., Berg, O.G., and von Hippel, P.H. (1981) Diffusion-driven mechanisms of protein translocation on nucleic acids. 3. The *Escherichia coli* lac repressor – operator interaction: kinetic measurements and conclusions. *Biochemistry* 20, 6961–6977.

14. Vant-Hull, B., Payano-Baez, A., Davis, R.H. et al. (1998) The mathematics of SELEX against complex targets. *J Mol Biol* 278, 579–597.

15. Stormo, G.D., and Fields, D.S. (1998) Specificity, free energy and information content in protein-DNA interactions. *Trends Biochem Sci* 23, 109–113.

16. Schneider, D., Gold, L., and Platt, T. (1993) Selective enrichment of RNA species for tight binding to *Escherichia coli* rho factor. *FASEB J* 7, 201–207.

17. Stormo, G.D. (2000) DNA binding sites: representation and discovery. *Bioinformatics* 16, 16–23.

18. Fields, D.S., He, Y., Al-Uzri, A.Y. et al. (1997) Quantitative specificity of the Mnt repressor. *J Mol Biol* 271, 178–194.

19. Takeda, Y., Sarai, A., and Rivera, V.M. (1989) Analysis of the sequence-specific interactions between Cro repressor and operator DNA by systematic base substitution experiments. *Proc Natl Acad Sci USA* 86, 439–443.

20. Sarai, A., and Takeda, Y. (1989) Lambda repressor recognizes the approximately 2-fold symmetric half-operator sequences asymmetrically. *Proc Natl Acad Sci USA* 86, 6513–6517.

21. Benos, P.V., Bulyk, M.L., and Stormo, G.D. (2002) Additivity in protein-DNA interactions: how good an approximation is it? *Nucleic Acids Res* 30, 4442–4451.

22. Man, T.K., and Stormo, G.D. (2001) Non-independence of Mnt repressor operator interaction determined by a new quantitative multiple fluorescence relative affinity (QuMFRA) assay. *Nucleic Acids Res* 29, 2471–2478.

23. Bulyk, M.L., Johnson, P.L., and Church, G.M. (2002) Nucleotides of transcription factor binding sites exert interdependent effects on the binding affinities of transcription factors. *Nucleic Acids Res* 30, 1255–1261.

24. O'Flanagan, R.A., Paillard, G., Lavery, R. et al. (2005) Non-additivity in protein-DNA binding. *Bioinformatics* 21, 2254–2263.

25. Djordjevic, M., Sengupta, A.M., and Shraiman, B.I. (2003) A biophysical approach to transcription factor binding site discovery. *Genome Res* 13, 2381–2390.

26. Wingender, E., Chen, X., Hehl, R. et al. (2001) The TRANSFAC system on gene expression regulation. *Nucleic Acids Res* 29, 281–283.

27. Salgado, H., Gama-Castro, S., Martinez-Antonio, A. et al. (2004) RegulonDB (version 4.0): transcriptional regulation, operon organization and growth conditions in *Escherichia coli* K-12. *Nucleic Acids Res* 32, D303–D306.

28. Frech, K., Quandt, K., and Werner, T. (1997) Finding protein-binding sites in DNA sequences: the next generation. *Trends Biochem Sci* 22, 103–104.

29. Robison, K., McGuire, A.M., and Church, G.M. (1998) A comprehensive library of DNA-binding site matrices for 55 proteins applied to the complete *Escherichia coli* K-12 genome. *J Mol Biol* 284, 241–254.

30. Cui, Y., Wang, Q., Stormo, G.D. et al. (1995) A consensus sequence for binding of Lrp to DNA. *J Bacteriol* 177, 4872–4880.

31. Liu, J., and Stormo, G.D. (2005) Combining SELEX with quantitative assays to rapidly obtain accurate models of protein-DNA interactions. *Nucleic Acids Res* 33, e141.

32. Velculescu, V.E., Zhang, L., Vogelstein, B. et al. (1995) Serial analysis of gene expression. *Science* 270, 484–487.

33. Schuster, S. (2008) Next generation sequencing transforms today's biology. *Nat Methods* 5, 16–18.

34. Phillips, C.M., Meng, X., Zhang, L. et al. (2009) Identification of chromosome sequence motifs that mediate meiotic pairing and synapsis in *C. elegans. Nat Cell Biol* 11, 934–942.

35. Jagannathan, V., Roulet, E., Delorenzi, M. et al. (2006) HTPSELEX – a database of high-throughput SELEX libraries for transcription factor binding sites. *Nucleic Acids Res* 34, D90–D94.

36. Ponomarenko, J.V., Orlova, G.V., Ponomarenko, M.P. et al. (2000) SELEX_DB: an activated database on selected randomized DNA/RNA sequences addressed to genomic

sequence annotation. *Nucleic Acids Res* 28, 205–208.

37. Lawrence, C.E., Altschul, S.F., Boguski, M.S. et al. (1993) Detecting subtle sequence signals: a Gibbs sampling strategy for multiple alignment. *Science* 262, 208–214.

38. Bailey, T.L., and Elkan, C. (1994) Fitting a mixture model by expectation maximization to discover motifs in biopolymers. In: *Proceedings of the Second International Conference on Intelligent Systems for Molecular Biology.* AAAI Press, Menlo Park, CA, pp. 28–36.

39. Sengupta, A.M., Djordjevic, M., and Shraiman, B.I. (2002) Specificity and robustness of transcription control networks. *Proc Natl Acad Sci USA* 99, 2072–2077.

40. Fletcher, R. (1987) Practical methods of optimization. Wiley, New York, NY.

41. Irvine, D., Tuerk, C., and Gold, L. (1991) SELEXION. Systematic evolution of ligands by exponential enrichment with integrated optimization by non-linear analysis. *J Mol Biol* 222, 739–761.

42. Eckert, K.A., and Kunkel, T.A. (1990) High fidelity DNA synthesis by the *Thermus aquaticus* DNA polymerase. *Nucleic Acids Res* 18, 3739–3744.

43. Nagaraj, V.H., O'Flanagan, R.A., and Sengupta, A.M. (2008) Better estimation of protein-DNA interaction parameters improve prediction of functional sites. *BMC Biotechnol* 8, 94.

44. Thompson, W., Rouchka, E.C., and Lawrence, C.E. (2003) Gibbs recursive sampler: finding transcription factor binding sites. *Nucleic Acids Res* 31, 3580–3585.

Chapter 13

Kernel-Based Identification of Regulatory Modules

Sebastian J. Schultheiss

Abstract

The challenge of identifying *cis*-regulatory modules (CRMs) is an important milestone for the ultimate goal of understanding transcriptional regulation in eukaryotic cells. It has been approached, among others, by motif-finding algorithms that identify overrepresented motifs in regulatory sequences. These methods succeed in finding single, well-conserved motifs, but fail to identify combinations of degenerate binding sites, like the ones often found in CRMs. We have developed a method that combines the abilities of existing motif finding with the discriminative power of a machine learning technique to model the regulation of genes (Schultheiss et al. (2009) *Bioinformatics* 25, 2126–2133). Our software is called KIRMES, which stands for kernel-based identification of regulatory modules in eukaryotic sequences. Starting from a set of genes thought to be co-regulated, KIRMES can identify the key CRMs responsible for this behavior and can be used to determine for any other gene not included on that list if it is also regulated by the same mechanism. Such gene sets can be derived from microarrays, chromatin immunoprecipitation experiments combined with next-generation sequencing or promoter/whole genome microarrays. The use of an established machine learning method makes the approach fast to use and robust with respect to noise. By providing easily understood visualizations for the results returned, they become interpretable and serve as a starting point for further analysis. Even for complex regulatory relationships, KIRMES can be a helpful tool in directing the design of biological experiments.

Key words: Kernel methods, support vector machines, machine learning, string kernels, regulatory modules, transcription factor binding motifs, eukaryotic gene regulation, motif finding.

1. Introduction

Understanding transcriptional regulation of eukaryotic cells is a very important challenge for computational biology. We present a method called KIRMES that aims at predicting transcription factor target genes based on their regulatory regions. These regions contain binding sites for transcription-regulating proteins, i.e.,

I. Ladunga (ed.), *Computational Biology of Transcription Factor Binding*, Methods in Molecular Biology 674,
DOI 10.1007/978-1-60761-854-6_13, © Springer Science+Business Media, LLC 2010

transcription factors, and are often located immediately upstream of a gene's transcription start site. Often, a binding site is characterized by a conserved motif that is specific to a certain transcription factor, while most of these proteins can recognize several distinct motifs.

Motif-finding approaches typically try to identify motifs based on a sample of regulatory regions that have been selected for a common reaction to external or internal perturbations, e.g., co-occurring expression change, or because of binding signals in chromatin immunoprecipitation experiments. The method proposed here is no exception; sets of conjointly reacting genes are exactly the kind of input data that are expected, alongside a list of genes that are thought not to be regulated by the same mechanism.

We use support vector machines (SVMs), a kernel-based, discriminative, and supervised machine learning method (1, 2). A kernel is a distance measure function that has to fulfill certain mathematical properties and can essentially calculate how similar two input vectors (e.g., sequences) are. Discriminative means the method will return a class label for each gene: in the positive case, whether it belongs to a class of genes that contain similar regulatory elements and is thus regulated by the same (combination of) transcription factors as other members of this class, or not, in the negative case. Supervised means that our newly developed SVM kernel has to be trained on input data for which the correct classification is already known or at least strongly suspected. After training, the kernel can be applied to data where the classification is not yet known. The SVM output will then consist of an assigned class for each input vector: positive if the gene is controlled by the same regulatory mechanism as the input data and negative if not.

In addition to the classification, KIRMES returns the user-specified number of sequence logos of the modules with the highest discriminative power for the positive class. This automatically excludes strong but abundant motifs that occur in both classes and levels background distributions of nucleotides.

A standard method to identify overrepresented oligomers in a sample of co-expressed genes is Gibbs sampling, which tries to capture motifs as position weight matrices (3). While being successful for prokaryotes and even in yeast, this motif-centered approach tends to fail in eukaryotic gene sets, where regulatory regions are much larger, motifs are often degenerate, and a combination of several binding sites is often required for a transcription factor to bind (cooperatively).

We incorporate comparative genomic information from related organisms and model homotypic or heterotypic combinations of binding sites, known as *cis*-regulatory modules (CRMs), for this method (4). CRMs are defined as a set of transcription factor binding sites in a region of up to a few hundred bases in the vicinity of the gene they regulate (5). Due to the size of a genome and the fact that binding patterns are often degenerate, putative sites can be found all over the genome. Biological experiments like chromatin immunoprecipitation show that only a select few of these are actually bound by transcription factors in vivo (6). Since transcription factors often bind cooperatively, a combination of similarly spaced binding sites, even if they are degenerate, is much more improbable to occur by chance than a single binding site. An additional redeeming factor is the conservation of important regulatory elements in related organisms (6). Finding and modeling conserved CRMs thus allows much more accurate predictions.

To capture these modules, we include three types of features: positional data of binding sites relative to each other and to the transcription start or end, sequence, and conservation (**Fig. 13.1**). By using SVM kernels instead of zero-order Markov chains (position weight matrices), we can model higher order sequence information and high-dimensional positional interdependence of motifs.

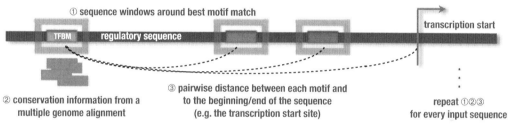

Fig. 13.1. The data used by the regulatory modules kernel: Overrepresented motifs present in a majority of training data are located in all sequences. (1) The best matching location (*highlighted bars*) serves as an anchor point for the sequence window that is excised (*boxed*) and used in the feature vector. (2) Conservation information for this window is retrieved from a previously computed multiple genome alignment. (3) Additionally, the pairwise distance of each window to another and to the start of the sequence is used in the feature vector (*dashed lines*).

To obtain these three types of features from gene sets, we developed the following procedure (*see* also **Fig. 13.2** and the **Section 3**).

First, a third-party motif finder, such as the INCLUSive MotifSampler (7) or PRIORITY (8), can be used to find overrepresented motifs. Alternatively, we implemented a simple oligomer counting algorithm that takes into account repeating nucleotide sequences of length k. Additional parameters include the number of such k-mers to be considered or a threshold on the minimum number of times an oligomer has to occur in the sequences in order to be considered. This simple approach has proven to be rather powerful as it usually returns a larger number of putative binding sites than motif finders, which often return fewer distinct motifs.

The second step begins by combining the three feature types for every motif into input vectors for the regulatory modules

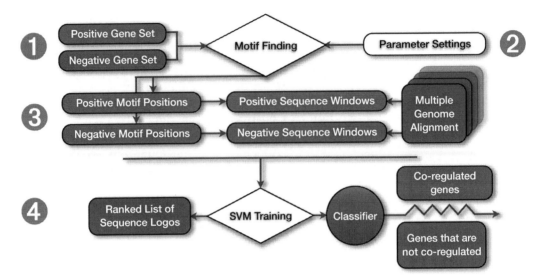

Fig. 13.2. Workflow of the KIRMES method: (1) The first step requires two sets of sequences, with the positives suspected to be co-regulated and the negatives in some way confirmed to be unaffected by the same regulator(s). (2) Parameters can then be adjusted and a round of motif finding begins. (3) The resulting positions, sequence windows, and optionally a multiple genome alignment for conservation information are used to construct feature vectors to train the RM kernel. (4) Along with a trained classifier, a ranked list of sequence logos is returned. The trained classifier then can be used on any other regulatory region to determine if it is regulated by the same mechanism as the ones in the positive training data set.

(RM) kernel, one for each input sequence. The kernel is trained on the positive and negative data sets, determines the most significant modules, and returns them. The trained classifier can then be applied to any gene set from the same organism to predict whether a gene is regulated by the same mechanism as the ones from the positive input set. We define "the same mechanism" as the dominant signature present in the positive training examples. Co-regulation of two genes observed in a small number of experiments may well be the result of a very different combination of transcription factors for each of them. Our approach is able to correctly predict several different CRMs in the positive set as long as they are absent from the negative set.

In our own experiments with the regulatory network of stem cells in *Arabidopsis thaliana*, we were able to show that KIRMES outperforms a Gibbs sampler on its own and even other SVM kernels (9). We used several publicly available knockdown and overexpression microarray experiments of genes involved in the regulation of the organizing center of the shoot apical meristem. Here, the plant maintains stem cells throughout its life and we are interested which transcription factors are involved in keeping some of these cells undifferentiated. A major player is the transcription factor *WUSCHEL*: it is expressed in cells of the organizing center immediately surrounding the stem cells and is critical in keeping them undifferentiated. It also seems to promote the expression of the genes CLAVATA3 and AGAMOUS. From microarray data, we constructed positive and negative gene sets to train KIRMES, in which we considered the regulatory region 1,500 base pairs upstream from the annotated gene start and 500 base pairs downstream from the gene. KIRMES was able to identify a putative binding site for *WUSCHEL* and confirmed it using two independent biological assays: gel shift and SELEX (9). The site is present in almost all genes we had previously suspected as regulatory targets of WUSCHEL. It is a palindromic octamer, suggesting that *WUSCHEL* binds as a homodimer. This is confirmed with the previously mentioned SELEX assay, in which the monomeric *WUSCHEL* protein only binds one half of the sequence. With the central pair of nucleotides very degenerate, the binding site has been missed by conventional motif-finding methods. This illustrates the power of the KIRMES method.

2. Materials

There are several experimental techniques that can yield input data for the proposed method. Essentially all that is needed are the regulatory regions of two sets of genes, a positive and a negative training data set.

Finding positive data sets is straightforward: For microarray experiments, the sequence regions can be selected from anywhere around each gene's locus where regulatory elements are expected, which varies from one organism to another (**Fig. 13.3**). In general, well-designed biological experiments will be the key to obtain meaningful results from the KIRMES method. The set of differentially expressed genes (between experiment and control) can then be selected as the positive training data.

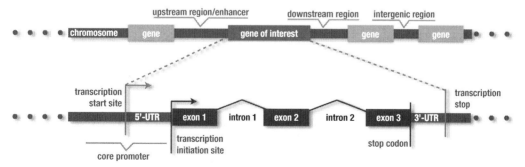

Fig. 13.3. General eukaryotic gene structure. Any part that contains regulatory elements can be used as input data for KIRMES.

Depending on the experimental technique, there are also other ways in which the regulatory region can be selected. For chromatin immunoprecipitation (10), followed either by hybridization to a microarray chip (11) or by ultra-high-throughput sequencing (12), large bound regions that contain the actual binding sites can be determined from the experimental data. These regions can be used directly as the positive training data.

Finding negative training data can be more challenging, because it is often not known with certainty if a particular gene is not regulated by a transcription factor under any circumstances. *See* **Note 1** for more details on selecting training data.

When researching the regulatory network of the response to an external stimulus such as heat stress or drugs, time-series experiments are very helpful. An individual positive data set can be created for each step in the series.

For experimental data from complex diseases, KIRMES will not be able to elucidate the complete regulatory network at once. A stepwise isolation of key players and dominant signatures in the sequences can be performed, ideally followed by another round of carefully designed biological experiments with the newly identified transcription factors and subsequent analysis with KIRMES. In this manner, more complex regulatory mechanisms can be untangled.

3. Methods

KIRMES is written in Python, and its source code has been released under the GNU General Public License. The RM kernel developed for KIRMES has become a part of the large-scale machine learning toolbox SHOGUN (13). A Web service version of KIRMES is available publicly at http://galaxy.fml.mpg.de/, our Galaxy analysis workbench for genomic data. Galaxy is an open-source, scalable workbench for tool and data integration (14). A downloadable version with a command line interface is also available. The following procedure is applicable to both interfaces but assumes a general knowledge of either the command line or a Web service (**Fig. 13.2**).

1. The program expects two FASTA files of sequences to be (up-)loaded for initial training of the classifier and determination of the sequence elements that are most helpful in discriminating between the positive and negative data sets. Sequence conservation information is not supported in the online version, as it would require a considerably larger infrastructure. There is no upper limit to the amount of

input sequences, but every data set should contain at least five sequences for cross-validation to work.

2. Several parameters can be adjusted; these are described in detail in the documentation of the Web and command line versions. Most importantly, the number of motifs to be considered and reported can be selected. Increasing the number of motifs increases processing time.

3. The motifs are returned in a list ranked by their discriminative power and are good starting points for further analysis of downstream regulatory targets or for in vitro binding experiments. The ranking is calculated by performing cross-validation of a classifier trained on the set of all motifs except one. The average difference in prediction accuracy – measured as the area under the receiver operating characteristic curve (15) – versus the accuracy with the complete set of motifs is the basis of the ranking. Motifs are returned as sequence logos for easier interpretation (16). Internally, we use the SVM kernel to calculate a positional oligomer importance matrix, which is described in **Note 2**, from which the sequence logos are derived (17).

4. After training of the classifier, a third data set can be uploaded containing sequences where the classification is not known. KIRMES will predict the class of each sequence based on the presence or absence of CRMs learned from the training data sets.

The underlying machine learning method, SVM, uses a similarity measure known as a kernel function to determine how similar two input vectors are. For KIRMES, we developed a new string kernel, the RM kernel, which is able to use information from sequences of any length, and at the same time incorporate positional and conservation information for the sequence. The RM kernel is based on the weighted degree kernel with shifts (18), with added capabilities to evaluate conservation information per nucleotide. It uses the locations of overrepresented motifs to excise 20 base pair sequence windows from the input data. A set of 20–200 such motifs is generated for any training data set, using either the oligomer counting method or a Gibbs sampler. The best matching position of each motif is determined in every input sequence, allowing for mismatches. This position is the center of the 20 base pair window that is excised and added to the feature vector for this sequence. The SVM can then determine the similarity between any two sequences by calculating the kernel function (which is exactly equivalent to the scalar product) of the two vectors representing the sequences. During training, the SVM adjusts the kernel's weight vector in such a way that it can optimally distinguish between the members of the two classes of input data.

4. Notes

1. Experimental design: For noisy data sources such as expression microarrays, well-designed experiments are key to predict regulation relatively accurately. Ideally, several experimental conditions should be tested with as many replicates as feasible, cf., (19). Time-series experiments make it possible to distinguish first-order responses from downstream reactions to the experimental condition. If some genes are already suspected to be transcription factors and have a large number of genes they regulate, overexpression and knockdown experiments are invaluable. This also applies to chromatin immunoprecipitation techniques, where precipitating a knockdown control can identify promiscuous binding of the antibody, which can be subtracted in a downstream data preparation step.

2. Regulatory region selection: Any region putatively containing transcription factor binding sites can be used for this method (**Fig. 13.3**). This includes the promoter region; the larger enhancer region; any non-coding, untranslated sequence, either upstream or downstream from the exons; the first intron or all introns; and even coding regions. For instance, in our experiments with microarray data from the plant *A. thaliana*, we use 1,500 base pairs upstream from the annotated gene start and another 500 base pairs downstream from the last exon. This will vary for other organisms.

3. Negative training data: To obtain a good negative training data set from microarray experiments, use the same regulatory regions as for the positive data set. Select those genes that exhibit a uniformly high expression level (significantly above the detection threshold of the array) and change little between experiment and control arrays. Reasonable differences within the limits of expected variation of microarrays may be acceptable. This will not exclude one or the other gene that shares a binding site with many of the positive genes, but SVMs are quite robust against mislabeled examples.

   Equally balancing the positive and negative data sets is not necessary; in fact, when the expected distribution of positives and negatives in the prediction data is far from the one in the training data, balancing is counterproductive. The distribution of positives and negatives in the training data should be as similar as possible to the one in the prediction data set used subsequently.

4. Prediction data sets: A data set for prediction with a trained classifier can for instance be comprised of the regulatory

regions of all genes of the organism you work with that have been annotated so far. This can be especially helpful if the microarray chip used for expression experiments is outdated compared to the current genome annotation of that organism. This way, even genes without expression data can be classified. For genes that are consistently expressed below a reliable detection threshold that can thus not be readily included in either the positive or the negative data set, a prediction is possible as well.

5. Contribution of vector features: It is of interest to know which parts of the input vector contribute most to the discriminative power. We used a representative gene set from *A. thaliana* microarray experiments and considered different combinations of the three feature types: sequence windows, conservation, and position. For the complete set of these features, we achieve an area under the receiver operating characteristic curve of 0.89 (1.0 is the maximum). Omitting conservation, performance is reduced slightly to 0.85 and omitting positional information, prediction accuracy is impaired more significantly, at 0.73. Using only the sequence windows, we get an area of only 0.69, which drops even more sharply, to 0.51, when using the positional information only (an area of 0.5 is equivalent to randomly guessing the classification, and thus not better than a random classifier).

   Sequence windows are consistently the most important feature, while their position can sometimes make a big difference, as in the data set discussed here. For other data sets (data not shown), its contribution is marginal. Positional preference of transcription factors – or lack thereof– has been studied previously (20). Conservation typically boosts performance by about 5 percentage points. There is work in progress to include other types of data as features, such as position-specific histone modification or nucleosome positions.

6. Positional oligomer importance matrices (POIMs) (17): POIMs can be calculated from a trained RM kernel. They contain information on which part of the sequence the kernel is used to distinguish between the positive and negative training data. The idea behind this is that these sequence peculiarities are the same ones that transcription factors recognize in vivo. POIMs are difficult to visualize in a meaningful way and thus KIRMES converts them to the more familiar sequence logos. This is not a lossless conversion; a lot of information contained in a POIM cannot be represented in a sequence logo. For instance, the length of the most discriminative sequence cannot be shown. It may usually be seen implicitly by the positions of the logo that are more clearly

defined compared to others with fewer information, but the exact length can only be estimated. Users interested in the actual POIM of the trained kernel can instruct KIRMES to return it in a separate results file.

7. Interpretation of results: Even though SVMs are rather robust when it comes to mislabeled training data, small data sets can still yield misleading results, when many of the genes are mislabeled. Generally, both negatives and positives should contain as many sequences as are available, while remaining as stringent as possible with the criteria for the positive data set. The sequence logos that are returned may not match well in all of the sequences, as can be determined by a run with a program like INCLUSive MotifLocator (2); this is an indicator of mislabeled sequences.

A change in returned motifs from one time point in a series to another is indicative of a downstream reaction; most probably one of the positive genes of the previous time points is a regulator that binds to this new motif.

References

1. Boser, B., Guyon, I., and Vapnik, V. (1992) A training algorithm for optimal margin classifiers. *ACM Press Proceedings COLT' 92* , 144–152.
2. Noble, W.S. (2006) What is a support vector machine? *Nat Biotechnol* 24, 1565–1567.
3. Lawrence, C.E., Altschul, S.F., Boguski, M.S. et al. (1993) Detecting subtle sequence signals: a Gibbs sampling strategy for multiple alignment. *Science* 262, 208–214.
4. Gupta, M., and Liu, J. (2005) De novo *cis*-regulatory module elicitation for eukaryotic genomes. *Proc Natl Acad Sci USA* 102, 7079–7084.
5. Howard, M.L., and Davidson, E.H. (2004) *cis*-Regulatory control circuits in development. *Dev Biol* 271, 109–118.
6. Blanchette M., Bataille, A.R., Chen, X. et al. (2006) Genome-wide computational prediction of transcriptional regulatory modules reveals new insights into human gene expression. *Genome Res* 16, 656–668.
7. Thijs, G., Lescot, M., Marchal, K. et al. (2001) A higher order background model improves the detection of regulatory elements by Gibbs sampling. *Bioinformatics* 17, 1113–1122.
8. Gordân, R., Narlikar, L., and Hartemink, A. (2008) A fast, alignment-free, conservation-based method for transcription factor binding site discovery. *LNCS RECOMB Springer, Heidelberg* 4955, 98–111.
9. Schultheiss, S. J., Busch, W., Lohmann, J. U. et al. (2009) KIRMES: kernel-based identification of regulatory modules in euchromatic sequences. *Bioinformatics* 25, 2126–2133.
10. Das, P.M., Ramachandran, K., van Wert, J., and Singal, R. (2004) Chromatin immunoprecipitation assay. *Biotechniques* 37, 961–969.
11. Buck, M.J., and Lieb, J.D. (2004) ChIP-chip: considerations for the design, analysis, and application of genome-wide chromatin immunoprecipitation experiments. *Genomics* 83, 349–360.
12. Barski, A., and Zhao, K. (2009) Genomic location analysis with ChIP-seq. *J Cell Biochem* 107, 11–18.
13. Sonnenburg, S., Rätsch, G., Schäfer, C., and Schölkopf, B. (2006) Large-scale multiple kernel learning. *J Mach Learn Res* 7, 1531–1565.
14. Giardine, B., Riemer, C., Hardison, R.C. et al. (2005) Galaxy: a platform for interactive large-scale genome analysis. *Genome Res* 15, 1451–1455.
15. Davis, J., and Goadrich, M. (2006) The relationship between precision-recall and ROC curves. *Proceedings ICML* 23, 233–240.
16. Schneider, T.D., and Stephens, R.M. (1990) Sequence logos: a new way to display consensus sequences. *Nucleic Acids Res* 18, 6097–6100.

17. Sonnenburg, S., Zien, A., Philips, P., and Rätsch, G. (2008) POIMs: positional oligomer importance matrices – understanding support vector machine-based signal detectors. *Bioinformatics* 24, i6–i14.

18. Rätsch, G., Sonnenburg, S., and Schölkopf, B. (2005) RASE: recognition of alternatively spliced exons in *C. elegans. Bioinformatics* 21(Suppl. 1), i369–i377.

19. Hughes, T.R., Marton, M.J., Jones, A.R. et al. (2000) Functional discovery via a compendium of expression profiles. *Cell* 102, 109–126.

20. Smith, B., Fang, H., Pan, Y. et al. (2007) Evolution of motif variants and positional bias of the cyclic-AMP response element. *BMC Evol Biol* 7(Suppl. 1), S15.

Chapter 14

Identification of Transcription Factor Binding Sites Derived from Transposable Element Sequences Using ChIP-seq

Andrew B. Conley and I. King Jordan

Abstract

Transposable elements (TEs) form a substantial fraction of the non-coding DNA of many eukaryotic genomes. There are numerous examples of TEs being exapted for regulatory function by the host, many of which were identified through their high conservation. However, given that TEs are often the youngest part of a genome and typically exhibit a high turnover, conservation-based methods will fail to identify lineage- or species-specific exaptations. ChIP-seq has become a very popular and effective method for identifying in vivo DNA–protein interactions, such as those seen at transcription factor binding sites (TFBS), and has been used to show that there are a large number of TE-derived TFBS. Many of these TE-derived TFBS show poor conservation and would go unnoticed using conservation screens. Here, we describe a simple pipeline method for using data generated through ChIP-seq to identify TE-derived TFBS.

Key words: Transposable elements, ChIP-seq, gene regulation, gene expression, transcription factors, CTCF.

1. Introduction

Transposable elements (TEs) are segments of DNA that possess the ability to 'transpose,' meaning that they can move themselves to distant locations of the host genome and replicate when they do so. TEs are present in all domains of life and are abundant in the genomes of many sequenced eukaryotes accounting for a large portion of non-coding DNA and the genomes as a whole (nearly 50%, ~1.4 Gb of the human genome) (1). Broadly speaking, there are two types of TEs. Type I TEs, or retroelements, transpose by a copy and paste mechanism via an RNA intermediate, generating a new insertion. Type II TEs, or DNA transposons,

I. Ladunga (ed.), *Computational Biology of Transcription Factor Binding*, Methods in Molecular Biology 674, DOI 10.1007/978-1-60761-854-6_14, © Springer Science+Business Media, LLC 2010

move by a 'cut-and-paste' mechanism where the actual insertion is moved (2). Most TEs harbor their own promoters and regulatory sequences, and many active elements encode genes for their own transposition. Active elements are a small minority, however, and most TE insertions are unable to transpose.

1.1. Exaptation of Transposable Elements

TEs exist solely to continue their own existence; they do not, simply by their replication, contribute anything to the host (3, 4). It is likely that many, if not the large majority of TE insertions, have little or no functional role for the host and are effectively under neutral or nearly neutral selection. However, given the very large number of TE insertions in eukaryotic genomes and the opportunistic nature of evolution, it is only reasonable to expect that some would be 'exapted' (5) over time to take on a functional role that benefits the host, a process that could have a wide variety of results (6, 7). A key factor in TE exaptation events is their ability to promote their own transcription; without this ability, they could not replicate themselves. Given this ability, it stands to reason that TEs could be exapted to provide alternative promoters for host genes; this has been seen a number of times (8, 9). Of most importance to this chapter, however, is the ability of TEs to provide new TFBS to the host. If there existed an active TE that contained a TFBS, then each new insertion that the TE generated would also contain the TFBS. If the TE were highly active, it could quickly spread the TFBS around the genome. Even if the TE simply had a sequence that was only close to the TFBS, it could still spread this 'progenitor sequence' around the genome. Over time, point mutations in individual insertions could alter the progenitor sequence so that it would now be bound by the TF (10). Either way, the TE could spread the TFBS around the genome over timer and create a network of TFBS, and in doing so alter the expression patterns of host genes. For example, it was recently shown that a large number of human c-myc binding sites are located in TE insertions, possibly creating a sub-network for c-myc control (11). For a comprehensive review of TE-derived regulatory networks, *see* (12).

1.2. Transposable Elements Evolve Rapidly

Transposable elements are generally the most rapidly evolving part of a genome; so long as their insertions are not too deleterious to the host, TEs can quickly increase in copy number and then are generally free to accumulate point mutations. The rapid activity of TEs relative to the host genome means that lineage-specific insertions can be accumulated in a very short time frame. In the 6 million years since the human–chimpanzee divergence, for example, there have been several thousand new TE insertions in each genome (13). There also appears to be very little selective pressure on the deletion of most insertions, which can result in their chance deletion from one lineage, while they are retained in others. Between human and mouse, there is generally very little conservation of non-coding regions in the genome, including TEs. Many insertions that appear to predate the human–mouse divergence are present in one genome, but have been lost in the other (**Fig. 14.1**) (14). The rapid insertion of TEs combined with their rapid loss means that two lineages can develop distinct TE complements in a relatively short time after divergence. Given that two lineages can have very different TE complements, it could be possible for a large number of lineage or even species-specific exaptation events (**Fig. 14.1**). If the exaptation events were the creation of new TFBS or promoters, then the spread of TEs could create species-specific patterns of gene expression (15, 16).

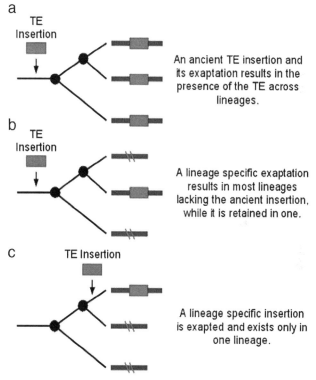

Fig. 14.1. Evolutionary scenarios related to TE exaptation events. **a** An ancient insertion is exapted and the resulting regulatory sequences are shared across multiple derived evolutionary lineages. **b** An ancient insertion is exapted but only selectively conserved in some of the derived evolutionary lineages. This could result in regulatory divergence between lineages. **c** A recent lineage-specific insertion is exapted resulting in regulatory differences between lineages. TEs are particularly prone to this scenario given how dynamic and rapidly evolving they are.

1.3. Detection of Functional TE-Derived Non-coding Sequences

There are three widely used methods to find TFBS in genomes. It should be noted that these approaches are not mutually exclusive; indeed, the methods are often combined to more rigorously predict and locate TFBS.

1.3.1. Phylogenetic Footprinting

The first approach, phylogenetic footprinting (17), can be done solely computationally via comparative sequence analysis. A phylogenetic screen attempts to find regions of different genomes that have been conserved over time and, in the case of TFBS, looking for conserved non-coding elements (CNEs). Screens looking for conserved non-coding elements (CNEs) represent a very successful technique for identifying the oldest and, due to their conservation most likely to be essential, non-coding parts of the genome. Shortly after the sequencing of the human and mouse genomes, it was shown that a larger than expected number of mouse MIR and L2 elements had human orthologs (14). Subsequently, several thousand insertions or insertion fragments near human genes were shown to be under purifying selection, suggesting their exaptation and possible involvement in transcriptional control (18). In recent years, a number of insertions have been shown to be enhancers for human and vertebrate genes, many identified with phylogenetic screens. An insertion from the CORE-SINE family was shown to be conserved across the mammalian lineage and to be an enhancer of the POMC gene in mice (19). The amniote SINE 1, AmnSINE1, family of TEs is a very old family that spread early in the amniote lineage. However, a number of conserved AmnSINE1 insertions exist in the human genome, two of which were shown to be enhancers involved in brain development (20–22). A mammalian interspersed repeat (MIR) was shown to have enhancer 'boosting' activity, in that its presence greatly increased the action of a nearby enhancer, while the MIR could not on its own be an enhancer (23). The problem with an approach based on conservation is that, while it will find many important regions, the screen will miss other regions that are also important, but also lineage specific. Lineage-specific TFBS, such as those that could be provided by lineage-specific TE insertions, could generate lineage-specific expression, and this would be missed by CNE screens (16). Another case in which older elements may be overlooked in CNE screens is one in which an old insertion has been lost, as many are, in several lineages, but exapted in one (**Fig. 14.1**). Such an insertion may well play some role in the lineage that kept it, but it will be completely missed in CNE screens. CNE screens will miss not only new TE exaptations but also other non-coding functional elements. It has been shown previously that sequences with low conservation can play important functional roles, such as rapidly evolving, long non-coding RNAs (24).

1.3.2. Motif Search

The second of the three methods to identify TFBS is also computational and involves scanning a genome for the sequence motif that the TF in question recognizes. REST, the RE1 silencing transcription factor, is known to repress neuronal genes in non-neuronal cells. Using experimentally identified REST

Identification of Transcription Factor Binding Sites   229

binding sites, which contain the RE1 motif, Johnson et al. (25) created a position-specific scoring matrix, PSSM, for the motif and used it to screen for possible REST binding sites in the human genome. Johnson et al. were able to show that there are a number of TE-derived REST binding sites that had the ability to bind REST in vitro, suggesting that TEs have helped to spread the REST network. When a PSSM is used to search for new TFBS in a genome, false positives are controlled by shuffling the sequence in the PSSM, re-scanning the genome with the shuffled sequence, and comparing the number of sites identified with the original PSSM to those found with the shuffled PSSM (26). This approach will not work, however, for TFs that recognize motifs smaller than the RE1 motif as there will likely be many false positives. In addition, the presence of a TFBS sequence motif does not guarantee that the sequence that bears it is actually bound by its corresponding TF, while sequences that lack similarity to the motif may in fact be bound by that factor. These challenges to the sequence-based computational approach necessitate an approach to identifying TFBS on a genome-wide scale that does not depend on the sequence of the TFBS, only the binding of the TF to the region.

1.3.3. ChIP-seq or ChIP-chip

The third major approach to finding TFBS is identifying in vivo protein–DNA interactions via chromatin immunoprecipitation (ChIP) followed by microarray analysis (ChIP-chip) or sequencing of the captured DNA (ChIP-seq, *see* **Chapters 9**, **10**, and **11**). Of the three approaches, this one offers the greatest sensitivity and potential specificity. ChIP is able to find genomic DNA that is bound by a transcription factor, not just those regions that are conserved or for which there exists a well-defined TFBS motif. ChIP is also distinguished from the other approaches in the sense that it identifies sequences that are experimentally characterized to be bound by transcription factors, i.e., not just computational predictions. Genome-wide ChIP assays such as ChIP-PET or ChIP-chip have been used successfully in the past; however, a newer and relatively inexpensive method, ChIP-seq, has quickly become the dominant method of experimentally identifying TFBS, and it is on ChIP-seq that we focus the rest of our discussion. The ChIP-seq method combines ChIP with massively parallel sequencing of the bound DNA (27). The sequencing is usually carried out on one of the currently available short-read sequencers: Illumina Genome Analyzer, ABI SOLiD, or Helicos HeliScope. ChIP-seq has a number of advantages over ChIP-chip and ChIP-PET. There is no cross-hybridization, as can occur in ChIP-chip, and the ChIP-seq signal is a digital count of reads mapping to the TFBS, rather than a fluorescence signal. ChIP-seq is also far less costly than ChIP-PET, which typically relied on capillary sequencing. Using several ChIP-based data sets, including one derived

230   Conley and Jordan

with ChIP-seq, Bourque et al. (28) identified a large number of TE-derived TFBS. The majority of TFBS they observed were not well conserved, with many being lineage specific. This strongly suggests that expansion of TEs within a genome can lead to the concurrent expansion of transcription regulatory networks. Below, we provide a specific example detailing how analysis of ChIP-seq data can be used to identify TE-derived TFBS.

2. Software

All the software we describe and recommend here is publicly available.

Bowtie (29) http://bowtie-bio.sourceforge.net/

MuMRescueLite (30) http://genome.gsc.riken.jp/osc/english/dataresource/

UCSC Genome Browser (31) http://genome.ucsc.edu

UCSC Table Browser (32) http://genome.ucsc.edu

3. Methods

This section describes our choice of tools for the identification of TFBS derived from TE insertions using ChIP-seq data, and we show how these tools can be assembled into an analytical pipeline. The tools presented were chosen for their speed, utility for analysis of TE-derived TFBS, ease of use, and good documentation. To illuminate the use of these tools, we first provide an overview of our analytical pipeline for the detection of TE-derived TFBS (**Fig. 14.2**) and then we give a specific example of how ChIP-seq data can be analyzed to yield genome-wide set of TE-derived TFBS.

3.1. Methods Basics

3.1.1. Mapping

The first step in finding TE-derived TFBS is to map reads generated by ChIP-seq back to the genome used. Massively parallel sequencers generate millions of reads in run of a ChIP-seq experiment. Mapping these reads in a genome as large as the human or mouse genomes with traditional techniques like BLAST (33) or BLAT (34) quickly becomes computationally overly expensive. Fortunately, a number of programs have been developed explicitly for the mapping of short-read data. The fastest of these

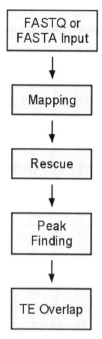

Fig. 14.2. Schematic of the analytical pipeline presented here for finding TE-derived TFBS with ChIP-seq. Each individual step is described in detail in the text along with important caveats, which are listed in 'Notes' section.

are those that employ the Burroughs–Wheeler transform (35) to build a very dense index of the genome, then map reads using the index. We recommend Bowtie for general mapping because of its speed and useful options (*see* **Note 1**). Bowtie is generally the fastest of these aligners, and it can utilize read quality information in the FASTQ format data generated from Illumina sequencing. However, it cannot currently use colorspace reads generated from SOLiD sequencing (*see* **Note 2**).

3.1.2. Read Rescue

Were genomes fully random sequences of the four bases, then almost any ChIP-seq read would be mappable to a unique region of the genome. However, due in large part to the vast number of TE insertions, this is not the case. There are numerous repeated sequences in eukaryotic genomes, and sequence tags derived from these regions may not map unambiguously back to the genome, i.e., they may map to multiple genomic regions with equal probability. The problem of such multiple-mapping ChIP-seq reads arises in part due to their short length. ChIP-seq reads must necessarily be short in order to provide good resolution protein binding locations in the genome; a 500 bp read from ChIP-seq would be easy to unequivocally map to the genome, but would give very little information about the exact location of the DNA–protein interaction. A shorter read, on the order of <50 bp, as most

ChIP-seq data sets contain, gives good resolution regarding the location of the DNA binding, but will have a much greater probability of mapping to multiple locations in the genome. If a TE insertion provides a TFBS, the insertion is very young, and there are many similar TEs in the genome, then it may not be possible to map the ChIP-seq reads from that insertion. For slightly older elements, there will be far fewer possible places to map the reads. Many studies have simply discarded multi-mapping reads for both simplicity of analysis and a desire to be conservative in their findings. However, this becomes an obvious problem when studying TEs, as this will result in the loss of many of the reads coming from TE insertions. To appropriately analyze ChIP-seq data in regard to TEs, some 'rescue' method must be used to resolve reads of the map to multiple locations.

3.1.3. Different Methods of Rescue

There are currently several different schools of thought regarding 'rescuing' reads that map to multiple genomic locations. MAQ (36) is a very commonly used mapping utility for short-read data. When it encounters reads that map to multiple locations with equal probability, it randomly chooses one of the locations to map the tag. This poses problems for TE-derived sequences, as it will dilute the signal from legitimate TFBS, potentially resulting in both false positives and false negatives. This method also ignores information on the local context of potential map positions given by uniquely mapping reads. MUMRescueLite (30, 37) takes this information into account and assumes that multi-mapping reads are more likely to come from regions which already have more uniquely mapping reads and probabilistically determines where a read most likely came from. We recommend that MuMRescueLite be used after the initial mapping to resolve multi-mapping reads.

3.1.4. Peak Calling

Quality mapping is critically important for downstream analysis, and once this has been achieved, the first step is often finding 'peaks' or, more generally speaking, regions that have a density of mapped ChIP-seq reads significantly higher than the background (*see* **Note 3**). These peaks are the regions bound by the TF that are being looked at in the ChIP assay and should contain the TFBS. Methods for peak calling, and indeed the area itself, are still new, and while there is work to be done in the area, there are several quality software choices available for identifying peaks in ChIP-seq data. Quantitative Enrichment of Sequence Tags (QuEST) is reviewed in **Chapter 10** and CisGenome in **Chapter 9**. PeakSeq (38) and SISSRs (39) are two widely used utilities, and in this review, we recommend SISSRs due to its good documentation.

Identification of Transcription Factor Binding Sites   233

3.1.5. Finding TE-Derived TFBS

SISSRs attempts, and in general is highly successful at, finding the TFBS to within a few tens of base pairs based on the strand orientations of reads forming the peak, as well as the density of reads in the region. Ideally, the TFBS would always be at the point of highest read density. In reality, it is very often co-located with the highest density or if not that then very near by, and SISSRs is correct in its predictions the large majority of the time. What this means, practically, is that finding those regions identified by SISSRs that are contained within TEs will tell us which TFBS are TE derived (*see* **Note 4**). This can be accomplished in a number of ways, the simplest being the creation of two BED-formatted custom tracks for the UCSC Genome Browser (31), one from the predicted TFBS and one from the TEs, and uploading them to the browser. Then, the table browser can be used to intersect the tracks (*see* **Note 5**). Below, we provide a specific step-by-step example of how this can be done using the software cited in **Section 2**.

3.2. Example

Here we provide an example using ChIP-seq data for the CCCTC-binding factor (CTCF) from the human ENCODE (ENCyclopedia of DNA Elements) project (40). CTCF is zinc finger binding protein with multiple regulatory functions including both transcriptional activation and repression as well as insulator and enhancer blocking activity (41). The ChiP-seq data for CTCF are available at http://hgdownload.cse.ucsc.edu/goldenPath/hg18/encodeDCC/wgEncodeChromatinMap/. For this example, we will be using the first repetition of CTCF and the control. The majority of the steps in this procedure are done from the command line in the Unix/Linux operating system environment.

3.2.1. Mapping

The program Bowtie requires an index for the genome that the user wishes to map the tags to. This is accomplished with the 'bowtie-build' utility. It takes as input a FASTA file that contains the genome in question, the human genome in our example:

```
$bowtie-build <human genome FASTA> <index name>
```

Building the index typically takes several hours depending on the machine, though once built there is no need to build it again for different samples. Bowtie takes as input a FASTQ file and the parameters to control the mapping (*see* **Note 1**), as well as the index to use for the mapping:

```
$bowtie -q -k 10 -m 10 --best --strata <index name>
<FASTQ> <bowtie output>
```

The mapping should be done for both the CTCF ChIP-seq set and the control set. Bowtie is capable of mapping several thousand reads per second, or far more, depending how many cores it is allowed to use (*see* **Note 1**).

234 Conley and Jordan

3.2.2. Multi-mapping Read Rescue

MuMRescueLite takes all of the information that the Bowtie output has, but the information needs to be rearranged to meet the requirements of MuMRescueLite:

```
$awk '/./ {print $1"\t"$7 + 1"\t"$3"\t"$2"\t"$4"\t"$4 +
length($5)"\t1"}' < <bowtie output> > <MuM Input>
```

While the above command may appear daunting, it is simply using awk to rearrange the columns of the Bowtie output and put tabs between them. MuMRescueLite is invoked with a much simpler command:

```
$MuMRescueLite.py <MuM Input> <MuM Output><Window Size>
```

Keeping the window size small will prevent distant reads from rescuing reads that do not really come from the location. We suggest keeping the window size under 100. MuMRescueLite produces output that is the same as the input, with an additional column that represents the calculated probability that the read in question is from that site. Using the desired probability cutoff for multi-mapping read, use awk to create a BED track from the MuMRescueLite output for analysis with SISSRs:

```
$awk '$8 > <cut off> {print $3"\t"$5"\t"$6"\t"$4}'
< <MuM Output> > <Mapping BED>
```

The output should then be sorted by chromosome, then start, then stop:

```
$sort -k 1,1 -k 2n,2n -k 3n,3n -o <Mapping BED>
<Mapping BED>
```

As with the mapping, the rescue should be done for both sets.

3.2.3. Peak Calling

SISSRs takes as input the two BED files created in the previous step and creates another file with peak calls:

```
$sissrs.pl -i <CTCF File> -b <Control File> -o
<Output File>
```

Use the -i option to specify the ChIP set as the input and the -b option to specify the control set as the background. The -o option tells SISSRs where to write the output. Formatting the output into a BED file will allow overlap of the identified TFBS with TEs in the UCSC genome browser:

```
$awk '/^chr/ {print 1,2,3}' < <Output File> >
<TFBS BED>
```

3.2.4. Identification of TE-Derived TFBS

The final step is to upload the SISSRs-identified TFBS, BED-formatted track to the UCSC genome browser as a custom track. The name of the track should be changed so as not to be overwritten by later tracks. Once that is done, create another custom track that will contain only TEs using the table browser.

This can be done by filtering the RepeatMasker track for only those repeats which have a 'repClass' of 'LINE,' 'SINE,' 'LTR,' or 'DNA.' Intersecting the track of CTCF TFBS with this TE-only track will give those TFBS that reside in TE insertions. If everything has gone right, then there should be examples like that shown in **Fig. 14.3**. Here, two distinct CTCF binding sites are shown for a solo long terminal repeat sequence from the endogenous retrovirus family K (ERVK). Although these particular binding sites were identified solely based on ChIP-seq data, they can also be seen to possess known CTCF binding site sequence motifs at the bound genomic intervals. Thus, a computational survey of TE sequences that possess TFBS motifs may have turned up this example.

Genome wide there are 326 CTCF-bound sites located within ERVK sequences, and ERVK elements show more than an order of magnitude greater likelihood to be bound by CTCF than members of other ERV families. The number of CTCF-bound ERVK sequences suggests that these TE-derived TFBS may play some role in regulating human genes, and in fact many ERVs

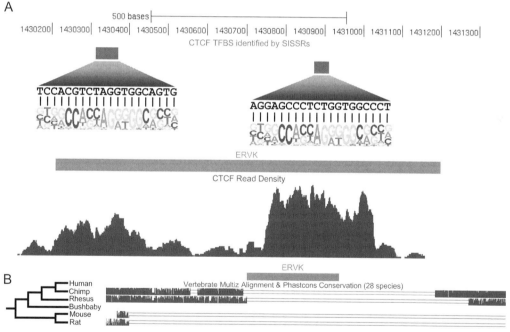

Fig. 14.3. An example of two TE-derived CTCF binding sites found using ChIP-seq data. **a** Two CTCF TFBS identified by the SISSRs program are found within the long terminal repeat sequence of an endogenous retrovirus TE (ERVK). The ChIP-seq read density shows two peaks in the ERVK that correspond to the CTCF-bound regions. Analysis of the bound regions with a CTCF position weight matrix (PWM) (45) using the program CLOVER (46) confirms the presence of two conserved CTCF binding site sequence motifs in the regions identified with the ChIP-seq data. The sequences of the binding sites are shown compared to the sequence logo representing position-specific variation in the CTCF PWM. **b** Regions orthologous to the ERVK insertion site from completely sequenced mammalian genomes were compared using the vertebrate Multiz alignment. Sequence regions conserved between species are shown. Regions flanking the ERVK element are conserved in other mammalian genomes, but the insertion itself is human specific.

are located in close proximity to genes. For instance, the CTCF-bound ERVK shown in **Fig. 14.3** is located in the 5′ regulatory region ~6 kb upstream of the ATAD3A gene.

ERV sequences in general and members of the ERVK family in particular are young lineage-specific elements that are poorly conserved across species. Phylogenetic analyses revealed that the ERVK family invaded the primate lineage subsequent to the diversification between New World and Old World monkeys (42). Consistent with their recent evolutionary origin in the human genome, ERVK sequences have a mean PhyloP (http://www.genome.ucsc.edu/cgi-bin/hgTrackUi?hgsid=147315896&c=chr1&g=phyloPCons28way) base-wise conservation score of 0.22, while the genome as a whole has a mean score of 0.47. Therefore, phylogenetic footprinting approaches, which identify regulatory sequences in non-coding DNA by virtue of their sequence conservation, would be exceedingly unlikely to turn up any cases of ERVK-derived TFBS. Indeed, comparison of the CTCF-bound ERVK insertion shown in **Fig. 14.3** with orthologous mammalian genomic regions indicates that this particular ERVK insertion is human specific and missing in all other mammals. Such lineage-specific TE-derived regulatory sequences may be of particular interest in the sense that they could be responsible for driving regulatory divergence between species (15, 16).

4. Notes

1. Bowtie is currently the fastest short-read aligner available and our preference for mapping short-read data, such as that generated by ChIP-seq or RNA-seq. It has many of the same advantages of MAQ, such as taking quality information into account, but also has other features useful for looking at TE-derived sequences that MAQ currently lacks. Bowtie is also quite memory efficient and it scales well with genome size. Bowtie can be run with the human genome on a computer with 4 GB of RAM, though on such a computer nothing else should be started in the meantime, as when Bowtie is forced out of memory it tends not to recover. Bowtie has a large number of options for controlling mapping and output, which can be listed by executing bowtie with no arguments. The more important options are listed and explained here:

   -k <integer> this option is critically important among those available. This option tells bowtie that it should report more than one mapping, as by default it reports only the first. At the current time, MAQ will not report more than one mapping. Currently, MAQ will use the quality scores to choose a location and assign the mapping a quality of 0. Output of multi-mapping reads and their possible location is essential for the rescue and analysis of TE-derived sequences.

--best giving this option will cause bowtie to report only those mappings which have the highest quality and is recommended if you have the FASTQ data and not just the FASTA data of base calls. This can greatly reduce the number of multi-mapping reads.

--strata This option is used along with the --best option and will cause bowtie to return only the highest quality mappings.

-m <integer> will eliminate reads that map more than m times. We suggest making it the same as k. This will remove reads that map to so many places in the genome that they could likely never be placed with confidence.

One major advantage of Bowtie is that it allows for the easy use of multiple cores, which every desktop shipped in the last ~3 years has. Speed will become increasingly important as the number of reads generated per run increases. On a dual-core machine, such as a machine with an Intel Core Duo, only one core is advisable. However, on a quad-core machine, it is generally advisable to use two or three cores. On an eight-core machine six cores are recommended. The number of cores (processors) is set with the -p option. In some unfortunate cases, FASTQ files from a ChIP-seq experiment are not available, and only the base calls are supplied. In this case, you would not supply the '-q' flag to indicate FASTQ format. It is in these cases that the rescue is especially important.

2. The ABI SOLiD sequencing platform does not produce base calls like the Illumina platform, but rather 'color' calls that represent transitions between two bases. Bowtie cannot currently map colorspace reads, and we suggest the SOCS program for this purpose (43). Like Bowtie, it has generally low memory requirements and is also capable of using multiple cores when available.

3. Though many peaks from ChIP-seq data will be quite large and obvious, others may be closer to the background noise. Complicating this is that the background in ChIP-seq is non-random and tends to form peaks of its own. Most peak-finding utilities will look for peaks with just the ChIP-seq data alone, but many also allow the use of both the ChIP-seq data and a control set. By comparing the control set and the experimental set, false positives that result from peaks not related to the ChIP can be removed.

4. While SISSRs and other peak finders do a very good job of finding the actual TFBS from ChIP-seq data, they may still be off on occasion. A more accurate way to find the exact TFBS is to scan the identified TFBS, along with their flanks, with a PSSM for the TFBS motif with a program such as MAST (44). This will give the exact location of the TFBS if it exists in the peak region.

5. In this chapter, we suggest using the UCSC Genome Browser and table browser for the overlap of the identified TFBS and transposable elements. This is very simple to do, but requires loading BED-formatted tracks to the browser and (relatively) lots of manual work. 'Kent Source Tree' is a large series of utilities, many of which form the back end of the browser. One such utility, 'bedOverlap,' will overlap two sets of tracks without having to upload them to the browser. Numerous other useful utilities include the 'bedItemOverlapCount' utility that can produce custom 'wiggle' tracks for the UCSC Genome Browser, which visualize the density of ChIP-seq reads, and hence protein binding intensity, along the genome. Compilation and installation of the Kent Source Tree is not always easy, but is recommended if possible.

Acknowledgments

King Jordan was supported by an Alfred P. Sloan Research Fellowship in Computational and Evolutionary Molecular Biology (BR-4839). King Jordan and Andrew B. Conley were supported by NIH HG000783 granted to Mark Borodovsky. The authors would like to thank Jianrong Wang for help with the CTCF analysis.

References

1. Lander, E.S., Linton, L.M., Birren, B., et al. (2001) Initial sequencing and analysis of the human genome. *Nature* 409, 860–921.
2. Wicker, T., Sabot, F., Hua-Van, A., et al. (2007) A unified classification system for eukaryotic transposable elements. *Nat Rev Genet* 8, 973–982.
3. Doolittle, W.F., and Sapienza, C. (1980) Selfish genes, the phenotype paradigm and genome evolution. *Nature* 284, 601–603.
4. Orgel, L.E., and Crick, F.H. (1980) Selfish DNA: the ultimate parasite. *Nature* 284, 604–607.

5. Gould, S.J., and Vrba, E.S. (1982) Exaptation; a missing term in the science of form *Paleobiology* 8, 4–15.
6. Jordan, I.K. (2006) Evolutionary tinkering with transposable elements. *Proc Natl Acad Sci USA* 103, 7941–7942.
7. Kidwell, M.G., and Lisch, D.R. (2001) Perspective: transposable elements, parasitic DNA, and genome evolution. *Evolution* 55, 1–24.
8. Cohen, C.J., Lock, W.M., and Mager, D.L. (2009) Endogenous retroviral LTRs as promoters for human genes: a critical assessment. *Gene* 448, 105–114.

9. Conley, A.B., Piriyapongsa, J., and Jordan, I.K. (2008) Retroviral promoters in the human genome. *Bioinformatics* 24, 1563–1567.

10. Zemojtel, T., Kielbasa, S.M., Arndt, P.F. et al. (2009) Methylation and deamination of CpGs generate p53-binding sites on a genomic scale. *Trends Genet* 25, 63–66.

11. Wang, J., Bowen, N.J., Chang, L. et al. (2009) A c-Myc regulatory subnetwork from human transposable element sequences. *Mol Biosyst* 5, 1831–1839.

12. Feschotte, C. (2008) Transposable elements and the evolution of regulatory networks. *Nat Rev Genet* 9, 397–405.

13. Chimpanzee Sequencing and Analysis Consortium (2005) Initial sequence of the chimpanzee genome and comparison with the human genome. *Nature* 437, 69–87.

14. Silva, J.C., Shabalina, S.A., Harris, D.G. et al. (2003) Conserved fragments of transposable elements in intergenic regions: evidence for widespread recruitment of MIR- and L2-derived sequences within the mouse and human genomes. *Genet Res* 82, 1–18.

15. Marino-Ramirez, L., and Jordan, I.K. (2006) Transposable element derived DNaseI-hypersensitive sites in the human genome. *Biol Direct* 1, 20.

16. Marino-Ramirez, L., Lewis, K.C., Landsman, D. et al. (2005) Transposable elements donate lineage-specific regulatory sequences to host genomes. *Cytogenet Genome Res* 110, 333–341.

17. Zhang, Z., and Gerstein, M. (2003) Of mice and men: phylogenetic footprinting aids the discovery of regulatory elements. *J Biol* 2, 11.

18. Lowe, C.B., Bejerano, G., and Haussler, D. (2007) Thousands of human mobile element fragments undergo strong purifying selection near developmental genes. *Proc Natl Acad Sci USA* 104, 8005–8010.

19. Santangelo, A.M., de Souza, F.S., Franchini, L.F. et al. (2007) Ancient exaptation of a CORE-SINE retroposon into a highly conserved mammalian neuronal enhancer of the proopiomelanocortin gene. *PLoS Genet* 3, 1813–1826.

20. Nishihara, H., Smit, A.F., and Okada, N. (2006) Functional noncoding sequences derived from SINEs in the mammalian genome. *Genome Res* 16, 864–874.

21. Sasaki, T., Nishihara, H., Hirakawa, M. et al. (2008) Possible involvement of SINEs in mammalian-specific brain formation. *Proc Natl Acad Sci USA* 105, 4220–4225.

22. Hirakawa, M., Nishihara, H., Kanehisa, M. et al. (2009) Characterization and evolutionary landscape of AmnSINE1 in Amniota genomes. *Gene* 441, 100–110.

23. Smith, A.M., Sanchez, M.J., Follows, G.A. et al. (2008) A novel mode of enhancer evolution: the Tal1 stem cell enhancer recruited a MIR element to specifically boost its activity. *Genome Res* 18, 1422–1432.

24. Pang, K.C., Frith, M.C., and Mattick, J.S. (2006) Rapid evolution of noncoding RNAs: lack of conservation does not mean lack of function. *Trends Genet* 22, 1–5.

25. Johnson, R., Gamblin, R.J., Ooi, L. et al. (2006) Identification of the REST regulon reveals extensive transposable element-mediated binding site duplication. *Nucleic Acids Res* 34, 3862–3877.

26. Thornburg, B.G., Gotea, V., and Makalowski, W. (2006) Transposable elements as a significant source of transcription regulating signals. *Gene* 365, 104–110.

27. Johnson, D.S., Mortazavi, A., Myers, R.M. et al. (2007) Genome-wide mapping of in vivo protein-DNA interactions. *Science* 316, 1497–1502.

28. Bourque, G., Leong, B., Vega, V.B. et al. (2008) Evolution of the mammalian transcription factor binding repertoire via transposable elements. *Genome Res* 18, 1752–1762.

29. Langmead, B., Trapnell, C., Pop, M. et al. (2009) Ultrafast and memory-efficient alignment of short DNA sequences to the human genome. *Genome Biol* 10, R25.

30. Hashimoto, T., de Hoon, M.J., Grimmond, S.M. et al. (2009) Probabilistic resolution of multi-mapping reads in massively parallel sequencing data using MuMRescueLite. *Bioinformatics* 25, 2613–2614.

31. Kuhn, R.M., Karolchik, D., Zweig, A.S. et al. (2009) The UCSC genome browser database: update 2009. *Nucleic Acids Res* 37, D755–D761.

32. Karolchik, D., Hinrichs, A.S., Furey, T.S. et al. (2004) The UCSC table browser data retrieval tool. *Nucleic Acids Res* 32, D493–D496.

33. Altschul, S.F., Madden, T.L., Schaffer, A.A. et al. (1997) Gapped BLAST and PSI-BLAST: a new generation of protein database search programs. *Nucleic Acids Res* 25, 3389–3402.

34. Kent, W.J. (2002) BLAT – the BLAT-like alignment tool. *Genome Res* 12, 656–664.

35. Burrows, M., and Wheeler, D.J. (1994) A block-sorting lossless data compression algorithm. *Digital Systems Research Center.*

36. Li, H., Ruan, J., and Durbin, R. (2008) Mapping short DNA sequencing reads and

calling variants using mapping quality scores. *Genome Res* 18, 1851–1858.

37. Faulkner, G.J., Forrest, A.R., Chalk, A.M. et al. (2008) A rescue strategy for multimapping short sequence tags refines surveys of transcriptional activity by CAGE. *Genomics* 91, 281–288.

38. Rozowsky, J., Euskirchen, G., Auerbach, R.K. et al. (2009) PeakSeq enables systematic scoring of ChIP-seq experiments relative to controls. *Nat Biotechnol* 27, 66–75.

39. Jothi, R., Cuddapah, S., Barski, A. et al. (2008) Genome-wide identification of in vivo protein-DNA binding sites from ChIP-Seq data. *Nucleic Acids Res* 36, 5221–5231.

40. Birney, E., Stamatoyannopoulos, J.A., Dutta, A. et al. (2007) Identification and analysis of functional elements in 1% of the human genome by the ENCODE pilot project. *Nature* 447, 799–816.

41. Gaszner, M., and Felsenfeld, G. (2006) Insulators: exploiting transcriptional and epigenetic mechanisms. *Nat Rev Genet* 7, 703–713.

42. Sverdlov, E.D. (2000) Retroviruses and primate evolution. *Bioessays* 22, 161–171.

43. Ondov, B.D., Varadarajan, A., Passalacqua, K.D. et al. (2008) Efficient mapping of Applied Biosystems SOLiD sequence data to a reference genome for functional genomic applications. *Bioinformatics* 24, 2776–2777.

44. Bailey, T.L., and Gribskov, M. (1998) Combining evidence using p-values: application to sequence homology searches. *Bioinformatics* 14, 48–54.

45. Kim, T.H., Abdullaev, Z.K., Smith, A.D. et al. (2007) Analysis of the vertebrate insulator protein CTCF-binding sites in the human genome. *Cell* 128, 1231–1245.

46. Frith, M.C., Fu, Y., Yu, L. et al. (2004) Detection of functional DNA motifs via statistical over-representation. *Nucleic Acids Res* 32, 1372–1381.

Chapter 15

Target Gene Identification via Nuclear Receptor Binding Site Prediction

Gabor Varga

Abstract

In spite of numerous advances in recent years, the complete list of direct target genes for nuclear receptors remains elusive. The integrated application of new computational and experimental methods reviewed in this chapter provides insight into the complex network of regulatory pathways mediated by nuclear receptors which is expected to improve the understanding of the physiology and the pathology of metabolism, development, homeostasis, and other fundamental processes.

Key words: Predictive modeling, nuclear receptor binding site, target gene identification, integrative informatics, liver X receptor, ChIP-chip, composite binding element, cooperative binding.

1. Introduction

Nuclear receptors form a superfamily of proteins which act as sensors of hormone, vitamin, or cholesterol levels and other important molecular signals (1). They play key roles in the transcriptional regulation of many gene pathways implicated in prevalent diseases. In addition, some nuclear receptors can respond to elevated levels of xenobiotics, such as pollutants and foreign chemicals, and mediate the response via increasing the transcription of detoxifying enzymes. The cognate gene programs regulated by nuclear receptors affect virtually all aspects of cellular processes, including embryogenesis, homeostasis, reproduction, cell growth, and death (1). This makes them appealing targets for drug discovery. Thirteen percent of FDA-approved drugs interact with nuclear receptors (2), making them the second most targeted

I. Ladunga (ed.), *Computational Biology of Transcription Factor Binding*, Methods in Molecular Biology 674, DOI 10.1007/978-1-60761-854-6_15, © Springer Science+Business Media, LLC 2010

class of proteins. Examples of widely used therapeutics interacting with nuclear receptors include rosiglitazone for type II diabetes, cortisol for topical inflammation, the oncology drugs tamoxifen and raloxifene for osteoporosis.

Nuclear receptors are classified as transcription factors because they have the ability to directly bind DNA, although some receptors can also exert their activity via tethering to other nuclear receptors. They are multidomain proteins which often form homo or heterodimers. The N-terminal domain of nuclear receptors is the least conserved evolutionarily. This region is subject to alternative splicing and differential promoter usage. The majority of known nuclear receptor isoforms differ in this region. The C-terminal domain acts as the ligand binding (often referred to as hormone binding) domain. The N- and C-terminal domains have been described for several nuclear receptors as transcription activation domains based on structure–function studies (1). The DNA binding domain is located in between the above two domains and confers sequence-specific recognition of DNA binding sites (also referred to as response elements or REs). X-ray and NMR experiments have shown that this region contains four zinc-finger motifs and helical regions which are involved in RE selection. Downstream of the DNA binding domain is the variable length hinge region which is thought to allow the more conserved regions to adopt multiple conformations without creating steric hindrance problems. For some receptors the hinge region also contains elements of the nuclear compartmentalization signal.

1.1. Computational Models

The computational models constructed for target gene identification make use of the known transactivation mechanisms of nuclear receptors. The transactivation involves the transcription factor protein binding to an RE in the promoter or enhancer region of the target genes. The binding event triggers the recruitment of co-activator proteins and initiates the transcription of the genes (1).

Structures of the RE sequence motifs share common characteristics across the four types of nuclear receptors. Steroid receptors (i.e., estrogen, androgen, and glucocorticoid receptors) form homodimers (3, 4). Dimeric orphan receptors [i.e., hepatocyte nuclear factor 4 (HNF-4)] also form homodimers. The third type of nuclear receptors [i.e., liver X receptor (LXR) or vitamin D receptor (VDR) or peroxisome proliferator-activated receptor gamma (PPARG)] can form heterodimers with the RXR protein, but in some instances can form homodimers with different binding specificity (5). The dimers then bind to RE sites consisting of two half-sites separated by a variable length spacer DNA (typically one to four nucleotides long DNA with low levels of sequence conservation). The core recognition sequence of the half-site can be repeated in three different ways to form either a *direct repeat* (DR, the orientation of the two half-sites is the same), a *palindromic inverted repeat* (IR, the orientation of the second half-site is reversed), or an *everted repeat* (ER, the orientation of the first half-site is reversed). The fourth type of nuclear receptors, the monomeric/tethered orphan receptors [i.e., RAR-related orphan receptor alpha (RORA) and Rev-Erb], also form heterodimers with the RXR protein, but bind to an RE that consists of only a half-site.

2. Methods

2.1. Generalized Modeling Approaches for Nuclear Hormone Receptor Binding Site Recognition

The half-site sequence for the nuclear receptor REs is often represented as a hexamer with a canonical sequence of 5'-AG(G|T)TCA-3'. The variations in the half-site sequence, their relative orientation, and the length of the spacer sequence allow for a wide range of RE configurations. This enables the different nuclear receptor dimers to bind with different specificities and to modulate the transcription of a unique set of target genes. However, when multiple nuclear receptors have affinity for an RE, receptors compete for the binding site creating cross talk between signaling pathways.

The simplest method for nuclear receptor binding site recognition involves the matching of the canonical sequence (in terms of regular expressions or the IUPAC ambiguous nucleotide code) to the DNA sequence being examined. However, binding site sequences which differ from the canonical sequence are not identified by this approach as only a match/no match decision can be made at each position. Consequently, such regular expression-type pattern matching methods produce low selectivity and are not used for the genome scale analysis of binding sites.

Multiple approaches have been developed to improve the modeling and the detection of nuclear receptor binding sites in genome sequences. One popular approach is to represent the nucleotide preferences, in a given binding motif, using a position weight matrix (PWM) (6, 7) profile where the profile is derived from a set of nucleotide sequences experimentally demonstrated to bind the transcription factor (**Fig. 15.1**).

When multiple sites can be derived from the experimental binding data then multiple profiles can be defined to represent them. For example, the MatInspector database (7) contains two separate PWM profiles for PPARG, the Pal3 motif (same as the Pal3 motif in the JASPAR database that was referenced earlier) and the DR1 (direct repeat motif with a single nucleotide spacer) binding site, both of which were previously confirmed using gel

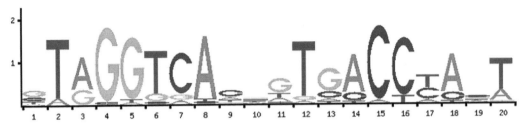

Fig. 15.1. Sequence logo representation of the PPARG nuclear receptor profile (MA0066.1) from the JASPAR database (6). The profile is based on the Pal3 palindromic binding motif for the PPARG homodimer (5) consisting of two half-sites (5′-AGGTCA-3′ and 5′-TGACCT-3′) separated by a three-nucleotide spacer with low level of sequence conservation.

shift assay experiments (5). As new experimental data become available, the PWM profile can be easily extended. For instance, the PPARG DR1 motif published by Okuno et al. (5) was based on an alignment of 24 binding sites. Lemay et al. (8) later refined the profile using an alignment of 73 confirmed sites.

JASPAR and MatInspector represent PWM profile-based approaches that provide a rich resource for covering a broad range of protein–DNA binding interactions, but are not optimized for the sensitive and selective recognition of the unique characteristics of nuclear receptors. For example, the weight matrix models are not able to represent the variable spacing and inversions of half-sites. The NUBIScan algorithm (9) provides improved specificity for the recognition of nuclear receptor binding sites compared to the PWM profile-based method by utilizing the *two hexamer half-site separated by spacer* structure. NUBIScan uses weighted nucleotide distribution matrices to recognize and score single half-sites in the DNA sequence of interest. The product of the individual matrix scores gives the overall score for the desired arrangement and spacing of the half-sites.

Moehren et al. (10) identified selective androgen receptor elements (sAREs) using an extended version of the NUBIScan algorithm (the new features are available now as options when running the program). The first extension allows the definition of different positional weight matrices for each of the two hexamer half-sites. The second extension introduces the notion of scaled positional weight factor to enable the highly conserved positions to contribute more to the score than the variable positions. Androgen receptor (AR) is known to interact with two different sets of AREs. One set known as the classical steroid-hormone-response elements (cAREs) are recognized by the other receptors that belong to the same subgroup of nuclear receptor superfamily. AR also interacts with a second set of REs, referred to as sAREs that are three-nucleotide-spaced partial direct repeats of the 5′-TGTTCT-3′ monomer binding element. Other receptors that belong to the same subgroup of nuclear receptors, such as the glucocorticoid receptor (GR), do not bind these elements.

The extended version of the NUBIScan algorithm was configured using the sARE binding site information and applied to screen the regulatory regions of 85 known human androgen-responsive genes. The most promising hits were followed up by in vitro DNA binding (band-shift) assay experiment. The result of the combined in silico prediction and experimental validation was that the elements found in two genes, the aquaporin-5 and the Rad9 genes, showed selective AR versus GR binding. This conferred the ability of the extended NUBIScan algorithm to identify REs which mediate androgen, but not glucocorticoid, responsiveness in reporter gene expression.

NUBIScan utilizes a single half-site model using weight matrix methodology and evaluates pairs of half-sites with a preselected configuration (for example, DR1) that must be preselected by the user. NUBIScan reports predictions which exceed a Z-score threshold, where the Z-scores are derived from the distribution of all scores generated from the input sequence. This makes the predictions for a given threshold dependent on the length and composition of the input sequence (11). An alternative approach was introduced by the NHR-scan model (11). NHR-scan uses a hidden Markov model (HMM) framework to recognize DR, IR, and ER elements. The model is capable of high-sensitivity recognition of DNA binding sites while maintaining the selectivity comparable to weight matrix profile approaches (11).

2.2. Application of a Specialized Binding Site Model to Liver X Receptor Target Gene Discovery

While both NUBIScan and NHR-scan improve the sensitivity and maintain the selectivity of identifying the presence of an RE compared to earlier models, they cannot distinguish between nuclear receptor subtypes or identify which receptor is most likely to bind to the predicted site. The extended version of NUBIScan takes an important step in this direction by allowing the definition of different positional weight matrices for each of the two hexamer half-sites. To combine the strengths and to improve upon the shortcomings of previously reported models, Varga et al. (12) developed a library of hidden Markov models as part of the LXRE.HMM algorithm with special focus on recognizing the REs of a single nuclear receptor, the liver X receptor (LXR).

LXR is a steroid nuclear hormone receptor that plays a pivotal role in the regulation of fatty acid, cholesterol, and glucose metabolism. Studies identifying LXR target genes gave insights into the regulatory pathways affecting metabolic diseases, such as diabetes and atherosclerosis. The canonical consensus sequence for the LXR response element (LXRE) half-sites is 5′-AGGTCA-3′. However, due to the diversity of the LXRE sequences, neither the pattern matching approach using the canonical sequence nor the available LXRE weight matrix model in the MatInspector tool is able to recognize experimentally verified LXRE sites with accuracy. These findings exacerbated the need for the development of an improved LXRE model (12).

As the first step toward building an improved model, a set of 35 non-redundant LXREs were examined. The analysis revealed three LXRE types: DR with a single-nucleotide spacer (DR1), DR with a four-nucleotide spacer (DR4), and IR with a single-nucleotide spacer (IR1). Subsequent phylogenetic analysis of the 28 DR4 LXREs showed five distinct DR4 subtypes and four singleton groups based on the evolutionary distance. To enable the classification of LXREs into types and subtypes, a total of 11 models were built forming a library of models available via the LXRE.HMM algorithm.

The predictive accuracy of LXRE.HMM was measured using cross-validation, while selectivity was evaluated against the EPD vertebrate promoter database (13). Using the optimal cutoff score, the sensitivity of the predictions was 96.7% and the selectivity was 93.8%. The LXRE.HMM method markedly improved the sensitivity of identifying LXR binding sites compared to previously available tools while maintaining a high level of selectivity. These results enabled the in silico screening of large data sets for LXR target gene identification.

2.3. Functional Characterization of Binding Events

Because of the importance of known LXR target genes in metabolic diseases, it is of great value to identify novel LXR target genes that can be used as therapeutic targets or disease biomarkers to support personalized medicine approaches. Stayrook et al. (14) applied LXRE.HMM predictions in conjunction with an LXR alpha ChIP-chip (chromatin immunoprecipitation followed by promoter or genome-wide tiling microarray) experiment to identify a novel LXR target. First, regions of the genome with significant LXR occupancy sites, as determined by chromatin immunoprecipitation followed by hybridization to genome-wide microarray chips (ChIP-chip, *see* **Chapters 9, 10,** and **11**) active regions, were screened using the LXRE.HMM method. The active regions then were mapped to the human genome to identify genes in 1 kb proximity as potential LXRalpha target genes. The screen identified a putative binding site in the promoter region of the human 3alpha-hydroxy steroid dehydrogenase (AKR1C4) gene for the LXRalpha isoform, which is the physiological receptor for oxidized cholesterol metabolites (oxysterols) (15). Next, follow-up experiments showed that LXRalpha binds this LXRE and increases the transcription and protein expression of AKR1C4. Since AKR1C4 has a key role in bile acid synthesis (14), the results suggest that LXRalpha may modulate the bile acid biosynthetic pathway, and other AKR1C4 pathways, such as the metabolism of steroid hormones and xenobiotics (14).

Wang et al. (15) compared the gene list obtained by the LXRE.HMM screen of the occupancy regions from the LXRalpha ChIP-chip experiment, as described previously, with microarray data obtained from mice liver treated with LXR-directed antisense oligonucleotide. This allowed the identification of genes with LXRalpha occupancy, at least one predicted LXRE and with significantly altered liver gene expression in the LXR-directed antisense oligonucleotide knockdown experiment. Two of these genes, squalene synthase (FDFT1) and lanosterol 14-alpha-demethylase (CYP51A1), showed reduced gene expression in liver. This led to the novel finding that LXRalpha can act as a gene silencer. Furthermore, both genes encode for key proteins in the cholesterol biosynthesis pathway (15). Since oxysterols, the endogenous ligands of LXRalpha, are the end products of cholesterol biosynthesis, a negative feedback regulation of cholesterol biosynthesis exists and it is mediated by LXRalpha. Taken the results together, the identification of novel LXREs combined with gene expression data led to intriguing findings which indicate that LXRalpha plays an important role in end product inhibition of cholesterol biosynthesis, in addition to its role in cholesterol reverse transport and elimination.

Further analysis of the data set obtained via the LXRE.HMM screen of occupancy sites from the LXRalpha ChIP-chip experiment revealed a putative LXRE in the selective Alzheimer's disease indicator-1 (Seladin-1/DHCR24) gene. Wang et al. (16) characterized the novel LXRE in the second intron of the gene and found that it confers LXR-specific ligand responsiveness in a reporter gene assay. This finding, taken together with results from follow-up experiments, suggests that Seladin-1 is an LXR direct target gene and that LXR may regulate lipid raft formation via modulation of Seladin-1. Lipid raft formation is known to be associated with Alzheimer's disease. Thus the findings provide a link between the LXR and the processes associated with Alzheimer's disease (16).

The above results illustrate the strategy of the integrated application of new computational and experimental methods for the genome-wide identification of LXR direct target genes. To avoid the need to scan the entire genome, only regions of the genome with significant LXR occupancy sites from the ChIP-chip experiment were screened. The screening was performed with a specialized model (LXRE.HMM) which improved both sensitivity and selectivity of recognizing putative LXR binding sites and associated genes. This was followed by a selection of a set of genes based either on results from an independent gene expression experiment or on biological significance. Follow-up experiments further validated that the genes are direct targets of LXR and provided insights into novel aspects of LXR regulation of key processes associated with disease biology.

3. Notes

1. Single RE prediction does not take cooperative binding sites into consideration. Transcriptional activation is a dynamic process involving a complex of proteins and also depends on events other than a single transcription factor binding to an RE site. The sequence features surrounding a given binding site may provide important clues of cooperative binding by other proteins which influence the activation of target genes. Recent advances of developing algorithms for the detection of cooperative (also referred to as composite) binding site recognition (17, 18) may enable the recognition of nuclear receptor binding sites with increased accuracy. Furthermore, cooperative binding site recognition may provide insights into tissue-specific binding preferences of nuclear receptors and the complex cross talk between regulatory pathways.

2. Pitfalls of mapping occupancy sites to genes. In the method described above ChIP-chip active regions were screened using the LXRE.HMM algorithm to predict LXR occupancy sites in the genome. The predicted sites were then mapped to genes in 1 kb proximity to predict LXRalpha target genes. This mapping identifies LXREs adjacent to the transcription initiation site (i.e., in the promoter region) of the putative target genes. However, enhancer regions which contain REs and regulate target gene expression can be located much further away. Thus, the occupancy site to gene mapping based on 1 kb proximity will not detect enhancer regions associated with target genes. Improved algorithmic and experimental methods will be needed in the future to identify enhancers and their corresponding genes.

References

1. Locker, J. (2001) Transcription factors. *Humana Molecular Genetics. Academic Press* 167–214.
2. Overington, J.P., Al-Lazikani, B., Hopkins, A.L. (2006) How many drug targets are there? *Nat Rev Drug Discov* 5, 993–996.
3. Olefsky, J.M. (2001) Nuclear receptor minireview series. *J Biol Chem* 276, 36863–36864.
4. He, J., Cheng, Q., Xie, W. (2009) Minireview: nuclear receptor-controlled steroid hormone synthesis and metabolism. *Mol Endocrinol* 23, 1–11.
5. Okuno, M., Arimoto, E., Nishihara, T., Imagawa, M. (2001) Dual DNA-binding

specificity of peroxisome-proliferator-activated receptor gamma controlled by heterodimer formation. *Biochem J* 353, 193–198.
6. Sandelin, A., Alkema, W., Engstrom, P. et al. (2004) JASPAR: an open-access database for eukaryotic transcription factor binding profiles. *Nucleic Acids Res* 32, D91–D94.
7. Cartharius, K., Frech, K., Grote, K. et al. (2005) MatInspector and beyond: promoter analysis based on transcription factor binding sites. *Bioinformatics* 21, 2933–2942.
8. Lemay, D., and Hwang, D. (2006) Genome-wide identification of peroxisome prolif-

erator response elements using integrated computational genomics. *J Lipid Res* 47, 1583–1587.

9. Podvinec, M., Kaufmann, M.R., Handschin, C., and Meyer, U.A. (2002) NUBIScan, an in silico approach for prediction of nuclear receptor response elements. *Mol Endocrinol* 16, 1269–1279.

10. Moehren, U., Denayer, S., Podvinec, M. et al. (2008) Identification of androgen-selective androgen-response elements in the human aquaporin-5 and Rad9 genes. *Biochem J* 411, 679–686.

11. Sandelin, A., and Wassermen, W.W. (2005) Prediction of nuclear hormone receptor response elements. *Mol Endocrinol* 19, 595–606.

12. Varga, G., and Su, C. (2007) Classification and predictive modeling of liver X receptor response elements *BioDrugs* 21, 117–124.

13. Schmid, C.D., Perier, R., Praz, V., and Bucher, P. (2006) EPD in its twentieth year: towards complete promoter coverage of selected model organisms *Nucleic Acids Res* 34, D82–D85.

14. Stayrook, K.R., Rogers, P.M., Savkur, R.S. et al. (2008) Regulation of human 3 alpha-hydroxysteroid dehydrogenase (AKR1C4) expression by the liver X receptor alpha. *Mol Pharmacol* 73, 607–612.

15. Wang, Y., Rogers, P.M., Su, C. et al. (2008) Regulation of cholesterologenesis by the oxysterol receptor, LXRalpha. *J Biol Chem* 283, 26332–26339.

16. Wang, Y., Rogers, P.M., Stayrook, K.R. et al. (2008) The selective Alzheimer's disease indicator-1 gene (Seladin-1/DHCR24) is a liver X receptor target gene. *Mol Pharmacol* 74, 1716–1721.

17. Fogel, G.B., Porto, V.W., Varga, G. et al. (2008) Evolutionary computation for discovery of composite transcription factor binding sites. *Nucleic Acids Res* 36, e142.

18. Pape, U.J., Klein, H., Vingron, M. (2009) Statistical detection of cooperative transcription factors with similarity adjustment *Bioinformatics* 15, 2103–2109.

Chapter 16

Computing Chromosome Conformation

James Fraser, Mathieu Rousseau, Mathieu Blanchette, and Josée Dostie

Abstract

The "Chromosome Conformation Capture" (3C) and 3C-related technologies are used to measure physical contacts between DNA segments at high resolution in vivo. 3C studies indicate that genomes are likely organized into dynamic networks of physical contacts between genes and regulatory DNA elements. These interactions are mediated by proteins and are important for the regulation of genes. For these reasons, mapping physical connectivity networks with 3C-related approaches will be essential to fully understand how genes are regulated. The 3C-Carbon Copy (5C) technology can be used to measure chromatin contacts genome-scale within (*cis*) or between (*trans*) chromosomes. Although unquestionably powerful, this approach can be challenging to implement without proper understanding and application of publicly available bioinformatics tools. This chapter explains how 5C studies are performed and describes stepwise how to use currently available bioinformatics tools for experimental design, data analysis, and interpretation.

Key words: Chromosome Conformation Capture (3C), 3C-Carbon Copy (5C), microarrays, high-throughput DNA sequencing, spatial genome organization, three-dimensional modeling, gene expression.

---

1. Introduction

Understanding the transcriptional responses triggered by cellular pathways is a fundamental question in biology. Although significant advances were recently made in describing transcription networks involved in different processes such as cellular differentiation (1–3), the coordination of genes in response to environmental cues remains poorly understood. Gene regulation is known to involve both local and long-range changes in

---

I. Ladunga (ed.), *Computational Biology of Transcription Factor Binding*, Methods in Molecular Biology 674,
DOI 10.1007/978-1-60761-854-6_16, © Springer Science+Business Media, LLC 2010

chromatin structure (**Fig. 16.1**) (4–9). Local gene regulation primarily involves changes in post-translational histone modifications, in the association of proteins, and the position of nucleosomes at promoters and regulatory DNA elements (4, 10). In contrast, changes in long-range chromatin architecture also involves altering physical contacts between genes and/or DNA elements. Long-range interactions between genes and elements located either on the same (*cis*) or different (*trans*) chromosomes were found to be essential for the regulation of genes from various cellular pathways (11–29). Interestingly, long-range contacts have also been found between co-regulated genes (30, 31). Together these studies indicate that long-range contacts represent a general mechanism to control the expression of genes and that genomes are likely organized into dynamic networks of physical contacts, which are required for coordinated transcriptional responses. Therefore, mapping the three-dimensional

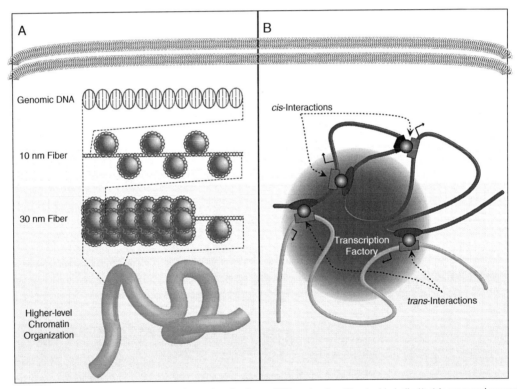

Fig. 16.1. Genomic DNA organization in the nucleus in vivo. **a** DNA packaging. The double helix (*top*) is wrapped around nucleosomes, which are shown as spheres, to form the 10 nm fiber. This fiber is further packaged into a "solenoid" or 30 nm fiber by wrapping onto itself. The mostly uncharacterized higher level in vivo chromatin organization is represented by a thick coil (*bottom*). **b** Functional DNA interactions in the genome. *Curved lines* represent different chromosomes featuring both *cis* and *trans* interactions. *Rectangles* represent promoters, and the *arrows* represent their transcriptional start sites. Physical interactions between promoters and enhancers (*ellipse*) or silencers (*hexagon*) are mediated by protein complexes illustrated by *circles*. Actively transcribed genes are further organized into transcription factories, represented by a *shaded area* at the *center* of the diagram.

organization of genomes will be essential to fully understand gene regulation.

1.1. Mapping Three-Dimensional Genome Organization with 3C Technology

The "Chromosome Conformation Capture" (3C) technology remains the method primarily used to measure changes in both local and long-range chromatin contacts (32–34). The 3C approach is shown in **Fig. 16.2** and involves first generating a library of DNA contacts captured in vivo by chemical cross-linking. Detailed protocols describing how to prepare high-quality 3C libraries have been published elsewhere (34, 35). Briefly, 3C libraries are generated by treating cells with formaldehyde, digesting with a restriction enzyme and ligating with T4 DNA ligase in dilute conditions (**Fig. 16.2A**). Thus, 3C libraries contain entire genomes digested and re-ligated into a wide variety of "head-to-head," "tail-to-tail," "tail-to-head," and "head-to-tail" products between restriction fragments. Since these four configurations exist at equimolar ratios in libraries, only one (e.g.,

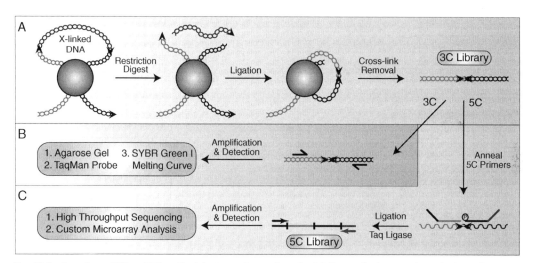

Fig. 16.2. Overview of 3C and 5C methodologies. **a** Generation of a 3C library. A *cis* genomic contact, mediated by a protein complex represented by a *circle* is shown as an example. *Arrowheads* indicate both the location of restriction cut sites and the directionality of restriction fragments. Formaldehyde cross-linked chromatin is first digested with a restriction enzyme, ligated under dilute conditions to favor intermolecular ligation of cross-linked fragments, and deproteinated to remove cross-links. Resulting 3C libraries contain DNA contacts derived from the entire genome, where the amount of each ligation product is inversely proportional to the in vivo spatial proximity between DNA segments. A "head-to-head" ligation product is illustrated as an example of in vivo 3C contact. **b** Conventional analysis of 3C products by semi-quantitative PCR detection, TaqMan quantitative real-time PCR, or melting curve analysis (36). The junctions of 3C ligation products are usually PCR-amplified individually with specific 3C primer pairs. **c** 5C detection of 3C products. 5C primers are first annealed to 3C libraries in a highly multiplexed setting to detect up to millions of different chromatin contacts simultaneously. 5C analysis required both forward and reverse 5C primers to detect "head-to-head" 3C junctions. Forward and reverse 5C primers anneal immediately next to each other on the same DNA strand and are then quantitatively ligated by Taq DNA ligase to generate 5C libraries containing "Carbon-Copies" of existing 3C contacts. This 5C library is finally amplified by PCR with primers specific to universal T3 and T7 5C tail sequences. 5C libraries can be analyzed on either custom microarrays or by high-throughput DNA sequencing.

the "head-to-head") is typically measured during 3C. An important consideration while selecting a restriction enzyme to produce 3C libraries is whether the presence of SDS and Triton X-100 substantially affects enzymatic activity. Commonly used restriction enzymes that work well under these experimental conditions include EcoRI, BglII, or HindIII. DNA ligation is conducted under dilute conditions to favor intermolecular ligation of cross-linked DNA fragments, and cross-links are finally removed by treating with proteinase K and phenol–chloroform DNA extraction.

The key to 3C detection lies at the ligation step. By diluting the reaction tenfold, the enzyme preferentially ligates genomic DNA fragments held together through protein cross-links. Fragments located close to each other in the nuclear space in vivo will be cross-linked more frequently and form a larger number of 3C products than fragments located very far apart. The level of these ligation products in a 3C Library is approximately inversely proportional to the original three-dimensional distance separating the pair of genomic regions in vivo. 3C products are conventionally measured semi-quantitatively by individual PCR amplification of predicted "head-to-head" ligation junctions and agarose gel detection or melting curve analysis (**Fig. 16.2B**) (36). Although more costly, predicted 3C products may also be more accurately quantified with TaqMan probes (37).

1.2. Limitations to the 3C Approach

A major limitation to the 3C technology is the lack of scalability. During 3C, a minimum of three PCR reactions must be prepared individually for each predicted contact to be interrogated and for each tested cellular condition. Also, since 3C relies on the individual PCR amplification of predicted ligation junctions with specific primers, differences in primer pair efficiency must be corrected. This normalization step is achieved by generating PCR triplicates in control 3C libraries containing equimolar ratios of all predicted contacts as previously described (34). Thus, 3C requires that at least six PCR reactions be resolved and quantified on agarose gels or measured by quantitative real-time PCR with TaqMan probes. For this reason, obtaining detailed information about the three-dimensional organization of transcriptional networks or even single genomic regions is tedious and can be very expensive with the conventional 3C approach. Other important limitations to the 3C method include occasional low specificity of the PCR amplification, the frequent need to re-design 3C primers, and the identification of amplified 3C junctions based on size rather than sequence.

1.3. The 5C Technology

The "3C-Carbon Copy" (5C) technique was more recently developed to increase the throughput of 3C (38–41). 5C analysis combines 3C library production with highly multiplexed ligation-mediated amplification (LMA) to quantify up to millions of ligation junctions simultaneously (**Fig. 16.2C**). To characterize three-dimensional chromatin organization with 5C, 5C primers are first annealed to the 3C library in a highly multiplexed setting. 5C primers are designed to include both a specific sequence corresponding to the 3′ end of restriction fragments and a universal primer sequence used for PCR amplification of 5C libraries. Since 5C also detects predicted "head-to-head" 3C ligation products, one 5C forward and one 5C reverse primer must be used to quantitatively detect 3C junctions. Only 5C primers annealed immediately next to each other on the same strand at 3C junctions can be ligated together by the Taq DNA ligase. This process quantitatively converts existing 3C contacts into "3C-Carbon Copies", which are then amplified with universal primers in a single PCR step. This "5C library" can be analyzed either on custom microarrays or by ultra-high-throughput DNA sequencing (38).

1.4. The Challenge of 5C Technology

In contrast to 3C, 5C analysis cannot be performed manually and requires the use of several computer programs for experimental design, analysis, and data interpretation. Understanding the function and limitations of these programs is key for successful 5C analysis. First, forward and reverse 5C primers must be designed for selected networks or throughout the regions of interest. Since 5C is typically used to generate large genomic organization data sets, hundreds to thousands of primers must be predicted for experimental design. This task can be performed using the "5CPrimer" program as described in **Section 3.2** below. 5C products can be detected by deep sequencing or on custom microarrays. While ultra-high-throughput DNA sequencing offers the advantage of greater linear detection range, microarray analysis remains the most affordable analysis method to characterize low-complexity 5C libraries. However, since the complexity of 5C libraries can increase exponentially with increasing primer numbers, the only viable detection method to characterize high-complexity libraries remains deep sequencing. Custom 5C microarrays can be designed with our "5CArrayBuilder" program. This program uses 5C primer lists generated with 5CPrimer and is described in **Section 3.3**. Raw microarray signals must then be normalized for background levels, the inherent signal variability due to array synthesis, and signal saturation. Corrected data from cellular and control 5C libraries must also be processed to calculate interaction frequencies (IFs) for each predicted contact. These corrected IFs can be estimated with our "IFCalculator" program described in **Section 3.4**. Although IFs are usually catalogued in the form of heatmaps that may be analyzed systematically with programs such as Excel, 5C data sets can also be interpreted with several bioinformatics tools, including "5C3D" and "Microcosm." The 5C3D program integrates

256   Fraser et al.

all 5C data in the form of a three-dimensional model, which can then be analyzed further to measure defined sets of metrics useful for comparing different genomic features. The program Microcosm uses 5C3D output models to estimate local DNA base densities surrounding given genomic features. These programs are described in **Sections 3.5** and **3.6**.

2. Software

The software described in this chapter can be found on our website under the "Tools" section at the following address: http://dostielab.biochem.mcgill.ca. All software is available for download as command line applications and should run well on any computer capable of running C (5CPrimer, 5CArrayBuilder) and Java (5C3D, Microcosm) programs. Additionally, 5CPrimer is available through an easy to use web interface on our website.

3. Methods

This section describes the computational steps required to conduct 5C analysis.

3.1. Genomic Annotation

The first step in conducting any 5C study involves selecting the restriction enzyme to generate 3C libraries. The chosen enzyme should preferentially yield restriction fragments that are uniform in size throughout the region of interest. The enzyme should also generate few very small or very large fragments. For example, when a 6-cutter enzyme is selected, fragments should be between 500 and 8,000 bp in length. Restriction digest patterns can be predicted with any DNA analysis software and for hypothesis-based projects, we highly suggest creating a reference file cataloging restriction sites, primers, genes, regulatory DNA elements, and other important genomic features. These reference files are usually updated throughout the project and routinely consulted to verify or optimize experimental design.

3.2. 5CPrimer

Once a restriction enzyme has been selected and a reference file created, the next computational step in 5C analysis is to design forward and reverse 5C primers throughout the region of interest. 5C primers can be designed with the "5CPrimer" computer program that we developed recently (42). This program is written in the C programming language and has a web interface easily accessible from our website (http://dostielab.biochem.mcgill.ca).

5CPrimer predicts both forward and reverse 5C primers for any given region digested with most restriction enzymes. The program first locates cutting sites in a DNA sequence for a restriction enzyme selected by the user and designs 5C primers iteratively from the center of each cut site. Primer length is set at a minimum of 18 bp, although the final length is affected by both the primer's melting point (Tm) and the number of cycles required to synthesize corresponding products onto microarrays. Both parameters are user customizable, and default settings represent values used in our laboratory. The Tm is calculated using Nearest-Neighbor thermodynamics as described by Breslauer et al. (43). We restrict the length of our primers to the maximum number of corresponding cycles required to generate the full-length feature on arrays to harmonize 5C primer and array design. After selecting the specific genomic region of primers, the 5CPrimer program attaches the universal T7 tail sequence on forward primers and the complementary T3 (T3c) sequence on reverse primers (**Fig. 16.3A**). These common sequences are used to PCR-amplify 5C libraries.

5CPrimer also offers the option of using the Repeat-Masker program to identify repetitive DNA sequences and low-complexity regions within primers and eliminate potentially problematic primers (44). These sequences introduce high background levels and therefore must be avoided (38). 5CPrimer does not perform a BLAST search yet to verify the uniqueness of primers. Therefore these searches have to be done manually. 5C primers sharing homology with other regions of the human genome such as conserved gene-coding sequences cannot be used in order to avoid cross-hybridization. The final output of 5CPrimer is a tab-delimited text file containing all the information necessary to place an order with vendors.

3.3. 5CArrayBuilder

The "5CArrayBuilder" computer program can be used to design custom arrays from 5CPrimer output files. This program is written in the C programming language and is currently only available for command line, although a web interface is currently under development. To design custom 5C arrays, the user specifies a list or a range of both forward and reverse primers from each of the 5CPrimer output files, and 5CArrayBuilder predicts every possible 5C product from the selected primers. Specifically, 5CArrayBuilder generates eight different probes of varying length centered at the restriction cut site, for each predicted 5C product (sense and anti-sense strand) (**Fig. 16.3B**). The first predicted feature is typically 30 bp long and used as a background signal when calculating interaction frequencies. The remaining seven probes vary in length from 36 bp to 48 bp in 2 bp increments. The output of 5CArrayBuilder is simply a text file where every line consists of a user-specified tag used for extracting array information, followed by the sequence of the probe. This file can

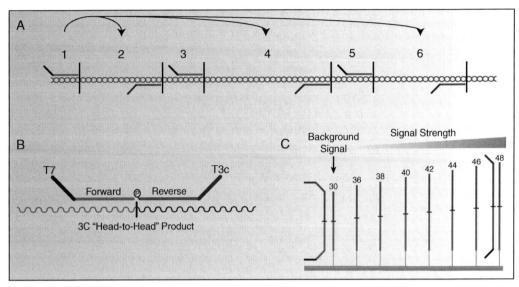

Fig. 16.3. 5C analysis using custom microarrays. **a** 5C experimental design. Forward and reverse 5C primers are illustrated at the *top* and *bottom* of the DNA double helix, respectively. Both forward and reverse 5C primers are designed to anneal at the 3′ end of restriction fragments immediately upstream of restriction sites, which are represented by *vertical lines*. Since forward and reverse 5C primers are complementary to each other, only one primer per fragment, either forward or reverse, can be used at one time to generate 5C libraries. For this reason, the maximum coverage possible per 5C library is 50%. Possible ligation products between forward 5C primer 1 with the reverse 5C primers in the remaining restriction fragments are indicated by *arrows* at the *top* of panel. **b** Anatomy of a "head-to-head" 5C ligation product. 5C primers are shown annealed to a "head-to-head" 3C ligation product. Forward primers typically include a T7 universal sequence at the 5′ end, which is shown, angled away from the sequence specific to genomic DNA. Reverse primers also feature specific genomic sequences corresponding to the reverse complement of the 3′ end of restriction fragments. Reverse primers also include a universal T3 complementary (T3c) tail and a 5′ phosphate used in the ligation reaction with Taq DNA ligase. **c** Hybridization of 5C ligation products on custom microarrays. Microarray probes generated with the "5CArrayBuilder" computer program are shown vertically attached to the array surface. Probe sizes usually range from 30 nt (background) to 48 nt (maximum signal intensity) as indicated *above* each feature. Partial hybridization of a 5C library product to the 30 nt feature is shown on the *left* and complete binding to a full-length array feature is illustrated on the *right*. Signal intensity and specificity are predicted to increase with increasing array feature length.

be manually modified to exclude select feature sets, combined to other files, or sent out directly to vendors for microarray synthesis.

3.4. IFCalculator

We recommend conducting 5C analysis with custom arrays featuring half-site probe lengths of 15, 18, 19, 20, 21, 22, 23, and 24 nt as described above in **Section 3.3** "5CArrayBuilder." The 15 nt half-site probe signal is representative of background noise and is used to determine which of the remaining probe values should be included to calculate the average interaction frequency (IF) of its corresponding fragment pair. The "IFCalculator" program automatically excludes points close to background signal. For each interaction, and starting from the longest half-site, IFCalculator first compares the signal of each probe to the value of the corresponding 15 bp probe. If a signal is found to be less than 150% of it's 15 bp value, that half-site signal is discarded along with all remaining shorter probe length values. Corresponding 15 bp signals are then subtracted from the remaining values to remove background from each entry. Corrected values are used to calculate IFs by dividing cellular and control 5C signals of corresponding feature lengths. Interaction frequencies are finally averaged and the variance, count, and 95% confidence interval are reported in the output 5C data set. If all probe length values are rejected for being too close to the background signal, an IF value of 0 is reported and is indicated as a missing data point.

3.5. 5C3D

5C IF data sets can be analyzed three dimensionally by generating average three-dimensional models with the 5C3D program. 5C3D converts all non-zero interaction frequencies to distances (D) as follows: $D(i, j) = 1/\text{IF}(i, j)$, where $\text{IF}(i, j)$ is the IF between points i and j and $D(i, j)$ is the three-dimensional Euclidean distance between points i and j ($1 \leq i, j \leq N$). Each point represents a single restriction fragment. Next, the program initializes a virtual three-dimensional DNA strand represented as a piecewise linear three-dimensional curve defined on N points distributed randomly in a cube. The program then follows a gradient descent approach to find the best conformation, aiming to minimize the Misfit (M) between the desired values in the distance matrix D and the actual Euclidean distance $E(i, j)$:

$$M = \sqrt{\sum_{1 \leq i, j \leq N, i \neq j} \left(\frac{D(i, j) - E(i, j)}{E(i, j)} \right)^2}$$

Each point is considered one-at-a-time and is moved in the inverse direction of the gradient, ∇, of the misfit function (for which an analytical function is easily obtained), using a step size equal to $d^* |\nabla|$. Small values of $d (d = 5 \times 10^{-5})$ were used to ensure convergence of the method but increase the number of iterations needed. The process of iteratively moving each point in the virtual DNA strand in order to decrease the misfit is repeated until convergence (change in misfit between successive iterations less than 0.001). The gradient descent approach converges relatively rapidly, with a running time of a few minutes on a desktop workstation. The resulting set of points is then considered to be the best fit for the experimental data and is represented as a piecewise linear three-dimensional curve. This curve is then annotated with differently colored transparent spheres centered at the transcription start sites of the genes present along the DNA sequence (or any other specified feature locations). An example of 5C3D output models highlighting spatial chromatin changes in the HoxA cluster during THP-1 differentiation is shown in **Fig. 16.4** (42). THP-1 cells are myelomonocytes that can be

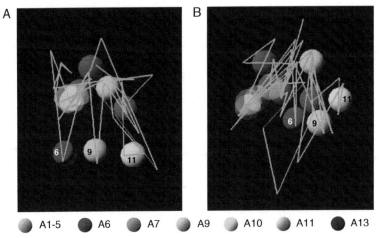

Fig. 16.4. 5C3D output models of the human HoxA cluster. 5C3D models were generated with 5C data from either undifferentiated **a** or differentiated **b** THP-1 cells. *Shaded spheres* represent transcriptional start sites (TSSs) of genes as indicated in the legend *below*. TSSs displaying the greatest change between the two states are labeled (HoxA 6, 9, and 11). The *lines* do not represent genomic DNA, but are simply used to connect the different vertices generated by 5C3D that represent the different DNA fragments.

terminally differentiated into macrophages in the presence of phorbol myristate acetate (PMA). These cells are often used as a model system to study acute myeloid leukemia (AML). In fact, the dynamic transcription factor network involved in differentiation and growth arrest was extensively described by the FANTOM4 Consortium. We previously showed that transcriptional silencing of the HoxA9, 10, 11, and 13 genes following THP-1 differentiation is accompanied by clustering of repressed genes (42). Interestingly, HoxA9 and 10 are oncogenes important in promoting cellular proliferation in these cells. Understanding the relationship between transcription repression and spatial chromatin remodeling may uncover important information about the mechanism by which these genes are regulated and regulate other genes.

3.6. Microcosm

The "Microcosm" program is designed to empirically analyze 5C3D structural models by measuring local base density along the structure's length. The input data includes the average IF values, variance, counts (or number of technical replicates), and 95% confidence intervals for each pair of points, as obtained from the IFCalculator program. To establish the robustness and significance of the observed measurements, Microcosm selects, for each fragment pair, an IF at random from a normal distribution with the mean and standard deviation corresponding to it. This process is repeated for each fragment pair to generate "randomly sampled" 5C array data sets based on original 5C data. Each randomly sampled data set is then used individually by 5C3D to infer the best fitting model.

The final models are analyzed to determine the local density of the environment surrounding each feature F. The local density is defined as the total number of DNA base pairs from any DNA segment that lies within the sphere of a fixed radius centered at F's genomic position. The radius should be chosen with respect to the overall size of the model. To remain significant, the radius should not be too large in order to avoid including too much of the model, or too small to exclude most surrounding DNA as indicated in **Section 4**. The process described above is repeated 100 times for each original 5C data set to generate 100 individual models and local density estimates around each feature. It is relatively computationally inexpensive to generate and analyze the 100 models required, with a running time of a couple of hours on a desktop workstation. The average local density, its variance, and 95% confidence interval for the mean are then calculated for each feature and reported in a graphical format called a local density plot. Local density plots can be compared between experimental states to identify features with significant differences in local density. If such a comparison is performed, then a p-value is calculated for each difference and corresponds to the probability of incorrectly predicting a difference in local densities assuming normality of the data. A low p-value therefore indicates that the difference in local densities of a feature's environment between two states is likely to be real. An example of Microcosm output is shown in **Fig. 16.5**. This figure presents a local base density plot of the human HoxA cluster in undifferentiated (solid) and differentiated (dashed) THP-1 cells. In agreement with 5C3D output models,

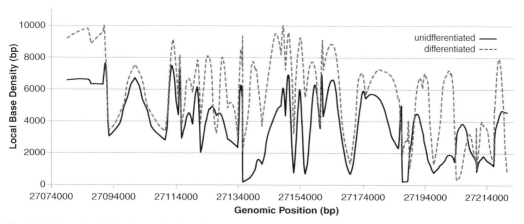

Fig. 16.5. Example of a local base density analysis by Microcosm. We previously examined the conformational changes in the HoxA cluster during differentiation of THP-1 cells. Microcosm was used to analyze 5C data from the HoxA gene cluster before differentiation (*solid line*) and after differentiation (*dotted line*). The horizontal axis shows the position along chromosome 7 (ENCODE, Mar. 2006 Assembly), while the *vertical* axis measures the density of DNA in the three-dimensional area.

3.7. Conclusion

this example shows that based density is considerably higher in differentiated cells where gene expression is repressed.

5C is a powerful tool for examining changes in three-dimensional chromatin architecture. The software described in this chapter should promote the use of 5C for hypothesis-based research and other functional chromatin studies. The identification of HoxA chromatin conformation signatures (CCSs) referred throughout this chapter represents the second 5C analysis of a genomic region. 5C was previously used to characterize the three-dimensional chromatin organization of the human beta-globin cluster in different cell lines (38). In that study, the locus control region (LCR) was shown to physically interact specifically with the transcribed genes. These interactions were recapitulated by conventional 3C analysis. Importantly, analysis of the beta-globin locus with 5C identified a new long-range interaction between the LCR and the transcribed pseudogene thereby demonstrating the tremendous power of 5C to identify physical contacts de novo. A flowchart summarizing the various molecular and computational steps involved in 5C analysis is presented in **Fig. 16.6**.

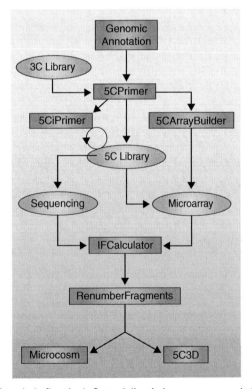

Fig. 16.6. 5C analysis flowchart. Computational steps are represented by *rectangles* and molecular biology experiments by *ovals*.

4. Notes

1. Using 5C iPrimers for quality control of 5C libraries

   Since 5C products are similar in size to primer dimer artifacts, we recommend verifying the identity and integrity of 5C products in libraries before proceeding with microarray hybridization or high-throughput DNA sequencing. Quality control of 5C libraries can be performed by serial dilution and semi-quantitative PCR amplification of representative DNA contacts with internal 5C primers as described previously (38). This approach offers the advantage of both distinguishing legitimate products from artifacts and verifying the linearity of 5C libraries. iPrimers correspond partially to universal primer sequence selected as tail and to specific sequences homologous to genomic DNA. These primers can be designed with our "5CiPrimer" computer program from 5C primer lists generated by 5CPrimer. 5C products can also be distinguished from primer dimers in "bulk" by digesting entire libraries with restriction enzymes cutting reconstituted restriction sites at the junction of ligated 5C forward and reverse primers.

2. Coverage limitation in individual 5C libraries

   Since forward and reverse 5C primers are complementary to each other, only one primer per restriction fragment can be used at one time, and a maximum interaction coverage of 50% per 5C library can be attained. Fifty percent coverage is achieved using alternating forward and reverse 5C primers for consecutive restriction fragments (**Fig. 16.3**). Although often sufficient to identify overall changes in three-dimensional genome organization, certain long-range looping contacts may not be resolved using this experimental design. To solve this problem, merging complementary 5C data sets was previously suggested in order to increase overall interaction coverage. For example, the typical R-F-R-F configuration could be complemented with R-R-F-R-R-F, and R-R-R-F-R-R-R-F configurations to achieve up to 87.5% coverage. However, since there are currently no available tools for merging complementary libraries and obtain higher coverage, this process must be done manually. Higher interaction coverage can also be achieved by analyzing 3C libraries generated with 4-cutter restriction enzymes.

3. Limitations in primer design

   Genomic regions are not always perfectly compatible with 5C analysis. For example, when a genomic domain contains a large number of repetitive DNA sequences, few 5C primers may be successfully designed for a given restriction pattern.

5C primers homologous to repetitive DNA sequences generate high background levels in 5C libraries and must be excluded. 5C primer design is restricted to regions immediately adjacent to restriction cut sites and for that reason, 5C primers may not be suitable if cut sites fall within repeats or within very low-complexity regions. Different restriction enzymes should be considered *before* producing 3C libraries when highly repetitive genomic domains are characterized in order to maximize coverage during 5C library production.

4. File formats

Interaction frequencies (IFs): The microarray signals used as input to the IFCalculator program need to be in a single tab-delimited file. The format of this file is specified in the readme file included with the IFCalculator program download.

Feature locations: The genomic positions of the features analyzed must be specified in a single input file. The format of this file is specified in the readme file included with the 5C3D program download.

Fragment lengths: The length of the genomic fragments (in base pairs) must be specified as start and stop base indices in a single tab-delimited file. The format of this file is specified in the readme file included with the 5C3D program download.

5. IFCalculator

IFCalculator considers that listed entries are consecutive. If some fragments were excluded from the analysis, gaps will be present in the numbering of fragments and the output data file by IFCalculator should be first converted with the "RenumberFragments" program before pursuing data analysis and interpretation. The output from RenumberFragments can then be used instead of the IFCalculator file for subsequent steps with the 5C3D and Microcosm programs.

Although IFCalculator typically applies a 150% cutoff for background as described above, the user can specify a different value as a command line parameter. For example a 150% cutoff may not be optimal when detecting changes between low IF signals. In this case, the cutoff may be reduced to include more noise but detect small changes. Conversely, a 150% background signal cutoff may not be appropriate for data sets with higher background levels. In that case, the cutoff may be increased to highlight higher confidence changes in IFs. Thus, increasing the value will increase the stringency of the filter and vice versa.

6. 5C3D

If the gradient descent process is not converging (oscillating) or is converging too slowly, the value of epsilon (the step size) is inappropriate. The user can specify a different value as a command line parameter (use a bigger value to speed up the process; use a smaller value to improve convergence problem).

7. Microcosm

The user can specify the radius of the sphere as a command line parameter. If the selected radius of the sphere is too large, the local base density measurements will include too much of the structure and will not correctly characterize the environment of genomic locations. Measurements will be very high and similar throughout the structure. In this case, the value for the radius of the sphere should be reduced.

If the selected radius of the sphere is too small, the local base density measurements will not include sufficient surrounding DNA and will not correctly reflect actual local base densities. Measurements will be very low and similar throughout the structure. In this case, the value for the radius of the sphere should be increased.

8. Interpretation of three-dimensional models

Several important considerations must be taken into account when interpreting three-dimensional models generated from 5C data sets. First, because of the inherent nature of 3C and 5C, the output models generated with 5C3D represent averaged structural models rather than true individual in vivo structures. Nonetheless, these models can be used to identify three-dimensional changes in chromatin organization by comparing the same genomic region under various cellular conditions. Also, three-dimensional modeling integrates all contacts into individual possible "structures" to generate averaged models of compatible IFs. This process reduces the incidence of false positives by considering each IF in context of entire data sets and by not favoring strong gain or loss of contacts.

Second, in 5C3D output models, straight lines represent the shortest path between the endpoints of each fragment rather than actual chromatin. Indeed, this information is excluded from our models since there is no actual information regarding the structure of chromatin between end points. Therefore, straight lines cannot be trusted as "true" structures.

Third, since small IF values are less accurate than large IF 5C values, points that are far away from others should be considered less importantly in overall models. Indeed, because we consider distances to be inversely proportional to IFs, small IF values are converted to large distances in models and points that are far away from everything else have very little constraints on their location in space in the three-dimensional model. Such points have a very high degree of flexibility in their position and should therefore not be interpreted as actually being located at any one point in space. A safer interpretation would be that such points are "far" away from the rest of the structure but the exact position of such points relative to the rest of the structure should not be trusted when considering the models.

9. Reproducibility of 5C data

Since 5C libraries derive from populations of cells, reproducibility between biological replicates is greatly influenced by growth conditions. We find that general trends such as long-range contacts and compaction levels are conserved between libraries generated from similarly treated cells, although relative IFs may vary between biological repeats. We are currently examining the variability and noise levels between 5C technical repeat in order to assess the reproducibility of 5C technology. To minimize biological noise, we recommend fixing cells at a fixed point of their growth curve. Synchronizing cells and collecting at a defined point after release may also reduce variability between preparations. To reduce technical noise between libraries, we recommend using the same reagents (provider, lot numbers, etc.) and a standard operating protocol each time to generate libraries.

References

1. Severin, J., Waterhouse, A.M., Kawaji, H. et al. (2009) FANTOM4 EdgeExpressDB: an integrated database of promoters, genes, microRNAs, expression dynamics and regulatory interactions. *Genome Biol* 10, R39.

2. Kawaji, H., Severin, J., Lizio, M. et al. (2009) The FANTOM web resource: from mammalian transcriptional landscape to its dynamic regulation. *Genome Biol* 10, R40.

3. Suzuki, H., Forrest, A.R., van Nimwegen, E. et al. (2009) The transcriptional network that controls growth arrest and differentiation in a human myeloid leukemia cell line. *Nat Genet* 41, 553–562.

4. Berger, S.L. (2007) The complex language of chromatin regulation during transcription. *Nature* 447, 407–412.

5. Heard, E., and Bickmore, W. (2007) The ins and outs of gene regulation and chromosome territory organisation. *Curr Opin Cell Biol* 19, 311–316.

6. Babu, M.M., Janga, S.C., de Santiago, I. et al. (2008) Eukaryotic gene regulation in three dimensions and its impact on genome evolution. *Curr Opin Genet Dev* 18, 571–582.

7. Miele, A., and Dekker, J. (2008) Long-range chromosomal interactions and gene regulation. *Mol Biosyst* 4, 1046–1057.

8. Kleinjan, D.A., and van Heyningen, V. (2005) Long-range control of gene expression: emerging mechanisms and disruption in disease. *Am J Hum Genet* 76, 8–32.

9. West, A.G., and Fraser, P. (2005) Remote control of gene transcription. *Hum Mol Genet* 14, R101–R111.

10. Kouzarides, T. (2007) Chromatin modifications and their function. *Cell* 128, 693–705.

11. Wright, M.M., Kim, J., Hock, T.D. et al. (2009) Human heme oxygenase-1 induction by nitro-linoleic acid is mediated by cyclic AMP, AP-1, and E-box response element interactions. *Biochem J* 422, 353–361.

12. Vakoc, C., Letting, D.L., Gheldof, N. et al. (2005) Proximity among distant regulatory elements at the beta-Globin locus requires GATA-1 and FOG-1. *Mol Cell* 17, 453–462.

13. Tsytsykova, A.V., Rajsbaum, R., Falvo, J.V. et al. (2007) Activation-dependent intrachromosomal interactions formed by the TNF gene promoter and two distal enhancers. *Proc Natl Acad Sci USA* 104, 16850–16855.

Computing Chromosome Conformation

14. Tolhuis, B., Palstra, R.J., Splinter, E. et al. (2002) Looping and Interaction between hypersensitive sites in the active beta-globin locus. *Mol Cell* 10, 1453–1465.

15. Spilianakis, C.G., Lalioti, M.D., Town, T. et al. (2005) Interchromosomal associations between alternatively expressed loci. *Nature* 435, 637–645.

16. Spilianakis, C.G., and Flavell, R.A. (2004) Long-range intrachromosomal interactions in the T helper type 2 cytokine locus. *Nat Immunol* 5, 1017–1027.

17. Pirozhkova, I., Petrov, A., Dmitriev, P. et al. (2008) A functional role for 4qA/B in the structural rearrangement of the 4q35 region and in the regulation of FRG1 and ANT1 in facioscapulohumeral dystrophy. *PLoS One* 3, e3389.

18. Palstra, R.J., Tolhuis, B., Splinter, E. et al. (2003) The beta-globin nuclear compartment in development and erythroid differentiation. *Nat Genet* 35, 190–194.

19. Ott, C.J., Suszko, M., Blackledge, N.P. et al. (2009) A complex intronic enhancer regulates expression of the CFTR gene by direct interaction with the promoter. *J Cell Mol Med* 13, 680–692.

20. Murrell, A., Heeson, S., and Reik, W. (2004) Interaction between differentially methylated regions partitions the imprinted genes Igf2 and H19 into parent-specific chromatin loops. *Nat Genet* 36, 889–893.

21. Liu, Z., and Garrard, W.T. (2005) Long-range interactions between three transcriptional enhancers, active Vkappa gene promoters, and a 3′ boundary sequence spanning 46 kilobases. *Mol Cell Biol* 25, 3220–3231.

22. Lim, J.H., Kim, H.G., Park, S.K. et al. (2009) The promoter of the Immunoglobulin J Chain gene receives its authentic enhancer activity through the abutting MEF2 and PU.1 sites in a DNA-looping interaction. *J Mol Biol* 390, 339–352.

23. Kabotyanski, E.B., Rijnkels, M., Freeman-Zadrowski, C. et al. (2009) Lactogenic hormonal induction of long-distance interactions between {beta}-casein gene regulatory elements. *J Biol Chem* 284, 22815–22824.

24. Jiang, H., and Peterlin, B.M. (2008) Differential chromatin looping regulates CD4 expression in immature thymocytes. *Mol Cell Biol* 28, 907–912.

25. Hakim, O., John, S., Ling, J.Q. et al. (2009) Glucocorticoid receptor activation of the Ciz1-Lcn2 locus by long range interactions. *J Biol Chem* 284, 6048–6052.

26. D'Haene, B., Attanasio, C., Beysen, D. et al. (2009) Disease-causing 7.4 kb cis-regulatory deletion disrupting conserved non-coding sequences and their interaction with the FOXL2 promotor: implications for mutation screening. *PLoS Genet* 5, e1000522.

27. Chavanas, S., Adoue, V., Mechin, M.C. et al. (2008) Long-range enhancer associated with chromatin looping allows AP-1 regulation of the peptidylarginine deiminase 3 gene in differentiated keratinocyte. *PLoS One* 3, e3408.

28. Duan, H., Xiang, H., Ma, L. et al. (2008) Functional long-range interactions of the IgH 3′ enhancers with the bcl-2 promoter region in t(14;18) lymphoma cells. *Oncogene* 27, 6720–6728.

29. Dhar, S.S., Ongwijitwat, S., and Wong-Riley, M.T. (2009) Chromosome conformation capture of all 13 genomic Loci in the transcriptional regulation of the multisubunit bigenomic cytochrome C oxidase in neurons. *J Biol Chem* 284, 18644–18650.

30. Brown, J.M., Leach, J., Reittie, J.E. et al. (2006) Coregulated human globin genes are frequently in spatial proximity when active. *J Cell Biol* 172, 177–187.

31. Barnett, D.H., Sheng, S., Charn, T.H. et al. (2008) Estrogen receptor regulation of carbonic anhydrase XII through a distal enhancer in breast cancer. *Cancer Res* 68, 3505–3515.

32. Dekker, J., Rippe, K., Dekker, M. et al. (2002) Capturing chromosome conformation. *Science* 295, 1306–1311.

33. Splinter, E., Grosveld, F., and de Laat, W. (2004) 3C technology: analyzing the spatial organization of genomic loci in vivo. *Methods Enzymol* 375, 493–507.

34. Miele, A., Gheldof, N., Tabuchi, T.M. et al. (2006) Mapping chromatin interactions by chromosome conformation capture (3C). In: *Current protocols in molecular biology* (Ausubel, F. M., R. Brent, R.E. Kingston, D.D. Moore, J.G. Seidman, J.A. Smith, and K. Struhl, Eds.) pp. 21.11.1–21.11-20, Wiley, Hoboken, NJ.

35. Miele, A., and Dekker, J. (2009) Mapping cis- and trans- chromatin interaction networks using chromosome conformation capture (3C). *Methods Mol Biol* 464, 105–121.

36. Abou El Hassan, M., and Bremner, R. (2009) A rapid simple approach to quantify chromosome conformation capture. *Nucleic Acids Res* 37, e35.

37. Hagege, H., Klous, P., Braem, C. et al. (2007) Quantitative analysis of chromosome conformation capture assays (3C-qPCR). *Nat Protoc* 2, 1722–1733.

38. Dostie, J., Richmond, T.A., Arnaout, R.A. et al. (2006) Chromosome conformation capture carbon copy (5C): a massively

parallel solution for mapping interactions between genomic elements. *Genome Res* 16, 1299–1309.

39. Dostie, J., Zhan, Y., and Dekker, J. (2007) Chromosome conformation capture carbon copy technology. *Curr Protoc Mol Biol* Chapter 21, Unit 21.14.

40. Dostie, J., and Dekker, J. (2007) Mapping networks of physical interactions between genomic elements using 5C technology. *Nat Protoc* 2, 988–1002.

41. van Berkum, N.L., and Dekker, J. (2009) Determining spatial chromatin organization of large genomic regions using 5C technology. *Methods Mol Biol* 567, 189–213.

42. Fraser, J., Rousseau, M., Shenker, S. et al. (2009) Chromatin conformation signatures of cellular differentiation. *Genome Biol* 10, R37.

43. Breslauer, K.J., Frank, R., Blocker, H. et al. (1986) Predicting DNA duplex stability from the base sequence. *Proc Natl Acad Sci USA* 83, 3746–3750.

44. Smit, A.F.A., Hubley, R., and Green, P. (1996–2004) RepeatMasker Open-3.0. http://wwwrepeatmaskerorg.

Chapter 17

Large-Scale Identification and Analysis of C-Proteins

Valery Sorokin, Konstantin Severinov, and Mikhail S. Gelfand

Abstract

The restriction-modification system is a toxin–antitoxin mechanism of bacterial cells to resist phage attacks. High efficiency comes at a price of high maintenance costs: (1) a host cell dies whenever it loses restriction-modification genes and (2) whenever a plasmid with restriction-modification genes enters a naïve cell, modification enzyme (methylase) has to be expressed prior to the synthesis of the restriction enzyme (restrictase) or the cell dies. These phenomena imply a sophisticated regulatory mechanism. During the evolution several such mechanisms were developed, of which one relies on a special C(control)-protein, a short autoregulatory protein containing an HTH-domain. Given the extreme diversity among restriction-modification systems, one could expect that C-proteins had evolved into several groups that might differ in autoregulatory binding sites architecture. However, only a few C-proteins (and the corresponding binding sites) were known before this study. Bioinformatics studies applied to C-proteins and their binding sites were limited to groups of well-known C-proteins and lacked systematic analysis. In this work, the authors use bioinformatics techniques to discover 201 C-protein genes with predicted autoregulatory binding sites. The systematic analysis of the predicted sites allowed for the discovery of 10 structural classes of binding sites.

Key words: Restriction-modification systems, C-proteins, DNA-binding proteins, bioinformatics, transcription regulation.

1. Introduction

1.1. Restriction-Modification Systems

The restriction-modification (RM) phenomenon was first discovered more than 50 years ago during the studies of bacterial anti-phage defense (1, 2). Of the three types of RM systems, type II is the simplest and most prevalent. Restriction endonucleases encoded by type II systems are widely used in molecular cloning. A typical type II RM system is essentially a two-component

I. Ladunga (ed.), *Computational Biology of Transcription Factor Binding*, Methods in Molecular Biology 674, DOI 10.1007/978-1-60761-854-6_17, © Springer Science+Business Media, LLC 2010

toxin–antitoxin system. A type II restriction endonuclease (toxin) cleaves unprotected or unmodified DNA at specific sites triggering DNA degradation and cell death. A methyltransferase (antitoxin) protects DNA from cleavage by methylating the same DNA sites (3, 4). Most RM systems' genes are plasmid borne and capable of horizontal spread through bacterial populations. A bacterial cell that acquires an RM system becomes resistant to infection by phages whose genomes contain unmethylated sites in their DNA. On the other hand, plasmids harboring RM systems' genes behave like selfish genetic elements, because a loss of an RM plasmid leads to cell death since methyltransferase (antitoxin) is shorter living than the corresponding restriction endonuclease (toxin).

The horizontal transfer of an RM system among naïve (i.e., lacking such genes) bacteria imposes constrains on RM genes expression. Since the corresponding genomic DNA sites in a naïve cell are unmethylated and therefore vulnerable to cleavage by the restriction endonuclease, the methylase must be synthesized first, while the appearance of restriction endonuclease activity must be delayed until all sites are methylated. On the other hand, once an RM system is established in the host cell, a steady-state ratio of restriction endonuclease and methyltransferase activity needs to be maintained. Too low endonuclease activity (or excessive methyltransferase activity) could lead to a loss of protection against bacteriophage infection and cell (and RM plasmid) death.

One of the most prevalent regulation strategies involves a dedicated transcription regulator encoded by a separate gene. These regulators are called control(C)-proteins and they influence the level of the endonuclease gene (and sometimes methylase gene) transcription in many RM systems. Since computational analysis of this type of regulation relies on the standard structure of C-containing RM-loci, we shall describe them in more detail.

1.2. Regulation of Transcription by C-Proteins

C-protein-dependent regulation was first described for the *Pvu*II system (5). The X-ray analysis of the C.*Ahd*I and C.*Bcl*I C-proteins revealed a 5-alpha-helical protein, which could be assigned to the Xre family of transcriptional regulators (6, 7). Of the five alpha helices, two represent a typical helix-turn-helix (HTH) domain. The remaining three alpha helices allow for effective dimerization of the protein. The similarity between C-proteins and the Xre family regulators (**Fig. 17.1**) suggests that like the latter, C-proteins activate transcription by directly interacting with the σ^{70} subunit of RNA polymerase.

The endonuclease gene is usually localized immediately downstream of the C-protein gene, forming a single CR (C-protein-restriction endonuclease) transcription unit (8). The methylase gene constitutes a separate transcription unit, either convergent or divergent with respect to the CR unit. In all

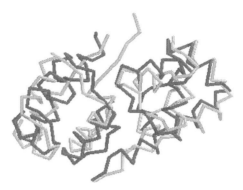

Fig. 17.1. Overlaid structures and protein alignment of C-protein C.AhdI (*black*) and cytosine regulator CylR2 (*gray*), an Xre-subfamily protein.

experimentally studied systems, the region upstream of the C-protein gene contains two C-protein binding sites. The distal site (located further from the C-protein translation start site) has a higher affinity. A weak promoter is usually positioned downstream of the two binding sites. Thus, the basal level of the CR unit transcription is low. After a sufficient amount of C-protein has been produced, the C-protein dimerizes and binds to the distal site activating the CR promoter. Upon further C-protein accumulation, a second dimer binds to the weaker proximal site and represses the further transcription of the CR unit. These positive and negative feedback loops allow for maintaining a steady level of the endonuclease transcript (9, 10).

In some RM systems, C-proteins also affect the transcription of the methylase gene. One (indirect) way of doing this operates when the *M* and *CR* transcription units are divergent. The distance between the transcription units is so small that the methylase gene promoter overlaps with the C-protein binding sites. Upon binding to the distal site, the C-protein dimer prevents the RNA polymerase from binding to the methylase promoter. This regulatory mechanism is operational in the *Eco*RV RM system (11).

Another direct way of affecting the methylase gene transcription has been recently described for the *Esp*1396I RM system (12), where the *M* and *CR* transcription units are convergent. A single, high-affinity C-protein binding site was found upstream of the methylase gene. Because of the high affinity, the C-protein dimer binds to this site early on, repressing the methylase gene transcription. As larger quantities of C-protein are produced, the C-protein binds first to the distal and then to the proximal sites located upstream of the C-protein gene, causing activation then repression of endonuclease gene transcription as described above. The schema of genetic organization of several C-protein-dependent RM system loci is shown in **Fig. 17.2**.

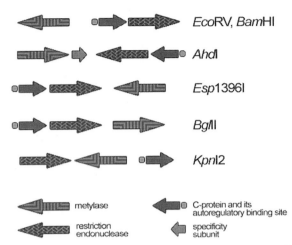

Fig. 17.2. The genetic organization of several known C-protein-dependent RM systems.

1.3. Computational Analysis of Transcription Regulation by C-Proteins

C-protein binding sites were first reported in 1995 (13). The analysis of upstream sequences of six then known C-protein genes revealed a conserved 12-bp region with the consensus sequence of ACTTATAGTCTG, later extended to 18 bp (aCTYATaGTCYGTNGNYt) (14). Newly discovered C-proteins changed the binding site motif (15). The authors argued that the 18-bp sites were unnecessarily long for monomer binding while lacking the dyad symmetry expected for dimer-binding sites. Based on these considerations, short binding sites (C.SmaI: AATGCTACT; C.NmeSI: TGCTACTTATAG; C.BglII: GATACTTATAGTC) were proposed. These sites were considered to interact with C-protein monomers.

As of July 2007, the main repository of RM systems, Rebase, contained as few as 48 C-proteins and 8 confirmed C-protein binding sites. The binding sites formed three structural groups, each named after its archetypical representative, with two groups containing only one member. The largest group with six binding sites was named after C.PvuII. It contained binding sites that lacked a pronounced palindromic structure expected for the dimeric form of C-proteins. The remaining binding sites of the C.EcoRV and C.EcoO109I group were palindromic.

The C.PvuII-like group of C-protein binding sites was subsequently extended to include 24 binding sites for known C-proteins from Rebase (16). A typical C.PvuII-like site was found to have a complex structure with a highly conserved tetranucleotide between two copies of the palindrome. Each palindrome arm is called a C-box, and the entire site thus contains four C-boxes. Interestingly, the proximal (3′) palindrome was less similar to the consensus, i.e., was "weaker"

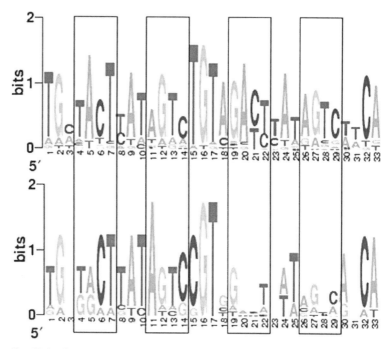

Fig. 17.3. Sequence logo showing the structure of C.*Pvu*II-like binding sites (16). Framed tetranucleotides represent C-boxes. A pair of C-boxes constitutes a palindrome. Additionally, the motifs contain highly conserved GT-core nucleotides and self-complementary dinucleotides at motif termini.

than the distal (5′) one, consistent with the above transcription regulatory mechanism. The observed palindromic structure (**Fig. 17.3**) likely reflects the dimeric form of C-proteins binding to the sites.

2. Methods

2.1. Large-Scale Identification and Analysis of C-Proteins

A systematic study of candidate C-proteins and their binding sites was reported in (17). All data related to the study can be found at http://iitp.bioinf.fbb.msu.ru/vsorokin.

Forty-six C-proteins from Rebase were used as queries in a BLAST (18) search against the non-redundant GenBank (19) nucleotide collection (tblastn, threshold: 1e-05, *see* **Note 1**). After manual curation, 245 unique hits were retained for further analysis.

The upstream sequences of candidate C-protein genes were analyzed in order to identify conserved regions that could correspond to the binding sites. Since there is no reason to expect a single conserved motif for all C-proteins, this was done separately in groups of closely related C-proteins. Specifically, a multiple alignment of all 291 proteins (46 known Rebase C-proteins and 245 putative C-proteins) was built using MUSCLE (20) (default parameters) and the maximum likelihood tree was constructed using the PROML procedure from the PHYLIP (21) package using the default parameter settings (*see* **Note 2**). Examination of the resulting phylogenetic tree revealed several large, separate branches. A slightly reduced variant of the original tree, containing only those proteins for which the putative binding motifs could be identified (see below), is shown in **Fig. 17.4**.

2.2. Identification of Candidate C-Protein Binding Motifs

Each major branch of the C-protein phylogenetic tree was analyzed independently. Hundred-base pairs long sequences upstream of most closely related genes were aligned using

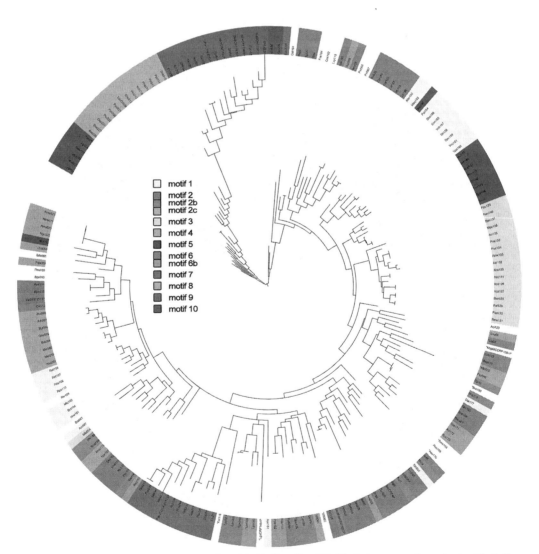

Fig. 17.4. The structure of motifs of predicted C-protein binding sites (17). Clustering was performed using ClusterTree-RS (22). C-boxes are *highlighted*, *arrows* represent palindromes formed by pairs of C-boxes.

MUSCLE (20) with the default parameter settings. If the alignment contained many highly conserved regions, the alignment was further extended by adding more distant members of the branch (*see* **Note 3**). The extension process terminated when alignments deteriorated completely. The resulting alignments were analyzed manually and the conserved islands that remained were considered to correspond to C-protein binding sites.

To identify additional binding sites, and, in particular, to account for a possibility of incorrectly annotated start codons of some predicted C-proteins, conserved sites obtained as described above were used to build HMM profiles (http://hmmer.wustl.edu, *see* **Note 4**), and the latter were used to scan by hmmer using default parameters for matches in regions from −100 to +50 relative to annotated start codons of all predicted C-protein genes. As expected, the majority of matches coincided with already predicted binding sites. However, several new matches were found and some matches were at a different location than the initially identified sites. A small overall number of corrections indicated the overall consistency of the prediction procedure. In total, 201 binding sites were predicted.

2.3. Validation of Predicted C-Protein Binding Sites

Candidate binding sites were predicted for 201 of the total of 291 C-proteins. As mentioned above, C-protein binding sites for eight Rebase RM systems have been identified experimentally and 24 more sites computationally (16). All these sites were present among the sites identified by our procedure.

2.4. The Structure of Binding Motifs

Previously known binding sites were assigned to one structural group named after its archetype C.*Pvu*II and two single-member groups: C.*Eco*RV and C.*Eco*O109I. The C.*Pvu*II-group sites had a structure of two short palindromes separated by highly conserved tetranucleotides (16). Two palindrome arms represent the so-called C-boxes, which are likely binding sites of individual C-protein monomers when they form dimers. Pairs of conserved, complementary positions outside the palindromes were disregarded in (16). Unlike the C.*Pvu*II-like sites, the binding sites of C.*Eco*RV and C.*Eco*O109I are single palindromes.

The set of newly predicted sites from (17) was split into clusters [ClusterTree-RS procedure (22)], which we will refer to as motifs. This procedure yielded 10 stable motifs (**Fig. 17.5**), which comprised 181 (90%) of the 201 predicted binding sites. While the remaining sites resembled motifs 1–6 (see below) the procedure failed to cluster them, probably due to the search stringency. The motifs logos (**Fig. 17.5**) are described below, while their main features are listed in **Table 17.1**.

Motifs 7 and 8

Motifs 7 and 8 correspond to the previously recognized C.*Eco*RV and C.*Eco*O109I groups, respectively. However, the

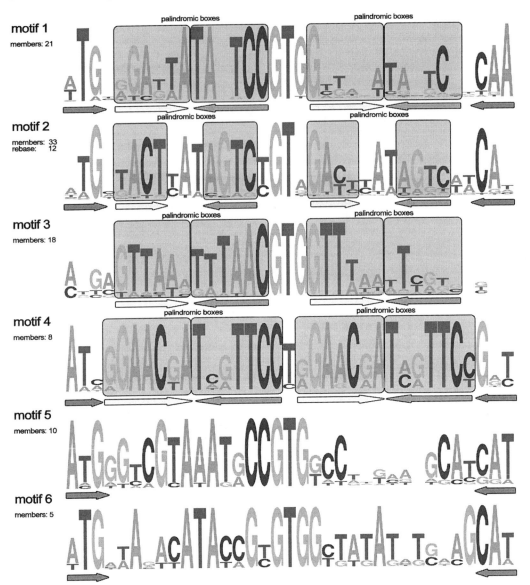

Fig. 17.5. The structure of motifs of predicted C-protein binding sites (17). Clustering was performed using ClusterTree-RS (22). C-boxes are highlighted. Arrows represent palindromes formed by pairs of C-boxes.

constructed motifs are longer due to additional conserved positions.

Motifs 1–6

Motifs 1–6 have the same length and share common structural features, described by the scheme

$$Z\text{-}X\text{-}N\text{-}X^*\text{-}[GT\text{-rich spacer}]\text{-}x\text{-}n\text{-}x^*\text{-}Z^*,$$

where Z denotes conserved complementary trinucleotides, X represents palindrome-forming C-boxes, and asterisks denote

Table 17.1
The classification of C-proteins and their binding motifs

Group 1 C.PvuII-like sites	Group 2 Palindromic sites	Group 3 Possible false positives
Motifs 1–4 • Length: 35 bp • GT-rich central area • Double palindromes (four C-boxes) • Conserved terminal complementary trinucleotides (except motif 3)	Motifs 7, 8, 10 • Single palindromes (two C-boxes) • Additional downstream palindromes • Form three separate, individual branches on the C-protein tree	Motif 9 • Unusually short • No pronounced palindromic structure • A separate branch on the C-protein tree
Motifs 5, 6 • Length: 35 bp • GT-rich central area • No pronounced C-boxes • Conserved terminal complementary trinucleotides		

complementary elements. The uppercase X-N-X* indicates that the 5′-copy is much closer to the overall palindromic consensus than the 3′-copy. The conserved spacer between the copies is unique for each motif.

Motifs 1, 2, and 4 fit the above scheme exactly. Motif 3 lacks external trinucleotides. Motifs 5 and 6 lack the palindromic structure but retain the highly conserved complementary external trinucleotides. All previously identified C.PvuII-like binding sites conform to the motif 2 structure.

Motif 9

Motif 9 is rather short (10 bp) but well conserved. It lacks any palindromic symmetry and reveals no other structural features. While these could be false positive, predicted binding sites for six C-proteins from Rebase belong to this motif. Hence, this prediction, while tentative, warrants experimental verification.

Motif 10

This single palindromic motif is not related to other motifs. Rebase contains no C-proteins predicted to bind to this motif. Again, this prediction requires experimental verification.

2.5. Downstream Sites

X-ray analysis revealed the dimeric nature of C-proteins C.AhdI and C.BclI (7, 8). Indeed, all experimentally studied C-protein binding events involved paired binding sites. Activation required their interaction with the high-affinity promoter-distal site. This is followed by repression through interaction with the low-affinity promoter-proximal site of the CR transcription unit.

Thus, if the predicted sites are functional, additional weaker sites could be expected in close vicinity and downstream of predicted "single" sites.

Motifs 1–4 already consisted of two palindromes representing promoter-distal and promoter-proximal binding sites. Consistent with the existing model, the consensus of the proximal palindrome was weaker than the consensus of the distal palindrome.

Motifs 7, 8, and 10 include single palindromes. A special procedure was applied to search for additional sites downstream of these motifs. First, HMM profiles were constructed for each motif (*see* **Note 4**). Next, upstream sequences of C-protein genes containing sites forming a motif were searched for profile matches. As expected, a strong match coinciding with already predicted binding site was observed in all cases. Importantly, all "second best" matches were downstream of primary matches and were predicted to be the downstream binding sites. The consensus of proximal (downstream) sites was weaker than that of distal sites. This result agrees with the established model of C-protein transcription regulation involving activation upon C-protein binding to the upstream site and subsequent repression following binding to the downstream site.

Unlike the case for motifs 1–4, the distance varies between the distal and proximal copies for motifs 7, 8, and 10. This suggests

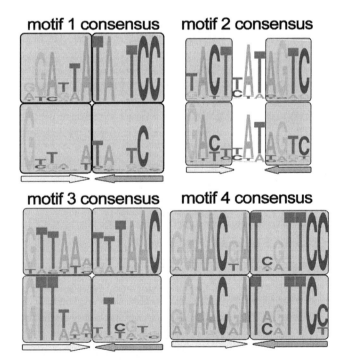

Fig. 17.6. Sequence logo of distal (*top*) and proximal (*bottom*) pairs of C-boxes from identified binding sites (17). Notation is as in **Fig. 17.5**.

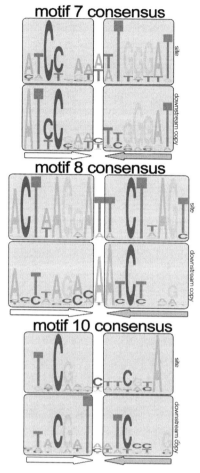

Fig. 17.6. (continued)

that C-protein dimers bound to these sites do not interact with each other, unlike C-proteins with constant distances.

No downstream copies were observed for C.*Pvu*II-like motifs 5 and 6 (lacking palindromic structure) as well as for non-palindromic motif 9. It is therefore unclear whether these binding sites, if real, operate using the same activation-repression mechanism as other C-proteins. The distal and proximal pairs of C-boxes are shown in **Fig. 17.6**.

3. Conclusions

Candidate C-proteins and predicted binding sites were consistently identified by a set of computational methods. Having started with known C-proteins from Rebase, we retrieved from

GenBank nucleotide sequences which could represent C-protein genes. A multi-step manual analysis was applied to remove possible false positives followed by the prediction of autoregulatory binding sites in upstream regions of C-protein genes. Manual analysis of the multiple alignments of closely related upstream sequences allowed for the identification of highly conserved islands, which were predicted to be the binding sites. Indeed, the predicted binding sites completely matched all known sites. The predicted binding sites were clustered into 10 motifs, of which only three had been known previously.

The work resulted in 201 predicted C-protein genes with predicted autoregulatory binding sites.

4. Notes

1. The use of such a liberal BLAST e-value threshold requires considerable care. On the other hand, since the typical C-protein length is only about 70 amino acids, stricter thresholds could yield loss of relevant hits. In (17), the authors controlled for several factors to avoid false positives. First, multiple alignment of all candidates was constructed and the presence of a pronounced HTH domain was verified. Second, manual analysis of hit annotations demonstrated that the set of candidates did not contain transcription factors with known, unrelated function. The third filter was the requirement of upstream regulatory sites, as described in the text.

2. The construction of phylogenetic trees requires high-quality multiple alignments. In case of distantly related proteins, computer-generated alignments need to be inspected manually. Alignments lacking a relatively conserved region (i.e., regions with a small number of mismatches and gaps) were refined as follows. The most distant sequences were removed until the conserved region appeared. Also, constructing a maximum likelihood tree of 300 proteins is a highly CPU-intensive task. If the purpose is to obtain a guide for iterative alignment of gene upstream regions, as here, the time may be decreased by temporarily removing very similar proteins. Subtrees for individual groups may then be constructed with complete data.

3. The "phylogenetic footprinting" approach (23) is used when candidate sites are expected to occur upstream of orthologous genes. This procedure is based on the assumption that the binding sites are more conserved than the

surrounding upstream regions. To achieve the sharpest contrast between the putative sites and the rest of the region, one needs to find the best set of sequences to be aligned. This is done heuristically by gradually adding more and more distant sequences until the alignment disintegrates. The alignment constructed at the last step before that is the one, where the conserved islands likely correspond to the binding sites. It is useful to constrain the alignment by retaining in the sequences to be aligned some part of the protein-coding region, which normally is sufficiently strongly conserved to be uniformly alignable at the nucleotide level.

Since for the RM-systems, due to frequent horizontal transfer, orthology relationships are challenging to establish, the evolutionary distance between the C-proteins (the regulators encoded by the regulated genes at the same time) was used as a guide for the progressive alignment of the upstream regions. The phylogenetic tree of C-proteins sets the order in which the upstream regions are considered. However, when the alignment starts to deteriorate, one should try to add several different sequences, since the exact topology of the tree is not very reliable.

4. Given the short extent of the binding sites (here, at most 35 bp), HMM profiles needs to be calibrated by the *hmmcalibrate* procedure. This allows one to increase precision of the *e*-value estimation when the e-value falls within the range of [1e-05, 1].

References

1. Bertani, G., Weigle, J.J. (1953) Host controlled variation in bacterial viruses. *J Bacteriol* 65, 113–121.
2. Lurlia, S.E., and Human, M.L. (1952) A nonhereditary, host-induced variation of bacterial viruses. *J Bacteriol* 64, 557–569.
3. Bickle, T.A., and Krueger, D.H. (1993) Biology of DNA restriction. *Microbiol Rev* 57, 434–450.
4. King, G., and Murray, N.E. (1994) Restriction enzymes in cells, not eppendorfs. *Trends Microbiol* 2, 465–469.
5. Knowle, D., Lintner, R., Touma, Y.M., and Blumenthal, R.M. (2005) Nature of promoter activated by C. PvuII, an unusual regulatory protein conserved among restriction-modification systems. *J Bacteriol* 187, 488–497.
6. Sawaya, M.R., Zhu, Z., Mersha, F. et al. (2005) Crystal structure of the restriction modification system control element C.BclI

and mapping of its binding site. *Structure* 13, 1837–1847.
7. McGeehan, J.E., Streeter, S.D., Papapanagiotou, I. et al. (2005) High-resolution crystal structure of the restriction-modification controller protein C.AhdI from *Aeromonas hydrophila. J Mol Biol* 346, 689–701.
8. Bart, A., Dankert, J., and van der Ende, A. (1999) Operator sequences for the regulatory proteins of restriction modification systems. *Mol Microbiol* 31, 1277–1278.
9. Bogdanova, E., Djordjevic, M., Papapanagiotou, I. et al. (2008) Transcription regulation of type II restriction-modification system AhdI. *Nucleic Acids Res* 36, 1429–1442.
10. Semenova, E., Minakhin, L., Bogdanova, E. et al. (2005) Transcription regulation of the EcoRV restriction modification system. *Nucleic Acids Res* 33, 6942–6951.
11. Zheleznaya, L.A., Kainov, D.E., Yunusova, A.K. et al. (2003) Regulatory C protein of

the EcoRV modification–restriction system. *Biochemistry (Moscow)* 68, 125–132.

12. Cesnaviciene, E., Mitkaite, G., Stankevicius, K. et al. (2003) Esp1396I restriction-modification system: structural organization and mode of regulation. *Nucleic Acids Res* 31, 743–749.

13. Rimseliene, R, Vaisvila, R, and Janulaitis, A. (1995) The eco72IC gene specifies a trans-acting factor which influences expression of both DNA methyltransferase and endonuclease from the Eco72I restriction-modification system. *Gene* 157, 217–219.

14. Anton, B.P., Heiter, D.F., Benner, J.S. et al. (1997) Cloning and characterization of the BglII restriction-modification system reveals a possible evolutionary footprint. *Gene* 187, 19–27.

15. Bart, A., Dankert, J., and van der Ende, A. (1999) Operator sequences for the regulatory proteins of restriction-modification systems. *Mol Microbiol* 31, 1275–1281.

16. Mruk, I., Rajesh, P., and Blumenthal, R.M. (2007) Regulatory circuit based on autogenous activation-repression: roles of C-boxes and spacer sequences in control of the PvuII restriction-modification system. *Nucleic Acids Res* 35, 6935–6952.

17. Sorokin, V., Severinov, K., and Gelfand, M.S. (2009) Systematic prediction of control proteins and their DNA binding sites. *Nucleic Acid Res* 37, 441–451.

18. Altschul, S.F., Madden, T.L., Schaffer, A.A. et al. (1997) Gapped BLAST and PSI-BLAST: a new generation of protein database search programs. *Nucleic Acids Res* 25, 3389–3402.

19. Benson, D.A., Karsch-Mizrachi, I., Lipman, D.J. et al. (2000) Genbank. *Nucleic Acids Res* 28, 15–18.

20. Edgar, R.C. (2004) MUSCLE: multiple sequence alignment with high accuracy and high throughput. *Nucleic Acids Res* 32, 1792–1797.

21. Felsenstein, J. (1989) PHYLIP – Phylogeny inference package (version 3.2). *Cladistics* 5, 164–166.

22. Stavrovskaia, E.D., Makeev, V.I., Mironov, A.A. (2006) ClusterTree-RS: the binary tree algorithm for identification of co-regulated genes by clustering regulatory signals. *Mol Biol (Moscow)* 40, 524–532.

23. Wasserman, W.W., and Fickett, J.W. (1998) Identification of regulatory regions which confer muscle-specific gene expression. *J Mol Biol* 278, 167–181.

Chapter 18

Evolution of *cis*-Regulatory Sequences in *Drosophila*

Xin He and Saurabh Sinha

Abstract

Cross-species comparison is an emerging paradigm for identifying *cis*-regulatory sequences and understanding their function and evolution. In this chapter, we review probabilistic models of evolution of transcription factor binding sites, which provide the theoretical basis for a number of new bioinformatics tools for comparative sequence analysis. We illustrate how important functional and evolutionary insights on binding site gain and loss can be acquired through sequence comparison. This includes the observation that binding site turnover follows a molecular clock and that its rate correlates with the strength of binding sites and the presence of other sites in the neighborhood. We also comment on emerging trends that go beyond individual binding sites to a more holistic study of regulatory evolution. We point out common technical challenges, such as reliable sequence alignment and binding site prediction, when doing comparative regulatory sequence analysis and note some potential solutions thereof.

Key words: *cis*-regulatory modules, regulatory evolution, transcription factor binding sites, probabilistic models, binding site turnover.

1. Introduction

The spatial–temporal expression patterns of genes are controlled by regulatory sequences often in the neighborhood of genes, through binding of transcription factors (TFs) to their binding sites (TFBSs) within these sequences. TFBSs tend to be clustered in ~1 kbp length sequences, forming "*cis*-regulatory modules" (CRMs) (1). Unlike coding sequences, CRM sequences are not surrounded by identifiable canonical elements, nor are their compositional rules as well understood, thus predicting their positions in genomes remains a difficult problem (2, 3). Despite intense efforts, we lack a good understanding of the organizational

I. Ladunga (ed.), *Computational Biology of Transcription Factor Binding*, Methods in Molecular Biology 674,
DOI 10.1007/978-1-60761-854-6_18, © Springer Science+Business Media, LLC 2010

principles of CRMs, e.g., how the arrangement of binding sites within a CRM affects their function. Cross-species comparisons provide a major opportunity to address these challenges: (i) CRM sequences tend to possess certain evolutionary signatures, which could be exploited to identify yet unknown CRMs (4), and (ii) the evolution of CRM sequences is constrained by functional requirements, so the study of CRM evolution should allow us to infer the underlying sequence–function relationships (5, 6). Finally, it has been argued that the change of regulatory sequences is a major source of evolutionary novelty in animal development, therefore, the study of CRM evolution will help to shed light on the path "from DNA to diversity" (7). The goal of this chapter is to introduce ideas and models pertaining to *cis*-regulatory evolution and illustrate how they can be leveraged to extract information from sequence comparison. We will focus on two aspects: probabilistic modeling of binding site evolution and the empirical study of binding site gain and loss during evolution.

Traditionally, TFBSs are identified by matching sequences with the binding specificities of TFs, represented as simple probabilistic objects called position weight matrices (PWMs) (8). For a TF with binding sites of length w, the PWM is defined as an ordered set of w multinomial distributions, each distribution representing the nucleotide frequencies of the corresponding position in the binding sites. A PWM is often visualized as a sequence logo, following the convention in (9) (**Fig. 18.1**). Since PWMs are generally short (usually less than 20 bp) and degenerate (one position of PWM may be occupied by diverse nucleotides with different probabilities), simply matching a sequence segment with a PWM tends to produce many false positives in large genomes (3). Furthermore, scoring by PWM implicitly makes the assumption that each position in a binding site is independent, thus ignoring

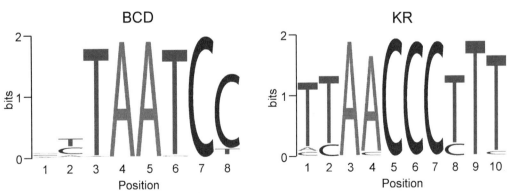

Fig. 18.1. The position weight matrices (PWMs) for the binding sites of Bicoid (Bcd) and Kruppel (Kr), two transcription factors involved in the segmentation of *D. melanogaster*. At each position, the height of each nucleotide is proportional to the frequency of this nucleotide at that position, and the total height of four nucleotides is equal to the information content.

the cases where different positions may be correlated (10). A key idea to improve the accuracy of prediction is to note that binding sites tend to be more evolutionarily constrained than neutral sequences, thus conservation across species enhances the signal for functional sites. Several methods have been developed to identify TFBSs based on their conservation, e.g., only sequences that match a PWM in two species will be reported as candidate sites (11–13). The formal way to utilize cross-species information is through a probabilistic model of evolution of TFBSs, which we explain below in detail. Because of the inherent randomness of evolutionary processes, a statistical framework based on such a model is essential for proper assessment of the statistical significance of binding site conservation or change.

Several studies have demonstrated that a functional binding site is not necessarily preserved throughout evolution; instead, a more complex and dynamic picture has emerged from these studies. Emberly et al. (14) found that binding sites are not substantially more conserved than their adjacent sequences in *Drosophila*. Gain and loss of TFBS were found to be common in promoters or known CRM sequences (14–16) and in regions experimentally found to be occupied by TFs in vivo (17, 18). However, what leads to this frequent loss and gain is far from being resolved. For example, are the binding sites with higher affinities more likely to be conserved in evolution? How does the local context of a site, i.e., the presence of other sites in the neighborhood, affect its likelihood of being lost during evolution? Understanding the causes of binding site turnover will be crucial to our understanding of the function and evolution of *cis*-regulatory sequences and to our efforts of building realistic quantitative models.

2. Methods

2.1. Evolutionary Models of Transcription Factor Binding Sites

A probabilistic model of sequence evolution estimates the probability of one sequence evolving to another during a specified time span. Comparisons of protein or DNA sequences across species have generally relied on such models. Nucleotide substitution models have been the cornerstone of many successful methods for reconstructing phylogenetic trees (19) and related tasks such as sequence alignment (20). The same class of models has found uses in regulatory sequence evolution as well. For example, a type of substitution model, commonly referred to as F81 model (21), was incorporated into several popular tools of binding site prediction (22, 23). The nucleotide substitution model uses a Markov chain of four states to represent the evolutionary transitions among four nucleotides at any position in a sequence.

The F81 model is parameterized by an equilibrium distribution π and a rate parameter α (19, 24). The "rate matrix" of the model is given by

$$
Q = \begin{bmatrix}
-(\pi_C + \pi_G + \pi_T)\alpha & \pi_C\alpha & \pi_G\alpha & \pi_T\alpha \\
\pi_A\alpha & -(\pi_A + \pi_G + \pi_T)\alpha & \pi_G\alpha & \pi_T\alpha \\
\pi_A\alpha & \pi_C\alpha & -(\pi_A + \pi_C + \pi_T)\alpha & \pi_T\alpha \\
\pi_A\alpha & \pi_C\alpha & \pi_G\alpha & -(\pi_A + \pi_C + \pi_G)\alpha
\end{bmatrix}
$$

[1]

where π_A, π_G, π_C, π_T specify the equilibrium distribution of nucleotides. The transition probability matrix of the Markov chain is $P(t) = \{P_{ij}(t)\} = e^{Q(t)}$, where $P_{ij}(t)$ is the probability that the descendent nucleotide is j after time t conditioned on the ancestral nucleotide i. To adopt the nucleotide substitution models to binding sites of a TF, one simply sets the equilibrium distribution at any position to be the distribution of that position in the PWM of the TF, while α is set to the neutral mutation rate inferred from other studies. In other words, the nucleotide at each position in a TFBS is assumed to evolve independently, with evolutionary dynamics dictated by the F81 model parameterized by the corresponding position of the PWM.

Halpern and Bruno (25) developed a model that also considers the evolution of binding sites as independent substitutions of nucleotides, but explicitly treats mutation and natural selection. Letting $\mu(a, b)$ denote the mutation rate of nucleotide a to b in the absence of selection (26), and N denote the population size, the substitution rate $u(a, b)$ can be written as the product of the total mutation rate in the population, $2N\mu(a, b)$, and the probability of fixation of the a to b mutation. According to population genetics theory (27), this rate is equal to

$$
u(a, b) = 2N\mu(a, b)\frac{1 - \exp[-2(F(b) - F(a))]}{1 - \exp[-4N(F(b) - F(a))]}
$$

[2]

where $F(\cdot)$ is the relative "fitness" of a nucleotide. The key idea in the Halpern–Bruno (HB) model is to relate the fitness of a nucleotide to the equilibrium distribution: the greater the fitness, the larger the probability in the equilibrium distribution. By formalizing this intuition, Halpern and Bruno were able to derive the following equation for functional sequences:

$$
u(a, b) = \mu(a, b)\frac{\log\dfrac{\pi(b)\mu(b, a)}{\pi(a)\mu(a, b)}}{1 - \dfrac{\pi(a)\mu(a, b)}{\pi(b)\mu(b, a)}}
$$

[3]

where $\pi(\cdot)$ is the equilibrium distribution of the functional nucleotide, which in the case of TFBSs is the nucleotide distribution of the corresponding position in the PWM (28).

A key assumption of the above models is that each position of a binding site evolves independently. Clearly, this may not always be true. The same mutation can have a very different effect on the functionality of a site depending on how strong the site was to begin with. A site that is close to optimal will probably remain a site even if an important nucleotide is mutated, thus this substitution is likely to be fixed in the population. On the other hand, the same nucleotide mutation inside a weak site may have a larger functional consequence (the site loses its binding functionality), thus will be less likely to be fixed. This intuition is captured in a model called the "Site-level Selection" or "SS" model that treats binding sites as single evolutionary units (29). The model assumes that the fitness of a binding site may take two values: 1 if the binding affinity of this site is below some threshold (non-functional) and $1 + s$ if the affinity is above this threshold (functional), for $s > 0$. (This approach to modeling genotypic variations with differing fitness consequences is standard practice in population genetics, for example (30).) By applying equation [2] where a and b now refer to sites, we have the substitution rate $u(a, b) = \mu(a, b)$, the neutral mutation rate, if both a and b are functional sites or both are non-functional sites. When a is non-functional and b is functional, we have

$$u(a, b) = \mu(a, b)\frac{4\,Ns}{1 - e^{-4\,Ns}} \qquad [4]$$

For the opposite situation, we have

$$u(a, b) = \mu(a, b)\frac{4\,Ns}{e^{4\,Ns} - 1} \qquad [5]$$

Note that N and s are inseparable in the above equations, so we will use the single quantity $4\,Ns$ to represent the intensity of selection. There are additional models of differing complexities that treat binding sites as evolutionary units, based on similar ideas. Interested readers should refer to (31–34).

The last two models of TFBS evolution differ in one key assumption: the HB model assumes that each position of a TFBS evolves independently, while the SS model assumes that the entire TFBS evolves as one unit. Is there a way to determine which of these two models (or assumptions) is more realistic? One way to answer this question is to simulate the evolution of TFBSs following each model and ascertain which model gives better agreement with real multi-species data. This test was performed in (29) as described next. First, a set of CRMs involved in blastoderm development of *D. melanogaster* was obtained. Next, predicted binding sites were collected for seven TFs – *Bicoid, Caudal, Dstat, Hunchback, Knirps, Kruppel,* and *Tailless* – in these CRMs in *D. melanogaster*, along with their respective aligned sequences (whether designated binding sites or not) in a closely related species (*Drosophila yakuba*). Let us (arbitrarily) call the sites in the former species as "ancestral" and the latter as "descendant." Assigning an "energy score" to each site based on its similarity to the PWM of the corresponding TF (35), the difference in energy scores was calculated between the ancestral and descendant sites and used as the statistic to represent binding site evolution. For each TF, the histogram of this "energy difference" statistic was computed. This formed the real data used in the test. Another data set was constructed by simulating TFBS evolution (by each model separately) for an amount of time equaling the divergence time of the two species and noting the difference in strengths between each pair of ancestral and evolved sites. (The evolutionary simulation procedure is described below.) Simulated data from the SS model was seen to provide a significantly better fit to the real data than the HB model, regardless of the TF whose sites were used in the test (**Fig. 18.2**). This result thus provides support to the notion that there are significant epistatic interactions among different positions in a TFBS and a proper model should view binding sites instead of individual positions as units of evolution.

Simulation of TFBS evolution. The simulation procedure consists of the following steps: (i) compute the rate of each substi-

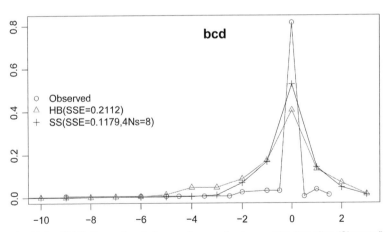

Fig. 18.2. Distributions of evolutionary changes in observed binding sites (Observed) and those simulated by the Halpern–Bruno (HB) and Site-level Selection (SS) models for the transcription factor Bcd in *D. melanogaster* and *D. yakuba* species pair. The *x*- and *y*-axes represent energy difference and frequency, respectively. SSE denotes the sum of squared errors between the observed and the simulation-based distributions and "4*Ns*" denotes the optimal value of this free parameter of the SS model. Reproduced from the authors' publication in *PLoS Genetics* (29).

Evolution of *cis*-Regulatory Sequences in *Drosophila*   289

tution event at each position according to equation [3] for HB model or equations [4] and [5] for SS model; (ii) choose a substitution event with probability proportional to the rate of the event; (iii) update the TFBS according to this event and increment the "clock" by an exponential random variable with mean equal to the inverse of the total rates of all events. The procedure is run until the "clock" reads a pre-specified time (the divergence between the species studied). This procedure simulates the evolution of a single TFBS, and repeating the procedure several times provides the data required.

2.2. Patterns of Binding Site Gain and Loss

The previous section presented models of TFBS evolution that try to quantitatively describe the dynamics of TFBS evolution. They do not attempt to explain the evolutionary forces underlying this phenomenon. For instance, we do not yet know what causes the gain and loss of TFBSs. They could be caused simply by chance (a process called random drift (30)) or be the consequences of changes in the forces of selection favoring gain of new sites or loss of existing ones (adaptive selection) (36). One way to investigate this issue is to test if gains and losses of TFBSs during evolution follow a "molecular clock" (29). According to the neutral theory of evolution, when mutations follow the molecular clock, this may suggest the absence of adaptive selection (26). To examine this hypothesis, let us calculate the fraction of binding sites in *D. melanogaster* that have an orthologous site (above the threshold) in a second species and plot this fraction as a function of evolutionary divergence from the second species. For all transcription factors, the fraction of conserved binding sites is seen to decrease linearly ($R^2 > 0.90$) as the divergence time increases, a clear sign of a molecular clock (**Fig. 18.3**). Even though we cannot exclude the presence of adaptive selection in individual cases, this result seems to suggest that selection mainly acts to maintain the existing functional sites during evolution, coupled with the occasional losses of binding sites due to random drift.

Next, let us look into the specific causes that influence the evolutionary fate of binding sites, as reported in (29). To begin with, we need a measure of TFBS turnover. For each set of orthologous TFBSs in 12 *Drosophila* species, let us construct a phylogenetic tree by labeling a leaf node as 1 if its corresponding species has the site and 0 otherwise. A subtree rooted at the least common ancestor of leaf nodes labeled 1 is then identified. The turnover rate of this TFBS is defined as the parsimony cost calculated for the subtree (i.e., the number of branches that carry a change in label, either 0 to 1 or 1 to 0) divided by the sum of branch lengths of the subtree (**Fig. 18.4**). The overall turnover rate across multiple sets of orthologous TFBSs is defined similarly, where summations of both parsimony costs and branch lengths are over all orthologous TFBS sets. With formal definitions of

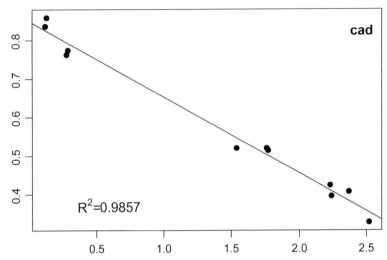

Fig. 18.3. The fraction of *D. melanogaster* TFBSs that are conserved in a related species (*y*-axis), as a function of the divergence time to that species (*x*-axis), for transcription factors Cad. Reproduced from the authors' publication in *PLoS Genetics* (29).

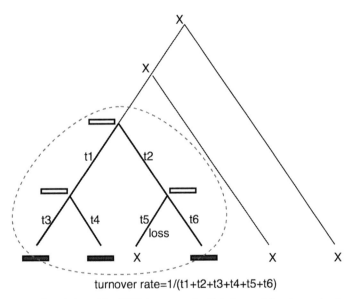

Fig. 18.4. Calculation of the TFBS turnover rate. Only three of the six species have a binding site (*rectangles* at the leaves). The subtree rooted at the least common ancestor of the binding sites is identified (in the *dashed circle*). There is one loss event in the subtree and, thus, the turnover rate is 1 (the number of events) divided by the sum of t_1 though t_6 (branch lengths in the subtree). Reproduced from the authors' publication in *PLoS Genetics* (29).

TFBS turnover in hand, we are now ready to see if their observed values correlate with measurements of other variables that represent potential causes of turnover.

One factor that may affect the rate of loss of existing sites is the strength of those sites, which may be related to their functional importance. To test the correlation between binding site strengths and their turnover rates, let us define the strength of a site as the degree of match of this site to the corresponding motif, as measured by a log-likelihood ratio (LLR) score (8). Using turnover rates as defined above, significant negative correlations are observed between TFBS strength and turnover, for six of the seven TFs analyzed (Table 5 in (29)). A simple explanation for this finding is that stronger sites are more likely to be important to CRM function, thus under stronger constraint (37). An alternative explanation is that the functionality of a site is determined by a strength threshold, and once a site drops below that threshold, it is impervious to selective forces. Assuming this is true, a weaker site is closer to the threshold than a stronger site and may thus be lost more easily as supported by (34).

Binding sites often interact with the sites of other factors in their neighborhood (38, 39). It is therefore natural to speculate that evolutionary constraints on a TFBS would depend on the presence of its interacting partner sites in close proximity. Since we do not know, in general, the interacting pairs of sites a priori, we may ask if there is a correlation between TFBS turnover and the presence of any other TFBS in close proximity. Following a procedure similar to that of Hare et al. (40), let us classify a TFBS as belonging to the proximal, distal, or overlap class depending on whether the closest site of another factor is within 10 bp, more than 10 bp away, or overlaps with this site. The sites in the overlap or proximal categories are found to be more conserved (present in all 12 species) than the sites in the distal category ($p = 4.39 \times 10^{-5}$, hypergeometric test). The same test is then repeated individually for each factor. Comparing the proximal and distal classes (Table 7 in (29), column "P versus D"), we find *Dstat* and *Tailless* sites are significantly more conserved when they have a proximal partner site ($p = 0.021$ and 0.027, respectively). In a similar comparison of the overlap class with its complement (proximal or distal) (Table 7 in (29), column "O versus NO"), *Caudal*, *Hunchback*, and *Kruppel* sites are found to be more conserved when having an overlapping partner ($p = 0.002$, 0.017, and 0.039, respectively). In summary, five of the seven TFs show a significant tendency to be conserved when they have a partner either overlapping with or proximal to them. To understand these results we note that different mechanisms of local interactions between TFBSs are known in developmental CRMs, e.g., cooperative binding between two factors (38, 41), short-range quenching (39, 42), competitive binding to overlapping sites (41). In all these cases, the loss of a single binding site may disrupt the interactions and create a large change of gene expression. As a consequence, these locally interacting pairs of binding

292   He and Sinha

sites may be under stronger selection. A recent paper reported similar results for four CRMs of the *even-skipped* gene (40).

Finally, we note that the validity of all evolutionary analyses presented above relies on the accuracy of the underlying bioinformatics methods used in evolutionary comparisons, primarily the alignment of orthologous sequences (Section **3.1**) and the prediction of TFBS (Section **3.2**).

2.3. Evolution of Entire cis-Regulatory Modules

We have seen above evidence that a realistic model of binding site evolution should treat sites as whole units. The individual binding sites, however, are only part of an intermediate level of organization of genome sequences. The function of regulatory sequences is to control the expression of their target genes, which requires coordination of multiple binding sites. Therefore, it is more natural to consider entire *cis*-regulatory modules as the appropriate functional units for evolution. Under this scenario, the evolutionary process of a CRM starts with a random mutation that may create a new site, or change the affinity of an existing site, or even destroy one. The fixation of this mutation in the population will depend on its functional consequence, in other words, the change of fitness of the new sequence. This mutation selection cycle is repeated over many generations. The critical component of such a model is the fitness function of a sequence, which may be defined as, for example, whether the sequence matches certain pre-specified constraints (e.g., it must contain certain combinations of binding sites) or the similarity between the expression pattern directed by this sequence and the desired pattern. This approach centered on entire functional sequences has led to several simulation-based studies which reveal unique insights of *cis*-evolution (43–45). We believe this line of work promises to reveal more intimate connections between function and evolution of regulatory sequences.

3. Notes

3.1. Alignment of Regulatory Sequences

Evolutionary comparison crucially depends on alignment of orthologous sequences, but in general, alignments cannot be perfectly determined and may be a source of biased conclusion. It has been shown that the alignment procedure may seriously affect the results of comparative genomic analysis including reconstruction of phylogenetic trees and detection of adaptive selection (46). We illustrate this problem with an example of sequence alignment on one CRM sequence using the popular tool, LAGAN (**Fig. 18.5a**). If one knows the relevant motifs of the regulatory sequences being studied, it may be possible to develop customized tools to take advantage of their unique structural and evolutionary properties. This is the aim of several recent studies.

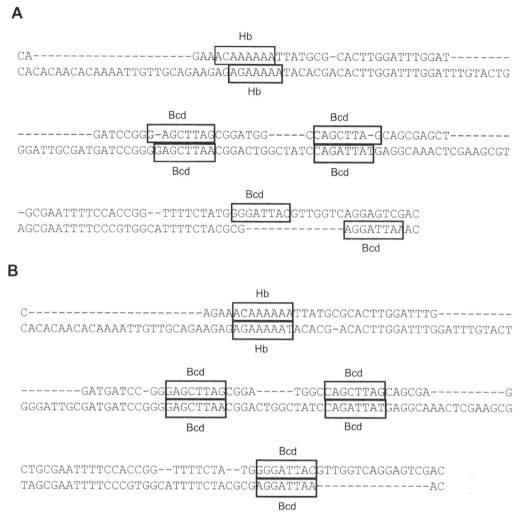

Fig. 18.5. *D. melanogaster–D. pseudoobscura* alignment of part of the CRM, "hb anterior activator." Shown in the *D. melanogaster* sequence (*top*) are the FlyReg sites of Bcd and Hb and shown in the *D. pseudoobscura* sequence are the predicted sites in this region. **a** Lagan; **b** EMMA. Reproduced from the authors' publication in *PLoS Computational Biology* (51).

Among them, the programs CONREAL (47), EEL (48), and SimAnn (49) align putative CRM sequences with scoring schemes that favor the alignment of multiple sites matching the same TFBS profiles. Developing on the same ideas, the more recent tools Morph (50) and EMMA (51) use evolutionary models of binding sites and allow binding sites to occur in only one species, but not the other (all these tools construct pairwise alignments). **Figure 18.5b** demonstrates the benefits of using TF motif information to improve alignment of regulatory sequences from EMMA. In the study on which much of the results in this chapter are based (29), a multiple alignment tool called ProbCons-Morph was used that combines the strengths of Morph in pairwise alignment with those of a probabilistic multiple alignment

3.2. Prediction of Transcription Factor Binding Sites

framework called ProbCons (52). The best strategy for dealing with alignment when analyzing regulatory sequences seems to be using customized tools such as EEL and EMMA when the motifs are known a priori and comparing a few different common alignment tools when the motifs are unknown so that the results are robust to the specific tool used.

Another important issue when doing regulatory sequence analysis across species is the prediction of TFBSs. Since this step relies on a PWM (TF binding specificity) that was characterized in one species, we first need to examine whether the PWM is conserved in different species. Researchers have found examples in yeast where the change of TF binding specificities is an important part of the evolutionary change of regulatory networks (53). In the context of the results shown in this chapter, there is prior computational evidence that the binding specificities of the TFs studied have not changed significantly between *D. melanogaster* and *Drosophila pseudoobscura* (54). The next issue to consider is the potentially high false-positive rate associated with prediction of binding sites using PWMs. We suggest several ideas to address this problem, while noting that the problem is difficult as at molecular level, the interactions between TFs and their binding sites are stochastic and quantitative, and it may not be appropriate to have a binary classification of sequences. (1) If possible, one should limit the PWM search regions to sequences that are likely to be functional. In (29), the sequences that were scanned for TFBSs had been experimentally determined previously to be regulatory sequences. The increasingly common data from genome-wide ChIP-chip or ChIP-seq experiments also provide excellent constraints for putative functional sequences (55, 56). A PWM match in sequences with such experimental validation is much more like to be a true functional binding site. (2) If applicable, one should use inter-species conservation to boost the signals for functional binding sites. Tools such as rVista (11), MONKEY (57), and EMMA (51) serve this purpose. If the goal is to study evolution of sites, in particular, their gain and loss, one reasonable strategy is to predict binding sites from a subset of species and analyze the evolution in other species not used for site prediction. This is meant to avoid "ascertainment bias," because most TFBS prediction tools already assume the binding sites tend to be conserved. (3) If none of these ideas is applicable, one will need to estimate the false-positive (FP) rates and make certain corrections in reaching conclusions. To estimate the FP rates, for example, one can treat all putative sites as a mixture of functional and neutral sites. If the relevant parameters of functional and neutral sites (e.g., substitution rates, the probability of being conserved across species, or the density of sites) are known, then the ratio of neutral sites can be inferred from the overall patterns (18, 29).

References

1. Howard, M.L., and Davidson, E.H. (2004) cis-Regulatory control circuits in development. *Dev Biol* 271, 109–118.
2. Papatsenko, D., and Levine, M. (2005) Computational identification of regulatory DNAs underlying animal development. *Nat Methods* 2, 529–534.
3. GuhaThakurta, D. (2006) Computational identification of transcriptional regulatory elements in DNA sequence. *Nucleic Acids Res* 34, 3585–3598.
4. Stark, A., Lin, M.F., Kheradpour, P. et al. (2007) Discovery of functional elements in 12 *Drosophila* genomes using evolutionary signatures. *Nature* 450, 219–232.
5. Wray, G.A., Hahn, M.W., Abouheif, E. et al. (2003) The evolution of transcriptional regulation in eukaryotes. *Mol Biol Evol* 20, 1377–1419.
6. Wittkopp, P.J. (2006) Evolution of cis-regulatory sequence and function in *Diptera*. *Heredity* 97, 139–147.
7. Carroll, S., Grenier, J., and Weatherbee, S. (2001) *From DNA to diversity: molecular genetics and the evolution of animal design*. Blackwell Science, Oxford.
8. Stormo, G.D. (2000) DNA binding sites: representation and discovery. *Bioinformatics* 16, 16–23.
9. Schneider, T.D., and Stephens, R.M. (1990) Sequence logos: a new way to display consensus sequences. *Nucleic Acids Res* 18, 6097–6100.
10. Sharon, E., Lubliner, S., and Segal, E. (2008) A feature-based approach to modeling protein-DNA interactions. *PLoS Comput Biol* 4, e1000154.
11. Loots, G. G., and Ovcharenko, I. (2004) rVISTA 2.0: evolutionary analysis of transcription factor binding sites. *Nucleic Acids Res* 32, W217–221.
12. Wasserman, W.W., Palumbo, M., Thompson, W. et al. (2000) Human-mouse genome comparisons to locate regulatory sites. *Nat Genet* 26, 225–228.
13. Doniger, S.W., Huh, J., and Fay, J.C. (2005) Identification of functional transcription factor binding sites using closely related *Saccharomyces* species. *Genome Res* 15, 701–709.
14. Dermitzakis, E.T., Bergman, C.M., and Clark, A.G. (2003) Tracing the evolutionary history of *Drosophila* regulatory regions with models that identify transcription factor binding sites. *Mol Biol Evol* 20, 703–714.
15. Ludwig, M.Z., Patel, N.H., and Kreitman, M. (1998) Functional analysis of eve stripe

2 enhancer evolution in *Drosophila*: rules governing conservation and change. *Development* 125, 949–958.
16. Doniger, S.W., and Fay, J.C. (2007) Frequent gain and loss of functional transcription factor binding sites. *PLoS Comput Biol* 3, e99.
17. Borneman, A.R., Gianoulis, T.A., Zhang, Z.D. et al. (2007) Divergence of transcription factor binding sites across related yeast species. *Science* 317, 815–819.
18. Moses, A.M., Pollard, D.A., Nix, D.A. et al. (2006) Large-scale turnover of functional transcription factor binding sites in *Drosophila*. *PLoS Comput Biol* 2, e130.
19. Yang, Z. (2006) *Computational Molecular Evolution*. Oxford University Press.
20. Miklos, I., Novak, A., Satija, R., Lingso, R., and Hein, J. (2009) Stochastic models of sequence evolution including insertion-deletion events. *Stat Methods Med Res* 18(5), 453–485.
21. Felsenstein, J. (1981) Evolutionary trees from DNA sequences: a maximum likelihood approach. *J Mol Evol* 17, 368–376.
22. Sinha, S., van Nimwegen, E., and Siggia, E.D. (2003) A probabilistic method to detect regulatory modules. *Bioinformatics* 19(Suppl. 1), i292–i301.
23. Siddharthan, R., Siggia, E.D., and van Nimwegen, E. (2005) PhyloGibbs: a Gibbs sampling motif finder that incorporates phylogeny. *PLoS Comput Biol* 1, e67.
24. Hasegawa, M., Kishino, H., and Yano, T. (1985) Dating of the human-ape splitting by a molecular clock of mitochondrial DNA. *J Mol Evol* 22, 160–174.
25. Halpern, A.L., and Bruno, W.J. (1998) Evolutionary distances for protein-coding sequences: modeling site-specific residue frequencies. *Mol Biol Evol* 15, 910–917.
26. Kimura, M. (1983) *The neutral theory of molecular evolution*, Cambridge University Press, Cambridge, MA.
27. Crow, J.F., and Kimura, M. (1970) *An introduction to population genetics theory*. Harper & Row Publishers, New York, NY.
28. Moses, A.M., Chiang, D.Y., Kellis, M. et al. (2003) Position specific variation in the rate of evolution in transcription factor binding sites. *BMC Evol Biol* 3, 19.
29. Kim, J., He, X., and Sinha, S. (2009) Evolution of regulatory sequences in 12 *Drosophila* species. *PLoS Genet* 5, e1000330.
30. Hartl, D.L., and Clark, A.G. (2006) *Principles of poulation genetics*. Sinauer Associates, Sunderland, MA.

31. Berg, J., Willmann, S., and Lassig, M. (2004) Adaptive evolution of transcription factor binding sites. *BMC Evol Biol* 4, 42.
32. Mustonen, V., and Lässig, M. (2005) Evolutionary population genetics of promoters: predicting binding sites and functional phylogenies. *Proc Natl Acad Sci USA* 102, 15936–15941.
33. Mustonen, V., Kinney, J., Callan, C.G., Jr., and Lassig, M. (2008) Energy-dependent fitness: a quantitative model for the evolution of yeast transcription factor binding sites. *Proc Natl Acad Sci USA* 105, 12376–12381.
34. Raijman, D., Shamir, R., and Tanay, A. (2008) Evolution and selection in yeast promoters: analyzing the combined effect of diverse transcription factor binding sites. *PLoS Comput Biol* 4, e7.
35. Stormo, G.D., and Fields, D.S. (1998) Specificity, free energy and information content in protein-DNA interactions. *Trends Biochem Sci* 23, 109–113.
36. Ludwig, M.Z. (2002) Functional evolution of noncoding DNA. *Curr Opin Genet Dev* 12, 634–639.
37. Papatsenko, D., and Levine, M. (2005) Quantitative analysis of binding motifs mediating diverse spatial readouts of the Dorsal gradient in the *Drosophila* embryo. *Proc Natl Acad Sci USA* 102, 4966–4971.
38. Struhl, K. (2001) Gene regulation. A paradigm for precision. *Science* 293, 1054–1055.
39. Gray, S., and Levine, M. (1996) Transcriptional repression in development. *Curr Opin Cell Biol* 8, 358–364.
40. Hare, E.E., Peterson, B.K., Iyer, V.N., Meier, R., and Eisen, M.B. (2008) Sepsid even-skipped enhancers are functionally conserved in *Drosophila* despite lack of sequence conservation. *PLoS Genet* 4, e1000106.
41. Small, S., Blair, A., and Levine, M. (1992) Regulation of even-skipped stripe 2 in the *Drosophila* embryo. *EMBO J* 11, 4047–4057.
42. Kulkarni, M.M., and Arnosti, D.N. (2005) *cis*-regulatory logic of short-range transcriptional repression in *Drosophila melanogaster*. *Mol Cell Biol* 25, 3411–3420.
43. Huang, W., Nevins, J.R., and Ohler, U. (2007) Phylogenetic simulation of promoter evolution: estimation and modeling of binding site turnover events and assessment of their impact on alignment tools. *Genome Biol* 8, R225.
44. MacArthur, S., and Brookfield, J.F. (2004) Expected rates and modes of evolution of enhancer sequences. *Mol Biol Evol* 21, 1064–1073.
45. Khatri, B.S., McLeish, T. C., and Sear, R. P. (2009) Statistical mechanics of convergent evolution in spatial patterning. *Proc Natl Acad Sci USA* 106, 9564–9569.
46. Wong, K.M., Suchard, M.A., and Huelsenbeck, J.P. (2008) Alignment uncertainty and genomic analysis. *Science* 319, 473–476.
47. Berezikov, E., Guryev, V., Plasterk, R.H., and Cuppen, E. (2004) CONREAL: conserved regulatory elements anchored alignment algorithm for identification of transcription factor binding sites by phylogenetic footprinting. *Genome Res* 14, 170–178.
48. Hallikas, O., Palin, K., Sinjushina, N. et al. (2006) Genome-wide prediction of mammalian enhancers based on analysis of transcription factor binding affinity. *Cell* 124, 47–59.
49. Bais, A.S., Grossmann, S., and Vingron, M. (2007) Simultaneous alignment and annotation of *cis*-regulatory regions. *Bioinformatics* 23, e44–e49.
50. Sinha, S., and He, X. (2007) MORPH: probabilistic alignment combined with hidden Markov models of *cis*-regulatory modules. *PLoS Comput Biol* 3, e216.
51. He, X., Ling, X., and Sinha, S. (2009) Alignment and prediction of *cis*-regulatory modules based on a probabilistic model of evolution. *PLoS Comput Biol* 5, e1000299.
52. Do, C.B., Mahabhashyam, M.S., Brudno, M., and Batzoglou, S. (2005) ProbCons: probabilistic consistency-based multiple sequence alignment. *Genome Res* 15, 330–340.
53. Tanay, A., Regev, A., and Shamir, R. (2005) Conservation and evolvability in regulatory networks: the evolution of ribosomal regulation in yeast. *Proc Natl Acad Sci USA* 102, 7203–7208.
54. Sinha, S., Schroeder, M.D., Unnerstall, U. et al. (2004) Cross-species comparison significantly improves genome-wide prediction of *cis*-regulatory modules in *Drosophila*. *BMC Bioinformatics* 5, 129.
55. Bulyk, M.L. (2006) DNA microarray technologies for measuring protein-DNA interactions. *Curr Opin Biotechnol* 17, 422–430.
56. Barski, A., and Zhao, K. (2009) Genomic location analysis by ChIP-Seq. *J Cell Biochem* 107, 11–18.
57. Moses, A.M., Chiang, D.Y., Pollard, D.A. et al. (2004) MONKEY: identifying conserved transcription-factor binding sites in multiple alignments using a binding site-specific evolutionary model. *Genome Biol* 5, R98.

Chapter 19

Regulating the Regulators: Modulators of Transcription Factor Activity

Logan Everett, Matthew Hansen, and Sridhar Hannenhalli

Abstract

Gene transcription is largely regulated by DNA-binding transcription factors (*TFs*). However, the TF activity itself is modulated via, among other things, post-translational modifications (*PTMs*) by specific modification enzymes in response to cellular stimuli. TF-PTMs thus serve as "molecular switchboards" that map upstream signaling events to the downstream transcriptional events. An important long-term goal is to obtain a genome-wide map of "regulatory triplets" consisting of a TF, target gene, and a modulator gene that specifically modulates the regulation of the target gene by the TF. A variety of genome-wide data sets can be exploited by computational methods to obtain a rough map of regulatory triplets, which can guide directed experiments. However, a prerequisite to developing such computational tools is a systematic catalog of known instances of regulatory triplets. We first describe PTM-Switchboard, a recent database that stores triplets of genes such that the ability of one gene (the TF) to regulate a target gene is dependent on one or more PTMs catalyzed by a third gene, the modifying enzyme. We also review current computational approaches to infer regulatory triplets from genome-wide data sets and conclude with a discussion of potential future research. PTM-Switchboard is accessible at http://cagr.pcbi.upenn.edu/PTMswitchboard/

Key words: Transcription factor, post-translational modification, modifying enzyme, regulatory network, computational biology.

1. Introduction

Gene transcription is regulated, in large part, by transcription factor (TF) proteins that bind to genomic *cis*-regulatory elements in a sequence-specific fashion. The activities of the TFs are often modulated by other proteins. One class of modulators consists of modifying enzymes that, through direct physical interaction with the TF, chemically modify specific residues of the

I. Ladunga (ed.), *Computational Biology of Transcription Factor Binding*, Methods in Molecular Biology 674, DOI 10.1007/978-1-60761-854-6_19, © Springer Science+Business Media, LLC 2010

TF protein, thereby altering its nuclear transport, protein degradation, DNA-binding, or co-factor interactions (1). Hundreds of distinct types of post-translational modification (PTM), such as phosphorylation, acetylation, methylation, glycosylation, have been reported (2, 3), at least a dozen of which are known to regulate TF activities, as reviewed in (1, 4–10). TF-PTMs therefore act as "molecular switchboards" that map inputs from cell signaling pathways to gene transcripts (5, 6). On one extreme, such regulation can be simple and binary – i.e., PTMs that serve as "on/off" switches for TF activity. More often, however, this regulation is highly complex (11), with multiple signaling inputs integrated into tightly regulated transcript levels, and each PTM affecting each target gene in a manner dependent on the larger promoter context (12).

The characterized instances of TF-PTMs are biased largely toward well-studied TFs, e.g., the cAMP-response element binding protein (CREB) (13), and PTM types, e.g., phosphorylation (8). For example, Ser133 phosphorylation on mammalian CREB is a heavily studied regulatory TF-PTM (**Fig. 19.1**). This modification has long been characterized as a key event in protein kinase A (PKA) signaling that results in the activation of target genes. The primary serine phosphorylation allows CREB to interact with its co-activator, CBP, thereby recruiting the core transcriptional machinery (5). Initially, this appeared to be a simple "on/off" switch, but further experimentation has revealed that other kinases also activate different downstream

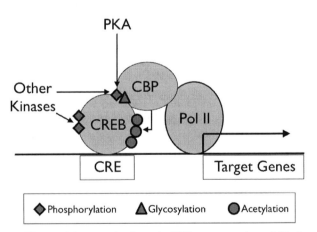

Fig. 19.1. The regulatory mechanism of cAMP-response element binding protein (CREB). CREB is a well-studied TF that exemplifies the complexity of TF-PTM regulatory circuits. Canonical CREB regulation begins with phosphorylation (*diamond*) of Ser133 by protein kinase A (PKA), which facilitates interaction with CREB binding protein (CBP) to recruit RNA polymerase II (RNAPII) and promote transcription of target genes. Other kinases can also regulate CREB through Ser133 and other phosphorylations, CBP can further regulate CREB activity through multiple acetylations (*circles*), and glycosylation (*triangle*) can disrupt the activating effect of the Ser133 phosphorylation.

Transcription Factor Modification 299

transcriptional programs through this same TF-PTM. Other PTMs that alter CREB activity have also been discovered, including additional phosphorylations, acetylations, and interestingly, an O-linked N-acetyl glycosylation that antagonizes the primary activating phosphorylation. These modifications are reviewed in further detail in (5). CREB exemplifies the potential complexity of TF-PTM regulatory programs beyond simple "on/off" switches, and the emerging appreciation for modifications other than phosphorylation.

Few TFs have been studied as extensively as CREB. Many TF-PTMs remain to be discovered and/or linked to specific modifying enzymes and downstream transcriptional programs. A complete mechanistic understanding of transcriptional regulation critically requires a map of the connectivity between TF-modifying enzymes, TFs, and target genes. The potential combinations of modifying enzymes, TFs, and target genes, even in relatively simple organisms such as *Saccharomyces cerevisiae*, are overwhelming. Experimental identification of these regulatory relationships on a genome-wide scale is currently not feasible. Fortunately, a variety of high-throughput data sources can be exploited via computational methods to predict the most likely regulatory relationships, which can then be used to prioritize experiments. Moreover, a comprehensive view of the TF-PTM landscape would provide clues to broader biological questions, such as the process-specific roles of individual types of PTMs, and the common functional and evolutionary principles underlying TF-PTM circuits.

In this chapter we focus on the problem of mapping what we term "regulatory triplets" where the regulation of a target gene by a TF is dependent on a third modulator gene. A special case of regulatory triplet is what we term a "modifier-factor-gene (MFG) triplet" (14), where the regulatory relationship between *F* and *G* is modulated by a modifying enzyme *M* via direct PTM of *F*. We first describe a publicly available database designed to catalog experimentally determined MFG triplets, which can serve as a reference set for validating new computational methods. We then review existing computational methods designed to detect regulatory triplets. We conclude with future ideas for integrative model-based approaches to detect MFG triplets.

2. Methods

2.1. PTM-Switchboard: A Database of MFG Triplets

While the current knowledge of TF-PTMs is limited to a subset of TFs and modifying enzymes, it is nevertheless highly valuable for the development of computational models. Thus, there is a need for a catalog of experimentally derived regulatory triplets, not

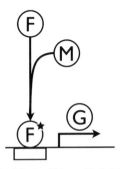

Fig. 19.2. MFG triplet example. A factor (*F*) is unable to bind the promoter of target gene (*G*) in its unmodified state. Modification is catalyzed by the enzyme (*M*), which transitions *F* to a new state (marked with a *star*) at which point it is able to bind the promoter and regulate *G*.

only to allow researchers to quickly assess what is already known about a given triplet but also to benchmark the performance of computational methods. Such a database system would further provide a framework in which to organize future computational predictions and experimental validations. Everett et al. (14) have developed *PTM-Switchboard*, a database that stores the regulatory relationships as triplets that include modulators of the TF-gene relationships. In this critical aspect PTM-Switchboard differs from previous molecular pathway databases (15–17), which support only pair-wise relationship between genes. PTM-Switchboard is specifically focused on MFG triplets, in which the modulator is a direct modifying enzyme of the TF (**Fig. 19.2**), and currently contains over 500 experimentally characterized triplets in the model organism *S. cerevisiae* – a sufficient set to train and benchmark computational methods.

The MFG triplet representation is designed specifically to support target-specific effects of a TF-PTM (*see* **Note 1**), and PTM-Switchboard includes a summary of additional experimental evidence gleaned from the literature. For instance, when the specific residues in the TF protein targeted by the modifier have been mapped, these are included in the database. The effect (activation or repression) of regulatory relationships is also provided. In some especially complex cases, such as Sko1, the TF can act as both an activator and a repressor, depending on the activity of the modifying enzyme, i.e., Hog1 (18). PTM-Switchboard provides several fields to describe the behavior in all cases (*see* **Note 2**).

As an exploratory tool for molecular biologists, PTM-Switchboard provides a considerable amount of supporting data for the curated instances of the MFG triplets, including links to external annotation resources (16, 19). All genes are recorded using both their gene symbol and ORF ID according to the *Saccharomyces* Genome Database (SGD) (20), thus directly linking each gene to its SGD annotation page. All information contained in PTM-Switchboard is annotated with links to the relevant literature or other knowledge bases from which it was derived.

PTM-Switchboard is also intended to seed a larger community effort to build a more comprehensive database of MFG triplets, as they are extremely laborious to search and curate from the literature. Text-mining approaches (21, 22) are currently limited to identifying pair-wise interactions from individual articles. MFG triplets are rarely studied together as part of a single paper, and therefore require the integration of knowledge from multiple literature sources (*see* **Note 3**). Furthermore, while genetic experiments alone can detect regulatory relationships, they cannot distinguish MFG triplets from other types of regulatory triplets. For example, a modifying enzyme that modulates a TF in a genetic experiment may be further upstream from the TF in a signaling pathway. Many kinases are also known to operate as co-factors, i.e., they bind TFs at promoters to help recruit, activate, or block the core transcriptional machinery (23–25). Thus, it is also essential to combine literature featuring both genetic and biochemical methods.

To date, Everett et al. have manually curated 519 experimentally validated MFG triplets (as opposed to more general regulatory triplets) covering 14 modifying enzymes, 15 TFs, and 212 target genes. The contents of the database can serve as an ideal "reference set" for training and testing computational models of MFG or regulatory triplets. For example, this database was used in part to validate and benchmark the Mimosa algorithm (26), discussed later in this chapter. In the future, computational predictions can be conveniently compiled in PTM-Switchboard and made accessible to molecular biologists for experimental validation. PTM-Switchboard is available on the web at: http://cagr.pcbi.upenn.edu/PTMswitchboard/

2.2. Computational Methods for Predicting Modulators of Transcription

The prediction of MFG triplets can be viewed as a special case of the more general problem of predicting TF modulators, i.e., genes that directly or indirectly affect the ability of a TF to regulate its target genes. Other examples of TF modulators include co-factors, chromatin modifying enzymes, and upstream signaling molecules. An even more general problem is the study of conditional TF regulation – inferring those conditions under which a TF is active. Conditional regulation is relevant to modulator prediction because the activity of modulator genes determines the conditions under which TF-target regulation occurs. Therefore, computational methods developed for these problems are relevant, and in some cases already applied, to the study of TF-PTMs and MFG triplets.

Most computational methods for studying TF activity primarily use gene expression profiles generated in targeted experiments, usually deposited in publicly available databases such as the Gene Expression Omnibus (GEO) (27). Many functionally related genes, including members of a pathway, biological process, or a protein complex, tend to have similar expression patterns (28, 29). Indeed, co-expression has been used extensively to infer functional relatedness (30–33). Various metrics have been proposed to quantify the correlated expression, such as Pearson and Spearman correlations (29, 31), and mutual information (MI) (32, 34–37). However, these measures are symmetric and they neither provide the causality relationships nor do they discriminate between direct and indirect relations. For instance, two co-expressed genes may be co-regulated, or one may regulate the other, either directly or indirectly. Despite this limitation, such methods have been used to successfully infer direct regulatory relationships. Other forms of data, such as TF binding site locations, can be used to strengthen the inference that a particular relationship is direct and regulatory in nature (38).

An initial motivation for studying conditional TF regulation is the observation that the activities of most TFs are likely restricted to specific cell types and/or experimental conditions (39). Thus the common practice of using large compendia of gene expression data to estimate functional relatedness is likely to include irrelevant expression samples that add noise to the co-expression signal. Furthermore, it has been observed that the conditional association between TFs and their target genes can be used to infer the modulator genes upon which these associations depend (40). The methods discussed below are therefore useful both for improving the inference of TF-gene relationships and for detecting relevant modulators of such relationships, including TF-modifying enzymes.

Previous computational methods developed to study the aforementioned problems fall broadly into two categories: *single-condition* methods, which model TF behavior separately in each experimental condition, and *partition-based* methods, which model TF behavior collectively over many experimental conditions. In one example of a single-condition method, McCord et al. (41) identified differentially expressed genes in specific experimental conditions, then searched the upstream promoter regions of those genes for enrichment of specific TF-binding motifs, thereby inferring which TFs are active in each condition. In an alternative approach, Boorsma et al. (42) used TF-binding data to first establish "regulons" for each TF, and test these regulons for differential expression in each individual experimental condition. These methods generally allow for the identification of conditions in which a TF is active, but can also be used to infer individual modulator genes in cases where the experimental conditions correspond to specific perturbations of those modulators. For instance, to infer the dependence of TF activity on histone modification enzymes, Steinfeld et al. (43) analyzed the expression of TF regulons in yeast samples where specific histone modification enzymes were knocked out. Cheng et al. (44) applied a similar method in yeast strains with knockouts of particular kinases related to life span, thereby inferring likely MFG triplets, although their approach does not guarantee direct interactions between the identified kinases and the TFs.

In contrast, partition-based methods search for broader patterns across many conditions. Such methods typically start by splitting a set of conditions into two or more partitions based on certain biological information. Once a partition structure has been established, these methods either calculate the differential association of gene pairs across partitions or attempt to fit each partition to a separate model of gene regulation. Hu et al. (45) have proposed a non-parametric test to detect differentially correlated gene pairs in two sets of expression samples from different disease classes. In a different study, Hudson et al. (46) analyzed two sets of expression data in cattle, a less-muscular wild-type and another with mutant TF myostatin. They found that the co-expression of myostatin with another gene, *MYL2*, was significantly different between the mutant and the wild-type sets of expression. This differential co-expression led them to detect a change in myostatin activity even though the expression of myostatin gene itself was not significantly different between the mutant and the wild type.

Larger compendia of expression profiles can also be partitioned based on the expression profile of a potential modulator gene, and then used to infer modulation of regulatory pairs or regulons that behave differently between the partitions. Zhang et al. (47) have proposed a method in which each potential modulator is first analyzed for bimodality in its expression profile, and this information is subsequently used to split the samples in the expression compendium. Regulatory pairs are then tested for a significant difference in correlation between the two partitions. Wang et al. (40) proposed a similar approach called MINDY, in which the expression compendium is partitioned into equal sizes according to the highest and lowest expression values of a selected modulator, and then regulatory pairs are tested for a significant difference in mutual information. This method has been applied to infer the kinases and other signaling molecules that directly or indirectly modulate TF activity in B cells (48). Segal et al. (49) proposed a related approach in which multiple partitions are learned according to a decision tree combining TFs and signaling molecules, and each partition is fit to a normal expression model for a particular module of genes. This method was applied in yeast to infer modules of genes regulated by a combination of TFs and upstream signaling genes. A different approach, termed Liquid Association, explicitly tries to detect gene triplets (X, Y, Z) where the change in correlation between X and Y varies continuously with the changes in the value of Z (50). However, methods that must exhaustively analyze many triplet combinations are generally inefficient when applied at a genome-wide scale. For example, the MINDY method is limited to a relatively small number of modulators and TFs of interest, rather than exhaustively searching all possible triplets. In the next section, we discuss a specific methodology that has been adapted from the partitioning paradigm to overcome this limitation.

2.3. Mimosa: A Mixture Model of Co-expression Data for Detecting TF Modulators

A major drawback of the partition-based methods reviewed above is that they require a priori partitioning of the expression data, based on a pre-selected modulator or condition. This typically limits the application of these methods to a particular list of known modulators and results in considerable combinatorial complexity when testing many regulatory triplets. It is not clear how to detect differentially co-expressed gene pairs with these methods when the appropriate partition of the expression samples is not provided and cannot be derived from the description of the experiments. This problem is an important practical challenge for large expression compendia that cover many diverse experimental conditions and for organisms with poor gene annotations or a large number of potential modulators.

The Mimosa algorithm (26) is a novel approach to mine expression data and detect potential modulators of regulatory interactions. In contrast to other methods that begin by selecting a modulator of interest, Mimosa begins with a known or predicted TF-target regulatory interaction, finds an appropriate partition structure in the co-expression data, and then infers a list of potential modulator genes, which are differentially expressed between these partitions (**Fig. 19.3**). A more detailed explanation of the Mimosa algorithm and preliminary results are provided below.

For a pair of genes with expression data across a set of conditions/samples, Mimosa assumes that the pair has correlated expression only in a subset of conditions and uncorrelated expression in all other conditions. Mimosa infers the hidden partition of the expression samples (the correlated and uncorrelated subsets) by fitting a mixture model to the pair of overall expression profiles using a maximum likelihood estimation (MLE) approach (*see* **Note 4**). At the end of this step, each sample is assigned a probability of originating from the correlated partition, which is more informative than an absolute partitioning of the data. The putative modulator genes, including TF-modifying enzymes, are inferred as those whose expression is significantly correlated with the vector of sample probabilities.

Mimosa has been validated on a number of expression data sets. The algorithm identifies with high accuracy the partition between wild type and mutant in a bovine data set from (46) using many of the known regulatory pairs. Specifically,

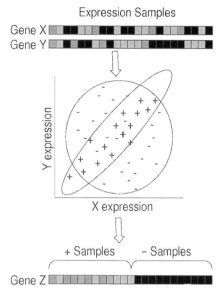

Fig. 19.3. The intuition behind Mimosa. Consider a TF gene X and a potential target gene Y. The expression values of X and Y for all expression samples are shown as a heat plot and as a scatter plot. We presume that X and Y expressions are correlated only in an unknown subset of samples (depicted by "+") and not in the remaining samples (denoted by "−"). Mimosa computes the maximum likelihood partition of samples. Then given the sample partition probabilities, a third gene Z with expression profile correlated to the partition structure may represent a potential modifier. More exactly, we assign a partition probability to each sample as opposed to a binary partition (adapted from ref. 26).

the detected sample probabilities were significantly correlated ($p \leq 0.05$) with the true sample partitions in 26% of the gene pairs tested. From a yeast expression compendium, Mimosa identified known MFG triplets from PTM-Switchboard (14) among the top 5% of predicted regulatory triplets. Mimosa was also applied to B-cell expression data (33) to predict modulators of the TF STAT1, and identified several known STAT1 modifiers as well as interesting candidates from pathways known to have essential roles in B-cell regulation and cross-talk with JAK-STAT signaling. It is always possible that a putative modulator predicted by this method is further upstream of the TF (indirect), or is actually another downstream target of the actual modulator (parallel). However, the list of putative modulators is still useful in that it can be filtered or used to infer undetected modulators through the analysis of other data types. Other relevant data sources and possible methods for integrating these data sources directly into the modeling procedure are discussed in the next section.

2.4. Looking Ahead: Integrative Modeling of MFG Triplets and TF-PTMs

The number of methods for studying conditional regulation continues to grow, but the specific problem of detecting MFG triplets remains to be addressed directly. MFG triplets differ from other regulatory triplets primarily in the physical interaction between the modifying enzyme and the TF, which yields a covalent chemical change in the TF protein. A wide range of additional data involving the detection and modeling of enzyme–substrate interactions and PTMs can be applied to the study of MFG triplets. In this section, we focus on the issue of integrating additional data sources into MFG triplet prediction.

The substrate specificity of protein-modifying enzymes is a field of intense study, with a vast and rapidly growing set of computational models available to predict the likelihood that a particular protein is modified by a particular enzyme in vivo. Below we briefly highlight some of the most recent advances and remaining challenges for PTM prediction. The most widely applicable methods predict the likelihood of modification at a particular residue based on the surrounding (proximal) protein sequence (*see* **Note 5**). Computational models range from basic statistical representations such as position-specific scoring matrices (51) to advanced machine learning classifiers (52). The sequence-based modeling of kinase specificity has been particularly successful due to decades of experimental study and the fact that the catalytic subunits have a large influence on the substrate profile (51). On the other hand, the substrate specificity of phosphatases is, in large part, determined by combinations of secondary adaptor proteins, and therefore phosphatase specificity models are not yet as effective as those for kinases (53, 54). Experimental investigation of other classes of modifying enzymes, such as acetyltransferases and methyltransferases, have primarily focused on a small set of substrates, e.g., histones (10), and the data are thus insufficient to build enzyme-specific substrate models (55–57).

Tandem mass spectrometry (MS/MS) data are applicable to PTM prediction in several important ways. Protein sequencing by mass spectrometry inherently identifies modified residues (58), and this information can be used to generate improved models of enzyme specificity. PTM sites identified by MS/MS without knowledge of the modifying enzymes responsible can also be used to limit the application of pre-existing computational models, thus lowering the false-positive rate compared to scanning entire protein sequences. For example, Linding (59) used a combination of sequence-based specificity models and protein–protein interaction networks to predict the most likely modifying enzymes for thousands of phosphorylation sites identified in MS/MS experiments.

Protein–protein interaction networks represent another class of experimental data potentially useful for inferring MFG triplets. These networks do not specify which residues in the protein sequence are modified, but can still be useful for the inference of MFG and other regulatory triplets because they provide evidence for a direct interaction between the modulator and the TF.

Common methods such as yeast-2-hybrid (Y2H) (60) can provide useful knowledge about regulatory triplets in general, but should be used with caution in the case of MFG triplets because they cannot distinguish enzyme–substrate interactions from non-catalytic protein–protein interactions. A more informative tool for detecting enzyme-substrate interactions is the protein microarray (61). For instance, Ptacek et al. (62) used an array of thousands of yeast proteins to identify the in vitro protein substrates of over 80 major yeast kinases. Lin et al. (63) used a similar method to identify in vitro substrates for the yeast acetyltransferase NuA4.

Many of the most promising data sources discussed here are in their infancy, and have not yet generated data at a scale comparable to gene expression profiles (27). It is also a non-trivial task to integrate such data properly with the current expression-based methods for predicting regulatory triplets. One solution to the problem of integrating heterogeneous data types is to use explicit Bayesian models that describe the expected behavior of the different experiments.

Chen et al. (38) proposed a regression model where target gene expression is presumed to be a function of TF-gene regulatory interactions (with prior probabilities determined by ChIP-chip and TF-binding motifs), synergistic TF–TF interactions, and TF transcript expression. Estimating the parameters of this model implicitly provides a probabilistic prediction of each TF's target genes as well as TF–TF interactions. This model can be naturally extended to the problem of predicting modifier–TF interactions by adding modifier–TF interaction terms akin to the TF–TF interaction terms. The additional data sources discussed above can be used to estimate the priors for the modifier–TF interaction variables. While it is conceptually straightforward to develop more complex and realistic models, estimating the model parameters remains a major challenge due to insufficient data and computational limitations (*see* **Note 6**).

3. Conclusions

Given the importance of PTMs in determining the TF activity and the eventual control of gene transcription, it is imperative that models of transcriptional regulation incorporate PTMs and TF-modifying enzymes. Current network reconstruction approaches rely primarily on expression data (26, 40, 47, 49, 50) and genetic manipulations (43, 44) to infer modulators based on perturbations to the TF–gene interactions. These methods are likely to predict indirect modulators, such as upstream signaling molecules. To extend the existing view of TF–gene connectivity to modifier–TF–gene connectivity, a more principled and integrative model will be needed, which directly incorporates the modifying enzymes. As new types of protein-level data become available on a scale comparable to that of transcript expression, there will be a growing need for computational techniques that can combine heterogeneous data in a practical and informed manner to develop a comprehensive view of transcriptional regulation.

4. Notes

1. A particular TF-PTM may affect only specific target genes rather than uniformly affecting all genes regulated by the TF. Therefore, triplets contained in PTM-Switchboard may share one or two members when a TF-PTM affects multiple target genes. For example, the kinase Hog1 regulates the overall transcriptional activity of Sko1 at a set of target gene promoters, and therefore a separate triplet is included in the database for each target.

2. In any MFG triplet, the modifier can either have a positive or negative effect on the activity of the TF, and likewise the TF can be an activator or repressor of each target gene. PTM-Switchboard summarizes the overall activity of the triplet by recording the influence of the TF on the target gene (positive, negative, or neutral) in each of two cases: when the modifying enzyme is active and when the modifying enzyme is inactive. For example, in **Fig. 19.2**, the relationship between F and G is neutral when M is inactive, and positive when M is active.

3. The experimental evidence for a particular MFG triplet is usually spread across multiple journal articles. For example, the overall effects of a PTM on a TF's cellular localization, degradation, or DNA-binding activity may be studied in one reference, while the gene targets of the TF are studied independent of any PTMs in another reference. In some cases MFG triplets can be inferred from these references together, but only with careful consideration of the molecular mechanisms involved – a task clearly beyond current text-mining methods. Therefore, manual curation remains the only reliable way to extract MFG triplets from the literature.

4. The mixture model used in Mimosa has two free parameters. The mixing parameter is the overall proportion of samples in the expression compendium that fit the correlated model and is denoted by f. The strength of the correlation within that model is denoted as α. These parameters

are estimated by maximizing the likelihood using a quasi Newton-Raphson method. The most informative models have f close to 0.5 (even partition sizes, indicative of a correlation model that is truly condition dependent) and $|\alpha|$ close to 1 (strong correlation in one partition). Therefore, only TF-gene mixture models meeting these criteria are used to search for potential modulators.

5. Some models of PTM prediction predict the likelihood of modification by a specific enzyme, while others simply predict the likelihood of modification in general. This is an important distinction in the context of MFG triplet prediction. Enzyme-specific models are of primary interest, because they can be used to strengthen an inferred connection between a TF and a particular modifier.

6. In general, models with increased numbers of parameters require a larger set of input data in order to maintain the statistical rigor of the estimates. Overly complex models can also be ill-behaved in typical estimation techniques such as Gibbs Sampling. Even so, based on preliminary studies, the model-based approach described in **Section 2.4** holds promise and provides a rational framework in which to integrate heterogeneous data from disparate experiments [Everett and Hannenhalli, 2009, unpublished results].

Acknowledgments

This work was supported by The National Institutes of Health grants R01GM085226 (MH, SH), and T32HG000046 (LE).

References

1. Tootle, T.L., and Rebay, I. (2005) Post-translational modifications influence transcription factor activity: a view from the ETS superfamily. *Bioessays* 27, 285–298.
2. Creasy, D.M., and Cottrell, J.S. (2004) Unimod: protein modifications for mass spectrometry. *Proteomics* 4, 1534–1536.
3. Garavelli, J.S. (2004) The RESID database of protein modifications as a resource and annotation tool. *Proteomics* 4, 1527–1533.
4. Berk, A.J. (1989) Regulation of eukaryotic transcription factors by post-translational modification. *Biochim Biophys Acta* 1009, 103–109.

5. Khidekel, N., and Hsieh-Wilson, L.C. (2004) A 'molecular switchboard' – covalent modifications to proteins and their impact on transcription. *Org Biomol Chem* 2, 1–7.
6. Brivanlou, A.H., and Darnell, J.E. (2002) Signal transduction and the control of gene expression. *Science* 295, 813–818.
7. Freiman, R.N., and Tjian, R. (2003) Regulating the regulators: lysine modifications make their mark. *Cell* 112, 11–17.
8. Holmberg, C.I., Tran, S.E., Eriksson, J.E., and Sistonen, L. (2002) Multisite phosphorylation provides sophisticated regulation of

9. Lee, D.Y., Teyssier, C., Strahl, B.D., and Stallcup, M.R. (2005) Role of protein methylation in regulation of transcription. *Endocr Rev* 26, 147–170.

10. Sterner, D.E., and Berger, S.L. (2000) Acetylation of histones and transcription-related factors. *Microbiol Mol Biol Rev* 64, 435–459.

11. Yang, X. (2005) Multisite protein modification and intramolecular signaling. *Oncogene* 24, 1653–1662.

12. Barolo, S., and Posakony, J.W. (2002) Three habits of highly effective signaling pathways: principles of transcriptional control by developmental cell signaling. *Genes Dev* 16, 1167–1181.

13. Mayr, B., and Montminy, M. (2001) Transcriptional regulation by the phosphorylation-dependent factor CREB. *Nat Rev Mol Cell Biol* 2, 599–609.

14. Everett, L., Vo, A., and Hannenhalli, S. (2009) PTM-Switchboard – a database of posttranslational modifications of transcription factors, the mediating enzymes and target genes. *Nucleic Acids Res* 37, D66–D71.

15. Kanehisa, M., and Goto, S. (2000) KEGG: Kyoto encyclopedia of genes and genomes. *Nucleic Acids Res* 28, 27–30.

16. Gough, N.R. (2002) Science's signal transduction knowledge environment: the connections maps database. *Ann NY Acad Sci* 971, 585–587.

17. Choi, C., Krull, M., Kel, A. et al. (2004) TRANSPATH-A high quality database focused on signal transduction. *Comp Funct Genomics* 5, 163–168.

18. Rep, M., Proft, M., Remize, F. et al. (2001) The *Saccharomyces cerevisiae* Sko1p transcription factor mediates HOG pathway-dependent osmotic regulation of a set of genes encoding enzymes implicated in protection from oxidative damage. *Mol Microbiol* 40, 1067–1083.

19. Pruitt, K.D., Tatusova, T., and Maglott, D.R. (2007) NCBI reference sequences (RefSeq): a curated non-redundant sequence database of genomes, transcripts and proteins. *Nucleic Acids Res* 35, D61–D65.

20. Hong, E.L., Balakrishnan, R., Qing, D. et al. (2008) Gene ontology annotations at SGD: new data sources and annotation methods. *Nucleic Acids Res* 36, D577–D581.

21. Yuan, X., Hu, Z.Z., Wu, H.T. et al. (2006) An online literature mining tool for protein phosphorylation. *Bioinformatics* 22, 1668–1669.

22. Saric, J., Jensen, L.J., Ouzounova, R., Rojas, I., and Bork, P. (2006) Extraction of regulatory gene/protein networks from Medline. *Bioinformatics* 22, 645–650.

23. Pascual-Ahuir, A., Struhl, K., and Proft, M. (2006) Genome-wide location analysis of the stress-activated MAP kinase Hog1 in yeast. *Methods* 40, 272–278.

24. Pokholok, D.K., Zeitlinger, J., Hannett, N.M., Reynolds, D.B., and Young, R.A. (2006) Activated signal transduction kinases frequently occupy target genes. *Science* 313, 533–536.

25. Bardwell, L., Cook, J.G., Voora, D. et al. (1998) Repression of yeast Ste12 transcription factor by direct binding of unphosphorylated Kss1 MAPK and its regulation by the Ste7 MEK. *Genes Dev* 12, 2887–2898.

26. Hansen, M., Everett, L., Singh, L., and Hannenhalli, S. (2010) Mimosa: mixture model of co-expression to detect modulators of regulatory interaction. *Algorithms Mol Biol* 5, 4.

27. Barrett, T., Troup, D.B., Wilhite, S.E. et al. (2009) NCBI GEO: archive for high-throughput functional genomic data. *Nucleic Acids Res* 37, D885–D890.

28. Tornow, S., and Mewes, H.W. (2003) Functional modules by relating protein interaction networks and gene expression. *Nucleic Acids Res* 31, 6283–6289.

29. Stuart, J.M., Segal, E., Koller, D., and Kim, S.K. (2003) A gene-coexpression network for global discovery of conserved genetic modules. *Science* 302, 249–255.

30. von Mering, C., Jensen, L.J., Kuhn, M. et al. (2007) STRING 7 – recent developments in the integration and prediction of protein interactions. *Nucleic Acids Res* 35, D358–D362.

31. Magwene, P.M., and Kim, J. (2004) Estimating genomic coexpression networks using first-order conditional independence. *Genome Biol* 5, R100.

32. Margolin, A.A., Nemenman, I., Basso, K. et al. (2006) ARACNE: an algorithm for the reconstruction of gene regulatory networks in a mammalian cellular context. *BMC Bioinformatics* 7(Suppl. 1), S7.

33. Basso, K., Margolin, A.A., Stolovitzky, G. et al. (2005) Reverse engineering of regulatory networks in human B cells. *Nat Genet* 37, 382–390.

34. Butte, A.J., and Kohane, I.S. (2000) Mutual information relevance networks: functional genomic clustering using pairwise entropy measurements. *Pac Symp Biocomput* 2000, 418–429.

35. Daub, C.O., Steuer, R., Selbig, J., and Kloska, S. (2004) Estimating mutual information using B-spline functions – an

36. Faith, J.J., Hayete, B., Thaden, J.T. et al. (2007) Large-scale mapping and validation of *Escherichia coli* transcriptional regulation from a compendium of expression profiles. *PLoS Biol* 5, e8.

37. Steuer, R., Kurths, J., Daub, C.O., Weise, J., and Selbig, J. (2002) The mutual information: detecting and evaluating dependencies between variables. *Bioinformatics* 18(Suppl. 2), S231–S240.

38. Chen, G., Jensen, S.T., and Stoeckert, C.J. (2007) Clustering of genes into regulons using integrated modeling-COGRIM. *Genome Biol* 8, R4.

39. Boorsma, A., Foat, B.C., Vis, D., Klis, F., and Bussemaker, H.J. (2005) T-profiler: scoring the activity of predefined groups of genes using gene expression data. *Nucleic Acids Res* 33, W592–W595.

40. Wang, K., Nemenman, I., Banerjee, N., Margolin, A., and Califano, A. (2006) Genome-wide discovery of modulators of transcriptional interactions in human b lymphocytes. *Res Comput Mol Biol* 3909, 348–362.

41. McCord, R.P., Berger, M.F., Philippakis, A.A., and Bulyk, M.L. (2007) Inferring condition-specific transcription factor function from DNA binding and gene expression data. *Mol Syst Biol* 3, 100.

42. Boorsma, A., Lu, X., Zakrzewska, A., Klis, F.M., and Bussemaker, H.J. (2008) Inferring condition-specific modulation of transcription factor activity in yeast through regulon-based analysis of genomewide expression. *PLoS ONE* 3, e3112.

43. Steinfeld, I., Shamir, R., and Kupiec, M. (2007) A genome-wide analysis in *Saccharomyces cerevisiae* demonstrates the influence of chromatin modifiers on transcription. *Nat Genet* 39, 303–309.

44. Cheng, C., Fabrizio, P., Ge, H., Longo, V.D., and Li, L.M. (2007) Inference of transcription modification in long-live yeast strains from their expression profiles. *BMC Genomics* 8, 219.

45. Hu, R., Qiu, X., Glazko, G., Klebanov, L., and Yakovlev, A. (2009) Detecting intergene correlation changes in microarray analysis: a new approach to gene selection. *BMC Bioinformatics* 10, 20.

46. Hudson, N.J., Reverter, A., and Dalrymple, B.P. (2009) A differential wiring analysis of expression data correctly identifies the gene containing the causal mutation. *PLoS Comput Biol* 5, e1000382.

47. Zhang, J., Ji, Y., and Zhang, L. (2007) Extracting three-way gene interactions from microarray data. *Bioinformatics* 23, 2903–2909.

48. Wang, K., Alvarez, M.J., Bisikirska, B.G. et al. (2009) Dissecting the interface between signaling and transcriptional regulation in human B cells. *Pac Symp Biocomput* 20, 264–275.

49. Segal, E., Shapira, M., Regev, A. et al. (2003) Module networks: identifying regulatory modules and their condition-specific regulators from gene expression data. *Nat Genet* 34, 166–176.

50. Li, K. (2002) Genome-wide coexpression dynamics: theory and application. *Proc Natl Acad Sci USA* 99, 16875–16880.

51. Fujii, K., Zhu, G., Liu, Y. et al. (2004) Kinase peptide specificity: improved determination and relevance to protein phosphorylation. *Proc Natl Acad Sci USA* 101, 13744–13749.

52. Yoo, P.D., Ho, Y.S., Zhou, B.B., and Zomaya, A.Y. (2008) SiteSeek: post-translational modification analysis using adaptive locality-effective kernel methods and new profiles. *BMC Bioinformatics* 9, 272.

53. Meiselbach, H., Sticht, H., and Enz, R. (2006) Structural analysis of the protein phosphatase 1 docking motif: molecular description of binding specificities identifies interacting proteins. *Chem Biol* 13, 49–59.

54. Virshup, D.M., and Shenolikar, S. (2009) From promiscuity to precision: protein phosphatases get a makeover. *Mol Cell* 33, 537–545.

55. Chen, H., Xue, Y., Huang, N., Yao, X., and Sun, Z. (2006) MeMo: a web tool for prediction of protein methylation modifications. *Nucleic Acids Res* 34, W249–W253.

56. Li, A., Xue, Y., Jin, C., Wang, M., and Yao, X. (2006) Prediction of N-epsilon-acetylation on internal lysines implemented in Bayesian Discriminant Method. *Biochem Biophys Res Commun* 350, 818–824.

57. Shao, J., Xu, D., Tsai, S., Wang, Y., and Ngai, S. (2009) Computational identification of protein methylation sites through bi-profile Bayes feature extraction. *PLoS ONE* 4, e4920.

58. Larsen, M.R., Trelle, M.B., Thingholm, T.E., and Jensen, O.N. (2006) Analysis of post-translational modifications of proteins by tandem mass spectrometry. *BioTechniques* 40, 790–798.

59. Linding, R., Jensen, L.J., Ostheimer, G.J. et al. (2007) Systematic discovery of in vivo phosphorylation networks. *Cell* 129, 1415–1426.

60. Ratushny, V., and Golemis, E. (2008) Resolving the network of cell signaling pathways using the evolving yeast two-hybrid system. *BioTechniques* 44, 655–662.

61. Zhu, H., Bilgin, M., Bangham, R. et al. (2001) Global analysis of protein activities using proteome chips. *Science* 293, 2101–2105.

62. Ptacek, J., Devgan, G., Michaud, G. et al. (2005) Global analysis of protein phosphorylation in yeast. *Nature* 438, 679–684.

63. Lin, Y., Lu, J., Zhang, J. et al. (2009) Protein acetylation microarray reveals that NuA4 controls key metabolic target regulating gluconeogenesis. *Cell* 136, 1073–1084.

Chapter 20

Annotating the Regulatory Genome

Stephen B. Montgomery, Katayoon Kasaian, Steven J.M. Jones, and Obi L. Griffith

Abstract

Determining the timing and molecular repertoire responsible for gene expression is fundamental to understanding a gene's function. Heritable differences in this character are increasingly regarded as explanatory for complex and common traits. For many known trait-predisposing genes, studies have sought to elucidate the associated logic behind gene regulation. However, there exist many challenges in deciphering these mechanisms. Among them, it is recognized that we have limited understanding of regulatory complexity, the current models of gene regulation have low specificity and any gene's regulatory logic is dependent on biological context. Addressing these limitations and defining the regulatory genome is an ongoing challenge for molecular biology. We discuss current efforts to define and annotate the regulatory genome by focusing on curation and text-mining activities. We further highlight the type of information and curation process for describing regulatory elements within the ORegAnno database (www.oreganno.org) and how the general standards for such information are changing.

Key words: Annotation, curation, gene regulation, database, open regulatory annotation, ORegAnno, transcription, transcription factor binding site.

---

1. Introduction

The timing and location of molecular interactions within an individual are the defining features of its biology. Differences in the character of these interactions during growth and development, homeostasis and metabolism and in response to stimuli are the basis of phenotypic diversity. The heritability of this character from parent to progeny is predominantly encoded within each individual's genome, and phenotypic differences have been considered the product of sequence variation within genes.

I. Ladunga (ed.), *Computational Biology of Transcription Factor Binding*, Methods in Molecular Biology 674, DOI 10.1007/978-1-60761-854-6_20, © Springer Science+Business Media, LLC 2010

However, it has been increasingly recognized that sequence variation affecting the functional elements that coordinate the expression of genes is a major determinant in trait aetiology. Such regulatory changes are posited to be as explanatory as protein-level changes for human and chimpanzee evolution since the time of our most recent common ancestor (1, 2). They are regarded as intrinsic to an organism's developmental program as has been relatively well characterized in the fruit fly, *Drosophila melanogaster*, and sea urchin, *Strongylocentrotus purpuratus* (3, 4). They are also robust in their ability to generate within species phenotypic diversity (5). Furthermore, the role of gene regulation and regulatory variation is crucial to our ability to predict and treat complex and common disease as increasing numbers of genome-wide association studies have implicated regulatory variants (6).

There are some major challenges in annotating the regulatory fraction of any complex genome. Compared to genic sequences, the rules defining regulatory sequences are degenerate and without functional testing or other a priori knowledge are likely to be specious. Estimates of the regulatory fraction have largely taken advantage of cross-species genome comparisons where evolutionary constraint has implied functional importance; initial sequencing of the mouse genome indicated that as much as 2.5–3.5% of the non-coding genome between human and mouse is constrained (7). Estimates using additional genomes have largely remained the same (8). Recent functional characterization of 1% of the human genome within the ENCODE project has suggested that at least 60% of these constrained sequences (including the coding fraction) have experimentally verifiable function (9). However, functional analysis of the regulatory activity of deeply constrained sequences has demonstrated limited success (10). Furthermore, an important consequence of using cross-species genome comparisons is that they are dependent on underlying evolutionary dynamics and as such will be hindered in detecting functional sites that are species-specific or tolerant of turnover (11). Estimates of the human-specific regulatory fraction have been harder to ascertain since many computational techniques have been reliant on limited information regarding the location of true regulatory elements and even scarcer information on the fraction that is species specific. The role of the species-specific fraction is likely not negligible as experimental characterization of unconstrained functional elements in ENCODE suggests that there is a similar proportion of constrained to unconstrained regulatory elements. Such studies further highlight that there is extensive intrinsic regulatory character to the genome and a salient goal is determining the regulatory code.

Currently, promoters and enhancers are thought to be the primary functional elements of the regulatory genome. The role of the promoter is to initiate transcription by positioning the core transcription machinery. Enhancers facilitate transcriptional activity for one or many genes in response to stimuli or as part of a developmental programme. Characterization of well-known regulatory regions has demonstrated that there are typically four to eight transcription factor binding sites per promoter or enhancer (12). However, some promoters (e.g. haemoglobin beta chain) can have as many as ∼50 binding sites. These sites are the protein–DNA interfaces that enhance or suppress the rate of gene transcription by recruiting individual transcription factors and transcription factor complexes. Characteristically, they are on the order of 5–30 bp, exhibit a range of sequence variability and are co-located and interact with neighbouring transcription factor binding sites. Adding further to regulatory complexity, in humans, there are an estimated 1,900 transcription factors, which evolved in families based on their protein–DNA binding domains (13).

Although there is limited information describing the organization of regulatory regions and their associated transcription factors within humans, genome-wide association studies have highlighted the biomedical relevance of variation in gene expression. Surveying genome-wide gene expression with respect to natural variation within families and populations has demonstrated that gene expression variation is heritable and widespread (14–22). Targeted analysis of non-coding polymorphisms has identified variants associated with conditions like cancer (23, 24), depression (25), systemic lupus erythematosus (26), perinatal HIV-1 transmission (27), and response to type 1 interferons (28). Systems biology-based approaches have used gene expression data to implicate gene regulatory pathways involved in obesity (29). While genetic prediction is possible by understanding the general impact of genetic variation on gene expression, therapeutic intervention will benefit from understanding the direct processes involved in the perturbation of expression. In this respect, systematic approaches to annotating the regulatory genome are as important as ongoing comprehensive gene and non-coding RNA annotation.

Systematic approaches have been undertaken to annotate and curate the regulatory genome. Databases such as Jaspar and Transfac have been designed to annotate transcription factor binding site recognition sequences and compute sequence specificity profiles (30, 31). Typically, these models are generated using in vitro experiments like SELEX (*see* also **Chapter 12**) such that in vivo transcription factor binding sites can be inferred (32). However, very low specificity and an incomplete knowledge of the cellular repertoire of transcription factors have made robust determination of in vivo sites challenging. Efforts to determine the set of transcription factors have taken several forms, from protein function prediction in the DBD database (33) to literature curation in the TFcat database (34). Several databases have been designed to independently organize the sites of promoter activity (35–40), transcription factor binding (31, 41–43) and regulatory variation (44–46). Unfortunately, the information recorded in many of these databases is disparate and has limited supporting information to be able to dissect the reliability and context of the information. New databases like ORegAnno (47) and PAZAR (48) have sought to improve the annotation of regulatory regions by providing richer semantic definitions, tools for mining the literature and support for curator activities. Central to this is also providing facilities to integrate large-scale gene regulation data sets from genome-scale assays like Chip-Seq. In this review, we will highlight activities central to mining the regulatory literature, archival of annotation and improvement of annotation standards. We also provide practical guides for regulatory annotation using the ORegAnno system.

2. Methods and Data

2.1. Strategies for Mining Regulatory Literature

A central challenge in curating regulatory elements from literature is determining a priori the scope of information available. Recently, we were involved in an effort to ascertain the extent of relevant information available (49). A training set of papers was selected from Pubmed using the query 'transcript AND regulation AND "binding site" AND (promoter OR enhancer)'. These papers were used to create a vocabulary to score 16 million Pubmed abstracts available at that time. From this, 200 'evenly spaced' papers (according to relevance score) were selected across the top 100,000 abstracts and were evaluated by an expert as to their 'curatability' (i.e. whether they would lead to a successful transcription factor binding site annotation). The point at which the positive predictive value of the text-mined corpus was equivalent to that achieved through annotation of papers recommended directly from experts (54.4%) was 58,000 abstracts. Therefore, at the time of the study, ~30,000 papers were deemed to have relevant information that would lead to one or more annotations. This is more than an order of magnitude greater than the current size of the RegulonDB and ORegAnno corpuses. Also, while many of these additional papers may yield redundant annotation, many will also likewise contain multiple annotation or further experimental support for regulatory activity in different conditions or biological contexts.

Considering that the regulatory corpus is larger than existing databases and that human curation is costly, one solution has been to determine whether a proportion of the regulatory genome can be confidently extracted from the literature using text mining techniques. One such effort used a rule-based text mining approach to extract regulatory networks from *Saccharomyces cerevisiae* and was able to achieve ~30% coverage and 83% accuracy in the evaluation corpus (6,640 abstracts) where the regulator and target protein and their direction of effect were correct when identified (50). Another approach also used rule-based text mining to mine regulatory interactions in *Escherichia coli* and was able to recover 45% of the human-curated RegulonDB network and identified 19% more regulatory interactions that had been missed during initial human-curation (51). This has highlighted that even human triage and curation of the regulatory literature is not perfect and that text-mining tools can help to refine evaluation and curation. We have supplemented ORegAnno with >54,000 abstracts that matched human triage accuracy from our rule-based text-mining of Pubmed (47). Each of these records has been added to our publication queue with an associated text-mining relevancy score to allow annotators to select papers on the presumptive relevance. However, as the utility of text-mining approaches has been dependent on the breadth of human-curated annotation, ORegAnno has attempted to distinguish which records were curated through text-mining triage or which were curated through human triage of the literature as well as the controlled reasons for why a recommended paper has failed curation. Despite these efforts, the information that defines success or failure for a potential annotation is dependent on the constraints adopted by the curation system.

To harmonize the essential features of what defines a regulatory annotation and consequently what information should be mined and what information can be safely ignored, several efforts have been established to build and use ontologies that define key features of gene regulation (47, 48, 52). The Gene Regulation Ontology has been designed to cover semantic relationships and elements (like promoters or transcription factors) related to gene regulation. By defining the key features and their relationships, text-mining activities can focus on and be evaluated by their ability to populate a consistent semantic representation of the data. Databases such as ORegAnno and PAZAR have begun to use such ontologies and have integrated the eVOC and Brenda ontologies respectively to describe cell type and tissues under study (53, 54). ORegAnno has further defined its own gene regulation experiment ontology, which encompasses the major types of protocols that validate gene regulatory elements. Such formal computable definitions support text-mining extraction, but even within these constraints the information that defines a successful annotation is heterogeneous. The definition of success for individual databases, independent of ontology, could be anything from confirming '*X regulates Y*' to '*X binds sequence W to positively regulate Y by N-fold induction in cell-type Z under conditions A, B and C confirmed by experiments E and F.'* A pragmatic question remains as how to proceed forward and capture the majority of the desired information with the minimum amount of complexity.

2.2. Archiving Regulatory Annotation

Archival of known gene regulatory element information has been provided through many databases (**Table 20.1**). The majority of these databases are curated internally and provided in-house experimental validation or links to literature. However, databases like ORegAnno and PAZAR provide facilities to support community annotation.

Community-based archival of regulatory element information requires managing the state of activity such that curators are not investing redundant effort discovering or curating the same publications. To address this, we have developed a publication queue that tracks user-recommended publications and their curation state. Publications can be added to the queue by a user supplying one or many Pubmed identifiers. All papers that do not exist in the queue already are added with the state of 'Pending.' 'Pending' papers can be checked out by individual curators. When a paper is checked out its state is set to 'Open.' While, the state is 'Open,' the curator has ownership of curation activities in the database for this publication. A limit of five publications can be checked out by any curator. These are also the only papers that the curator can provide annotation in the database for. Once associated annotation has been added to the database, the publication state can be set to 'Closed' where one of four different sub-states is provided describing the success or failure conditions for the publication. These sub-states are 'success' when the paper leads to a successful regulatory annotation or one of three 'failure' sub-states when the publication does not describe a regulatory element, a publication describes a regulatory element but there is not enough evidence to annotate it or the paper has been closed with no information. Alternatively, a publication can be set back to the state of 'Pending' if further work is required, allowing the curator to come back to it later or another curator to check it out. The advantages of this type of system are that researchers can suggest papers for annotation; these papers can be maintained in the system until a curator decides to annotate one; the process of checking out a paper and curating it is controlled to prevent wasted effort and the ultimate state of the curation is recorded with a controlled state. The latter specifically provides a control set for further text-mining validation activities. The current set of 'expert entry' papers in the queue was obtained from existing sources of curated publications including the *Drosophila* DNase I Footprint Database (42), REDfly (55), a catalogue of regulatory elements for muscle-specific regulation of transcription (56, 57), ABS (58), TRED (59), ooTFD (60), DBTGR (61), or added

Table 20.1
Experimentally determined gene regulatory element databases

Database	Description
The *Arabidopsis* gene regulatory information server (86)	AGRIS consists of *Arabidopsis* promoters, transcription factor binding sites and regulatory network information
Argonaute (87)	Literature curation of mammalian miRNA and their known or predicted targets
Database of tunicate gene regulation (61)	Describes 12 transcription factors and 140 promoters from published experimental work
EdgeDB (88)	EdgeDB is a *C. elegans* differential gene expression database. It includes information regarding protein–protein and protein–DNA interactions, including expression patterns conferred by regulatory elements
FlyTF (79)	FlyTF is a database of fruit fly transcription factors. It contains experimentally verified transcription factor binding sites and position weight matrices
JASPAR (30)	An alternative to TRANSFAC, this database is open access. JASPAR contains tightly controlled binding profiles with strict quality restrictions. Furthermore, it provides a programming API for ease of data access
The liver-specific gene promoter database (LSPD) (89)	LSPD contains liver-specific promoters and transcription factor binding sites. It contains 178 specific genes listed with 368 regulatory elements
ORegAnno (47)	ORegAnno is an open-access database of community curated regulatory regions, transcription factors and regulatory mutations
PAZAR (48)	PAZAR is an open-access, community-annotated database of transcription factor and regulatory sequence annotation
REDfly 2.0 (90)	REDfly is a curated collection of known *Drosophila* transcriptional *cis*-regulatory modules (CRMs). It contains 737 CRMs and 1,342 TFBSs for 83 transcription factors
RegulonDB (91)	Describes transcriptional network information for *Escherichia coli* *K12*. Release 6.4 contains 1,771 promoters, 1,584 transcription factor binding sites and 169 transcription factors
TFcat (34)	A curated catalogue of mouse and human transcription factors
TRANSFAC (92)	Curated transcription factor binding sites and position weight matrices
TRED (59)	TRED is a mammalian regulatory element database. It contains genome-wide predictions of core promoters for human, mouse and rat. It also contains expert-curated transcription factor binding sites for cell-cycle factors either computationally or experimentally determined
TRRD (41)	TRRD contains information on structural and functional organization of transcription regulatory regions of eukaryotic genes. It contains over 10,000 transcription factor binding sites and 3,490 regulatory regions curated from 7,609 references

manually by individual ORegAnno users from literature searches and review articles. The expert entry queue currently contains 4,458 gene regulation papers of which 3,491 are open or pending and 967 are closed. An additional 54,351 papers were added to the queue by text-mining Pubmed as described above and in greater detail by Aerts et al. (49).

Users in the gene regulation community can check out papers from the publication queue and begin manual curation. A typical record entry consists of species, sequence type, sequence (plus sufficient flanking sequence for genome alignment), target gene, binding factor, experimental outcome and one or more detailed lines of experimental evidence demonstrating function of the sequence. Records are cross-referenced to Ensembl (62) or Entrez Gene identifiers (63), Pubmed (63) and dbSNP (63) (for regulatory polymorphisms). Before committing a record to the database, ORegAnno performs a number of error checks (e.g. that the sequence has not been entered previously, the external database identifiers are valid and sufficient information has been provided to uniquely map the record) and asks the user to verify its contents before submission. Once submitted, the record is added to the database and an email is generated containing an XML representation of this record to members of the ORegAnno developers' mailing list (oreganno-guts@bcgsc.ca). A BLAST-based mapping agent then assigns genome coordinates to each sequence, allowing it to be viewed as a track in the Ensembl or UCSC genome browsers. Once finished with a paper, a user will then set the publication state to 'Closed' in the queue and assign an annotation result. Existing records can be commented by any registered user and scored (positive if verified as correct; negative if a problem is identified), updated or replaced by a 'validator' user. The complete database or any subset can be searched or downloaded in a number of formats or accessed programmatically. A detailed walkthrough of the annotation process is provided below (**Section 2.4**). In some cases a user will have a large set of regulatory sequences, which are too numerous to add to ORegAnno using the manual annotation pages. For example, they may have hundreds or thousands of binding sites from a high-throughput experiment such as Chip-Seq. In such cases, users may upload their data using the ORegAnno XML template. Typically sequences and their associated experimental details are converted from a database or flat file to XML using a parsing language such as Perl or Python. The resulting XML file can then be uploaded using the web-based XML uploader for small data sets (~100 records) or a Perl DBI uploader for larger data sets. The XML template as well as an explanation of all required and optional elements are provided in **Appendix 1**. Users wishing to submit data sets by XML should contact an ORegAnno administrator at oreganno@bcgsc.ca.

At the time of writing, the ORegAnno database held 52,027 records. This total includes 37,489 regulatory regions, 14,360 TFBSs, and 178 regulatory variants (polymorphisms and haplotypes) from 20 species (**Fig. 20.1**). Of these sequence records 50,543 have been mapped to one of 15 species, representing a mapping success rate of 97.1%. A large fraction of these sites was obtained from previous large-scale collections such as the FlyReg resource (42), a large set of muscle/liver-specific regulatory sites curated by Wasserman, Fickett and others (56, 57), rSNP_DB (64), a large set of human promoters (65), the REDfly resource (55), HBB and Erythroid modules (66, 67), the Vista Enhancer data set (68), ChIP-chip sites for CTCF (69), Esr1 (mouse) (70) and multiple yeast TFs (71, 72), ChIP-Seq sites for STAT1 (73), REST (74), ESR1 (75), RELA (76) and FOXA2 (77), the NFIRegulomeDB database and a set of ancient phylogenetically conserved non-coding elements (PCNEs) (78). However, extensive manual curation of the literature has produced an additional 1,322 original sequence records. In total, 959 publications have been curated by 46 contributing users (from >550 registered

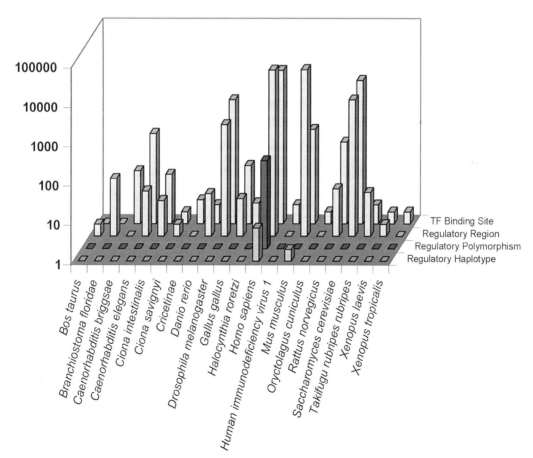

Fig. 20.1. Summary of ORegAnno data contents.

users). The complete set of records contains regulatory sequences for 4,093 genes and 545 TFs. The majority of records (99.3%) had positive experimental outcomes (i.e. the experiments demonstrated the sequence to be functional), but a small set of negative or neutral results have also been catalogued. Annotation activities follow a power-law relationship with the vast majority of contributions made by a minority of contributors. Over 90% of records (70% if high-throughput data sets excluded) were contributed by ~10% of the user base (**Fig. 20.2**). Indeed, the most prolific user contributed over 50% (20% if high-throughput data sets excluded) of all records. Contributions also tend to be sporadic with periods of intense activity followed by less active periods.

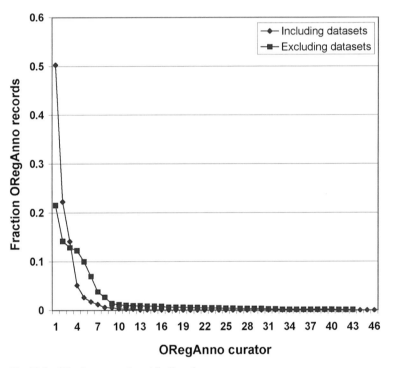

Fig. 20.2. ORegAnno record contributions by user.

The ORegAnno Database is freely available in a number of formats. Nightly XML data dumps and periodic flat file dumps are posted on the web site (http://www.oreganno.org/oregano/Dump.jsp). Human (hg18), fly (dm3) and yeast (sacCer2) records are available through the UCSC genome browser (http://genome.ucsc.edu/) as a standard track under the 'Regulation' or 'Expression and Regulation' tab. Programmatic interaction with ORegAnno is available through web services using the Perl SOAP modules. Requests for the entire database (e.g. a MySQL dump) or other formats can be addressed to the authors. The ORegAnno web application is available open-source under the Lesser GNU Public License at https://oreganno.dev.java.net/.

2.3. Improving Annotation Standards

Considerable efforts are being made to improve the standards for regulatory annotation. Of the databases that exist, there are disparate levels of supporting evidence regarding gene regulatory activity. Much of this is due to the lack of comprehensive and controlled sources of information regarding gene regulation. Perhaps central to this is the ensemble of transcription factor complexes involved in gene regulation and their biological contexts as these are known to act in tissue- and stage-dependent fashion. Effort has been undertaken to curate the ensemble of transcription factors native to each organism. Projects like TFcat and flyTF aim to address this through expert curation and controlled ontology regarding support for classification (34, 79). Other projects like TFdb and DPTF aim to combine known and predicted information (43, 80). Currently, only PAZAR integrates standardized TF information with regulatory element annotation features.

We conducted a workshop called RegCreative, which aimed to assess inter-annotator agreement in annotation for the ORegAnno database. Through analysis of redundant annotation, we identified a need for improved data standardization and development of associated regulatory ontologies. Specifically, this should include the open access development and integration of transcription factor naming conventions and sequence, cell type, cell line, tissue, and evidence ontologies. However, learning-based ontology development was widely regarded as an essential feature of the annotation process, such that annotators are not restricted from annotating based on the limitations of the controlled vocabulary and that these exceptions can be used to further develop the backbone ontologies. Furthermore, the development of these backbone ontologies should be decentralized from any one annotation system to provide an implementation-free core standard. It was also determined that current annotation systems were either too generic or too species-specific with no intermediate solution available.

A specific focus of the workshop was addressing the role of text-mining in facilitating regulatory annotation and improving annotation standards. It was recognized that beyond summarizing the regulatory corpus many text-mining activities could help facilitate discovery and then curation of regulatory information. Specifically, incorporating data from text-mining based efforts can help identify regulatory networks and transcription factor relationships (81–83). Furthermore, it was recognized that text-mining could improve inter-annotator agreement by extracting relevant key words that may be missed due to annotator error, such as tissue type, organism, sequence and experimental protocol.

324   Montgomery et al.

2.4. ORegAnno Annotation Walkthrough

Regulatory element records in ORegAnno are collected from users worldwide. Contributors can submit new records using the annotation functionality of the ORegAnno web interface. This section outlines the steps involved in annotating a publication and adding a new record to the database. The paper used as an example in this annotation walkthrough is from Schilling et al. (84).

2.4.1. Logging In

To start the annotation, existing users are asked to log in to ORegAnno using their usernames and passwords. New users will have to register for new accounts.

2.4.2. Adding a Paper to the ORegAnno Publication Queue

The first step in annotating a scientific publication is to add it to the publication queue. To do so, the user chooses the 'queue' option from the menu bar located on the right-hand side of the home page (http://www.oreganno.org/) and then from the Publication Queue page, the option 'Add paper(s) to publication queue (must be logged in)' (**Fig. 20.3**). This will take the user to a page to input the necessary information. The first required field is the PubMED Reference ID of the publication; for Schilling et al., it is 19615968. The second field specifies the criteria for the selection of the paper, 'Expert entry' or 'Predicted automatically by text mining entry'. In this case, the publication was selected by an ORegAnno user from a literature search and thus the option 'Expert entry' is chosen. Providing a transcription factor name and comment is optional (**Fig. 20.4**).

Publication Queue

Description: ORegAnno's work queue allows registered users to input relevant papers from scientific journals to a queue system for annotation. All that is required is a valid PubMED ID (PMID). Each added paper is set to PENDING. Any user can explicitly OPEN publications from the queue that are PENDING, to begin the annotation process. Once a paper has been completely annotated, the user will set the publication state to CLOSED. Otherwise, the user can revert the state back to PENDING. Comment fields are available for each change of state in the queue. A publication must be in the work queue before it can be annotated. VALIDATORS can set the state of a CLOSED paper back to PENDING, if they feel something was missed, otherwise they may correct the annotation directly.

Add to publication queue
Add paper(s) to publication queue (must be logged in)   ⬅

Search publication queue

	Search publications:		all ▾
(Optional filter)	only with state	ANY ▾	
(Optional filter)	for Transcription factor:	▾	
(Optional filter)	only include:	N/A ▾	
(Optional filter)	sort scores:	N/A ▾	

Fig. 20.3. The publication queue page.

2.4.3. Opening the Paper for Annotation

After adding the paper to the queue, the status of the paper in the queue is set to 'PENDING'. The next step is to search for

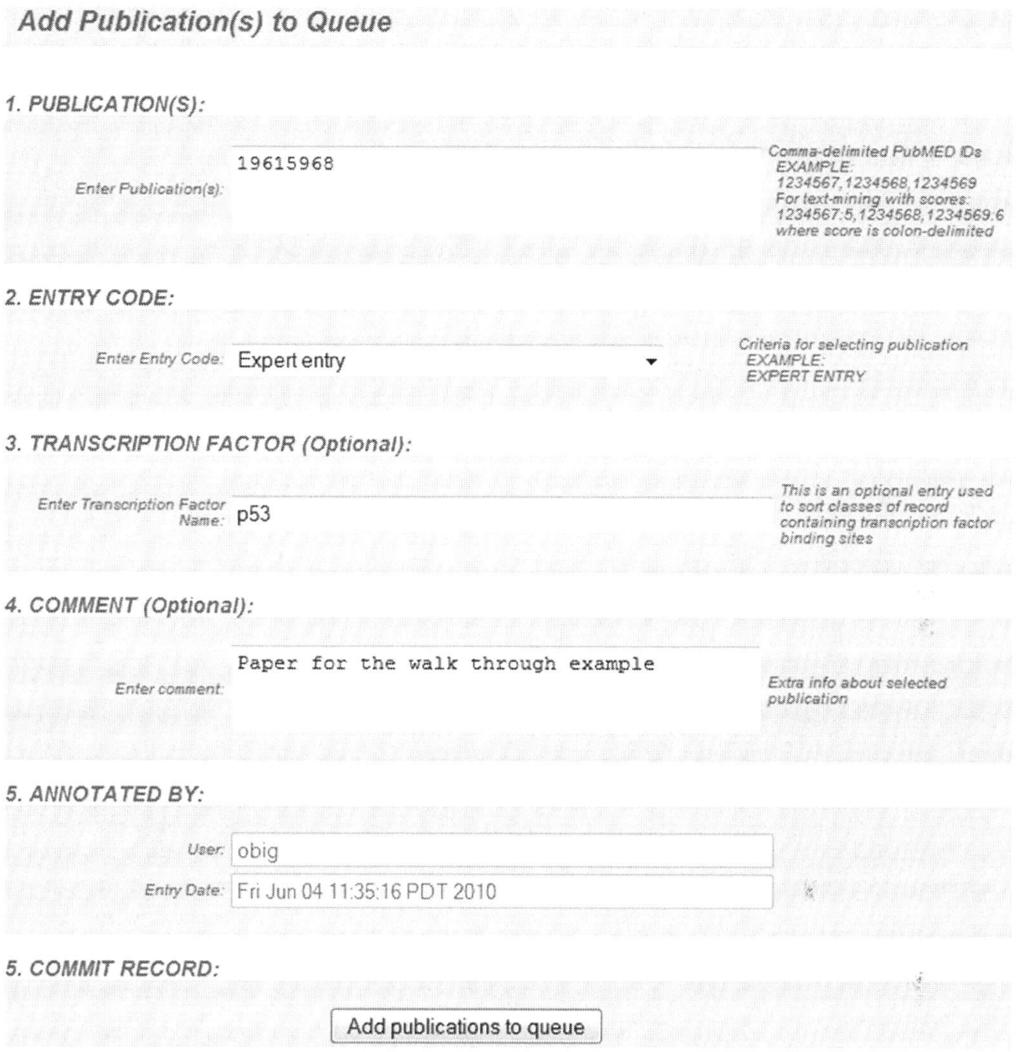

Add Publication(s) to Queue

1. PUBLICATION(S):

Enter Publication(s):

19615968

Comma-delimited PubMED IDs
EXAMPLE:
1234567,1234568,1234569
For text-mining with scores:
1234567:5,1234568,1234569:6
where score is colon-delimited

2. ENTRY CODE:

Enter Entry Code: Expert entry ▼

Criteria for selecting publication
EXAMPLE:
EXPERT ENTRY

3. TRANSCRIPTION FACTOR (Optional):

Enter Transcription Factor Name: p53

This is an optional entry used
to sort classes of record
containing transcription factor
binding sites

4. COMMENT (Optional):

Enter comment: Paper for the walk through example

Extra info about selected
publication

5. ANNOTATED BY:

User: obig

Entry Date: Fri Jun 04 11:35:16 PDT 2010

5. COMMIT RECORD:

[Add publications to queue]

Fig. 20.4. The web form for adding a publication to the queue.

the paper in the queue, using its PubMED ID for instance, and choose to 'OPEN/ANNOTATE' the paper (**Fig. 20.5**). Only the user who has opened the publication will then be allowed to annotate it.

2.4.4. Specifying the Type of the Annotation

Using the 'OPEN/ANNOTATE' option from the publication queue automatically opens the publication for annotation, takes the user to the annotation page and pre-fills the Pubmed ID. An alternative is to use the 'OPEN' option or identify a paper in the queue that the user has previously opened. To annotate such a paper, go directly to the annotation page by choosing 'annotate' from the user menu and then enter the correct Pubmed ID manually. The first step in annotating a regulatory sequence is to

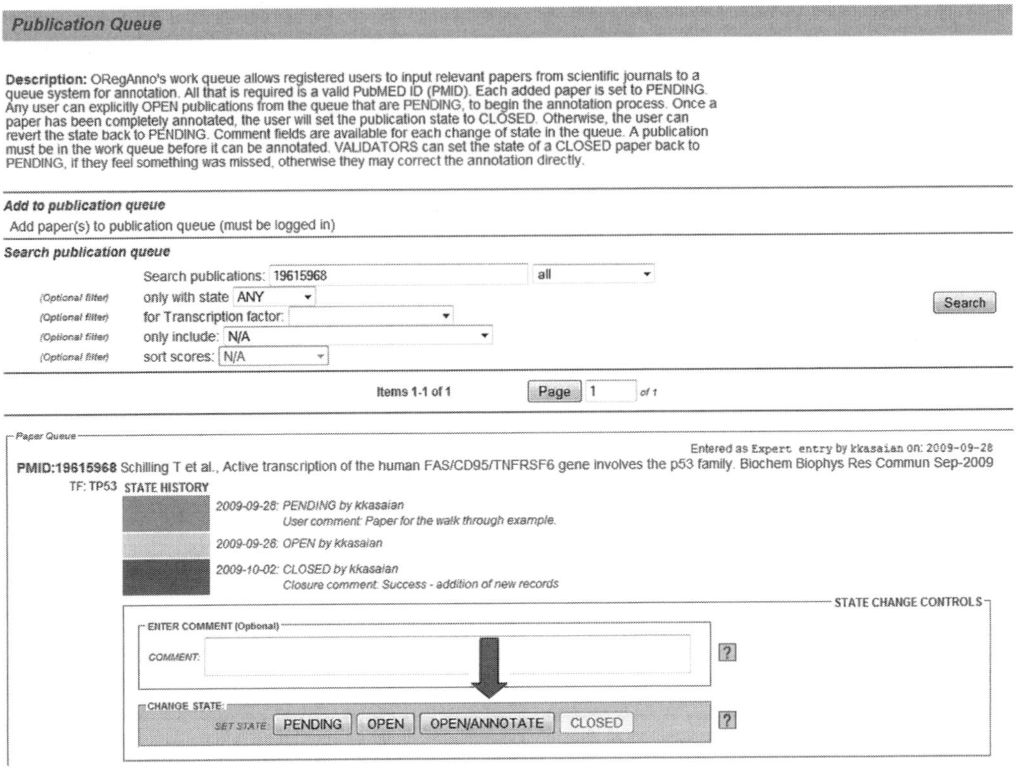

326   Montgomery et al.

Publication Queue

Description: ORegAnno's work queue allows registered users to input relevant papers from scientific journals to a queue system for annotation. All that is required is a valid PubMED ID (PMID). Each added paper is set to PENDING. Any user can explicitly OPEN publications from the queue that are PENDING, to begin the annotation process. Once a paper has been completely annotated, the user will set the publication state to CLOSED. Otherwise, the user can revert the state back to PENDING. Comment fields are available for each change of state in the queue. A publication must be in the work queue before it can be annotated. VALIDATORS can set the state of a CLOSED paper back to PENDING, if they feel something was missed, otherwise they may correct the annotation directly.

Add to publication queue

Add paper(s) to publication queue (must be logged in)

Search publication queue

	Search publications:	19615968		all	▾
(Optional filter)	only with state	ANY ▾			
(Optional filter)	for Transcription factor:		▾		
(Optional filter)	only include:	N/A	▾		
(Optional filter)	sort scores:	N/A ▾			

Search

Items 1-1 of 1   Page 1   of 1

┌ *Paper Queue* ───

Entered as Expert entry by kkasaian on: 2009-09-28

PMID:19615968 Schilling T et al., Active transcription of the human FAS/CD95/TNFRSF6 gene involves the p53 family. Biochem Biophys Res Commun Sep-2009

TF: TP53 STATE HISTORY

2009-09-28: PENDING by kkasaian
User comment: Paper for the walk through example.

2009-09-28: OPEN by kkasaian

2009-10-02: CLOSED by kkasaian
Closure comment: Success - addition of new records

STATE CHANGE CONTROLS

ENTER COMMENT (Optional)

COMMENT:

[?]

CHANGE STATE

SET STATE PENDING OPEN OPEN/ANNOTATE CLOSED [?]

Fig. 20.5. Opening a publication in the queue for annotation.

---

specify the type of annotation. Acceptable annotation types in the ORegAnno database are:

- Transcription factor binding site
- Regulatory region
- Regulatory polymorphism
- Regulatory haplotype

The regulatory sequence described in the Schilling et al. paper is a transcription factor binding site (**Fig. 20.6**). The results section of the publication states that the wild-type p53 transcription factor directly binds to the intronic binding site of the CD95 gene. The authors have confirmed their results by conducting electrophoretic mobility shift assays, reporter gene assays and mutagenesis analyses.

2.4.5. Adding the New Record to the Database

After specifying the annotation type, the user is asked to provide detailed information to add the new record to the database. The information in a typical record includes the gene of interest, the binding factor, the sequence of the regulatory region as well as its flanking sequence, the taxonomy ID of the species under study and the types of experiments and evidence the authors have used to validate their results. The information fields for adding a

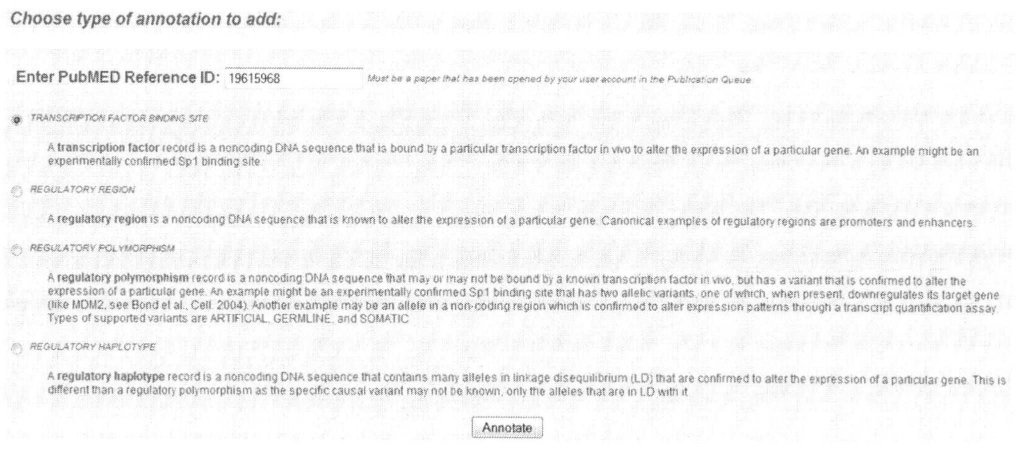

Annotating the Regulatory Genome   327

Choose type of annotation to add:

Enter PubMED Reference ID: 19615968   *Must be a paper that has been opened by your user account in the Publication Queue*

○ TRANSCRIPTION FACTOR BINDING SITE

A **transcription factor** record is a noncoding DNA sequence that is bound by a particular transcription factor in vivo to alter the expression of a particular gene. An example might be an experimentally confirmed Sp1 binding site.

○ REGULATORY REGION

A **regulatory region** is a noncoding DNA sequence that is known to alter the expression of a particular gene. Canonical examples of regulatory regions are promoters and enhancers.

○ REGULATORY POLYMORPHISM

A **regulatory polymorphism** record is a noncoding DNA sequence that may or may not be bound by a known transcription factor in vivo, but has a variant that is confirmed to alter the expression of a particular gene. An example might be an experimentally confirmed Sp1 binding site that has two allelic variants, one of which, when present, downregulates its target gene (like MDM2, see Bond et al., Cell 2004). Another example may be an allele in a non-coding region which is confirmed to alter expression patterns through a transcript quantification assay. Types of supported variants are ARTIFICIAL, GERMLINE, and SOMATIC.

○ REGULATORY HAPLOTYPE

A **regulatory haplotype** record is a noncoding DNA sequence that contains many alleles in linkage disequilibrium (LD) that are confirmed to alter the expression of a particular gene. This is different than a regulatory polymorphism as the specific causal variant may not be known, only the alleles that are in LD with it.

[Annotate]

Fig. 20.6. Specifying the type of the annotation.

transcription factor binding site record to the database are outlined here:

- *Field 1. Stable ID*
  This is a unique record identifier generated by ORegAnno for each record in the database. This ID is set automatically by ORegAnno; however, the user can regenerate it if they wish to do so.

- *Field 2. Data Set:*
  Several data sets have been annotated in ORegAnno (Refer to the ORegAnno home page for a list of these data sets, http://www.oreganno.org). If the new annotation is associated with any one of them, that data set can be selected; otherwise the data set 'OregAnno' can be chosen. This functionality allows external curators to manage particular sets of annotation using ORegAnno's curation tools. It is also possible to create new data sets. However, only ORegAnno administrator(s) can create a data set using information such as the description of the data set, the URL for the data source and the citation for the original publication describing the data set.

- *Field 3. Target Gene:*
  This field identifies the gene whose regulation is being studied. The gene ID and the source with which it is associated are specified here. The gene can either be user-defined or it can be one that has a record entry in Ensembl and/or NCBI. If Ensembl is chosen as the source, the database version needs to be specified as well. All entered information is cross-referenced against NCBI or Ensembl before a record is saved to the databases. The gene of interest in the Schilling et al. paper is *CD95*. Searching NCBI Entrez for this gene reveals that CD95 is also known as TNF receptor

superfamily, member 6 (FAS) and its NCBI gene ID is 355 (http://www.ncbi.nlm.nih.gov/sites/entrez?db=gene&term=355).

- *Field 3b. Transcription Factor:*
This field describes the transcription factor ID and its source, similar to the 'Target gene' field mentioned above. In the Schilling et al. paper, the transcription factor is p53 (TP53) and its NCBI gene ID is 7157 (http://www.ncbi.nlm.nih.gov/sites/entrez?db=gene&term=7157). The transcription factor does not have to be known for a record to be entered as a transcription factor binding site.

- *Field 4. Loci Name:*
Some regulatory regions have loci names associated with them; these can be user-defined or they can refer to other ORegAnno records. If the locus name is known, it should be entered as part of the annotation. In the case of the walk-through example, the locus name is unknown so the field is left blank.

- *Field 5. Target Species:*
This field identifies the species under study, using its taxonomy ID. From the title of the Schilling et al. paper, it is clearly human. It is also stated in the 'Materials and methods' section of the publication that the authors conducted their experiments using Hep3B cells, human liver carcinoma deficient in p53. The species *Homo sapiens* has the taxonomy ID of 9606 (NCBI, http://www.ncbi.nlm.nih.gov/Taxonomy).

- *Field 6. Sequence:*
Regulatory Sequence Entry: This entry refers to the sequence of the regulatory sequence being studied. In the example, the binding site of the p53 transcription factor is shown to be located in the first intronic region of the CD95 gene. The sequence of this region, composed of 20 base pairs, is GGACAAGCCCTGACAAGCCA (**Fig. 2C** of Schilling et al. paper). Before final entry of the sequence, a check against the current reference genome with sequence alignment programs should be preformed to confirm that the sequence aligns unambiguously and with no unexpected base discrepancies. The location of the p53 transcription factor binding site (TFBS) in the FAS gene was confirmed using Blat (http://genome.ucsc.edu/cgi-bin/hgBlat?command=start) (85). The 20-nucleotide long sequence was aligned to the human genome build hg19 assembly. The results showed only one match for the alignment of this sequence against the human genome. All 20 nucleotides were aligned to the reference and the match has a 100% identity, which was significant and highly likely to

indicate a p53 binding site. Navigating to the details of this match, we can see that the genomic location of the TFBS is on chromosome 10 at coordinates 90751066-90751085, and examining the genome browser confirms that these coordinates are located in the first intronic region of the FAS gene. BLAST could also be used to find the genomic coordinates of the sequence.

Sequence with Flank Entry: This field is the sequence of the regulatory region plus 35 to 50 base pairs flanking it (on each side). This field can be useful for genome mapping purposes. The regulatory region is usually entered in upper case and the flank in lower case. In the case of the p53 binding site in the CD95 gene, we can find the flanking sequence from the alignment of the p53 TFBS to the human genome. Thus, the sequence with flank entry is:

```
tttagggtcgctggaggggggaccccggttggagagaggagcgg
aactcctGGACAAGCCCTGACAAGCCAagccaaaggtccgctc
cggcgcgggtgggtgagtgcgcgccgccccgcgg
```

Search Space Entry: Search space is the region that has been assayed to find the regulatory sequence(s). In some cases, two or more regulatory regions might be discovered in the same search space. This field is typically used for experiments such as 'promoter-bashing' where a large sequence is systematically assayed for functional sub-sequences. If this sequence is not known, as is the case in the Schilling et al. paper, the field can be left blank.

Note: For the three sequence fields, an alternative to entering the nucleotide sequence is to provide chromosome, strand and genomic coordinates (start and end). If the exact coordinates for a specific Ensembl genome build are known, choose the 'Load from EnsEMBL' option.

- *Field 7. Reference:*
This field, the PubMED ID, is automatically entered by ORegAnno.

- *Field 8. Evidence:*
This field describes one or more pieces of evidence presented in the publication, supporting the claim about the gene under study and its regulatory element(s). The evidence classes, types and subtypes as well as the information on cell types (if known) must be recorded. Description of the evidence classes, types and subtypes can be found on the help documentation page under the heading 'EVIDENCE: Evidence in OregAnno' (http://www.oreganno.org/oregano/evidenceview.action). ORegAnno uses the eVOC cell-type ontology (http://www.evocontology.org/) to describe the different types of cells. Each line of evidence should also

be accompanied by an evidence comment providing in detail the specific implementation and results of the experiment.

The classes, types and subtypes of the evidence described in the paper as well as the cell types and evidence comments are as follows:

1. **Evidence Type:** Protein Binding Assay
   Evidence Subtype: Western Blot Assay
   Evidence Class: Transcription regulator
   Cell Type: Hepatocyte (EV:0200061)
   Evidence Comment: Paragraph 3 of the "Results" section states that "Western blotting shows p53 expression in nuclear extracts of Hep3B cells transduced with rAd-p53 (Replication-deficient adenoviral vectors encoding the complete human wt p53 cDNA together with GFP). Incubation with the ^{32}P-labeled oligonucleotide of the p53 intronic binding site resulted in the formation of a protein/DNA complex in the extracts of rAd-p53- but not of rAd-GFP (vectors with GFP alone)-transduced Hep3B cells." This experiment shows the expression of p53 protein in nuclear extracts of Hep3B cells following rAd-GFP transfer (refer to **Fig. 2A** of the paper).

2. **Evidence Type:** Electrophoretic mobility shift assay (EMSA)
   Evidence Subtype: Gel shift competition
   Evidence Class: Transcription regulator
   Cell Type: Hepatocyte (EV:0200061)
   Evidence Comment: Paragraph 3 of the "Results" section explains that "competition using a 100-fold molar excess of unlabeled wild type intronic p53-binding site has led to an inhibition of the DNA/protein complex, whereas 100× addition of the oligonucleotide of a mutated p53 intronic binding site or of the unspecific oligonucleotides Sp-1 and Oct-1 did not inhibit the complex". This observation shows the specific binding of wild type p53 to the intronic p53-binding site of the CD95 gene (refer to **Fig. 2B** of the paper).

3. **Evidence Type:** Electrophoretic mobility shift assay (EMSA)
   Evidence Subtype: Supershift
   Evidence Class: Transcription regulator
   Cell Type: Hepatocyte (EV:0200061)
   Evidence Comment: Paragraph 3 of the "Results" section states that "supershift of the complex upon addition of the p53 antibody DO-1 demonstrated p53-specificity. This result implies that activation of the CD95 gene is

mediated primarily via specific tight binding of wt p53 to the intronic p53-binding site" (refer to **Fig. 2B** of the paper).

4. **Evidence Type:** Reporter gene assay
Evidence Subtype: Transient transfection luciferase assay
Evidence Class: Transcription regulator site
Cell Type: Hepatocyte (EV:0200061)
Evidence Comment: Paragraph 1 of the "Results" section explains that "several luciferase-based reporter plasmids were constructed and assayed by transient transfection. The CD95 promoter alone was only minimally (up to 2-fold) activated by wt p53. This suggests that this promoter contains only weak p53-responsive elements. When the CD95 promoter was placed in conjunction with the p53-binding intronic CD95 DNA region, transcriptional activity became strongly (up to 80-fold) stimulated by wt p53." Thus, the reporter assay shows that the intronic p53 binding site is necessary for maximal transcription of the CD95 gene (*see* **Fig. 1** of the paper).

5. **Evidence Type:** Mutagenesis
Evidence Subtype: Site-directed
Evidence Class: Transcription regulator site
Cell Type: Hepatocyte (EV:0200061)
Evidence Comment: Paragraphs 4 and 5 in the "Results" section of the paper describe how the "p53-dependent transactivation of the CD95 gene was totally abrogated when the experimenters used CD95 luciferase constructs with mutated intronic p53-binding sites in transient transfection assays. The extent of stimulation by wt p53 was dramatically reduced from 50-fold to 2-fold by mutation of only one additional critical core nucleotide and decreased further by additional mutation of one or two crucial nucleotides. Mutation of two less important nucleotides for binding of wt p53 resulted in less but still highly significant reduction of CD95 gene transactivation (50-fold to 5-fold). These data imply that the first intron of the CD95 gene harbors a p53-responsive enhancer element that is essential for CD95 gene transactivation" (*see* **Figs. 2C** and **2D** of the paper).

- *Field 9. Experimental Outcome:*
Based on the experimental results, each record should be associated with a positive, neutral or negative outcome. If a sequence is shown to bind to a transcription factor, as in this example, then the record has a positive outcome. Records with uncertain value and inconclusive results will have a neutral outcome.

332 Montgomery et al.

- *Field 10. Comment (Optional):*
  This is an optional field for further comments and clarifications. It can be used to explain specific problems with annotation or important conclusions or data for which no appropriate field exists in the database.

- *Field 11. Meta data (Optional):*
  Each annotation in ORegAnno is optionally associated with several administrator-defined meta data types. This can be additional pieces of data relevant to a specific type of annotation or those captured in important data sets that would be useful to append to an ORegAnno annotation. Each added meta data element must match a pattern defined by the administrator for these meta data types.

- *Field 12. Annotated by:*
  The username of the annotator and the date of the annotation.

- *Field 13. Review Record (Before Commit):*
  This final field allows the review of all the completed fields before committing the record to the database. However, the "Review Record" page will not appear if the cross-referencing of the annotation against the necessary external databases is not successful. After the record is reviewed and added to the database, the user is informed of the ORegAnno stable ID (e.g. OREG0040664) associated with this record. A sample of a completed record (before committing to database) is shown in **Fig. 20.7**.

2.4.6. Setting the Status of the Publication to 'CLOSED'

Having finished the annotation, the user should set its status to 'CLOSED' in the publication queue and assign a reason for closure. In our example, the record was added to the database successfully; therefore the reason for closure is 'Success' (**Fig. 20.8**). This process prevents users from creating redundant records or annotating the same publication concurrently.

2.4.7. Final Notes

As mentioned before, the stable ID of a record can be used to uniquely identify it in the database. For instance, searching ORegAnno with the stable ID of OREG0040664 directs the user to the record on the Schilling et al. paper (**Fig. 20.9**).

Records in ORegAnno can be commented on by any user. These comments provide extra insight into the associated publication or function of these annotations. This also helps capture metadata that can be used by future validators in verifying annotation. Once a user has logged into ORegAnno, search results will return with the extra option of being able to 'Add comment' to any returned record (in addition to being able to 'View comments' for those not logged in).

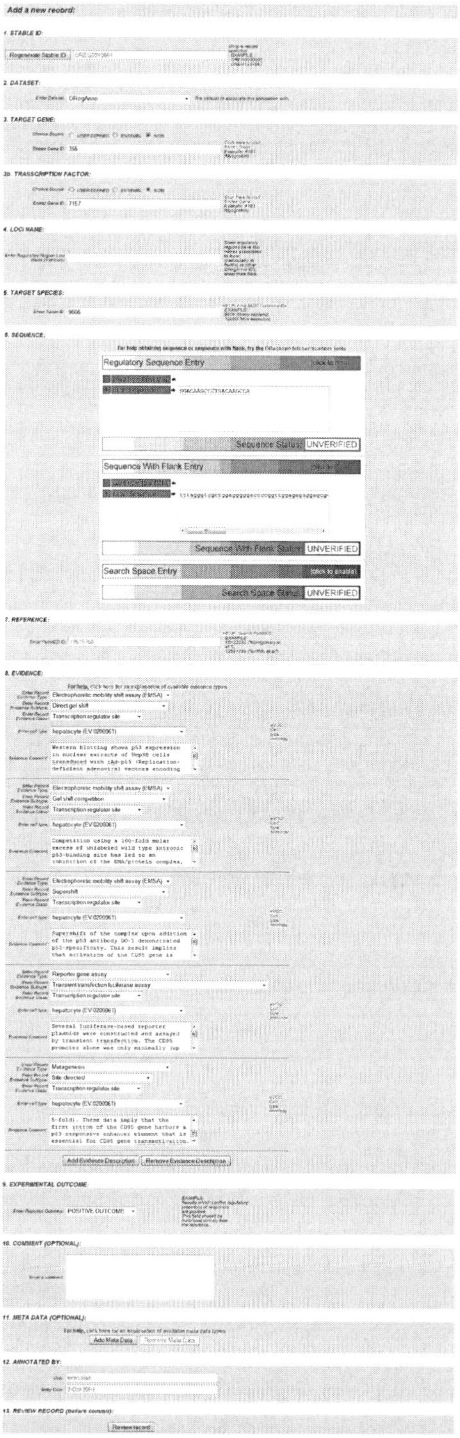

Annotating the Regulatory Genome   333

Fig. 20.7. The web form for adding a new record to the database.

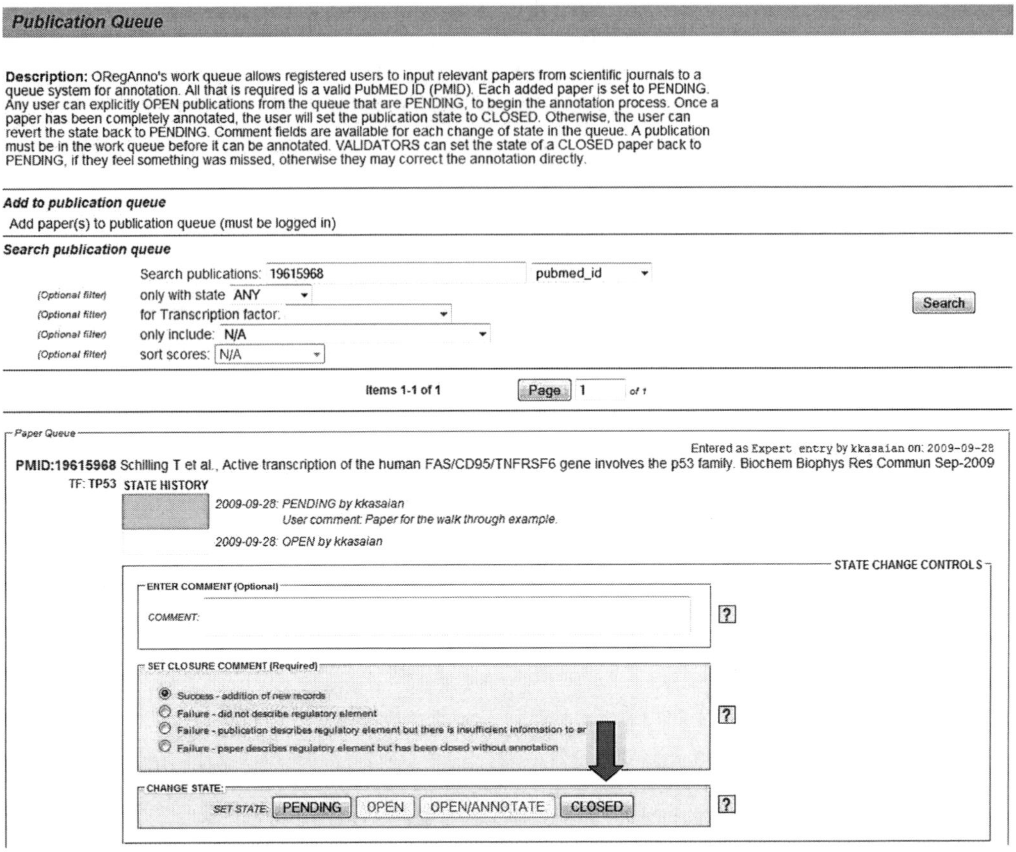

334 Montgomery et al.

Publication Queue

Description: ORegAnno's work queue allows registered users to input relevant papers from scientific journals to a queue system for annotation. All that is required is a valid PubMED ID (PMID). Each added paper is set to PENDING. Any user can explicitly OPEN publications from the queue that are PENDING, to begin the annotation process. Once a paper has been completely annotated, the user will set the publication state to CLOSED. Otherwise, the user can revert the state back to PENDING. Comment fields are available for each change of state in the queue. A publication must be in the work queue before it can be annotated. VALIDATORS can set the state of a CLOSED paper back to PENDING, if they feel something was missed, otherwise they may correct the annotation directly.

Add to publication queue

Add paper(s) to publication queue (must be logged in)

Search publication queue

	Search publications: 19615968	pubmed_id ▾	
(Optional filter)	only with state ANY ▾		Search
(Optional filter)	for Transcription factor: ▾		
(Optional filter)	only include: N/A ▾		
(Optional filter)	sort scores: N/A ▾		

Items 1-1 of 1   Page 1   of 1

Paper Queue

Entered as Expert entry by kkasaian on: 2009-09-28

PMID:19615968 Schilling T et al., Active transcription of the human FAS/CD95/TNFRSF6 gene involves the p53 family. Biochem Biophys Res Commun Sep-2009

TF: TP53 STATE HISTORY

2009-09-28: PENDING by kkasaian
User comment: Paper for the walk through example.

2009-09-28: OPEN by kkasaian

STATE CHANGE CONTROLS

ENTER COMMENT (Optional)

COMMENT: [?]

SET CLOSURE COMMENT (Required)

◉ Success - addition of new records
○ Failure - did not describe regulatory element   [?]
○ Failure - publication describes regulatory element but there is insufficient information to ar
○ Failure - paper describes regulatory element but has been closed without annotation

CHANGE STATE:
SET STATE: PENDING | OPEN | OPEN/ANNOTATE | CLOSED   [?]

Fig. 20.8. Closing a paper in the publication queue.

ORegAnno has three roles designed to help annotate and verify the data, users, validators and administrators. Users are allowed to annotate new records and make comments on existing ones. Validators are allowed to modify records in addition to annotating new records and commenting on them. For a record modification, a new record is created and the old record is marked as being deprecated by the newer record. Validators can also score records positive if verified as correct or negative if a problem is identified. Administrators have the additional responsibilities of adding new evidence classes, types and subtypes to the database as well as new data sets. They can also upload batch sets of records contributed from high-throughput experiments (*see* **Appendix 1**). To add batch sets of records, new data sets, evidence classes, types, or subtypes, or submit questions about the curation process, users can contact administrators through the ORegAnno mailing list at oreganno@bcgsc.ca.

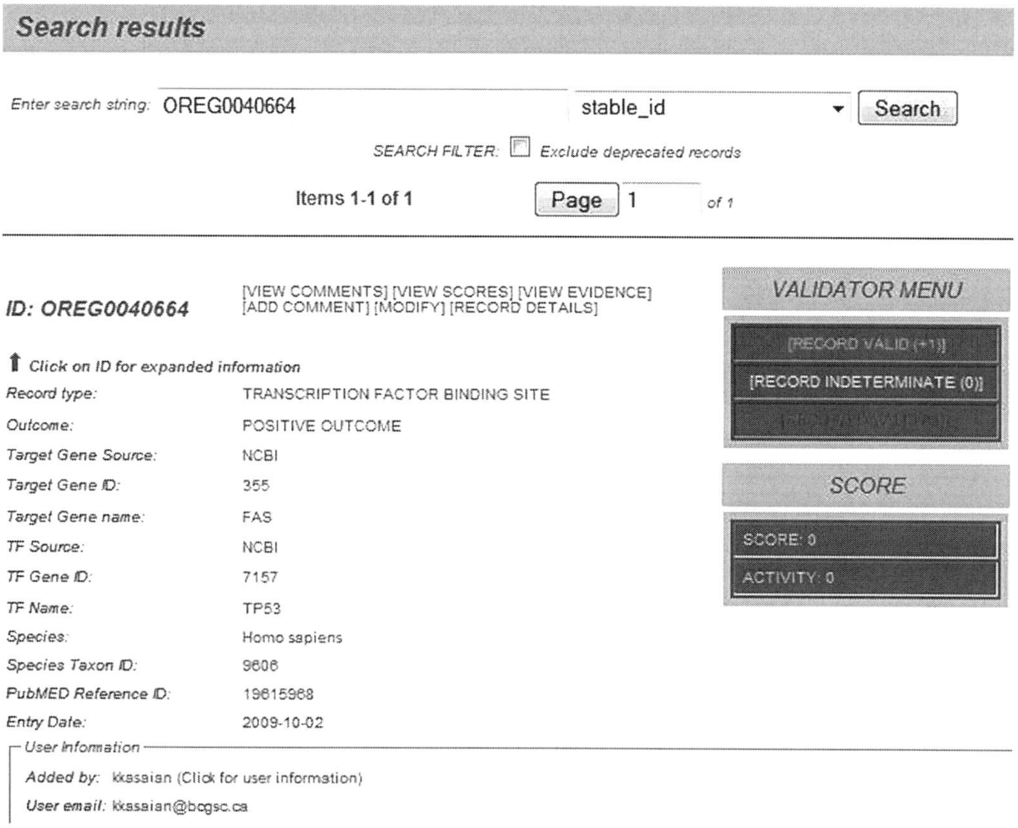

Search results

Enter search string: OREG0040664 | stable_id ▼ | Search

SEARCH FILTER: ☐ Exclude deprecated records

Items 1-1 of 1 | Page 1 | of 1

ID: OREG0040664

[VIEW COMMENTS] [VIEW SCORES] [VIEW EVIDENCE]
[ADD COMMENT] [MODIFY] [RECORD DETAILS]

VALIDATOR MENU

[RECORD VALID (+1)]
[RECORD INDETERMINATE (0)]

⬆ Click on ID for expanded information

Record type:	TRANSCRIPTION FACTOR BINDING SITE
Outcome:	POSITIVE OUTCOME
Target Gene Source:	NCBI
Target Gene ID:	355
Target Gene name:	FAS
TF Source:	NCBI
TF Gene ID:	7157
TF Name:	TP53
Species:	Homo sapiens
Species Taxon ID:	9606
PubMED Reference ID:	19615968
Entry Date:	2009-10-02

SCORE

SCORE: 0
ACTIVITY: 0

— User Information —

Added by: kkssaian (Click for user information)

User email: kkssaian@bcgsc.ca

Fig. 20.9. Result for searching the database with the stable ID OREG0040664.

3. Conclusions

Gene expression is a principal cellular function that is characteristic of the state and stage of the cell. However, the sites of transcription factor interactions and the downstream effects on expression are still largely uncharacterized. Effort has been made to dissect the role of expression in genetics and in the aetiology of disease. Likewise, effort has been made to understand the functional sites and constituent molecules that drive expression. New experimental methods such as Chip-Seq provide the ability to screen the genome for transcription factor binding sites and histone methylation marks characteristic of active promoter and enhancer activity. Integrating large amounts of new information with enhanced knowledge of the ensemble of transcription factors and with finer resolution regarding the sites of binding and their context-specific effect on expression will be central to determining the molecular mechanisms that define organismal biology and phenotypic diversity.

---

4. Notes

This section describes some of the problems that may occur during manual curation (**Section 2.4**) or batch uploading (**Appendix 1**) of larger data sets as well as tips on how to identify and overcome them.

1. Manual curation

1.1. General issues with the annotation process
When adding a new record to the database (**Section 2.4.5**), it is a good idea to assemble all the necessary information in a separate text document before entering them into ORegAnno. When ready, the data can be uploaded at once into the database. This prevents loss of information due to network timeout or unexpected hardware crashes. Another common problem while adding a new record to the database is encountered when a publication has identified a regulatory region (e.g. a promoter) composed of several TFBS. In these cases, separate ORegAnno records should be created for the regulatory region and each TFBS. Before a record can be created, the source publication must be entered into the publication queue and 'opened' for curation (**Section 2.4.3**). However, each ORegAnno user can have only five publications *open* for annotation at any time; opening a sixth publication will cause an error. In this case, one or more of the 'Open' papers should be reverted to 'Pending' or 'Closed'.

1.2. Problems with determining the 'type' of annotation
Every record in ORegAnno has a type associated with it. The types of record allowed in the database are 'Transcription factor binding site', 'Regulatory region', 'Regulatory polymorphism' and 'Regulatory haplotype' (**Section 2.4.4**). The description of these different types of annotation can be found on the help documentation page under the heading 'RECORDS: Understanding ORegAnno record types' (http://www.oreganno.org/oregano/Help.jsp). In some cases, the type of the regulatory sequence might be ambiguous. A 'Regulatory region' is a noncoding DNA sequence that is known to alter the expression of a particular gene. It may contain one or more specific binding sites or be bound by a specific binding factor whose exact binding site has not been fully localized (e.g. ChIP-Seq sites). Canonical examples of regulatory regions are promoters and enhancers. A 'Transcription factor binding site', on the other hand, is a noncoding DNA sequence that is bound by a particular transcription factor to alter the expression of a par-

ticular gene. A 'Regulatory polymorphism' is a noncoding DNA sequence that may or may not be bound by a known transcription factor in vivo, but has a variant that is confirmed to alter the expression of a particular gene. Statistical association between a non-coding variant and a condition is not sufficient evidence of a regulatory polymorphism as this may simply represent linkage disequilibrium (see 'regulatory haplotype'). The publication should specifically demonstrate the functional significance of the variant on gene expression, protein binding, etc. A 'regulatory haplotype' record is a noncoding DNA sequence that contains several alleles in linkage disequilibrium (LD), which are confirmed to alter the expression of a particular gene or binding of a particular protein factor. This is different from a regulatory polymorphism as the specific causal variant may not be known, only the alleles that are in LD with it.

1.3 Problems with identifying target gene or transcription factor
Entering target genes and transcription factors into ORegAnno requires specifying their Ensembl or Entrez gene IDs (**Section 2.4.5**, Fields 3 and 3b). However, curators also have the option of specifying a 'User-defined' gene as a last resort. For species where Ensembl or Entrez genes are not yet well defined or where another gene identification system is preferred (e.g. SGD ID for *S. cerevisiae*), using the 'User-defined' option is recommended. Other situations where the 'User-defined' option might be used include (1) cases where the transcription factor is an unknown member of a protein family or binding is non-specific for several family members; (2) a regulatory region such as an enhancer has more than one proximal gene that might be influenced by it.

1.4 Problems with identifying the species of interest
The taxonomy ID of the species under study must be specified when adding a new record to the database (**Section 2.4.5**, Field 5). The species under study should be explicitly stated in the source publication. However, if it is not, the species can be inferred with caution by the gene identifier or sequence. One source of problems with specifying a species is when a regulatory sequence for one species is tested in an experimental system of another species. In these cases, the species for the regulatory sequence should be specified. For example, a study that has tested human regulatory sequences in a mouse in vivo model system will have *Homo sapiens* as its target species. This allows correct mapping of the sequence to the appropriate genome. The experimental/model system should be fully explained in the evidence comment field.

There are also instances where the Entrez taxonomy IDs are ambiguous and both species and subspecies entry could be chosen as the target species. An example is *Takifugu rubripes* (taxonomy id=31033) and *Takifugu rubripes rubripes* (taxonomy id=47633). In these situations, the curator should use the most specific taxonomy entry that is accurate; if still in doubt, the convention of existing ORegAnno records should be followed.

1.5 Problems with identifying regulatory sequence

Each annotation should include the functional or bound sequence as reported in the publication (**Section 2.4.5**, Field 6). A problem often encountered by users is publications that do not clearly specify the identified regulatory sequence. A surprising number of studies describe experimental evidence for a binding site without providing any reliable means of locating that sequence in the appropriate genome. Using relative coordinates can help in finding the general search region; however, caution should be taken in using relative coordinates. They can be misleading if the exact genome version is not specified. The binding site then can be located in the search region. If site-specific mutagenesis or oligonucleotide competitions were performed, there may be PCR primers or oligonucleotide sequences that can be aligned to the genome in order to locate the binding site. There are also several tools at the ORegAnno web site (http://www.oreganno.org/oregano/Tools.jsp), which can help in locating the sequence of interest.

To facilitate mapping to the current genome, sufficient flanking sequence should be provided for each regulatory region. The minimum sequence length allowed by ORegAnno is 40 base pairs, but the recommended length is 100 base pairs or more. When choosing the amount of flank, the user should align the sequence using either Blat or BLAST to the appropriate genome to ensure a single unambiguous high-scoring alignment.

A sequence can also be specified using its coordinates. In that case, the correct Ensembl genome version, corresponding to the provided coordinates, must be selected. For example, if the sequence of interest was 'chr21:45316682-45316701' relative to the 'hg18' reference genome, this would correspond to the March 2006 human reference sequence (NCBI Build 36.1) (http://genome.ucsc.edu/cgi-bin/hgGateway?org=Human&db=hg18). Therefore, only an Ensembl version based on the human Build 36 genome would be appropriate. This is indicated by the last number in the Ensembl database version name (home_sapiens_core_47_**36**i). If the necessary Ensembl version is not available in the ORegAnno drop-down list, coordinates can be converted between genome builds using the UCSC LiftOver tool (http://genome.ucsc.edu/cgi-bin/hgLiftOver).

In some cases, a sequence in the publication does not match the sequence in the current reference genome version due to small differences between the reported sequence and the current reference genome. These might represent errors in an earlier assembly or different alleles at variant positions. These sequences are acceptable as long as the differences do not challenge the accuracy of the experimental results. However, it is recommended that the most current sequence is used in order to facilitate mapping. If it is believed that a particular allele (for a known variant) is of functional importance, the variant sequence can be used. As mentioned, sufficient flank for an unambiguous alignment to the correct genome location must be included.

Before final entry of the sequence, a check against the current reference genome with sequence alignment programs should be performed to confirm that the sequence aligns unambiguously, with no unexpected base discrepancies, to the expected location relative to the target gene. If the sequence is too short and thus Blat cannot be used to perform an alignment to the genome, BLAST, having more parameter options for short sequence alignments, should be used.

1.6 Problems with missing PubMED ID

Each annotation must reference a valid PubMED article (**Section 2.4.2**). This ensures that any record can be verified or validated by referring back to the original source. However, a single record or data set for a publication being prepared for submission or in press (with no PMID) can be added to the database by contacting ORegAnno administrators. The batch uploading method can be used to upload records with PMID set to 'PENDING'. The ORegAnno accession number(s) then can accompany a journal submission.

1.7 Problems with describing the experimental evidence

Each annotation specifies an evidence type, subtype and class describing the biological technique cited to discover the regulatory sequence. Each annotation can have multiple entries from any evidence class, type and subtype describing each piece of experimental evidence for the regulatory sequence and/or binding protein. As a minimum, a record must have at least one piece of in vivo or in vitro experimental evidence to be considered suitable for entry into ORegAnno. In silico or indirect evidence (e.g. evidence type: 'Sequence conservation') should be entered as supplemental evidence only (**Section 2.4.5**, Field 8).

Evidence classes are broken into two categories: the 'regulator' classes which describe evidence for the specific protein(s) that bind a site and the 'regulatory site' classes which describe evidence for the function of a regulatory sequence itself. These two categories are further divided into three levels of regulation (transcription, transcript stability and translation). Thus, a total of six evidence classes currently exist. For instance, 'transcription regulator site' describes evidence for the identity of a sequence that regulates transcription (e.g. transcription factor binding site) and 'transcription regulator' describes evidence for the identity of the protein that binds a transcription regulator sequence (e.g. transcription factor).

In many cases, the functional validation of a regulatory sequence depends on the context under which it was assayed. One important factor determining this context is the cell type. Therefore, wherever possible, the cell type in which experiments were conducted should be recorded for each piece of experimental evidence. If a particular experiment (e.g. a reporter gene assay) is performed in several different cell types (e.g. different cell lines), these can be considered multiple pieces of evidence (one for each cell type). ORegAnno currently uses the eVOC cell-type ontology for this purpose. If a specific cell type is not in the ORegAnno list but is present in eVOC, users can contact ORegAnno administrators regarding this. However, if the cell type is also missing from eVOC, the eVOC administrators should be contacted.

If there are any missing evidence types and subtypes, a request can be sent to ORegAnno administrators in order to add them to the evidence ontology. Evidence types describe the generic assay used while subtypes define specific implementations of these assays. For regulatory polymorphisms or haplotypes, association studies (evidence type: 'Association study') alone should not be considered sufficient evidence as these studies typically cannot distinguish a functional polymorphism from a non-functional polymorphism in linkage disequilibrium with the functional polymorphism (see discussion above). The evidence type 'Literature derived' should only be used in cases where sequences were manually curated by another group of experts adhering to standards materially equivalent to those outlined in this document but where specific experiments were not recorded or cannot be confidently mapped to the evidence ontology.

2. Batch upload

In the case of batch uploading, the most common problem is that the template is not followed closely enough or a data field is not used properly. Refer to **Appendix 1** for an example of the template and detailed explanation of each field.

Another issue is that the ORegAnno accession IDs are used for data set, evidence, and cell-type fields instead of their names (as in manual curation). The correct data set IDs can be found by following the link to the appropriate data set from the ORegAnno home page (http://www.oreganno.org). The correct evidence IDs can be determined from the ORegAnno evidence page (http://www.oreganno.org/oregano/evidenceview.action). The user must make certain that the evidence subtypes are correctly matched to their parent evidence types in the ontology.

Appendix 1.
ORegAnno XML
Sample

The method for manual annotation detailed in **Section 2.4** of the 'Methods and data' is recommended for individual low-throughput studies. In some cases, a user may wish to upload a large collection of previously annotated publications, an existing database, or the results of a high-throughput experiment. To allow this, an alternative 'batch upload' method is provided. This makes use of the ORegAnno XML template. Data may be converted into an XML format using the following example. Many of the required fields overlap with those explained in **Section 2.4**. However, each field is also explained in **Table 20.2**. Please contact the ORegAnno administrators (oreganno@bcgsc.ca) for help with creating and uploading an XML batch file.

```
<?xml version="1.0" encoding="ISO-8859-1"?>
<oreganno>
  <recordSet>
    <record>
      <id></id>
      <stabled></stabled>
      <type>REGULATORY REGION</type>
      <outcome>NEGATIVE OUTCOME</outcome>
      <geneId>ENSDARG00000062484</geneId>
      <geneName>ptprf</geneName>
      <geneSource>ENSEMBL</geneSource>
      <geneVersion>danio_rerio_core_42_6c</geneVersion>
      <tfId></tfId>
      <tfName></tfName>
      <tfSource></tfSource>
      <tfVersion></tfVersion>
      <lociName></lociName>
      <speciesName>Danio rerio</speciesName>
      <reference>19704032</reference>
      <date>5-Aug-2009</date>
      <sequence>
        <internalSequenceType>sequence</internalSequenceType>
        <sequence>GGTTAAGAGTGAAAAGAACCAACCTCCTCGAGGGTCTATGAGATGA
GGTGAGAGTTTGACCGGGTGATTTAATGGA</sequence>
        <ensembl_database_name>danio_rerio_core_42_6c</ensembl
_database_name>
        <sequence_region_name>2</sequence_region_name>
        <start>14342892</start>
        <end>14342967</end>
        <strand>1</strand>
        <verified>true</verified>
      </sequence>
      <sequenceWithFlank>
        <internalSequenceType>sequence_with_flank</internal
SequenceType>
        <sequence>ttgacacagataacaactagcctgaacgaaatataacattgctcttg
catctctttaatgcaggctcatgcaagtcacctgacacaacacattcagcctgaac
acaaaggtgaggggcggcataacgcagggagtgggattgata acaaggtctctga
ttaaagatggatccaggttggggtctgcaagcggcGGTTAAGAGTGAAAAGAACCAA
CCTCCTCGAGGGTCTATGAGATGAGGTGAGAGTTTGACCGGGTGATTTAATGGAgat
gaaattgaaagacagagacaaatggaaaacaagagaacatgaaaagacatttgtgaa
caatttcatggctgttagaaaaaaaaagaaacacaatggaaatttttaaaagacaga
```

```
          cacaaaagcataacattcacagaaaagtcggatattctaccatatttcatacatatt
          gcagcaacatccccaatg</sequence>
          <ensembl_database_name>danio_rerio_core_42_6c</ensembl_
          database_name>
          <sequence_region_name>2</sequence_region_name>
          <start>14342697</start>
          <end>14343159</end>
          <strand>1</strand>
          <verified>true</verified>
        </sequenceWithFlank>
        <searchSpace>
          <internalSequenceType>searchSpace</internalSequenceType>
          <sequence>ttgacacagataacaactagcctgaacgaaatataacattgctcttg
          catctcttttaatgcaggctcatgcaagtcacctgacacaacacattcagcctgaac
          acaaaggtgagggcggcataacgcagggagtgggattgataacaagggtctctga
          ttaaagatggatccaggttggggtctgcaagcggcGGTTAAGAGTGAAAAGAACCAA
          CCTCCTCGAGGGTCTATGAGATGAGGTGAGAGTTTGACCGGGTGATTTAATGGAgat
          gaaattgaaagacagagacaaatggaaaacaagagaacatgaaaagacatttgtgaa
          caatttcatggctgttagaaaaaaaaagaaacacaatggaaattttaaaagacaga
          cacaaaagcataacattcacagaaaagtcggatattctaccatatttcatacatatt
          gcagcaacatccccaatg</sequence>
          <ensembl_database_name>danio_rerio_core_42_6c</ensembl_
          database_name>
          <sequence_region_name>2</sequence_region_name>
          <start>14342697</start>
          <end>14343159</end>
          <strand>1</strand>
          <verified>true</verified>
        </searchSpace>
        <dataset>OREGDS00016</dataset>
        <evidenceSet>
          <evidence>
            <evidenceClassStableId>OREGEC00001</evidenceClassStableId>
            <evidenceTypeStableId>OREGET00002</evidenceTypeStableId>
            <evidenceSubtypeStableId>OREGES00021
            </evidenceSubtypeStableId>
            <comment>Each candidate conserved regulatory region was
            amplified by PCR and co-injected with an EGFP reporter
            construct into zebrafish embryos produced from natural
            matings between the 1-4 cleavage stages. Embryos were
            then assayed for GFP expression  on the second day of
            development (approximately 24-16 hpf). The conserved
            region is recorded here as "sequence," and the entire
            tested PCR product is recorded as "searchSpace." This
            element contains the PCNE 67-Dr_ECR7_C2.</comment>
            <date>5-Aug-2009</date>
            <userName>hufton</userName>
          </evidence>
        </evidenceSet>
        <commentSet>
          <comment>
            <comment>Ancient phylogenetically conserved non-coding
            elements (PCNEs) were identified around gene families
            from mouse, zebrafish, fugu, and the invertebrate
            chordate amphioxus. 42 of these elements were tested
            for enhancer activity in transgenic zebrafish embryos,
            including 22 amphioxus elements and 20 fish elements.
            Results for each of these elements, and 9 randomly
            chosen negative control elements, are described in
            this dataset.</comment>
            <date>5-Aug-2009</date>
            <userName>hufton</userName>
          </comment>
        </commentSet>
        <scoreSet></scoreSet>
      <variationSet></variationSet>
```

Annotating the Regulatory Genome 343

```
      <metaDataSet></metaDataSet>
      <deprecatedByDate></deprecatedByDate>
      <deprecatedByStableID></deprecatedByStableID>
      <deprecatedByUser></deprecatedByUser>
    </record>
  </recordSet>
  <speciesSet>
    <species>
      <name>Danio rerio</name>
      <taxonId>7955</taxonId>
    </species>
  </speciesSet>
  <userName>hufton</userName>
</oreganno>
```

Table 20.2
Explanation of data fields in ORegAnno XML template (In order of appearance)

XML tag	Description
oregano[a]	The root element of an ORegAnno xml file. Contains all other child elements detailed below
recordSet[a]	Contains one or more ORegAnno record elements
record[a]	The main parent element for an ORegAnno record
Id	The internal ID assigned to each record. For initial upload, leave empty and one will be assigned automatically
stableId	The stable display ID assigned to each record (e.g. OREG0040664). For initial upload, leave empty and one will be assigned automatically
type[a]	Type of regulatory sequence. Currently one of: 'REGULATORY REGION', 'TRANSCRIPTION FACTOR BINDING SITE', 'REGULATORY POLYMORPHISM' or 'REGULATORY HAPLOTYPE'
outcome[a]	Indicates whether the experimental evidence confirms or refutes regulatory function of the sequence. 'NEGATIVE OUTCOME' for confirmed negative control sequences (demonstrated lack of regulatory function). 'POSITIVE OUTCOME' for sequences with experimental evidence for regulatory function. 'NEUTRAL OUTCOME' for sequences with uncertain experimental outcome
geneId[a]	Entrez Gene Id (e.g. 7157) or Ensembl Gene Id (e.g. ENSG00000141510). If Entrez/Ensembl is unknown or unavailable, a user-defined Gene ID can be used from any another gene identification system
geneName[a]	Entrez Gene Name (e.g. TP53) or Ensembl Gene Name (e.g. TP53). If Entrez/Ensembl is unknown or unavailable, a user-defined Gene Name can be used from any another gene identification system
geneSource[a]	Indicates source of geneId and geneName (NCBI, ENSEMBL or USER DEFINED)
geneVersion	If geneSource is ENSEMBL, provide Ensembl database version (e.g. danio_rerio_core_42_6c)
tfId tfName tfSource tfVersion	Same as for gene details (i.e. geneId, geneName, geneSource and geneVersion), except referring to transcription factor (if known) that has been shown to bind the regulatory sequence being described

(continued)

344 Montgomery et al.

Table 20.2 (continued)

XML tag	Description
lociName	Name for gene locus
speciesName[a]	Name of species to which regulatory sequence belongs (e.g. *Danio rerio*)
reference[a]	A valid Pubmed ID (PMID) for the source publication describing the regulatory sequence and its experimental evidence. If submitting sequences as prerequisite for an unpublished but accepted publication, use 'PENDING' and then update ORegAnno admin upon assignment of PMID
date[a]	Date of record creation. Specified as dd-mmm-yyyy (e.g. 5-Aug-2009)
sequence[a] sequenceWithFlank[a] searchSpace	Each record must be accompanied by the functional/bound sequence plus sufficient flanking sequence to permit unambiguous alignment to the genome. The sequence of interest should be in upper case and the flanking sequence in lower case (e.g. aagatggatctgcaTTAATG-GAgatgaaatttgg). The sequenceWithFlank must be at least 40 bp in length, but ~100 bp is recommended to ensure mapping success. If a larger sequence region was assayed before narrowing down to a smaller functional region (e.g. promoter bashing experiment), then this sequence can be entered as the searchSpace. Sequences will be aligned to current genome builds of the specified species for the record. However, you can also provide position details explicitly (genome version, chr, start, end, strand) with the optional elements (details below)
internalSequenceType[a]	One of 'sequence', 'sequenceWithFlank' or 'searchSpace'
sequence[a]	Nucleotide sequence
ensembl_database_name	If providing sequence coordinates, also provide the appropriate Ensembl database version to establish which genome build these coordinates correspond to (e.g. danio_rerio_core_42_6c)
sequence_region_name	Chromosome or contig name
start	Genomic base position of start of sequence
end	Genomic base position of end of sequence
strand	Strand of sequence (1 or −1)
verified[a]	For XML upload, this can be set to 'true'
dataset	ORegAnno dataset stable ID for the dataset that the record belongs to (e.g. OREGDS00016). Typically used only if record belongs to a defined external dataset (e.g. an existing database of regulatory sequences or a logical collection of sequences such as ChIPseq sites for a particular transcription factor). If the uploader wishes to organize their records as a dataset, they can contact the ORegAnno administrators to have a new dataset created for them
evidenceSet[a]	Contains one or more evidence elements. Each ORegAnno record must have at least one experimental evidence element
evidence[a]	The main parent element for each piece of evidence in the evidence-Set. Contains several child elements describing details of the experimental evidence including cell type, comment, class, type and subtype. The complete ORegAnno evidence ontology can be found at: http://www.oreganno.org/oregano/evidenceview.action

(continued)

Table 20.2 (continued)

XML tag	Description
cellType	The cell type in which experiments were conducted if known. Uses the eVOC cell type ontology (e.g. EV:0200034)
comment	A detailed comment describing the experimental evidence and how it demonstrates the identity or function of a regulatory sequence
date[a]	Date of evidence entry. Specified as dd-mmm-yyyy (e.g. 5-Aug-2009)
evidenceClassStableId[a]	Evidence classes are broken into two categories: the 'regulator' classes describe evidence for the specific protein(s) that bind a site. The 'regulatory site' classes describe evidence for the function of a regulatory sequence itself. These two categories are further divided into three levels of regulation (transcription, transcript stability and translation). The most common class 'Transcription regulator site' (OREGEC00001) describes evidence for the identity of a sequence (e.g. transcription factor binding site) that regulates transcription. The 'Transcription regulator' (OREGEC00002) class describes evidence for the identity of the protein (e.g. transcription factor) that binds a transcription regulator site
evidenceTypeStableId[a]	Stable ID for an ORegAnno evidence type describing a type of biological assay. This is a generic type of experiment that may have several subtypes of experimentation associated with it, e.g. OREGET00002 - Reporter gene assay
evidenceSubtypeStableId[a]	Stable IDd for an ORegAnno evidence subtype describing a subtype of biological assay. This is the specific biological assay that was used in the associated literature, e.g. OREGES00004 – Transient transfection luciferase assay
userName[a]	ORegAnno user ID for evidence entry (e.g. kkasaian)
commentSet	Contains one or more comments. These can be added at record creation and also appended to a record at a later date by any user with sufficient permissions
comment (child)	Each comment (child) element includes a comment (sub-child) element as well as date and user elements
comment (sub-child)	The actual descriptive text for the comment
date	Date of comment creation specified as dd-mmm-yyyy (e.g. 5-Aug-2009)
userName	ORegAnno user ID for comment
scoreSet	Normally empty for a new entry. Records history of scores entered through ORegAnno's voting-based validation system
variationSet	Used only for records of type 'REGULATORY POLYMORPHISM' or 'REGULATORY HAPLOTYPE'. Details sequence variants (with respect to reference) shown to affect regulatory function of a sequence
metaDataSet	Typically used for import of existing databases of regulatory sequences with data fields not represented in the ORegAnno data model
deprecatedByDate	Date of deprecation of record. Used only in cases where a record has been updated or replaced. Not used for initial data upload. Record deprecation is handled internally by the ORegAnno web application

(continued)

Table 20.2 (continued)

XML tag	Description
deprecatedByStableID	ORegAnno Stable ID of record that the record has been deprecated by
deprecatedByUser	Name of user (userName) who initiated deprecation of record
speciesSet[a]	Contains one or more species elements for each species represented in the XML upload
species[a]	The main parent element for each species in the speciesSet. Contains child elements for name and taxon ID of each species
name[a]	Standard nomenclature for species (e.g. *Danio rerio*). Matches the 'speciesName' element associated with each record
taxonId[a]	NCBI Entrez taxonomy ID for the species. Can be found at: http://www.ncbi.nlm.nih.gov/sites/entrez?db=Taxonomy
userName[a]	ORegAnno user ID. Before an XML file can be uploaded, the uploader must create an ORegAnno user account: (http://www.oreganno.org/oregano/createuserpre.action). The userName element is specified once for the entire file, and also for each comment and evidence element. In this way, multiple users can comment and describe evidence for the same record

[a]Indicates required elements for upload of new dataset to ORegAnno. Note, optional unused elements should still be specified as either '<dataset><dataset/>' or '<dataset />'.

References

1. Khaitovich, P., Hellmann, I., Enard, W. et al. (2005) Parallel patterns of evolution in the genomes and transcriptomes of humans and chimpanzees. *Science* 309, 1850–1854.
2. King, M.C., and Wilson, A.C. (1975) Evolution at two levels in humans and chimpanzees. *Science* 188, 107–116.
3. Davidson, E.H., and Levine, M.S. (2008) Properties of developmental gene regulatory networks. *Proc Natl Acad Sci USA* 105, 20063–20066.
4. Levine, M., and Davidson, E.H. (2005) Gene regulatory networks for development. *Proc Natl Acad Sci USA* 102, 4936–4942.
5. Giurumescu, C.A., Sternberg, P.W., and Asthagiri, A.R. (2009) Predicting phenotypic diversity and the underlying quantitative molecular transitions. *PLoS Comput Biol* 5, e1000354.
6. Hardy, J., and Singleton, A. (2009) Genomewide association studies and human disease. *N Engl J Med* 360, 1759–1768.
7. Waterston, R.H., Lindblad-Toh, K., Birney, E. et al. (2002) Initial sequencing and comparative analysis of the mouse genome. *Nature* 420, 520–562.
8. Cooper, G.M., Stone, E.A., Asimenos, G. et al. (2005) Distribution and intensity of constraint in mammalian genomic sequence. *Genome Res* 15, 901–913.
9. Birney, E., Stamatoyannopoulos, J.A., Dutta, A. et al. (2007) Identification and analysis of functional elements in 1% of the human genome by the ENCODE pilot project. *Nature* 447, 799–816.
10. Attanasio, C., Reymond, A., Humbert, R. et al. (2008) Assaying the regulatory potential of mammalian conserved non-coding sequences in human cells. *Genome Biol* 9, R168.
11. Dermitzakis, E.T., and Clark, A.G. (2002) Evolution of transcription factor binding sites in Mammalian gene regulatory regions: conservation and turnover. *Mol Biol Evol* 19, 1114–1121.
12. Arnone, M.I., and Davidson, E.H. (1997) The hardwiring of development: organization and function of genomic regulatory systems. *Development* 124, 1851–1864.
13. Messina, D.N., Glasscock, J., Gish, W. et al. (2004) An ORFeome-based analysis of human transcription factor genes and the construction of a microarray to interrogate their expression. *Genome Res* 14, 2041–2047.

14. Cheung, V.G., Conlin, L.K., Weber, T.M. et al. (2003) Natural variation in human gene expression assessed in lymphoblastoid cells. *Nat Genet* 33, 422–425.
15. Frazer, K.A., Ballinger, D.G., Cox, D.R. et al. (2007) A second generation human haplotype map of over 3.1 million SNPs. *Nature* 449, 851–861.
16. Monks, S.A., Leonardson, A., Zhu, H. et al. (2004) Genetic inheritance of gene expression in human cell lines. *Am J Hum Genet* 75, 1094–1105.
17. Petretto, E., Mangion, J., Dickens, N.J. et al. (2006) Heritability and tissue specificity of expression quantitative trait loci. *PLoS Genet* 2, e172.
18. Price, A.L., Patterson, N., Hancks, D.C. et al. (2008) Effects of cis and trans genetic ancestry on gene expression in African Americans. *PLoS Genet* 4, e1000294.
19. Schadt, E.E., Monks, S.A., Drake, T.A. et al. (2003) Genetics of gene expression surveyed in maize, mouse and man. *Nature* 422, 297–302.
20. Spielman, R.S., Bastone, L.A., Burdick, J.T. et al. (2007) Common genetic variants account for differences in gene expression among ethnic groups. *Nat Genet* 39, 226–231.
21. Storey, J.D., Madeoy, J., Strout, J.L. et al. (2007) Gene-expression variation within and among human populations. *Am J Hum Genet* 80, 502–509.
22. Stranger, B.E., Nica, A.C., Forrest, M.S. et al. (2007) Population genomics of human gene expression. *Nat Genet* 39, 1217–1224.
23. Miao, X., Yu, C., Tan, W. et al. (2003) A functional polymorphism in the matrix metalloproteinase-2 gene promoter (−1306C/T) is associated with risk of development but not metastasis of gastric cardiac adenocarcinoma. *Cancer Res* 63, 3987–3990.
24. Bond, G.L., Hu, W., Bond, E.E. et al. (2004) A single nucleotide polymorphism in the MDM2 promoter attenuates the p53 tumor suppressor pathway and accelerates tumor formation in humans. *Cell* 119, 591–602.
25. Caspi, A., Sugden, K., Moffitt, T.E. et al. (2003) Influence of life stress on depression: moderation by a polymorphism in the 5-HTT gene. *Science* 301, 386–389.
26. Prokunina, L., Castillejo-Lopez, C., Oberg, F. et al. (2002) A regulatory polymorphism in PDCD1 is associated with susceptibility to systemic lupus erythematosus in humans. *Nat Genet* 32, 666–669.
27. Kostrikis, L.G., Neumann, A.U., Thomson, B. et al. (1999) A polymorphism in the reg-ulatory region of the CC-chemokine receptor 5 gene influences perinatal transmission of human immunodeficiency virus type 1 to African-American infants. *J Virol* 73, 10264–10271.
28. Saito, H., Tada, S., Ebinuma, H. et al. (2001) Interferon regulatory factor 1 promoter polymorphism and response to type 1 interferon. *J Cell Biochem* 81, 191–200.
29. Emilsson, V., Thorleifsson, G., Zhang, B. et al. (2008) Genetics of gene expression and its effect on disease. *Nature* 452, 423–428.
30. Bryne, J.C., Valen, E., Tang, M.H. et al. (2008) JASPAR, the open access database of transcription factor-binding profiles: new content and tools in the 2008 update. *Nucleic Acids Res* 36, D102–D106.
31. Matys, V., Kel-Margoulis, O.V., Fricke, E. et al. (2006) TRANSFAC and its module TRANSCompel: transcriptional gene regulation in eukaryotes. *Nucleic Acids Res* 34, D108–D110.
32. Roulet, E., Busso, S., Camargo, A.A. et al. (2002) High-throughput SELEX SAGE method for quantitative modeling of transcription-factor binding sites. *Nat Biotechnol* 20, 831–835.
33. Wilson, D., Charoensawan, V., Kummerfeld, S.K. et al. (2008) DBD – taxonomically broad transcription factor predictions: new content and functionality. *Nucleic Acids Res* 36, D88-D92.
34. Fulton, D.L., Sundararajan, S., Badis, G. et al. (2009) TFCat: the curated catalog of mouse and human transcription factors. *Genome Biol* 10, R29.
35. Lescot, M., Dehais, P., Thijs, G. et al. (2002) PlantCARE, a database of plant cis-acting regulatory elements and a portal to tools for in silico analysis of promoter sequences. *Nucleic Acids Res* 30, 325–327.
36. Pohar, T.T., Sun, H., and Davuluri, R.V. (2004) HemoPDB: hematopoiesis promoter database, an information resource of transcriptional regulation in blood cell development. *Nucleic Acids Res* 32, D86–D90.
37. Grienberg, I., and Benayahu, D. (2005) Osteo-Promoter Database (OPD) – promoter analysis in skeletal cells. *BMC Genomics* 6, 46.
38. Schmid, C.D., Perier, R., Praz, V. et al. (2006) EPD in its twentieth year: towards complete promoter coverage of selected model organisms. *Nucleic Acids Res* 34, D82–D85.
39. Shahmuradov, I.A., Gammerman, A.J., Hancock, J.M. et al. (2003) PlantProm: a database of plant promoter sequences. *Nucleic Acids Res* 31, 114–117.

40. Zhu, J., and Zhang, M.Q. (1999) SCPD: a promoter database of the yeast *Saccharomyces cerevisiae. Bioinformatics* 15, 607–611.

41. Kolchanov, N.A., Ignatieva, E.V., Ananko, E.A. et al. (2002) Transcription Regulatory Regions Database. (TRRD): its status in 2002. *Nucleic Acids Res* 30, 312–317.

42. Bergman, C.M., Carlson, J.W., and Celniker, S.E. (2005) Drosophila DNase I footprint database: a systematic genome annotation of transcription factor binding sites in the fruitfly, *Drosophila melanogaster. Bioinformatics* 21, 1747–1749.

43. Kanamori, M., Konno, H., Osato, N. et al. (2004) A genome-wide and nonredundant mouse transcription factor database. *Biochem Biophys Res Commun* 322, 787–793.

44. Tahira, T., Baba, S., Higasa, K. et al. (2005) dbQSNP: a database of SNPs in human promoter regions with allele frequency information determined by single-strand conformation polymorphism-based methods. *Hum Mutat* 26, 69–77.

45. Stenson, P.D., Ball, E.V., Mort, M. et al. (2003) Human Gene Mutation Database (HGMD): 2003 update. *Hum Mutat* 21, 577–581.

46. Zhao, T., Chang, L.W., McLeod, H.L. et al. (2004) PromoLign: a database for upstream region analysis and SNPs. *Hum Mutat* 23, 534–539.

47. Griffith, O.L., Montgomery, S.B., Bernier, B. et al. (2008) ORegAnno: an open-access community-driven resource for regulatory annotation. *Nucleic Acids Res* 36, D107–D113.

48. Portales-Casamar, E., Kirov, S., Lim, J. et al. (2007) PAZAR: a framework for collection and dissemination of cis-regulatory sequence annotation. *Genome Biol* 8, R207.

49. Aerts, S., Haeussler, M., van Vooren, S. et al. (2008) Text-mining assisted regulatory annotation. *Genome Biol* 9, R31.

50. Saric, J., Jensen, L.J., Ouzounova, R. et al. (2006) Extraction of regulatory gene/protein networks from Medline. *Bioinformatics* 22, 645–650.

51. Rodriguez-Penagos, C., Salgado, H., Martinez-Flores, I. et al. (2007) Automatic reconstruction of a bacterial regulatory network using Natural Language Processing. *BMC Bioinformatics* 8, 293.

52. Beisswanger, E., Lee, V., Kim, J.J. et al. (2008) Gene Regulation Ontology (GRO): design principles and use cases. *Stud Health Technol Inform* 136, 9–14.

53. Kelso, J., Visagie, J., Theiler, G. et al. (2003) eVOC: a controlled vocabulary for unifying gene expression data. *Genome Res* 13, 1222–1230.

54. Schomburg, I., Chang, A., Ebeling, C. et al. (2004) BRENDA, the enzyme database: updates and major new developments. *Nucleic Acids Res* 32, D431–D433.

55. Gallo, S.M., Li, L., Hu, Z. et al. (2006) REDfly: a Regulatory Element Database for *Drosophila. Bioinformatics* 22, 381–383.

56. Wasserman, W.W., and Fickett, J.W. (1998) Identification of regulatory regions which confer muscle-specific gene expression. *J Mol Biol* 278, 167–181.

57. Ho Sui, S.J., Mortimer, J.R., Arenillas, D.J. et al. (2005) oPOSSUM: identification of over-represented transcription factor binding sites in co-expressed genes. *Nucleic Acids Res.* 33, 3154–3164.

58. Blanco, E., Farre, D., Alba, M.M. et al. (2006) ABS: a database of Annotated regulatory Binding Sites from orthologous promoters. *Nucleic Acids Res* 34, D63–D67.

59. Jiang, C., Xuan, Z., Zhao, F. et al. (2007) TRED: a transcriptional regulatory element database, new entries and other development. *Nucleic Acids Res* 35, D137–D140.

60. Ghosh D. (2000) Object-oriented transcription factors database (ooTFD). *Nucleic Acids Res* 28, 308–310.

61. Sierro, N., Kusakabe, T., Park, K.J. et al. (2006) DBTGR: a database of tunicate promoters and their regulatory elements. *Nucleic Acids Res* 34, D552–D555.

62. Hubbard T.J., Aken B.L., Ayling S. et al. (2009) Ensembl 2009. *Nucleic Acids Res* 37, D690–D697.

63. Sayers E.W., Barrett T., Benson D.A. et al. (2009) Database resources of the National Center for Biotechnology Information. *Nucleic Acids Res* 37, D5–D15.

64. Ponomarenko, J.V., Merkulova, T.I., Vasiliev, G.V. et al. (2001) rSNP_Guide, a database system for analysis of transcription factor binding to target sequences: application to SNPs and site-directed mutations. *Nucleic Acids Res* 29, 312–316.

65. Trinklein, N.D., Aldred, S.J., Saldanha, A.J. et al. (2003) Identification and functional analysis of human transcriptional promoters. *Genome Res* 13, 308–312.

66. King, D.C., Taylor, J., Elnitski, L. et al. (2005) Evaluation of regulatory potential and conservation scores for detecting cis-regulatory modules in aligned mammalian genome sequences. *Genome Res* 15, 1051–1060.

67. Wang, H., Zhang, Y., Cheng, Y. et al. (2006) Experimental validation of predicted

68. Visel, A., Minovitsky, S., Dubchak, I. et al. (2007) VISTA Enhancer Browser – a database of tissue-specific human enhancers. *Nucleic Acids Res* 35, D88–D92.

69. Kim T.H., Abdullaev Z.K., Smith A.D. et al. (2007) Analysis of the Vertebrate Insulator Protein CTCF-Binding Sites in the Human Genome. *Cell* 128, 1231–1245.

70. Gao, H., Falt, S., Sandelin, A. et al. (2008) Genome-wide identification of estrogen receptor alpha-binding sites in mouse liver. *Mol Endocrinol* 22, 10–22.

71. Harbison, C.T., Gordon, D.B., Lee, T.I. et al. (2004) Transcriptional regulatory code of a eukaryotic genome. *Nature* 431, 99–104.

72. MacIsaac, K.D., Wang, T., Gordon, D.B. et al. (2006) An improved map of conserved regulatory sites for *Saccharomyces cerevisiae*. *BMC Bioinformatics* 7, 113.

73. Robertson, G., Hirst, M., Bainbridge, M. et al. (2007) Genome-wide profiles of STAT1 DNA association using chromatin immunoprecipitation and massively parallel sequencing. *Nat Methods* 4, 651–657.

74. Johnson D.S., Mortazavi A., Myers R.M. et al. (2007) Genome-wide mapping of in vivo protein-DNA interactions. Science 316, 1497–1502.

75. Lin, C.Y., Vega, V.B., Thomsen, J.S. et al. (2007) Whole-genome cartography of estrogen receptor alpha binding sites. *PLoS Genet* 3, e87.

76. Lim, C.A., Yao, F., Wong, J.J. et al. (2007) Genome-wide mapping of RELA(p65) binding identifies E2F1 as a transcriptional activator recruited by NF-kappaB upon TLR4 activation. *Mol Cell* 27, 622–635.

77. Wederell, E.D., Bilenky, M., Cullum, R. et al. (2008) Global analysis of in vivo Foxa2-binding sites in mouse adult liver using massively parallel sequencing. *Nucleic Acids Res* 36, 4549–4564.

78. Hufton, A.L., Mathia, S., Braun, H. et al. (2009) Deeply conserved chordate non-coding sequences preserve genome synteny but do not drive gene duplicate retention. *Genome Res.* 19, 2036–2051.

79. Adryan, B., and Teichmann, S.A. (2006) FlyTF: a systematic review of site-specific transcription factors in the fruit fly *Drosophila melanogaster*. *Bioinformatics* 22, 1532–1533.

80. Zhu, Q.H., Guo, A.Y., Gao, G. et al. (2007) DPTF: a database of poplar transcription factors. *Bioinformatics* 23, 1307–1308.

81. Maier, H., Dohr, S., Grote, K. et al. (2005) LitMiner and WikiGene: identifying problem-related key players of gene regulation using publication abstracts. *Nucleic Acids Res* 33, W779–W782.

82. Yang, H., Nenadic, G., and Keane, J.A. (2008) Identification of transcription factor contexts in literature using machine learning approaches. *BMC Bioinformatics* 9 Suppl 3, S11.

83. Steele, E., Tucker, A., 't Hoen, P.A. et al. (2009) Literature-based priors for gene regulatory networks. *Bioinformatics* 25, 1768–1774.

84. Schilling, T., Schleithoff, E.S., Kairat, A. et al. (2009) Active transcription of the human FAS/CD95/TNFRSF6 gene involves the p53 family. *Biochem Biophys Res Commun* 387, 399–404.

85. Kent, W.J. (2002) BLAT – the BLAST-like alignment tool. *Genome Res* 12, 656–664.

86. Palaniswamy, S.K., James, S., Sun, H. et al. (2006) AGRIS and AtRegNet. a platform to link cis-regulatory elements and transcription factors into regulatory networks. *Plant Physiol* 140, 818–829.

87. Shahi, P., Loukianiouk, S., Bohne-Lang, A. et al. (2006) Argonaute – a database for gene regulation by mammalian microRNAs. *Nucleic Acids Res* 34, D115–D118.

88. Barrasa, M.I., Vaglio, P., Cavasino, F. et al. (2007) EDGEdb: a transcription factor-DNA interaction database for the analysis of *C. elegans* differential gene expression. *BMC Genomics* 8, 21.

89. LSPD. (2006) http://rulai.cshl.edu/LSPD/.

90. Halfon, M.S., Gallo, S.M., and Bergman, C.M. (2008) REDfly 2.0: an integrated database of cis-regulatory modules and transcription factor binding sites in *Drosophila*. *Nucleic Acids Res* 36, D594–D598.

91. Gama-Castro, S., Jimenez-Jacinto, V., Peralta-Gil, M. et al. (2008) RegulonDB (version 6.0): gene regulation model of *Escherichia coli* K-12 beyond transcription, active (experimental) annotated promoters and Textpresso navigation. *Nucleic Acids Res* 36, D120–D124.

92. Wingender, E. (2008) The TRANSFAC project as an example of framework technology that supports the analysis of genomic regulation. *Brief Bioinform* 9, 326–332.

Chapter 21

Computational Identification of Plant Transcription Factors and the Construction of the PlantTFDB Database

Kun He, An-Yuan Guo, Ge Gao, Qi-Hui Zhu, Xiao-Chuan Liu, He Zhang, Xin Chen, Xiaocheng Gu, and Jingchu Luo

Abstract

Transcription factors (TFs) play an important role in gene regulation. Computational identification and annotation of TFs at genome scale are the first step toward understanding the mechanism of gene expression and regulation. We started to construct the database of *Arabidopsis* TFs in 2005 and developed a pipeline for systematic identification of plant TFs from genomic and transcript sequences. In the following years, we built a database of plant TFs (PlantTFDB, http://planttfdb.cbi.pku.edu.cn) which contains putative TFs identified from 22 species including five model organisms and 17 economically important plants with available EST sequences. To provide comprehensive information for the putative TFs, we made extensive annotation at both the family and gene levels. A brief introduction and key references were presented for each family. Functional domain information and cross-references to various well-known public databases were available for each identified TF. In addition, we predicted putative orthologs of the TFs in other species. PlantTFDB has a simple interface to allow users to make text queries, or BLAST searches, and to download TF sequences for local analysis. We hope that PlantTFDB could provide the user community with a useful resource for studying the function and evolution of transcription factors.

Key words: Transcription factors, database construction, plant genome, HMMER search, Ortholog.

1. Introduction

Transcription factors (TFs) bind selectively to specific DNA sequences in order to turn on or off the transcription of their target genes. Eukaryotes have a much more sophisticated transcription regulation mechanism than prokaryotes. For example,

I. Ladunga (ed.), *Computational Biology of Transcription Factor Binding*, Methods in Molecular Biology 674, DOI 10.1007/978-1-60761-854-6_21, © Springer Science+Business Media, LLC 2010

eukaryotic RNA polymerase complexes cannot turn on the gene transcription by themselves. Instead, TFs are needed to recognize and bind to the *cis*-regulatory elements that primarily reside in the promoter region of the target genes and to recruit the RNA polymerase complex for the initiation of transcription. The special sequence region in a TF that interacts directly with DNA is called the DNA-binding domain (DBD). Based on the sequence patterns and structural features of DBDs, TFs can be classified into diverse families (1).

Multicellular eukaryotes have to deal with cell differentiation with the aid of more sophisticated regulatory mechanism using a larger number of TFs (2). Larger genomes usually have higher numbers of TFs. It has been recently reported that a normal human somatic cell was turned into a fully functional stem cell by introducing four TFs into it (3). Therefore, deciphering the binding relationship and regulatory network is a key step to the understanding of the process of development and many other biological phenomena (4). Identification of TFs at the genome level and construction of knowledge databases for the TFs using computational approach have therefore become the fundamental step in studies on the regulation of gene expression.

TRANSFAC is one of the earliest attempts to build a TF knowledgebase with experimentally proven TFs, their binding sites, and regulated target genes (5). The January 2009 release contains 12,183 factors and 24,745 binding sites from various taxa including plants, mammals, fungi, and bacteria. It started to collect data identified by high-throughput technologies, such as chromatin immunoprecipitation on chip (ChIP-chip). Several databases of plant TFs have been developed after the completion of the *Arabidopsis* genome sequencing in 2000 as well as several other plant model organisms such as rice and poplar in the following years (**Table 21.1**).

AtTFDB is the first *Arabidopsis* TF databases hosted by the *Arabidopsis* gene regulatory information server (AGRIS) at the Ohio State University [*see* **Chapter 2** and ref. (6)]. It classified TFs into different families based on the type of the corresponding DBDs. Recently, AtTFDB further integrated information about the potential regulatory relationship among TFs which makes it a useful resource for regulatory network analysis of *Arabidopsis*. The *Arabidopsis* TF database (RARTF) hosted by the RIKEN BioResource Center in Japan applies PSI-BLAST and InterProScan to the identification of all putative TFs in *Arabidopsis* (7). Sequence information, InterPro domains, and links to public *Arabidopsis* genome databases such as TAIR, MIPS, and TIGR are provided for each predicted TF. Recently, a database of tobacco TFs (TOBFAC) has been built by the University of Virginia (8). TOBFAC contains about 2,500 putative tobacco TF genes predicted from the data source of both gene-space reads

Table 21.1
Databases of plant transcription factors

Name	Data source	Website and institution	References
TRANSFAC	Mainly *Arabidopsis*	http://www.gene-regulation.com/ BIOBASE, Germany	(5)
AtTFDB	*Arabidopsis*	http://arabidopsis.med.ohio-state.edu/AtTFDB/ The Ohio State University, USA	(6)
RARTF	*Arabidopsis*	http://rarge.psc.riken.jp/rartf/ RIKEN BioResource Center, Japan	(7)
TOBFAC	Tobacco	http://compsysbio.achs.virginia.edu/ tobfac/ University of Virginia, USA	(8)
LegumeTFDB	*Glycine max, Lotus japonicus, Medicago truncatula*	http://legumetfdb.psc.riken.jp/ RIKEN BioResource Center, Japan	
PlnTFDB	Genome sequences (19 species)	http://plntfdb.bio.uni-potsdam.de/ University of Potsdam, Germany	(9)
PlantTFDB	Genome sequence (5 species) EST sequence (17 species)	http://planttfdb.cbi.pku.edu.cn/ Peking University, China	(10)

and EST sequences. LegumeTFDB, a database of three legume plants (*Glycine max*, *Lotus japonicus*, *Medicago truncatula*) has been available online at the RIKEN BioResource Center, Japan. Interestingly, the number of predicted soybean (*Glycine max*) TFs listed in LegumeTFDB exceeds the number of predicted tobacco TFs by a factor of 2. Up to date, the two most comprehensive plant TF databases are the PlnTFDB developed by University of Potsdam, Germany (http://plntfdb.bio.uni-potsdam.de/) (9) and the PlantTFDB constructed by Peking University, China (http://planttfdb.cbi.pku.edu.cn/) (10). Both PlnTFDB and PlantTFDB attempt to collect plant TFs from all available data sources and provide comprehensive annotation at both the family and gene level.

In this chapter, we describe our computational approaches to the genome-wide identification of plant TFs, construction of the TF databases, and annotation of TF genes. In 2002, several research groups from China and the United States initiated a collaborative project for the genome-wide ORFeome cloning and analysis of *Arabidopsis* genes which encode TFs (11). As the only bioinformatics group involved in this project, we started to construct the database of *Arabidopsis* TFs (DATF) which became publicly available in 2005 (12). With the available genome sequences of two rice sub-species (*Oryza sativa*, ssp.

japonica and *Oryza sativa*, ssp. *indica*) and poplar (*Populus trichocarpa*), two databases of rice TFs (DRTF) (13) and poplar TFs (DPTF) (14) were further built up in the following years.

Based on the experiences obtained from the construction of these three plant TF databases, we have built an in silico pipeline for the systematic identification of the putative TFs from various plant genomes and developed a comprehensive plant TF database PlantTFDB. To provide a comprehensive data source for plant biologists, PlantTFDB contains TFs identified from 5 model organisms with whole genome sequences and 17 economically important plants with abundant transcripts (**Table 21.2**).

Table 21.2
TFs and ortholog numbers in PlantTFDB

Data source (version)	Name		Species	TFs[a]	TFs with orthologs
TAIR (v6)	*Arabidopsis*		*Arabidopsis thaliana*	2290	1346
JGI (v1.1)	Poplar		*Populus trichocarpa*	2576	2042
TIGR (v4.0)	Rice		*Oryza sativa* (ssp. indica)	2025	1763
			Oryza sativa (ssp. japonica)	2384	2124
JGI (v1.1)	Moss		*Physcomitrella patens*	1170	524
JGI (v3.0)	Green alga		*Chlamydomonas reinhardtii*	205	64
PlantGDB (v155a)	Crops	Barley	*Hordeum vulgare*	618	595
		Maize	*Zea mays*	764	734
		Sorghum	*Sorghum bicolor*	397	372
		Sugarcane	*Saccharum officinarum*	1177	1157
		Wheat	*Triticum aestivum*	1127	1074
	Fruits	Apple	*Malus x domestica*	1025	938
		Grape	*Vitis vinifera*	867	793
		Orange	*Citrus sinensis*	599	541
	Trees	Pine	*Pinus taeda*	950	644
		Spruce	*Picea glauca*	440	383
	Economic plants	Cotton	*Gossypium hirsutum*	1567	1430
		Potato	*Solanum tuberosum*	1340	1243
		Soybean	*Glycine max*	1891	1774
		Sunflower	*Helianthus annuus*	513	435
		Tomato	*Lycopersicon esculentum*	998	917
		Lotus	*Lotus japonicus*	457	434
		Medicago	*Medicago truncatula*	1022	914

[a]The TF numbers of *Arabidopsis* and rice *japonica* are the gene model numbers including alternative splicing.

Computational Identification of Plant Transcription Factors 355

TFs predicted from newly sequenced genomes are being added to PlantTFDB.

PlantTFDB attempts to provide comprehensive information for the identified TFs both at the family and at the gene level. A brief introduction can be found for each family. In addition to common sequence features derived from well-known domain database (15, 16) and Gene Ontology (http://www.geneontology.org/) (17), expression profiling data derived from UniGene and NCBI GEO repository are also available for each predicted TF. Moreover, automatically annotated homologs in related species can also be found for each TF. With a user-friendly Web interface, all sequences and annotation information are freely available online (http://planttfdb.cbi.pku.edu.cn/).

2. Materials

2.1. Sequence Data

Whole proteome sequences of five model organisms with completed genomes were downloaded from genome sequencing centers (**Table 21.3**). The *Arabidopsis* Information Resource (TAIR) maintains a database of genomic data for the model plant *Arabidopsis thaliana* (http://www.arabidopsis.org/). The genome sequence of the rice sub-species *japonica* was originally hosted at The Institute of Genome Research (TIGR) and moved to Michigan State University in 2007 (http://rice.plantbiology.msu.edu/), while another rice sub-species *indica*

Table 21.3
Data source of PlantTFDB

Species and data type	Website and institution[a]
Arabidopsis genome sequence	http://www.arabidopsis.org/ The Arabidopsis Information Resource (TAIR), USA
Rice (*japonica*) genome sequence	http://rice.plantbiology.msu.edu/ Michigan State University, USA
Rice (*indica*) genome sequence	http://rise.genomics.org.cn/ Beijing Genome Institute, China
Poplar, Moss, Green Algae genome sequence	http://www.jgi.doe.gov/ DOE Joint Genome Institute (JGI), USA
Plant unique transcripts (PUTs) of 17 plants assembled by PlantGDB	http://www.plantgdb.org/ Plant Genome Database (PlantGDB), USA

[a]Rice (*japonica*) was originally hosted at The Institute for Genomic Research (TIGR) and moved to Michigan State University in 2007.

was sequenced by the Beijing Genome Institute (BGI), China (http://rise.genomics.org.cn/). All other genome sequences are obtained from the Joint Genome Institute (JGI), the US Department of Energy (http://www.jgi.doe.gov/).

For the 17 plants whose genomic sequences were not available in 2007 when we started to construct PlantTFDB, we downloaded the assembled transcripts from the Plant Genome Database (PlantGDB, http://www.plantgdb.org/). By assembling mRNA and EST sequences available in public databanks, PlantGDB predicted a set of plant unique transcripts (PUTs) for each organism. Based on those transcripts, we further identified open reading frames and derived protein sequences using the framefinder program (http://www.ebi.ac.uk/~guy/estate/).

2.2. Software Tools

All data and information are stored in a MySQL relational database on a Linux server. MySQL is an open source database management system widely used for both small and large database applications (http://dev.mysql.com). Queries to the database are implemented in PHP scripts running in an Apache/PHP environment. Graphics are drawn using the PHP module of the GD graphics library. Three-dimensional structure illustration was created using Molscript (http://www.avatar.se/molscript/) (18).

The NCBI BLAST tool kit (19) is installed locally for sequence similarity search. The HMMER package (http://hmmer.janelia.org/) (20) is used for Hidden Markov Model (HMM) profile search. HMM profiles of known DNA-binding domains are obtained from the Pfam database (http://pfam.sanger.ac.uk/) (16). The ClustalW (21) program is used for multiple sequence alignment. The Phylip package (http://evolution.genetics.washington.edu/phylip.html) is implemented for the construction of the phylogeny trees. The InterProScan (15) program is employed to identify protein domains and assign Gene Ontology (GO) (17) terms to the putative TFs.

3. Methods

3.1. Classification of TF Families

Like most other homologous proteins, TFs are usually grouped into families based on the conserved sequences of their DBDs. DBDs are responsible for the recognition of the *cis*-regulatory elements in the promoter and other regulatory regions of target genes. As stated above, the DBD is used to determine whether or not a protein could be considered as a putative TF. However,

Computational Identification of Plant Transcription Factors 357

domain shuffling and horizontal gene transfer events have been abundant during evolution. Some of the TFs may contain more than one type of DBD so the classification of TFs into families is not always straightforward.

Richemann et al. (1) systematically compared families among *Arabidopsis* and other species and summarized the relationship between DBDs and TF families. Based on the convention they proposed, we classify all known *Arabidopsis* TFs into 64 families (**Fig. 21.1**). **Table 21.4** lists the TF number of each family in *Arabidopsis* and the other four model plant species.

There are three different relationships between TF families and their DNA-binding domains: required, possible, or forbidden domain. A required domain means that a TF in a certain family must contain the corresponding domain. Using two TF families CCAAT-Dr1 and CCAAT-HAP3 as examples (bottom left in **Fig. 21.1**), the existence of a Dr1 or HAP3 domain in either family is required to classify this TF as a CCAAT-Dr1 or CCAAT-HAP3 family member. A possible domain is defined such that a TF from a certain family might contain this domain in addition to the required domain. For instance, a member of CCAAT-HAP3 family might contain a Dr1 domain besides the required HAP3 domain. Finally, a forbidden domain means that it should not be contained in the TF of a certain family. In the above example, a CCAAT-Dr1 family member should not contain a HAP3 domain, otherwise it will be classified as a CCAAT-HAP3 member. In summary, if a protein contains a HAP3 domain, it is classified into the CCAAT-HAP3 family no matter whether it contains a Dr1 domain or not. On the other hand, a protein containing only the Dr1 domain but not the HAP3 domain will be classified into the CCAAT-Dr1 family.

3.2. Prediction of TFs

We combine automated search with manual curation for the identification of *Arabidopsis* TFs (**Fig. 21.2**). A list of 64 TF families was obtained based on the fundamental work by Riechmann et al. (1) as well as from a literature survey. HMM Profiles (20) are statistical models of multiple sequence alignments and contain position-specific information about the occurrence probabilities of all possible residues for each column in the alignment. HMMER (http://hmmer.janelia.org/) is an implementation of profile HMMs for biological sequence analysis. HMMER can be used to construct profile according to multiple sequence alignments, and it can also use a given profile to search for sequences belonging to the same family with the given profile. Pfam (16) is a database of protein domains represented by multiple sequence alignments and HMM profiles built using HMMER. HMM profiles of 48 TF families can be found in Pfam and are used in HMMER search. For the remaining 16 families where HMM profiles were not available at the time when we started to construct

358   He et al.

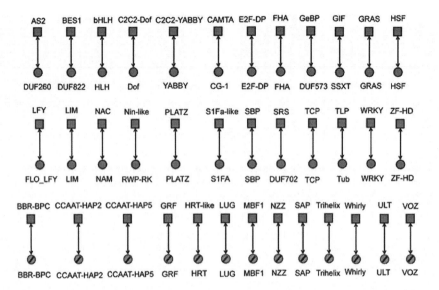

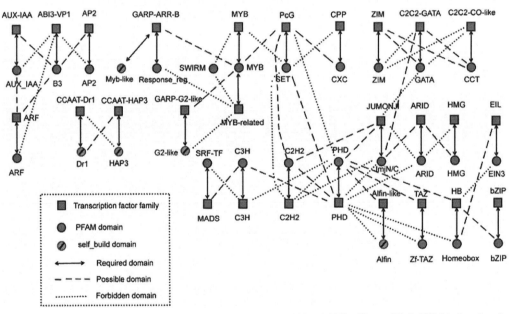

Fig. 21.1. Schematic representation of the relationship between *Arabidopsis* TF families and their DNA-binding domains (DBDs). *Squares* denote TF families, *circles* show DBDs obtained from the Pfam database, *circles* with latches inside indicate the DBDs we constructed. A double-headed *arrow* connects the DBD which must be contained in the corresponding TF family, *dashed lines* link one or more DBDs which might exist within this TF family, *dotted lines* demonstrate one or more DBDs which should not be contained in this family.

the database of *Arabidopsis* TFs, seed sequences were retrieved from the public protein sequence databases such as GenPept and Swiss-Prot and were taken as query sequence for BLAST search against the protein sequences of the *Arabidopsis* genome. With more plant genome sequences available, we have accumulated more data for each of the above 16 families and built HMM

Table 21.4
The number of predicted TF family members in five plant genomes

TF family	Description	At	Os	Pt	Pp	Cr	Function	TFBS
ABI3-VP1	Abscisic acid insensitive 3,viviParous1	60	52	108	37	1	ABA response	
Alfin(Zn)	Alfalfa protein	7	10	9	7	2	Salt response	
AP2-EREBP	APETALA2 and ethylene-responsive element binding proteins	146	165	212	156	12	Flower development, ethylene response, drought, cold response	GCC box
ARF	Auxin response factors	23	26	37	13		Auxin response	TGTCTC
ARID	AT-rich interaction domain	10	5	13	8	1	Chromatin remodeling	AT-rich
AS2	Asymmetric leaves2	42	36	57	29		Leaf venation establishment	
AUX-IAA	Auxin (indole-3-acetic acid)	29	29	33	2		Auxin response	
BBR-BPC	Barley B recombinant protein, basic pentacysteine	7	4	16			Ovule identity control	(GA/TC)8
BES1	Bri1-Ems-suppressor1	8	6	12	6		Brassinosteroid regulation	
bHLH	Basic helix loop helix	127	151	148	98	3	Secondary metabolism, cell proliferation	G-box E-box
bZIP	Basic leucine zipper	72	84	85	38	9	Flower and leaf development, hormone response	ACGT

(continued)

Table 21.4 (continued)

TF family		Description	At	Os	Pt	Pp	Cr	Function	TFBS
C2C2(Zn)	CO-like	Constans	37	39	39	6	6	Meristem identity control	
	Dof	DNA binding with one finger	36	30	42	1	1	Carbohydrate metabolism, defense, germination, hormone response	
	GATA	GATA	26	21	32	11	11	Light response	[TA]GATA[GA]
	YABBY		5	8	13			Abaxial identity control	
C2H2(Zn)		Cys2-His2 motif	134	94	81	49	5	Development	
C3H(Zn)		Cys3-His motif	59	57	78	37	12	Development	
CAMTA		Calmodulin-binding transcription activators	6	6	7	1		Calcium pathway	
CCAAT	Dr1	DRAP1	2	1	2	1			CCAAT
	HAP2	Heme activator protein 2	10	11	11			Embryo and flower development, circadian rhythm control, light signaling	CCAAT
	HAP3	Heme activator protein 3	11	12	19	2		Flowering control	CCAAT
	HAP5	Heme activator protein 5	13	16	19	2			CCAAT
CPP(Zn)		Cell shape control protein phosphatase	8	11	13	6	2	Cell proliferation, leghemoglobin gene regulation	
E2F-DP		Electro acoustic 2 factor (E2F)-DRTF1 polypeptide (DP)	8	8	10	11	3	Cell cycle regulation	
EIL		Ethylene insensitive 3 like	6	9	6	2		Ethylene response	
FHA		ForkHead associated	16	16	19	15	11	Cell cycle regulation	

(continued)

Table 21.4 (continued)

TF family	Description	At	Os	Pt	Pp	Cr	Function	TFBS
GARP								
ARR-B	*Arabidopsis* response regulator-B	10	8	15	1		Cytokinin signal transduction	
G2-like	GLK2-like	43	46	67	4		chloroplast development	
GeBP	Glabrous1 enhancer binding protein	21	15	7			Leaf cell development	
GIF	GRF-interacting factor	3	3	5	4	1	Leaf growth, pattern formation	
GRAS	Acronym for three genes: Gai, RgA, Scr	33	55	96	39		Meristem development, gibberellin response	
GRF	Growth regulating factor	9	12	9	2		Leaf and cotyledon growth	
HB	Homeo Box	87	82	106	40	1	Development	CAATNATTG
HMG	High mobility group	11	9	12	9	8		AT-hook
HRT-like(Zn)	Hordeum repressor of transcription (HRT)	2	1	1	3		Hormone response	
HSF	Heat shock transcription factor	23	25	31	8	2	Heat shock response	
JUMONJI		17	15	20	10	7	Flowering control	
LFY	LeaFY	1	1	1	2		Flower development	
LIM(Zn)	Acronym for three genes: Lin11, Isl-1, Mec-3	13	10	21	11		Lignin synthesis	Pal-box
LUG	LeUniG	2	6	6			Flower development	
MADS	Acronym for four genes: MCM1, AGAMOUS, DEFICIENS, SRF	104	63	111	22	1	Flower development	CC[A/T]6GG
MBF1	Multiprotein bridging factor 1	3	2	3	3	1		
MYB	MYeloBlastosis viral oncogene homolog	150	129	216	64	14	Cell proliferation, secondary metabolism, defense, ABA response	GGTTTAG

(continued)

Table 21.4 (continued)

TF family	Description	At	Os	Pt	Pp	Cr	Function	TFBS
MYB-related		49	60	84	31	7	Circadian rhythm control, cell proliferation	
NAC	Acronym for three genes: NAM, ATAF1, CUC2	107	130	172	32		Meristem development, hormone response, defense	
Nin-like	Nodule INception like	14	13	18	9	8	Root nodule development	
NZZ	Nozzle	1	1	2	3		Flower development	
PcG	PolyComb group	34	33	45	31	21	Seed development	
PHD(Zn)	Plant homeo domain	56	63	86	68	13	Light response, auxin signaling, leaf polarity	
PLATZ(Zn)	Plant AT-rich sequence and zinc-binding protein	10	16	20	13	4	Transcription repression	
S1Fa-like		3	2	2	1			
SAP	Sterile apetala	1	0	1			Flower development	
SBP(Zn)	Squamosa promoter binding protein	16	20	29	14	21	Flower and fruit development	TNCGTACAA
SRS(Zn)	Shi-related sequence	10	5	10	2		Gibberellin response	
TAZ(Zn)	Transcriptional adaptor zinc-binding domain	9	6	7	5	2	Calcium binding, stress response	
TCP	Acronym for three genes: TB1, CYC, PCFs	23	21	34	6		Flower development, cell division	GGNCCC
TLP	Tubby-like proteins	11	14	11	6	3	ABA pathway	
Trihelix		26	20	47	28		Light response	box II
ULT	Ultrapetal	2	2	3			Apical meristem development	

(continued)

Table 21.4 (continued)

TF family	Description	At	Os	Pt	Pp	Cr	Function	TFBS
VOZ(Zn)	Vascular plant transcription factors with one zinc finger	2	2	4	2		Pollen development	GCGTNx7ACGC
Whirly		2	1	2		1	Stress response	TGACAnnnnTGTCA
WRKY(Zn)	WRKY sequence motif	72	98	104	37	1	Defense	TGAC
ZF-HD(Zn)	Zinc finger homeo domain	16	15	25	8		Flower development	
ZIM(Zn)	Zinc finger motif	18	18	22	16		Leaf development	

Names with underlines are plant-specific families; (Zn): zinc finger protein; At: *Arabidopsis thaliana*; OS: *Oryza sativa japonica*; Pt: *Populus trichocarpa*; Cr: Pp: *Physcomitrella patens*

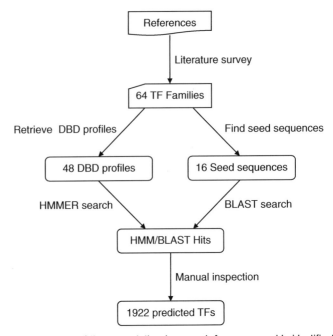

Fig. 21.2. Flowchart of the computational approach for genome-wide identification of transcription factors from *Arabidopsis thaliana*. Literature survey through published references is the first step to find all TFs characterized by experimental studies. A list of 64 TF families was constructed. HMM profiles of 48 DBDs were retrieved from the Pfam database and used in HMMER search. Seed sequences of DBDs for 16 families without HMM profiles were retrieved from the public protein sequence databases for BLAST search. Both HMMER and BLAST search results were manually checked to remove false positives. A total of 1,922 putative TFs were predicted from the *Arabidopsis* genome.

profiles for each DBD which can be used in the prediction of TFs in other species with either whole genome sequence or EST data.

3.3. Annotation

To provide sufficient information about the putative TFs, we made various annotations at both the family and gene levels. For each TF family, PlantTFDB gives a brief introduction including the potential function, the three-dimensional structure of the DBD, the characterization of the *cis*-regulatory element bound by the DBD of the family. For individual TFs, PlantTFDB shows general information such as database identifier, gene name, DNA sequence of both genomic and coding region, and protein sequence. The database of *Arabidopsis* TFs has the most comprehensive annotations benefiting from the rich published results of genetic and functional investigations of this model organism (**Fig. 21.3**). In addition to the general information, DATF includes the unique information as to whether a TF has been cloned (11) which can be browsed and searched in the "clone information" field. BLAST search was performed against well-known public databases and cross-references are linked to various public databases. Putative functional domains are identified

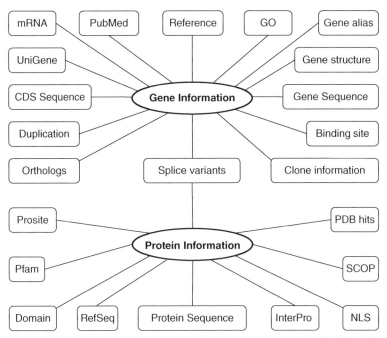

Fig. 21.3. Schematic demonstration of the annotation for individual TF genes in the PlantTFDB. Two ellipses show the key fields of "Gene Information" and "Protein Information" linked by "Splice Variants." *Round squares* with lines connected to either "Gene Information" or "Protein Information" show the annotation retrieved from various resources or cross-references to related databases.

and annotated by InterProScan, and Gene Ontology annotations are further extracted. In addition, the expression profiles collected from UniGene EST/cDNA information are also available.

3.4. Ortholog Identification

Recent genomic studies have already discovered several TF families that are specifically expanded in the plant kingdom. The ortholog information among different species is predicted using the BLAST score ratio (BSR), which is widely adopted by ENSEMBL (http://www.ensembl.org/) and other studies. An all-against-all BLASTP search with a strict cutoff E-value $<1 \times 10^{-20}$ was performed, and the BSR value was calculated for each hit. After comparing results at different BSR values, we chose the BSR value 0.4 as the cutoff and we retrieved the top sequences in a species with the largest BSR value as the putative ortholog(s). Using the bZIP family TF HY5 as an example, we identified orthologs of HY5 in 13 species (**Table 21.5**).

HY5 is an important activator for photomorphogenesis, which is essential for most of the land plants. As expected, all four land plants with completed genome data have HY5 orthologs. It was reported that the closest HY5 homologs in green algae has no COP complex interaction site detected, suggesting that HY5 might not be essential for this aquatic plant. The fact that HY5 orthologs have not been found in green algae could also be

Table 21.5
The orthologs of AtHY5 among different plant species

Species	PlantTFDB ID	Score ratio	Coverage	Identity	*E*-value
Physcomitrella patens	213225	0.43	0.82	0.58	4×10^{-35}
Populus trichocarpa	fgenesh4_pm.C_LG_XVIII000127	0.78	0.99	0.78	2×10^{-68}
Oryza sativa (japonica)	LOC_Os02g10860.1	0.58	0.98	0.69	7×10^{-49}
Oryza sativa (indica)	OsIBCD000496	0.53	0.95	0.65	2×10^{-44}
Citrus sinensis	PTCs00574.1	0.77	0.98	0.81	1×10^{-68}
Glycine max	PTGm01858.1	0.59	0.93	0.65	6×10^{-51}
Helianthus annuus	PTHa00512.1	0.54	0.71	0.74	3×10^{-46}
Lycopersicon esculentum	PTLe00971.1	0.69	0.94	0.77	4×10^{-60}
Lotus japonicus	PTLj00452.1	0.59	0.92	0.64	6×10^{-51}
Malus x domestica	PTMx01004.1	0.76	0.98	0.78	2×10^{-67}
Picea glauca	PTPg00432.1	0.43	0.55	0.76	3×10^{-35}
Solanum tuberosum	PTSt01300.1	0.64	0.86	0.79	7×10^{-56}
Vitis vinifera	PTVv00842.1	0.59	0.8	0.76	4×10^{-51}

due to our overly stringent criteria and the sequence difference between green alga and other plants. For the species from which HY5 orthologs were not detected, either genome sequence or more EST data should be used in the future.

4. Notes

1. Data source and database updating

   With the rapid progress of next-generation sequencing technology, more and more plant genomes have been already or are being sequenced. Sequence data of both genomic DNA and mRNA transcripts of different plant lineages become a rich source for computational identification of plant TFs at the genome level. We shall update PlantTFDB with the new release of sequence data of the five model organisms, as well as other newly sequenced genomes.[1]

---

[1] The PlantTFDB was updated to version 2.0 in July 2010, with predicted TFs from more species, and a new interface.

2. Family name and classification

There is no standard nomenclature for the name of TF families. Some of the families were named after the biological processes in which they are involved, and others by the three-dimensional structures of their DBDs. For example, the family ARF is referring to a group of auxin responsive factors, while bHLH is given to a group of TFs with conserved DBDs that form basic helix loop helix structures. For the former group, it is reasonable to expect most if not all of the members are involved in auxin responses-related processes.

Although distinct functions might be seen in different TF families, the family classification implemented in PlantTFDB should not be taken as the reflection of their biological functions. Functions of the TFs from the same family could be dramatically different. On the other hand, TFs from different families can recognize similar or even identical binding sites and be involved in the same biological process. For example, many bZIP family TFs share similar binding motifs with the bHLH family TFs, and MYB12 was found to cooperate with HY5 to regulate the expression of genes from the anthocyanin pathway.

3. Prediction

During the construction procedure of DATF when the HMM profiles for some families were not available, we used DBDs rather than the full-length protein sequence as seed sequences in BLAST search since members of the same TF family may share sequence conservation only at the DBD regions. On the other hand, non-TF proteins which do not contain DBDs may share sequence similarity with TFs in other regions in the flanking region of DBDs. We built the HMM profiles for those families and used them to predict TFs in other genomes. The best score ratio method used for ortholog prediction may result in both false positives and false negatives. On the other hand, phylogenetic methods reported for small-scale analysis are computationally too expensive for large TF families with dozens or even hundreds of members. New approaches such as comparative genomics are being investigated and hopefully can be successfully implemented in the future.

4. Annotation

The gene regulatory system is so complex that no single current available technology is sufficient to decipher the mechanism behind the complex networks. The simple model of one-to-one relationship between TFs and their binding sites may not reflect the real world of the regulatory network. Not only the same binding motif could be recognized by different TFs from the same or even different families, but also the same TF could bind to more than one known *cis*-regulatory element. The nonlinear relationship between TFs and their target genes is the basis of co-regulation underlying the very complex life processes. However, most of the current TF databases are using relational database systems such as MySQL. More complex schema or object-oriented database systems are required to handle more sophisticated regulatory network information.

In conclusion, the information provided by PlantTFDB for the predicted TFs should not be taken as a unique reference. Rather, it may serve as a starting point for further biological investigations using experimental approaches of genetic and molecular biology.

References

1. Riechmann, J.L., Heard, J., Martin, G., Reuber, L. et al. (2000) *Arabidopsis* transcription factors: genome-wide comparative analysis among eukaryotes. *Science* 290, 2105–2110.
2. Badis, G., Berger, M.F., Philippakis, A.A. et al. (2009) Diversity and complexity in DNA recognition by transcription factors. *Science* 324, 1720–1723.
3. Yu, J.Y., Vodyanik, M.A., Smuga-Otto, K. et al. (2007) Induced pluripotent stem cell lines derived from human somatic cells. *Science* 318, 1917–1920.
4. Qu, L.J., and Zhu, Y.X. (2006) Transcription factor families in *Arabidopsis*: major progress and outstanding issues for future research. *Curr Opin Plant Biol* 9, 544–549.
5. Wingender, E., Dietze, P., Karas, H. et al. (1996) TRANSFAC: a database of transcription factors and their DNA binding sites. *Nucleic Acids Res* 24, 238–241.
6. Davuluri, R.V., Sun, H., Palaniswamy, S.K. et al. (2003) AGRIS: *Arabidopsis* gene regulatory information server, an information resource of *Arabidopsis* cis-regulatory elements and transcription factors. *BMC Bioinformatics* 23, 25.
7. Iida, K., Seki, M., Sakurai, T., Satou, M. et al. (2005) RARTF: database and tools for complete sets of *Arabidopsis* transcription factors. *DNA Res* 12, 247–256.
8. Rushton, P.J., Bokowiec, M.T., Laudeman, T.W. et al. (2008) TOBFAC: the database of tobacco transcription factors. *BMC Bioinformatics* 9, 53.
9. Riaño-Pachón, D.M., Ruzicic, S., Dreyer, I. et al. (2007) PlnTFDB: an integrative plant transcription factor database. *BMC Bioinformatics* 8, 42.
10. Guo, A.Y., Chen, X., Gao, G. et al. (2008) PlantTFDB: a comprehensive plant transcription factor database. *Nucleic Acids Res* 36, D966–D969.
11. Gong, W., Shen, Y.P., Ma, LG. et al. (2004) Genome-wide ORFeome cloning and analysis of *Arabidopsis* transcription factor genes. *Plant Physiol* 135, 773–782.
12. Guo, A., He, K., Liu, D. et al. (2005) DATF: a database of *Arabidopsis* transcription factors. *Bioinformatics* 21, 2568–2569.
13. Gao, G., Zhong, Y., Guo, A., Zhu, Q. et al. (2006) DRTF: a database of rice transcription factors. *Bioinformatics* 22, 1286–1287.
14. Zhu, Q.H., Guo, A.Y., Gao, G. et al. (2007) DPTF: a database of poplar transcription factors. *Bioinformatics* 23, 1307–1308.
15. Quevillon, E., Silventoinen, V., Pillai, S. et al. (2005) InterProScan: protein domains identifier *Nucleic Acids Res* 33, W116–W120.
16. Finn, R.D., Tate, J., Mistry, J. et al. (2008) The Pfam protein families database. *Nucleic Acids Res* 36, D281–D288.
17. The Gene Ontology Consortium. (2000) Gene ontology: tool for the unification of biology. *Nat Genet* 25, 25–29.
18. Kraulis, P.J. (1991) MOLSCRIPT: a program to produce both detailed and schematic plots of protein structures. *J Appl Cryst* 24, 946–950.
19. Altschul, S.F., Madden, T.L., Schäffer, A.A. et al. (1997) Gapped BLAST and PSI-BLAST: a new generation of protein database search programs. *Nucleic Acids Res* 25, 3389–3402.
20. Durbin, R., Eddy, S., Krogh, A. et al. (1998) *Biological sequence analysis: probabilistic models of proteins and nucleic acids.* Cambridge University Press, Cambridge, MA.
21. Larkin, M.A., Blackshields, G., Brown, N.P. et al. (2007) Clustal W and Clustal X version 2.0. *Bioinformatics* 23, 2947–2948.

Chapter 22

Practical Computational Methods for Regulatory Genomics: A *cis*GRN-Lexicon and *cis*GRN-Browser for Gene Regulatory Networks

Sorin Istrail, Ryan Tarpine, Kyle Schutter, and Derek Aguiar

Abstract

The CYRENE Project focuses on the study of *cis*-regulatory genomics and gene regulatory networks (GRN) and has three components: a *cis*GRN-Lexicon, a *cis*GRN-Browser, and the Virtual Sea Urchin software system. The project has been done in collaboration with Eric Davidson and is deeply inspired by his experimental work in genomic regulatory systems and gene regulatory networks. The current CYRENE *cis*GRN-Lexicon contains the regulatory architecture of 200 transcription factors encoding genes and 100 other regulatory genes in eight species: human, mouse, fruit fly, sea urchin, nematode, rat, chicken, and zebrafish, with higher priority on the first five species. The only regulatory genes included in the *cis*GRN-Lexicon (*CYRENE genes*) are those whose regulatory architecture is validated by what we call the Davidson Criterion: *they contain functionally authenticated sites by site-specific mutagenesis, conducted in vivo, and followed by gene transfer and functional test.* This is recognized as the most stringent experimental validation criterion to date for such a genomic regulatory architecture. The CYRENE *cis*GRN-Browser is a full genome browser tailored for *cis*-regulatory annotation and investigation. It began as a branch of the Celera Genome Browser (available as open source at http://sourceforge.net/projects/celeragb/) and has been transformed to a genome browser fully devoted to regulatory genomics. Its access paradigm for genomic data is zoom-to-the-DNA-base in real time. A more recent component of the CYRENE project is the Virtual Sea Urchin system (VSU), an interactive visualization tool that provides a four-dimensional (spatial and temporal) map of the gene regulatory networks of the sea urchin embryo.

Key words: *cis*-regulatory architecture, gene regulatory networks, transcription factors, *cis*GRN-Lexicon, *cis*GRN-Browser, virtual sea urchin.

1. Introduction

When the GRN context is clear, we will use at times the shorthands "*cis*-Lexicon" and "*cis*-Browser". The *cis*-Lexicon and the *cis*-Browser are conceptually two separate entities, a database

I. Ladunga (ed.), *Computational Biology of Transcription Factor Binding*, Methods in Molecular Biology 674,
DOI 10.1007/978-1-60761-854-6_22, © Springer Science+Business Media, LLC 2010

and a visualization tool; however, from the user's point of view, they are intertwined into one integrated software environment for regulatory genomics.

The *cis*GRN-Lexicon is a database containing the regulatory architecture (the genomic regulatory region) of a set of transcription factor-encoding genes as well as of a number of other regulatory genes. This architecture is presented with full genomic structure known to date, including transcription factor binding site sequences, the organization into *cis*-regulatory modules (CRMs), and various other types of functional genomics (e.g., logic functions) annotations of the DNA regulatory region revealed by *cis*-regulatory analyses and systematic experimental perturbations of gene regulatory networks. The *cis*GRN-Lexicon annotations, accessible though the *cis*GRN-Browser, include the transcription factor binding site, the *trans* acting factor, the protein family to which the *trans* acting factor belongs, the *cis*-Regulatory Module (CRM) boundaries, the spatial and temporal functionality of the CRM, and the molecular function of the encoded protein. The *cis*GRN-Lexicon is embedded in and accessed through the *cis*GRN-Browser and is supported by various software libraries of tools for *cis*GRN-Lexicon annotators. One such system under development is CLOSE (*cis*-Lexicon Search Engine), a set of algorithmic strategies for literature extraction of *cis*-regulation articles to speed identification of new CYRENE genes and estimate the "dimension" of the CYRENE gene universe.

We describe the current state of the CYRENE *cis*GRN-Browser: its detailed architecture and its planned improvement in partnership with scientists from the Davidson Lab at the California Institute of Technology, where the *cis*-Browser has been in use for the past few years (18).

The Virtual Sea Urchin software system aims at giving a three-dimensional representation of the embryo's cellular anatomy stages in which gene expression is represented in time and within specific cell types. It will be integrated with the *cis*GRN-Browser and BioTapestry (19, 20) to present a *View from the GRNs*.

1.1. Algorithms for CRM Regulatory Architecture Prediction

The object of the *cis*-Lexicon is to create a data set that makes possible the prediction of *cis*-regulatory elements in DNA sequences of unknown function. To achieve a better prediction algorithm, the data set used must not be contaminated by low-quality data. Furthermore, the categories for annotation of a gene into the lexicon must be based on a relevant model of evolution and biological function.

Helpful surveys describing the state of the art of algorithms for prediction of sites, modules, and organization of regulatory regions are refs. 21–24. Many algorithms presented in the literature have aimed at addressing various aspects of the computational prediction problems related to regulatory genomics

(25–44, 19, 45–47). The present work aims at providing a database of *cis*-regulatory architectures, experimentally validated at the highest level as a basis for the design of the next generation of regulatory prediction algorithms.

1.2. CYRENE Genes and GRNs

The *cisGRN*-Lexicon presently contains the regulatory architecture of 200 transcription factors encoding genes and 100 other regulatory genes in eight species: human, mouse, fruit fly, sea urchin, nematode, rat, chicken, and zebrafish. The regulatory architecture of each of these CYRENE genes contains only functionally authenticated sites by site-specific mutagenesis, conducted in vivo, and followed by gene transfer and functional test. As the objective is to determine how genes are regulated in vivo, it follows that only in vivo *cis*-regulation studies should be admitted to the *cis*-Lexicon.

This database differs from other databases of gene regulation in several ways. It will be displayed in an interactive way that presents on one page the whole genome, a workspace for cis-regulatory analysis, and all the relevant gene functions. Many annotation categories and functions are unique to the *cis*-Lexicon. Experimental evidence required for admittance to the lexicon is stringently examined.

1.3. The Need for Integrated Cell Models

Combining GRN inference experiments (identification of regulatory genes (41, 48–50), perturbation experiments (51–52, 49, 53), *cis*-regulatory analysis (11, 13, 54, 55), etc.) with recent developments in systems biology imaging (56, 57) will make possible the construction of a full 4D spatiotemporal map of the sea urchin embryo. Such a map will fully describe the intra- and intercellular interactions of the GRN in the developing sea urchin embryo. The analysis and visualization tools needed to interpret these data seem to have lagged far behind experiments. Thus, we have been developing a natural visualization and analysis environment – the Virtual Sea Urchin (VSU) – that allows researchers to interrogate the developmental atlas at any time and at any position in the developmental process.

2. Materials

2.1. cisGRN-Browser: Software

The CYRENE *cis*-Browser was developed in the Eclipse IDE for Java Developers (http://www.eclipse.org). The foundation of the *cis*-Browser is the Celera Genome Browser (58), whose source code is available free on Source-Forge.net (http://sourceforge.net/projects/celeragb/). The

372   Istrail et al.

cis-Lexicon is stored as an Apache Derby Database (http://db.apache.org/derby/).

2.2. Virtual Sea Urchin: Software

Our system is composed of OGRE (an open-source 3D graphics engine), a C++ visualization application, and data describing the GRN such as transcription factor binding site affinity and products, cell-signaling pathways, etc. Three-dimensional sea urchin embryonic models were created using Blender. Future directions will include the integration of the regulatory network simulator BioTapestry (19, 20) and the *cis*GRN-Browser Cyrene. Portions of the embryo viewing application were provided by OgreMax (http://www.ogremax.com).

3. Methods

3.1. Cis-Lexicon

3.1.1. Anatomy of the Lexicon

The clues given by biology for the rules of *cis*-regulation are copious; the difficulties lie in creating criteria to represent the clues in a meaningful way. We must thus develop a classification system that neither oversimplifies biology, so that categories lose their physical meaning, nor overcomplicates the issues by creating more categories than necessary. While biology is inherently resistant to precise definitions, categories are necessary for the sake of high-throughput data clustering. The many challenges in creating controlled vocabularies are discussed shortly. The vocabularies of the *cis*-Lexicon are based on two guiding principles: (1) vocabularies should represent phenomena in a way that fits their physical interaction and (2) vocabularies should facilitate comparison. Where possible, the vocabularies are externally linked in such a way that when the vocabulary is updated, so is the *cis*-Lexicon.

Cis-Lexicon annotations include the transcription factor binding site (TFBS), the function of the TFBS, the *trans* acting factor, the protein family to which the *trans* acting factor belongs, the *cis*-regulatory module (CRM), the spatial and temporal functionality of the CRM, and the molecular function of the encoded protein. These categories in the *cis*-Lexicon were developed for useful data clustering. When a gene could not be annotated accurately within the constraints of our chosen vocabulary, new categories were created in order to categorize the *cis*-regulatory architecture in a biologically relevant manner.

3.1.2. cis-Regulatory Ontology

Transcription factor binding sites (TFBS) usually span 6 to 8 bp, though sometimes many more. Every TFBS in the lexicon is annotated as performing one or more of the following functions (more functions may exist in the natural world, but this is the set

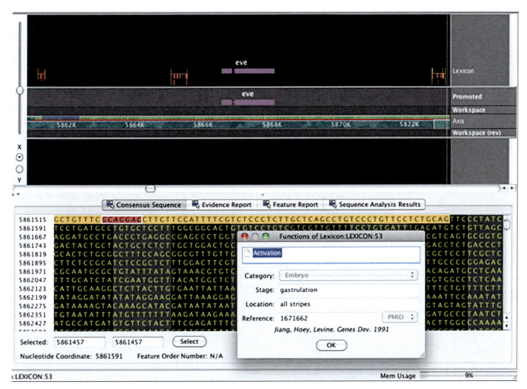

Fig. 22.1. Screenshot from the cis-Browser of the Drosophila transcription factor encoding gene, eve.

of all functions encountered so far) [See **Fig. 22.1** for an example of cis-regulatory function in the cisBrowser]:

- Repression – Indicates that mutating the TFBS increases gene expression or produces ectopic expression. Repressors may act "long range," when the repression effect may target more than one enhancer, or "short range," when repression affects only neighboring activators (59, 60). The function of repression applies in cases where the repressors interact with the basal transcription apparatus either directly or indirectly (61).

- Activation – Indicates that mutation decreases gene expression. An activator TFBS may act over a large genomic distance or short. See Latchman (62) for further discussion of some of the many ways a transcription factor can accomplish activation.

- Signal response – Indicates that the transcription factor has been shown to be activated by a ligand such as a hormone (phosphorylation is not included) (63).

- DNA looping – Indicates that the binding factor is involved in a protein–protein interaction with another binding factor some distance away that causes the DNA to form one or more loops. This looping brings distant regulatory elements closer to the basal transcription apparatus (64).

- Booster – Indicates that the TFBS does not increase gene expression on its own but can augment activation by other TFBSs.

- Input into AND logic – Indicates that the TFBS can activate gene expression only when two or more cooperating TFBSs are bound. Assigned to one of at least two TFBSs (14).

- Input into OR logic – Indicates that the TFBS can activate gene expression when either or both of two or more cooperating TFBSs are bound. Assigned to one of at least two TFBSs (14).

- Linker – Indicates that a TFBS is responsible for communicating between CRMs.

- Driver – Indicates that this TFBS is the primary determining factor of gene expression. The binding factor appears only in certain developmental situations and thus is the key input for directing gene expression. TFBSs that are not drivers usually bind ubiquitous factors (65).

- Communication with BTA (basal transcription apparatus).

- Insulator – Indicates that the TFBS causes *cis*-regulatory elements to be kept separate from one another. Insulators can separate the *cis*-regulatory elements of different genes as well as act as a barricade to keep active segments of DNA free of histones and remain active (66).

3.1.3. The trans Acting Factor

The transcription factor binding to the regulatory DNA is annotated as the gene name given in NCBI rather than the name given to the factor in the literature. For example, while Inagaki refers to human TF *c-Jun* (67), this is annotated in the *cis*-Lexicon as *JUN* for consistency. Each transcription factor in the lexicon is also assigned to a leaf of a transcription factor hierarchy adapted from TRANSFAC (68) (*see* **Figs. 22.2** and **22.3**)). More closely-related transcription factors may behave more similarly, so that when the data in the *cis*-Lexicon are clustered, patterns may be found by grouping transcription factors according to their evolutionary origins.

3.1.4. cis-Regulatory Modules

TFBSs occur in groups and each grouping usually directs gene expression in one temporal and spatial location. The CRM (1) includes the binding sites responsible for gene expression as well as the neighboring sequence established to enable the TBFSs to function correctly. Each CRM in the lexicon is annotated as functioning in a specific spatial and temporal location. There is currently no associated ontology for annotating this location in the lexicon. Exhaustively naming all locations and time points of

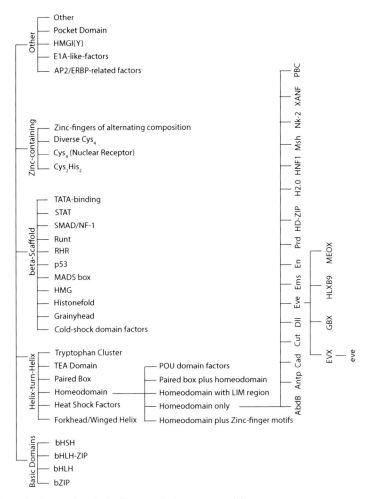

Fig. 22.2. Transcription factor hierarchy (homeodomain family expanded as an example).

an organism's development is beyond the scope of this project, though a controlled vocabulary of body parts and stages would be useful.

3.1.5. Gene Functions

Each gene whose *cis*-regulatory architecture is annotated in the lexicon is assigned to one of seven gene function categories. Many ontologies of gene functions already exist, such as the Gene Ontology (69) and Panther Classification System (70), but the *cis*-Lexicon Gene Functions ontology was created with the specific intent of grouping gene functions so that genes with similar *cis*-regulatory architecture are grouped together. The hypothesis is that housekeeping genes have *cis*-regulatory architecture distinct from transcription factor-encoding genes or signaling genes. GO annotations for each gene indicated in brackets show similarities and differences in gene function annotation between GO and the *cis*-Lexicon. *See* **Fig. 22.4** for gene functions in the *cis*-Lexicon.

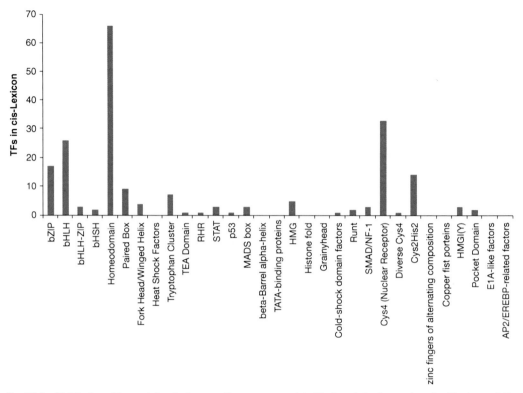

Fig. 22.3. Distribution of transcription factor encoding genes annotated in the *cis*-Lexicon categorized by transcription factor family (68).

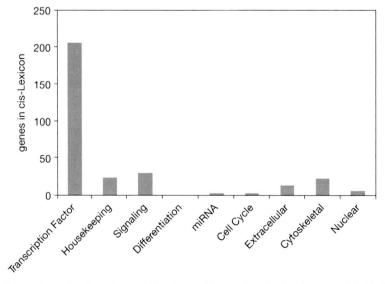

Fig. 22.4. Distribution of gene functions for which the *cis*-regulatory architecture has been annotated in the *cis*-Lexicon.

- *Cell cycle* – Genes involved in cell division that decide when to replicate DNA, when to divide the cell, etc., especially before cells terminally differentiate. Example: cyclins run in the *cis*-Lexicon: IL3 (*Homo sapiens*) (71) [extracellular space, cytokine activity, growth factor activity, etc.]; BCL2 (*H. sapiens*) (72) [apoptosis, etc.].

- *Cytoskeletal* – Genes involved in maintaining the structure of the cell, for example, actin and tubulin. In the *cis*-Lexicon: myh4 (*M. musculus*) (73) [actin binding, myosin filament, etc.; ASL (*G. gallus*) (74) [structural constituent of eye lens].

- *Differentiation* – Genes "expressed in the final stages of given developmental processes… They receive rather than generate developmental instructions" (2). In the *cis*-Lexicon: malpha (*Drosophila melanogaster*) (75) [cell fate specification, notch signaling pathway, sensory organ development].

- *Extracellular* – Genes whose product is released from the cell, such as hormones. These do not include membrane proteins, which could be categorized as housekeeping or signaling. In the *cis*-Lexicon: Col5A2 (*H. sapiens*) (76) [eye morphogenesis, skin development, extracellular matrix structural constituent, etc.]; SOD3 (*H. sapiens*) (77) [cytoplasm, extracellular region, zinc ion binding, etc.].

- *Housekeeping* – Genes continuously expressed that regulate processes inside the cell such as transcription apparatus, ribosomes, degradation proteins, and many enzymes. In the *cis*-Lexicon: btl (*D. melanogaster*) (78, 79) [endoderm development, glial cell migration, negative regulation of axon extension, etc.]; BACE1 (*Homo sapiens*) (80) [proteolysis, peptidase activity, etc.].

- *Transcription factor* – Genes whose product binds to DNA in a sequence-specific manner to affect gene expression. Does not include basal transcription factors. In the *cis*-Lexicon: twi (*D. melanogaster*) (81–84), [specific RNA polymerase II transcription factor activity, etc.]; Hoxa4 (*H. sapiens*) (85); Foxa2 (*Mus musculus*) (86, 87) [RNA polymerase II transcription factor activity, enhancer binding, etc].

- *miRNA* – Genes encoding micro RNAs. In the *cis*-Lexicon: DmiR-1 (*D. melanogaster*) (88) [cardiac cell differentiation, regulation of notch signaling].

- *Signaling* – Genes acting as part of a signaling pathway such as hormones, hormone receptors, and kinases. In the *cis*-Lexicon: IL4 (*M. musculus*) (89) [extracellular space, B cell activation, interleukin-4 receptor binding, etc.]; ins2 (*R. norvegicus*) (90) [cytoplasm, extracellular space, hormone activity, etc.].

378   Istrail et al.

3.1.6. Quintessential Diagram Problem

The categories used in the literature to classify the function of a TFBS are often described in simple terms that prevent the annotator from fully describing the function according to the *cis*-regulatory ontology (*see* **Section 3.1.2**). In the literature, a TFBS is generally declared an "activator" if deletion lowers output and a "repressor" if deletion increases output. Other more complex mechanisms may cause increased or decreased expression, such as DNA looping, communication with basal transcription apparatus, etc., but these are often unreported in the literature. Ideally, these more complex mechanisms would be known for each *trans* acting factor in the *cis*-Lexicon. Such biochemical clues would make possible effective data clustering, thus presenting clues for predicting *cis*-regulation. For example, DNA looping between two TFBSs cannot occur at less than a certain minimum distance, while Su(H), a transcription factor activated by signaling, may have a maximum distance from the transcription start site while still being able to direct gene expression of the sea urchin gene, *gcm* (9). Our lexicon is designed to handle more complex fields, but this information is not always available.

Perhaps more *cis*-regulatory information can be derived from the quantitative data obtained by mutating a TFBS (**Fig. 22.5**). Most literature containing data that meet the criteria of the *cis*-Lexicon contains a bar chart quantitatively describing gene expression as a result of mutating each of the TFBSs individually and in combination (18). Gene expression is thus a function of each of the inputs. Gene expression, the output of the function, is the combined effect of each of the individual inputs. This function is not simply the sum of the effect of each individual input; rather, the output depends on the interaction of the inputs. Thus, describing the gene expression requires a more complicated function than summing the effects of mutating each TFBS individually. A generalized mathematical function has been suggested, but applying the function to a broad range of mutational studies is difficult (15).

The *cis*-Lexicon currently does not handle the annotation of the quantitative data from literature. Knowing the relative impact of each TFBS on gene expression is important in properly describing *cis*-regulatory architecture. Thus a format for collecting these data that describe the biology effectively needs to be implemented from quantitative experimental data. Since gene expression can depend on which nucleotides are mutated within a TFBS, there is a relatively low certainty associated with these quantitative data, adding further complexity to the problem.

3.1.7. Examples in the Lexicon

See **Figs. 22.6** and **22.7**.

Practical Computational Methods for Regulatory Genomics 379

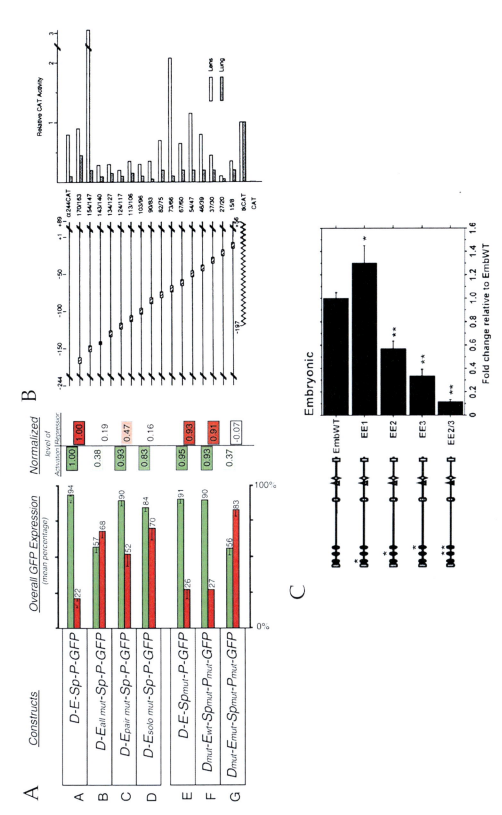

Fig. 22.5. Quintessential diagrams of cis-regulatory architecture. **a** sp-gcm in *Strongylocentrotus purpuratus* (9); **b** alphaA-crystallin in *Gallus gallus* (91); **c** myh3 in *Mus musculus* (92).

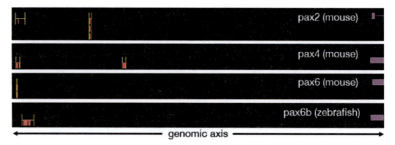

Fig. 22.6. Comparison of four pax genes in the *cis*-Lexicon (axis not to scale). The *purple* square represents the first exon, the *yellow* double-ended bar represents the CRM, and the *orange* blocks represent the TFBS.

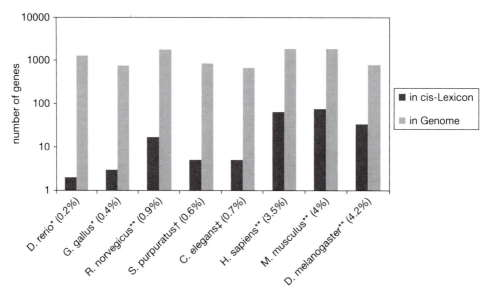

Fig. 22.7. Annotation progress for transcription factor encoding genes. The percent in parentheses is the percentage of transcription factor encoding genes in the corresponding genome that have been annotated in the *cis*-Lexicon. Genome-wide transcription factor encoding gene totals were taken from the following sources: * – DBD (93), ** – Panther (70), † – SpBASE (48), ‡ – (94).

3.2. cis-Browser

3.2.1. Development of the cis-Browser

The CYRENE *cis*-Browser is a genome browser tailored for *cis*-regulatory annotation and investigation. It began as a branch of the Celera Genome Browser, which is available as open source at http://sourceforge.net/projects/celeragb/. The features of the original Celera Genome Browser centered on viewing and annotating gene transcripts, so many new capabilities were added to address our new focus.

First, support for *cis*-regulatory modules (CRMs) and transcription factor binding sites (TFBSs) was added. Each of these new genomic features possesses several unique properties and

Practical Computational Methods for Regulatory Genomics 381

associated information. Unlike gene transcripts, whose borders are determined solely by their exons, the boundaries of CRMs can extend beyond the known binding sites contained inside (e.g., if evidenced by sequence conservation). It is often known whether or not whole CRMs or individual binding sites are conserved across species, and this information can be added and viewed via the *cis*-Browser. Each TFBS has a specific factor (or, occasionally, a family of related factors) that binds there. The NCBI GeneID or name for this factor and its effect on gene expression can be annotated and viewed in the *cis*-Browser. Support for new types was added by creating new Java classes. For example, the class CuratedCRM was created to represent CRMs, and the existing class CuratedTranscript (from the Celera Genome Browser) was used as a reference, since transcripts and CRMs share key traits.

The focus of the *cis*-Browser is on annotations that are supplemental to known genes, rather than on discovering transcripts. Therefore, instead of requiring annotators to input the genes themselves, the capability was added to download genes directly from NCBI. Within the *cis*-Browser application, the user can search for genes (*see* **Fig. 22.8**) just as in the NCBI Entrez Gene web site. When a gene is selected from the results, the genomic sequence of the region is downloaded and all the gene's transcripts and exons are automatically displayed. Data are accessed via the NCBI Entrez Programming Utilities service (http:// eutils.ncbi.nlm.nih.gov/entrez/query/static/eutils_help.html).

In the Celera Genome Browser, properties of genomic entities could be of two types: plain text (e.g., names) or a choice from a list of options (e.g., evidence type: *cis*-mutation, footprinting, etc). Properties could be nested, so that a single (parent) property could contain inside it several additional (child) properties. For *cis*-regulatory annotation, we required accurate recording of complex properties. First, we needed to support properties containing

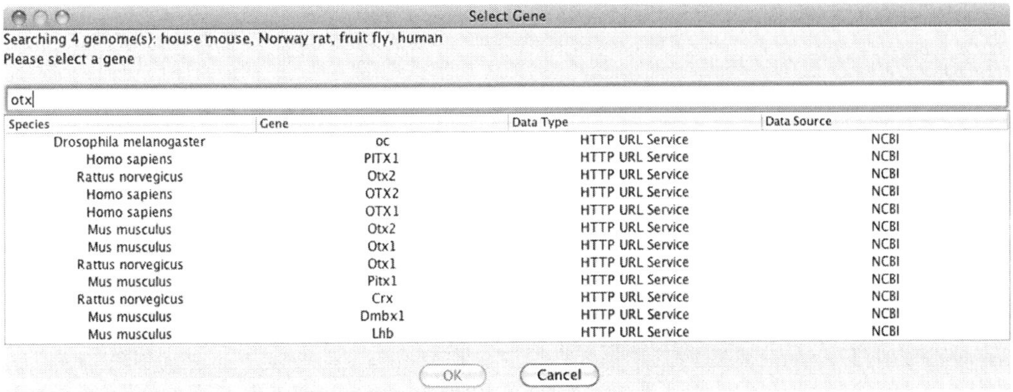

Fig. 22.8. Searching NCBI Entrez Gene within the *cis*-Browser.

multiple interdependent parts: for example, when annotating the factor that binds at a certain site, we must keep track of the factor name, its NCBI GeneID, and any synonyms mentioned in the literature, so that they do not fall out of sync. Second, we needed to support multiple values for a single property: multiple synonyms, multiple *cis*-regulatory functions, and conservation in multiple species.

For properties with multiple parts, we created rich dialog boxes ensuring that the user enters correct information. It would be tedious to ask the user to flip back and forth between the *cis*-Browser and the NCBI Entrez Gene web site to find GeneIDs for each binding factor, and it would be error-prone to make the user type in the factor names and GeneIDs, especially when the same factor binds at several sites for a single target gene. Therefore, the *cis*-Browser provides a special window for annotating the factor that binds to each site (*see* **Fig. 22.9**), so that the user can search the Entrez Gene site from within the browser; the search is automatically restricted to the species being annotated. If the same factor binds at multiple sites, for the second and later sites the user can select the gene from a menu rather than re-entering the information. There are similar windows for annotating conserved species (which searches NCBI for the correct scientific names and NCBI ID) and *cis*-regulatory functions (which ensures that the *cis*-regulatory ontology is followed in naming the regulatory

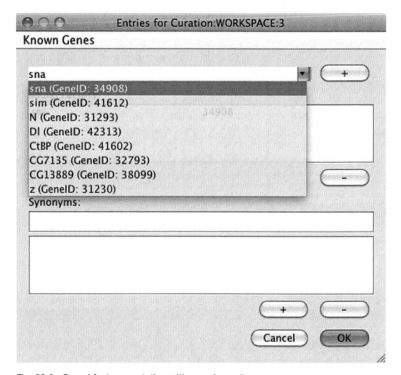

Fig. 22.9. Bound factor annotation with search results.

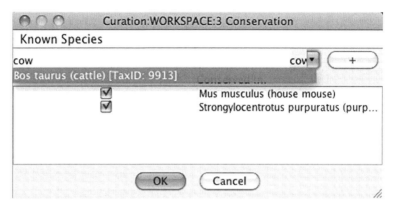

Fig. 22.10. Conserved species annotation with search results.

functions and verifies that PMIDs are typed correctly) – *see* **Figs. 22.10** and **22.11**.

Three different mechanisms were tested to support properties with multiple values; the first two are described in Notes section below. The problem was that in the Celera Genome Browser data model, only one property value can be assigned to a given name. In the GAME XML format supported by the browser, this is represented by the tag <property name="property-name" value= "property-value"/>. It is not valid to give the same property more than one value; e.g., <property name="synonym" value="gcm"/><property name="synonym" value= "spgcm"/>. We ultimately decided to represent a set of values for a single property as one property containing multiple child properties, one for each of the desired values, where each child has a unique name, for example <property name="synonyms" value="gcm, spgcm"><property name="synonym1" value="gcm"/><property name= "synonym2" value="spgcm"/></property> (the child property names do not matter, since only their values are used). The value of the parent property is generally a human-readable summary of the contents, for convenience of display, but again this does not matter, since only the values of the children are used in computations and searches.

The Celera Genome Browser was one part of a three-tiered architecture and communicated with an application server to access a relational database back end. It supported loading genomic features from files, but this was meant to supplement the database (with, for example, output from bioinformatics tools), not replace it. Initially, our *cis*-Lexicon was simply a collection of these XML files. Searching the lexicon required the *cis*-Browser to open, read, and process every one of these files – and this was repeated for every individual search request.

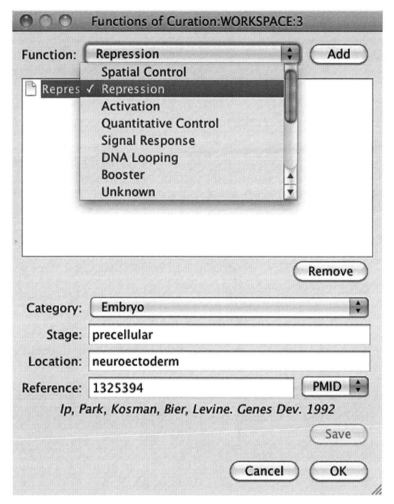

Fig. 22.11. Regulatory function annotation.

At one point we created an ad hoc index as an interim measure to allow certain restricted queries (e.g., list all genes in the lexicon or ask whether a given gene is in the lexicon). Later, we implemented the *cis*-Lexicon as a true database using Apache Derby, an open-source relational database engine. Since Apache Derby is implemented entirely in Java, the *cis*-Browser remains entirely cross-platform. Derby can be run either in embedded mode, where the database is stored and accessed locally, or as a network client, where the database is stored remotely and accessed via a server. This allows the *cis*-Lexicon to be packaged with the *cis*-Browser, for ease of access, or to be stored in one central location, for ease of updating. Relational databases such as Derby support automatic indexing of specific fields in a database record. By indexing the NCBI GeneID field of gene records, for example, searching for genes by GeneID immediately becomes fast. Relational databases also support foreign keys, which ensure that values intended to reference other entities are actually valid. For example, if the bound factor for a binding site is recorded as gene 373400, then the database either ensures that gene 373400 is present in the *cis*-Lexicon or rejects the annotation.

3.2.2. The CYRENE cis-Browser Interface

The *cis*-Browser interface has the same organization as the original Celera Genome Browser. The *cis*-Browser application window is split into four regions (clockwise from top left): the Outline View, the Annotation View, the Subview Container, and the Property Inspector View (*see* **Fig. 22.12**). The Outline View displays in a hierarchical tree format the species, chromosomes, and sequences loaded by the *cis*-Browser and ready for analysis. The Annotation View displays the locations of genomic features (e.g., transcripts, CRMs) on the sequence currently being examined. The Subview Container shows the user a set of views specific to the currently selected feature, and the Property Inspector View shows the properties of the selected feature in textual form.

The Annotation View allows real-time zooming from a chromosome-wide view down to the individual nucleotide level. When the user clicks on a genomic feature, information specific to the feature is visible in the Subview Container and the Property Inspector View. The Annotation View displays genomic features in tiers (horizontal rows grouping features according to their source) so that information from multiple sources is not intermixed and confused; in **Fig. 22.13**, for example, mapped Solexa

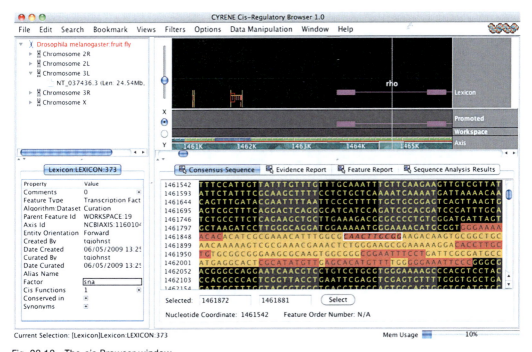

Fig. 22.12. The *cis*-Browser window.

386   Istrail et al.

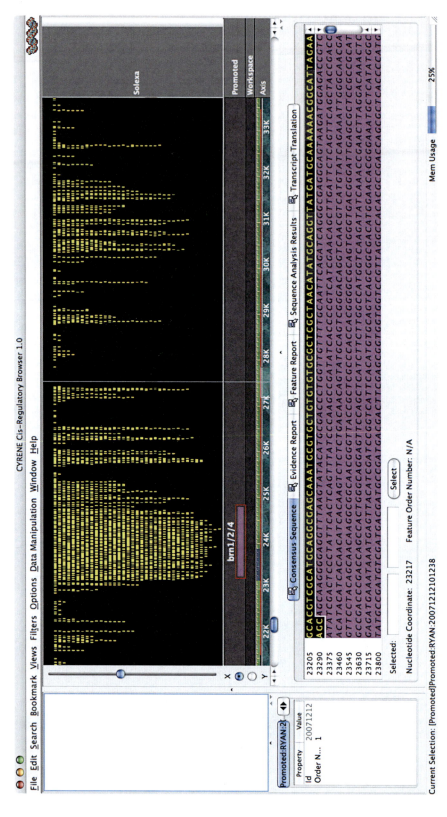

Fig. 22.13. The gene brn1/2/4 appears in the "Promoted" tier while mapped Solexa reads appear in the "Solexa" tier.

reads are grouped separately from genomic transcripts. One special tier in the Annotation View is the workspace, which is the tier that contains genomic features currently being edited. Features loaded from data sources such as XML files or the *cis*-Lexicon are considered immutable, so a copy must be made and placed in the workspace before it can be modified. New features added by the annotator, such as CRMs and binding sites, are always in the workspace.

Each type of genomic feature has particular traits that distinguish it from other types. For example, transcripts are translated into proteins and BLAST hits are the result of comparisons between different sequences. Therefore, viewing the translation of codons into amino acids is relevant only for transcripts, while examining the differences between the current sequence and a sequence that was searched against it is relevant only for BLAST hits. The Subview Container is the location for views such as these. When a feature is selected in the Annotation View, only the views relevant to that type of feature are shown in the Subview Container. Only one such view is shown at a time, to maximize the visible area; the rest are shown as tabs that the user may click on to switch to that view.

Every genomic feature has certain associated properties, such as name, NCBI accession number, or date of curation. The Property Inspector View displays these properties as a two-column table giving the name of each property on the left and the value on the right. Properties can be edited by double-clicking the current value. If the value is a simple string, then it can be edited in place; if it is more complex, such as binding factors, then a dialog box appears. Property changes affecting how the feature appears in the Annotation View are reflected in real time: when the name of a gene or CRM is modified, the new name appears immediately.

One subview (i.e., view appearing in the Subview Container) of critical importance is the Consensus Sequence View, which displays the sequence of the selected feature and the surrounding region and is also used to mark the location of new features. The user simply clicks and drags to select a sequence. Right-clicking shows a menu with options to create a transcript, CRM, or TFBS. The seqFinder feature quickly locates the exact coordinates of a sequence appearing in a published paper. Given a region of sequence to search within (e.g., a gene and its flanking sequence), the seqFinder lets the user type in only the minimum number of nucleotides to uniquely find the paper's sequence. For each letter the user types, the seqFinder tells whether the sequence typed so far is found more than once (i.e., multiple ambiguous matches, so more input is necessary), exactly once (i.e., a perfect match; no more typing needed), or never (i.e., a typo or possibly a true mismatch between the paper's sequence and the reference genome).

The user need only type a few letters from the beginning and then from the end to uniquely identify the entire sequence. When the start and end coordinates are known, the seqFinder automatically selects the sequence within the Consensus Sequence View. Typing the minimum sequence necessary lets the user locate the precise coordinates quickly yet accurately.

3.2.3. Annotation with the cis-Browser

To enter a genomic feature, the annotators first input the coordinates by locating them with the seqFinder. The relevant properties such as names, binding factors, *cis*-regulatory functions, and sequence conservation are set via the Property Inspector View. The Annotation View lets one do quick sanity checks – are the binding sites located upstream, downstream, or within introns of the regulated gene, as is usually the case? Are the CRMs of a reasonable size?

The annotators' work is saved as XML files in the GAME format, rather than directly input into the *cis*-Lexicon. This allows easy backup and sharing of past work and also prevents cluttering the database with half-finished or faulty annotations. A special software tool is required to move the annotations from these intermediate files into the *cis*-Lexicon; forcing the use of intermediate files and preventing unauthorized annotators from modifying the *cis*-Lexicon directly lets us keep the database at a strict high quality. An experienced annotator can verify the work of a trainee before it is entered into the *cis*-Lexicon.

3.3. Virtual Sea Urchin

We have worked closely with the Davidson laboratory at CalTech to produce a Virtual Sea Urchin prototype. The VSU uses spatial models and a graphics engine to simulate the four-dimensional sea urchin embryo, allowing the researcher to probe the GRN at levels of granularity from the multicellular embryo to the gene-regulatory network of an individual cell type. The embryo models were created by extrapolating to three dimensions cross-sectional color-coded tracings from photomicrographs (17).

The Virtual Sea Urchin currently provides models for the *Strongylocentrotus purpuratus* embryo at 6, 10, 15, 20, and 24 h. Cell types are defined by ambient and diffuse coloring as well as shape. Gene expression data are visible at a glance on an embryonic cell type using emission coloring (intensity of coloring is proportional to intensity of expression).

The VSU model of embryonic development will eventually be configurable, featuring realistic cell models and dynamics simulators. *In toto* imaging (56, 57) of the sea urchin embryo will enhance the model's accuracy and resolution, letting researchers probe the regulatory network activity per cell. We will ultimately combine the *cis*-regulatory sequence-analysis capabilities of CYRENE and the network building, visualization, and simulation capabilities of BioTapestry with the temporal and spatial

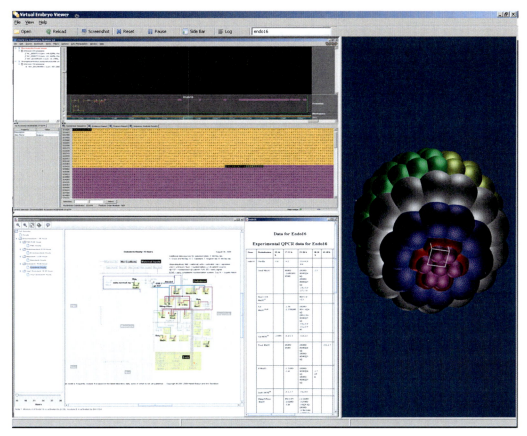

Fig. 22.14. Virtual Sea Urchin, BioTapestry, and Cyrene integration. In order to model the gene regulatory network of the developing sea urchin embryo completely, tools that specialize in analyzing different perspectives of the network must be integrated. BioTapestry operates at the network level, allowing users to manipulate and simulate network interactions. Cyrene operates at the DNA level, providing *cis*-regulatory module and binding site definition and other DNA sequence analysis. The Virtual Sea Urchin maps regulatory network information into space and time. The interoperability among these three views is a central component of future work.

analysis of the 4D Virtual Sea Urchin to yield a complete characterization of the GRN (*see* **Fig. 22.14**).

4. Notes

1. **Gene naming**

   Creating a controlled vocabulary starts with the name of the gene, as discussed previously (18), although there has been a large effort to establish common names across species (95). We would like to know the primary ortholog of some gene we have annotated to observe the similarities of *cis*-regulatory architecture across species. Our efforts toward

Table 22.1
Examples for gene name translation between mouse and
Drosophila

Mouse	Drosophila	References
Beta-catenin	Armadillo	van Noort et al. (2002) (103)
Cux-1	Cut	Sharma et al. (2004) (104)
TLE-4	Groucho	Sharma et al. (2004) (104)
six3, six6	So	López-Ríos et al. (2003) (105)

such a gene name translation table are only in the nascent stages. Examples for gene name translation between mouse and *Drosophila* are shown in **Table 22.1**. A gene occurring in two species is more likely to have a similar *cis*-regulatory architecture if the two genes are related by evolution *and* have a conserved function. Sets of genes meeting these criteria we have termed *Davidson Orthologs*. The method usually employed for determining homology is by sequence rather than conserved function. *orthoMCL* (96) and *inparanoid* (97) are examples of this kind of ortholog table. While the latter definition of ortholog is more easily searched in a database of genes automatically, the definition is not as stringent.

2. **CRM boundaries**

TFBSs are usually well described and require no guesswork by the annotator, but CRMs are often not well defined in the literature. Some of this confusion stems from the lack of a precise definition of the function of a CRM; additionally, many research groups are not interested in finding the boundaries of CRMs and limit their scope to discovery of TFBSs. When annotating the *cis*-regulatory architecture of a gene, the annotator often must make certain assumptions about the boundaries of a CRM that can be classified as follows (examples are referenced): (1) CRMs are not discussed in the literature and the annotator defines the CRM by the minimal sequence that correctly directs gene expression, usually approximated to within 100 bp (96). (2) CRMs are not discussed in the literature, but a graph in the paper shows sequence similarity to the same gene in other species (98). The annotator defines the CRM as the sequence most conserved in other species. If a sequence remains highly similar over a great evolutionary distance, there must be selective pressure to conserve the sequence, and therefore the sequence probably plays an important role in the organism. (3) CRMs are not discussed in the literature and the annotator defines the CRM by most extreme TFBSs determined to act in a specific location (99). That is, if three TFBSs are found to drive gene expression in a cell, the CRM annotation goes from the first nucleotide of the first TFBS to the last nucleotide of the last TFBS. Overall uncertainty and lack of consistency in CRM annotation reduce the quality of data on CRM boundaries.

3. **Caveats in Davidson criteria**

While the *cis*-Lexicon seeks to collect only the most reliable data, many uncertainties remain. Mutational studies show that a certain TFBS is important for correct gene expression, but the factor that binds to the site is not immediately certain. The sequence probably contains a transcription factor binding motif, so the factor that binds can often be guessed. In some annotations, the experimentalist has shown that a particular transcription factor binds to the TFBS in an assay (100) or that knockdown of the transcription factor also causes a change in gene expression. Such confirmations of *trans* acting factor are not always reported in the literature.

4. **Uncertainty in identifying the *trans* acting factor**

Often in the literature the exact transcription factor binding to a TFBS is not known, but the transcription factor family to which the *trans* acting factor belongs is reported; Shen and Ingraham (101) report an E-Box *trans* activator. Sometimes the authors report that the *trans* acting factor could be one of several; for instance, Clark et al. (102) report that either an RAR/RAR or RAR/RXR dimer activates transcription. The TFBS cannot be compared to others in the *cis*-Lexicon since the *trans* acting factor is not known, and the TFBS loses its usefulness in the data set for clustering and prediction. The transcription factor hierarchy described above (**Fig. 22.2**) was added as an annotation tool to combat this problem. *Trans* acting factors can be clustered at different levels of the transcription factor hierarchy.

5. ***cis*-Lexicon search engine**

A great challenge in the annotation process has been finding literature relevant to building the *cis*-Lexicon. So far the literature has been located by PubMed or Google Scholar searches or by browsing references describing previously annotated genes (**Fig. 22.15** shows journals cited in the *cis*-Lexicon). A formalized search process will rapidly uncover relevant literature; in addition, it will help determine the number of genes studied according to the *Davidson criteria* and give an estimate of *cis*-Lexicon completeness. When the *cis*-Lexicon is declared complete, searches will have to be performed continually to find new data. To accomplish these goals, the CLOSE Project (*cis*-Lexicon Ontology Search

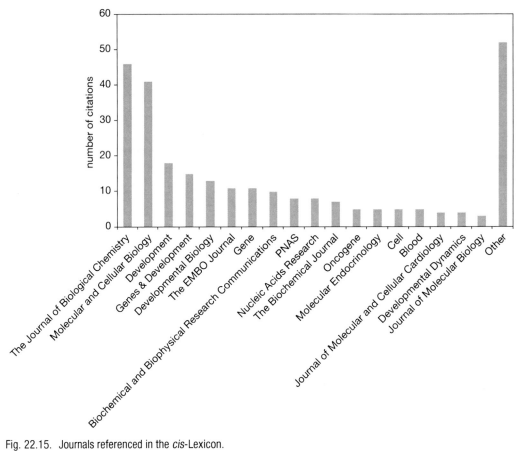

Fig. 22.15. Journals referenced in the *cis*-Lexicon.

Engine) aims to create a set of algorithmic strategies for literature extraction of *cis*-regulation articles.

PubMed cannot perform unrestricted phrase searching of citations and abstracts; only phrases in the PubMed Index are found. If a phrase is not in the Index, then a PubMed search cannot return exact matches, even if it appears in citations or abstracts. Instead, the query is treated as a standard non-phrase search, yielding almost entirely irrelevant results (http://www.nlm.nih.gov/bsd/disted/pubmedtutorial/020_450.html). This prevents the use of PubMed for advanced text-based queries. MeSH terms are not assigned consistently enough to make possible comprehensive searches by keyword alone (*see*, for example, Yuh et al. (13), which is not assigned a term for Base Sequence). Google does not yet offer an API to access Google Scholar, and the results of searching PubMed via standard search engines (Google, Yahoo, etc.) lack known important papers (unpublished

Practical Computational Methods for Regulatory Genomics   393

data). All of these point to the need for a specialized search engine to find the most relevant papers.

6. **Complex annotations in the *cis*-Browser**
Three different mechanisms for making complex annotations with multiple values were tested. The first was to extend the GAME XML format with a new tag for each new type of data. This method was explored first because supporting multiple complex values appeared analogous to storing annotator comments, which was already supported by the Celera Genome Browser. According to the Genome Browser data model, comments are completely distinct from properties. In the XML format, each comment was stored in a `<comment>` tag containing custom author and date attributes. It therefore seemed reasonable to add a `<cisreg_function>` tag with function location and time attributes. Multiple `<cis_reg_function>` tags can be associated with a single genomic entity, allowing multiple functions to be annotated. This required significant work in modifying not only the XML parser itself but all of the loading code governing the interactions between the XML parser (and other possible storage implementations) and the browser. A more general solution was deemed necessary with the prospect of additional complex annotations, since it would be infeasible to make similar changes multiple times.

The second mechanism was the addition of "facts," a new data type differing from properties in that multiple facts can be associated with the same name. Custom attributes could be replaced by child facts. This led to a straightforward conversion of `<cisreg_function>` tags to nested facts: `<cisreg_function func-tion="booster" location="mesoderm" time="24 h"/>` became `<fact name="cisreg_function" value = "booster"><fact name = "location" value="mesoderm"/><fact name="time" value ="24 h"/></fact>`. New facts could be added to store interspecies conservation and bound factors without modifying the parser or loading code. However, the format was still not standard GAME XML, and therefore could not be read by the Celera Genome Browser (which, being open source, is still under development) or other applications using the GAME format.

The desire to keep compatibility with the Genome Browser led us to consider the third option of using only nested properties, described in **Section 3**.

7. **Java XML parsing**
Two significant issues arose in XML parsing with Java, the first concerning external DTD loading. XML data received

from the NCBI Entrez Programming Utilities service always refer to DTDs located on NCBI's servers. The XML language requires parsers always to access the DTD, and the extra time required to download the DTD whenever such an XML file is parsed led to difficult-to-trace slowdowns in the *cis*-Browser application. The correct solution was a custom org.xml.sax.EntityResolver implementation returning cached copies of NCBI files. Once given to an XML-Reader object, it will utilize the cached copies rather than fetching the remote originals.

Second, computers with an old version of Java 6 may run out of memory when parsing large XML files, even when using efficient parsing methods (see http://bugs.sun.com/bugdatabase/view_bug.do?bug_id= 6536111). The easiest solutions are to use Java 5 if available or to upgrade to a more recent version of Java 6.

8. **Virtual Sea Urchin embryo modeling**
Producing the three-dimensional models for the Virtual Sea Urchin's 3D graphics engine is currently laborious; an embryology domain expert works with a 3D computer graphics and modeling software expert to create the embryonic models using a software suite such as Maya or Blender. These models are then exported and integrated into a format compatible with the VSU.

We can streamline this process by animating the anatomical structure using a hierarchy of cells and cell types that completely categorize the embryo at all relevant time slices. This hierarchical tree representation, in which each tree level defines the embryo at a specific time and which is defined by the experimentalist, can be visualized by meiotically splitting cells into appropriate cell types. With this new embryo representation, experimentalists can interact directly with the VSU and easily define embryo development without needing outside expertise.

Acknowledgments

The support of the National Science Foundation under grant DBI 0645955 is acknowledged with gratitude. We would also like to acknowledge the tremendous impact on this work of our collaborator, Eric H. Davidson of the California Institute of Technology, who has guided every step of our efforts. This work would not have been possible without the contributions of three generations of annotators, most notably Tim Johnstone, Jake Halpert, and David Moskowitz. (The first generation was David Moskowitz, Rohan Madamsetti, and Sanjay Trehan; the second generation was Tamar Melman, Mark Grabiner, and Kyle Schutter; the third generation is Tim Johnstone, Jake Halpert, Mei Cao, Kenneth Estrellas, Nicole Noronha, and Daniel Yang.) We would also like to thank Andy Ransick, Andy Cameron and Russell Turner for many discussions and valuable suggestions. Last but not least, many thanks go to Erin Klopfenstein for her outstanding work and many valuable contributions to the CYRENE Project.

References

1. Davidson, E.H. (2001) Genomic regulatory systems: In *Devel and evol,* Academic Press, San Diego, CA.
2. Davidson, E.H., and Erwin, D. (2006) Gene regulatory networks and the evolution of animal body plans. *Science* 311, 796–800.
3. Davidson, E.H. (1968) *Gene activity in early development.* Academic Press, New York, NY.
4. Sea Urchin Genome Consortium. (2006) The genome of the sea urchin *Strongylocentrotus purpuratus. Science* 314, 941–952.
5. Samanta, M.P., Tongprasit, W., Istrail, S. et al. (2006) The transcriptome of the sea urchin embryo. *Science* 314, 960–962.
6. Erwin, D.H., and Davidson, E.H. (2009) The evolution of hierarchical gene regulatory networks. *Nature Rev Gen* 10, 141–148.
7. Davidson, E.H., Rast, J.P., Oliveri, P. et al. (2002) A genomic regulatory network for development. *Science* 295, 1669–1678.
8. Britten, R.J., and Davidson, E.H. (1969) Gene regulation for higher cells: a theory. *Science* 165, 349–357.
9. Ransick, A., and Davidson, E. (2006) cis-regulatory processing of Notch signaling input to the sea urchin glial cells missing gene during mesoderm specification. *Dev Biol* 297, 587–602.
10. Oliveri, P., Tu, Q., and Davidson, E.H. (2008) Global regulatory logic for specification of an embryonic cell lineage. *Proc Natl Acad Sci USA* 105, 5955–5962.
11. Yuh, C.H., and Davidson, E.H. (1996) Modular cis-regulatory organization of Endo16, a gut-specific gene of the sea urchin embryo. *Development,* 122, 1069–1082.
12. Yuh, C.H., Bolouri, H., and Davidson, E.H. (1998) Genomic cis-regulatory logic: experimental and computational analysis of a sea urchin gene. *Science* 279, 1896–1902.
13. Yuh, C.H., Dorman, E.R., Howard, M.L. et al. (2004) An otx cis-regulatory mod-

ule: a key node in the sea urchin endomesoderm gene regulatory network. *Dev Biol* 269, 536–551.
14. Istrail, S., De-Leon, S.-T., and Davidson, E. (2007) The regulatory genome and the computer. *Dev Biol* 310, 187–195.
15. Istrail, S., and Davidson, E. (2005) Logic functions of the genomic cis-regulatory code 2005. *Proc Natl Acad Sci USA* 102, 4954–4959.
16. Levine, M., and Davidson, E.H. (2005) Gene regulatory networks for development. *Proc Natl Acad Sci USA* 102, 4936–4942.
17. Davidson, E.H. (2006) *The regulatory genome: gene regulatory networks in development and.* Academic Press, San Diego, CA.
18. Tarpine, R., and Istrail, S. (2009) On the concept of Cis-regulatory information: from sequence motifs to logic functions. *Algorithmic Bioprocesses* In (Condon, A., Harel, D., Kok, J.N., Salomaa, A., and Winfree, E. Eds.) pp. 731–742 Springer-Verlag, Berlin Heidelberg.
19. Longabaugh, W.J.R., Davidson, E.H., and Bolouri, H. (2005) Computational representation of developmental genetic regulatory networks. *Dev Biol* 283, 1–16.
20. Longabaugh, W.J.R., Davidson, E.H., and Bolouri, H. (2009) Visualization, documentation, analysis, and communication of large-scale gene regulatory networks. *Biochem Biophys Acta* 1789, 363–374.
21. Stormo, G.D. (2000) DNA binding sites: representation and discovery. *Bioinformatics* 16, 16–23.
22. Wasserman, W.W., and Sandelin, A. (2004) Applied bioinformatics for the identification of regulatory elements. *Nature Rev Gen* 5, 276–287.
23. Sandelin, A. (2004) In silico prediction of *cis*-regulatory elements. Karolinska Institutet. Stockholm, Sweden, 4–130.

24. Tompa, M., Li, N., Bailey, T.L. et al. (2005) Assessing computational tools for the discovery of transcription factor binding sites. *Nature Biotechnol* 23, 137–144.

25. Hannenhalli, S., and Levy, S. (2001) Promoter prediction in the human genome. *Bioinformatics* 17, S90–S96.

26. Hannenhalli, S., and Levy, S. (2002) Predicting transcription factor synergism. *Nucleic Acids Res* 30, 1–8.

27. Hannenhalli, S., Putt, M.E., Gilmore, J.M. et al. (2006) Transcriptional genomics associates FOX transcription factors with human heart failure. *Circulation J Am Heart Assoc* 114, 1269–1276.

28. Singh, L.N., Wang, L.S., and Hannenhalli, S. (2007) TREMOR—a tool for retrieving transcriptional modules by incorporating motif covariance. *Nucleic Acids Res* 35, 7360–7371.

29. Markstein, M., Markstein, P., Markstein, V. et al. (2002) Genome-wide analysis of clustered dorsal binding sites identifies putative target genes in the *Drosophila* embryo. *Dev Biol* 99, 763–768.

30. Linhart, C., Halperin, Y., and Shamir, R. (2008) Transcription factor and microRNA motif discovery: the Amadeus platform and a compendium of metazoan target sets. *Genome Res* 18, 1180–1189.

31. Tompa, M. (1999) An exact method for finding short motifs in sequences, with application to the ribosome binding site problem. *7th International Conference Intelligent Systems for Molecular Biology*, 262–271.

32. Sinha, S., and Tompa, M. (2003) Performance comparison of algorithms for finding transcription factor binding sites. *Proceedings of the 3rd IEEE Symposium on Bioinformatics and Bioengineering*, 213.

33. Blanchette, M., Schwikowski, B., and Tompa, M. (2002) Algorithms for phylogenetic footprinting. *J Comput Biol* 9, 211–223.

34. Wasserman, W.W., and Fickett, J.W. (1998) Identification of regulatory regions which confer muscle-specific gene expression. *J Mol Biol* 278, 167–181.

35. Benos, P.V., Bulyk, M.L., and Stormo, G.D. (2002) Additivity in protein-DNA interactions: how good an approximation is it? *Nucleic Acids Res* 30, 4442–4451.

36. Keich, U., and Pevzner, P.A. (2002) Subtle motifs: defining the limits of motif finding algorithms. *Bioinformatics* 18, 1382–1390.

37. Ng, P., Nagarajan, N., Jones, N. et al. (2006) Apples to apples: improving the performance motif finders and their significance analysis in the Twilight Zone. *Bioinformatics* 22, e393–e401.

38. Badis, G., Berger, M., Philippakis, A. et al. (2009) Diversity and complexity in DNA recognition by transcription factors. *Science* 324, 1720–1723.

39. Berger, M., Badis, G., Gehrke, A. et al. (2008) Variation in homeodomain DNA binding revealed by high-resolution analysis of sequence preferences. *Cell* 133, 1266–1276.

40. Noyes, M.B., Christensen, R.G., Wakabayashi, A. et al. (2008) Analysis of homeodomain specificities allows the family-wide prediction of preferred recognition sites. *Cell* 133, 1277–1289.

41. Cameron, R.A., Rast, J.P., and Brown, C.T. (2004) Genomic resources for the study of sea urchin development. *Methods Cell Biol* 74, 733–757.

42. He, X., Ling, X., and Sinha, S. (2009) Alignment and prediction of cis-regulatory modules based on a probabilistic model of evolution. *PLoS Comput Biol* 5, e100299.

43. Li, N., and Tompa, M. (2006) Analysis of computational approaches for motif discovery. *Algorithms Mol Biol* 1, 1–8.

44. Li, X., Zhong, S., and Wong, W.H. (2005) Reliable prediction of transcription factor binding sites by phylogenetic verification. *Proc Natl Acad Sci USA* 102, 16945–16950.

45. Papatsenko, D., and Levine, M. (2005) Quantitative analysis of binding motifs mediating diverse spatial readouts of the dorsal gradient in the *Drosophila* embryo. *Proc Natl Acad Sci USA* 102, 4966–4971.

46. Pilpel, Y., Sudarsana, P., and Church, G.M. (2001) Identifying regulatory networks by combinatorial analysis of promoter elements. *Nature Genet* 29, 153–159.

47. Zhu, Z., Pilpel, Y., and Churge, G.M. (2002) Computational identification of transcription factor binding sites *via* a transcription-factor-centric clustering (TFCC) algorithm. *J Mol Biol* 318, 71–81.

48. Howard-Ashby, M., Materna, S.C., Brown, C.T. et al. (2006) Identification and characterization of homeobox transcription factor genes in *Strongylocentrotus purpuratus*, and their expression in embryonic development. *Dev Biol* 300, 74–89.

49. Oliveri, P., Carrick, D.M., and Davidson, E.H. (2002) A regulatory gene network that directs micromere specification in the sea urchin embryo. *Dev Biol* 246, 209–228.

50. Calestani, C., Rast, J.P., and Davidson, E.H. (2003) Isolation of pigment cell specific genes in the sea urchin embryo by differen-

tial macroarray screening. *Development* 130, 4587–4596.

51. Imai, K.S., Levine, M., Satoh, N. et al. (2006) Regulatory blueprint for a chordate embryo. *Science* 312, 1183–1187.

52. Ransick, A., Rast, J.P., Minokawa, T. et al. (2002) New early zygotic regulators expressed in endomesoderm of sea urchin embryos discovered by differential array hybridization. *Dev Biol* 246, 132–147.

53. Stathopoulos, A., Van Drenth, M., Erives, A. et al. (2002) Whole-genome analysis of dorsal-ventral patterning in the *Drosophila* embryo. *Cell* 111, 687–701.

54. Revilla-i-Domingo, R., Minokawa, T., and Davidson, E.H. (2004) R11: a cis-regulatory node of the sea urchin embryo gene network that controls early expression of SpDelta in micromeres. *Dev Biol* 274, 438–451.

55. Lickert, H., and Kemler, R. (2002) Functional analysis of cis-regulatory elements controlling initiation and maintenance of early Cdx1 gene expression in the mouse. *Dev Dyn* 225, 216–220.

56. Megason, S., and Fraser, S. (2003) Digitizing life at the level of the cell: high-performance laser-scanning microscopy and image analysis for in toto imaging of development. *Mech Dev* 120, 1407–1420.

57. Megason, S., and Fraser, S. (2007) Imaging in systems biology. *Cell* 130, 784–795.

58. Turner, R., Chaturvedi, K., Edwards, N. et al. (2001) Visualization challenges for a new cyberpharmaceutical computing paradigm, *Proceedings of the Symposium on Large-Data Visualization and Graphics*, San Diego, CA.

59. Gray, S., Szymanski, P., and Levine, M. (1994) Short-range repression permits multiple enhancers to function autonomously within a complex promoter. *Genes Dev* 8(15), 1829–1838.

60. Courey, A., and Jia, S. (2001) Transcriptional repression: the long and the short of it. *Genes Dev* 15, 2786–2796.

61. Nakao, T., and Ishizawa, A. (1994) Development of the spinal nerves in the mouse with special reference to innervation of the axial musculature. *Anat Embryol* 189, 115–138.

62. Latchman, D. (2008) *Eukaryotic transcription factors*. Fifth Edition, Academic Press, London.

63. Barolo, S., and Posakony, J. (2002) Three habits of highly effective signaling pathways: principles of transcriptional control by developmental cell signaling. *Genes Dev* 16, 1167–1181.

64. Zeller, R., Griffith, J., Moore, J. et al. (1995) A multimerizing transcription fac-

tor of sea urchin embryos capable of looping DNA. *Proc Natl Acad Sci USA* 92, 2989–2993.

65. Smith, J., and Davidson, E. (2008) A new method, using cis-regulatory control, for blocking embryonic gene expression. *Dev Biol* 318, 360–365.

66. West, A., Gaszner, M., and Felsenfeld, G. (2002) Insulators: many functions, many mechanisms. *Genes Dev* 16, 271–288.

67. Inagaki, N., Maekawa, T., Sudo, T. et al. (1992) c-Jun represses the human insulin promoter activity that depends on multiple cAMP response elements. *Proc Natl Acad Sci* 89, 1045–1049.

68. Matys, V., Kel-Margoulis, O., Fricke, E. et al. (2006) TRANSFAC and its module TRANSCompel: transcriptional gene regulation in eukaryotes. *Nucleic Acids Res* 34, D108–D110.

69. Ashburner, M., Ball, C., Blake, J. et al. (2000) Gene ontology: tool for the unification of biology. The Gene Ontology Consortium. *Nat Genet* 25, 25–29.

70. Mi, H., Guo, N., Kejariwal, A. et al. (2007) PANTHER version 6: protein sequence and function evolution data with expanded representation of biological pathways. *Nucleic Acids Res* 35, D247–D252.

71. Gottschalk, L., Giannola, D., and Emerson, S. (1993) Molecular regulation of the human IL-3 gene: inducible T cell-restricted expression requires intact AP-1 and Elf-1 nuclear protein binding sites. *J Exp Med* 178, 1681–1692.

72. Regl, G., Kasper, M., Schnidar, H. et al. (2004) Activation of the BCL2 promoter in response to Hedgehog/GLI signal transduction is predominantly mediated by GLI2. *Cancer Res* 64, 7724–7731.

73. Wheeler, M., Snyder, E., Patterson, M. et al. (1999) An E-box within the MHC IIB gene is bound by MyoD and is required for gene expression in fast muscle. *Am J Physiol* 276, C1069–C1078.

74. Sekido, R., Murai, K., Funahashi, J. et al. (1994) The delta-crystallin enhancer-binding protein delta EF1 is a repressor of E2-box-mediated gene activation. *Mol Cell Bio* 14, 5692–5700.

75. Castro, B., Barolo, S., Bailey, A. et al. (2005) Lateral inhibition in proneural clusters: cis-regulatory logic and default repression by suppressor of hairless. *Development* 132, 3333–3344.

76. Penkov, D., Tanaka, S., Di Rocco, G. et al. (2000) Cooperative interactions between PBX, PREP, and HOX proteins modulate the activity of the alpha 2(V) colla-

77. Zelko, I., Mueller, M., and Folz, R. (2008) Transcription factors sp1 and sp3 regulate expression of human extracellular superoxide dismutase in lung fibroblasts. *Am J Respir Cell Mol Biol* 39, 243–251.

78. Murphy, A., Lee, T., Andrews, C. et al. (1995) The breathless FGF receptor homolog, a downstream target of Drosophila C/EBP in the developmental control of cell migration. *Development* 121, 2255–2263.

79. Ohshiro, T., and Saigo, K. (1997) Transcriptional regulation of breathless FGF receptor gene by binding of TRACHEALESS/dARNT heterodimers to three central midline elements in *Drosophila* developing trachea. *Development* 124, 3975–3986.

80. Christensen, M., Zhou, W., Qing, H. et al. (2004). Transcriptional regulation of BACE1, the beta-amyloid precursor protein beta-secretase, by Sp1. *Mol Cell Biol* 24, 865–874.

81. Pan, D., Huang, J., and Courey, A. (1991) Functional analysis of the *Drosophila* twist promoter reveals a dorsal-binding ventral activator region. *Genes Dev* 5, 1892–1901.

82. Thisse, C., Perrin-Schmitt, F., Stoetzel, C. et al. (1991) Sequence-specific transactivation of the *Drosophila* twist gene by the dorsal gene product. *Cell* 65, 1191–1201.

83. Jiang, J., Kosman, D., Ip, Y. et al. (1991) The dorsal morphogen gradient regulates the mesoderm determinant twist in early *Drosophila* embryos. *Genes Dev* 5, 1881–1891.

84. Akimaru, H., Hou, D., and Ishii, S. (1997) *Drosophila* CBP is required for dorsal-dependent twist gene expression. *Nature Genet* 17, 211–214.

85. Doerksen, L., Bhattacharya, A., Kannan, P. et al. (1996) Functional interaction between a RARE and an AP-2 binding site in the regulation of the human HOX A4 gene promoter. *Nucleic Acids Res* 24, 2849–2856.

86. Sasaki, H., Hui, C., Nakafuku, M. et al. (1997) A binding site for Gli proteins is essential for HNF-3beta floor plate enhancer activity in transgenics and can respond to Shh in vitro. *Development* 124, 1313–1322.

87. Yoon, J., Kita, Y., Frank, D. et al. (2002) Gene expression profiling leads to identification of GLI1-binding elements in target genes and a role for multiple downstream pathways in GLI1-induced cell transformation. *J Biol Chem* 277, 5548–5555.

88. Sokol, N., and Ambros, V. (2005) Mesodermally expressed *Drosophila* microRNA-1 is regulated by Twist and is required in muscles during larval growth. *Genes Dev* 19, 2343–2354.

89. Ho, I., Hodge, M., Rooney, J. et al. (1996) The proto-oncogene c-maf is responsible for tissue-specific expression of interleukin-4. *Cell* 85, 973–983.

90. Kajihara, M., Sone, H., Amemiya, M. et al. (2003) Mouse MafA, homologue of zebrafish somite Maf 1, contributes to the specific transcriptional activity through the insulin promoter. *Biochem Biophys Res Commun* 312, 831–842.

91. Matsuo, I., and Yasuda, K. (1992) The cooperative interaction between two motifs of an enhancer element of the chicken alpha A-crystallin gene, alpha CE1 and alpha CE2, confers lens-specific expression. *Nucleic Acids Res* 20, 3701–3712.

92. Belkin, D., Allen, D., and Leinwand, L. (2006) MyoD, Myf5, and the calcineurin pathway activate the developmental myosin heavy chain genes. *Dev Biol* 294, 541–553.

93. Wilson, D., Charoensawan, V., Kummerfeld, S., et al. (2008) DBD—taxonomically broad transcription factor predictions: new content and functionality. *Nucleic Acids Res* 36, D88–D92.

94. Haerty, W., Artieri, C., Khezri, N. et al. (2008) Comparative analysis of function and interaction of transcription factors in nematodes: extensive conservation of orthology coupled to rapid sequence evolution. *BMC Genomics* 9, 399.

95. Bult, C., Eppig, J., Kadin, J. et al. (2008) The mouse genome database (MGD): mouse biology and model systems. *Nucleic Acids Res* 36, D724–D728.

96. Chen, F., Mackey, A., Stoeckert, C. et al. (2006) OrthoMCL-DB: querying a comprehensive multi-species collection of ortholog groups. *Nucleic Acids Res* 34, D363–D368.

97. Berglund, A.-C., Sjölund, E., Ostlund, G. et al. (2008) InParanoid 6: eukaryotic ortholog clusters with inparalogs. *Nucleic Acids Res* 36(database issue), D263–D266.

98. Delporte, F., Pasque, V., Devos, N. et al. (2008) Expression of zebrafish pax6b in pancreas is regulated by two enhancers containing highly conserved cis-elements bound by PDX1, PBX and PREP factors. *BMC Dev Biol* 8, 53.

99. Warren, D., Simpkins, C., Cooper, M. et al. (2005) Modulating alloimmune responses with plasmapheresis and IVIG. *Curr Drug Targets Cardiovasc Haematol Disord* 5, 215–222.

100. Annicotte, J.-S., Fayard, E., Swift, G. et al. (2003) Pancreatic-duodenal homeobox 1 regulates expression of liver recep-

tor homolog 1 during pancreas development. *Mol Cell Biol* 23, 6713–6724.

101. Shen, J.-C., and Ingraham, H. (2002) Regulation of the orphan nuclear receptor steroidogenic factor 1 by Sox proteins. *Mol Endocrinol* (Baltimore, MD) 16, 529–540.

102. Clark, A., Wilson, M., London, N. et al. (1995) Identification and characterization of a functional retinoic acid/thyroid hormone-response element upstream of the human insulin gene enhancer. *Biochem J* 309, 863–870.

103. van Noort, M., van de Wetering, M., and Clevers, H. (2002) Identification of two novel regulated serines in the N terminus of beta-catenin. *Exp Cell Res* 276, 264–72.

104. Sharma, M., Fopma, A., Brantley, et al. (2004) Coexpression of Cux-1 and notch signaling pathway components during kidney development. *Dev Dyn* 231(4), 828–838.

105. López-Ríos, J., Tessmar, K., Loosli F. et al. (2003) Six3 and Six6 is moduated by members of the groucho family. *Development* 130, 185–195.

Chapter 23

Reconstructing Transcriptional Regulatory Networks Using Three-Way Mutual Information and Bayesian Networks

Weijun Luo and Peter J. Woolf

Abstract

Probabilistic methods such as mutual information and Bayesian networks have become a major category of tools for the reconstruction of regulatory relationships from quantitative biological data. In this chapter, we describe the theoretic framework and the implementation for learning gene regulatory networks using high-order mutual information via the MI3 method (Luo et al. (2008) *BMC Bioinformatics* 9, 467; Luo (2008) *Gene regulatory network reconstruction and pathway inference from high throughput gene expression data*. PhD thesis). We also cover the closely related Bayesian network method in detail.

Key words: Three-way mutual information, high-order mutual information, information theory, Bayesian network, systems biology, probabilistic graphical model, gene regulatory network (GRN), transcriptional regulation, microarray, gene expression data.

1. Introduction

Gene regulatory network (GRN) reconstruction is to infer GRN models from data. These models reveal the operational mechanism of biological systems at the molecular level. With the development of high-throughput technologies such as microarrays (1, 2), we can profile the expression of the whole genome at a time. This brings the potential to study biology at the whole transcriptome level. Correspondingly, we can reconstruct GRNs involving large number of genes in one study. Learning GRNs from high-throughput expression data (3) has become a major challenge in systems biology and a major focus of statistical learning.

Previous efforts to learn GRNs from gene expression data can be broadly divided into linear correlation and probability based

I. Ladunga (ed.), *Computational Biology of Transcription Factor Binding*, Methods in Molecular Biology 674,
DOI 10.1007/978-1-60761-854-6_23, © Springer Science+Business Media, LLC 2010

methods. Methods based on linear correlation, such as clustering (4, 5), correlation networks (6, 7) and graphical Gaussian models (8), are computationally fast and relatively easy to interpret. However, a key limitation with these methods is that they assume linear relationships between variables. While some components of any transcriptional regulatory network are linear, nonlinear events such as OR-, AND-, and XOR-type transcriptional regulations are relatively commonplace (9). These nonlinear interactions would not be captured with a linear model, leading to spurious relationships between variables.

Probability based methods form the second class of methods commonly used for reconstructing GRNs from biological data. Representative probability methods include Bayesian networks (10–13) and mutual information networks (14, 15). Probability based methods can capture both linear and nonlinear regulatory relationships and are relatively noise tolerant. Probabilistic graphical models use directed edges to represent causal relationship rather than correlative relationships. However, probabilistic methods require significantly more data than correlation based methods. They can be computationally slow.

A Bayesian network is a directed graphical probabilistic model that represents the joint probability distribution among variables in a decomposed form (16). A Bayesian network has two components: a directed acyclic graph (DAG), G, which encodes conditional independent relationships among nodes (variables); and parameter, θ, which specifies the conditional distribution for each variable given its parents. Bayesian networks have been widely used in modeling gene regulatory systems (10–13). These tools have several major advantages: (1) the ability to handle imperfect (incomplete and noisy) data sets; (2) the ability to identify causal relationships; (3) the ability to combine domain knowledge and data. A less obvious problem with Bayesian networks is that the joint probability score decomposes into local conditional probability terms. Conditional probability is still a generalized correlative metric for the two-way dependency between the target and the parent set, hence cannot effectively tell the real causal relationships from confounding ones based on observational data. This issue becomes severe when there are a large number of correlative variables, such as co-regulated genes in microarray data.

Mutual information is a probabilistic quantity defined in information theory to measure the similarity or dependency between two variables. Mutual information has been widely used to model gene networks (14, 15, 17). Mutual information captures both linear and nonlinear relationships between two genes, hence can replace correlation to learn more robust relevance or association networks. Mutual information may also work in place of conditional probability to capture dependency between target gene and parent gene set and learn directed causal networks (18).

Indeed, we will show below that mutual information is equivalent to conditional probability in local regulatory model learning (Eq. **14**). In both cases, mutual information measures the two-way dependency between two variables (genes or gene sets). Real biological systems frequently involve more complicated, higher order relationships, such as transcriptional regulation coordinated by multiple transcription factors, binding interactions in protein complexes, etc. The definition of mutual information has been extended to higher dimensional spaces to measure such high-order relationships among multiple variables (19–21).

To overcome the limitations of Bayesian networks and classic mutual information, we developed a novel statistical learning strategy, MI3 (22), which uses high-order mutual information scoring metric to capture high-order interactions. True causal relationships like genetic regulation feature positive higher order interactions (19, 20), the non-additive effect above the sum of the

Table 23.1

The non-additive property of high order interactions, i.e., $I(T;R1,R2) - I(T;R1) - I(T;R2) = I(T;R1,R2) > 0$ shown by common types of regulatory relationships involving two independent parents ($R1$ and $R2$) and a target (T). Entropies (H's) and mutual information (I's) are calculated according to definitions in Section 2.1. These are ideal cases. In reality, we do not always get positive high-order interactions due to the data quality and absence of real regulators in the data. Hence, we do not impose any threshold on high-order interaction alone. Table is copied, with permission, from (22)

Relationship	OR				AND				XOR			
Contigency table	p	$R1$	$R2$	T	p	$R1$	$R2$	T	p	$R1$	$R2$	T
	1/4	0	0	0	1/4	0	0	0	1/4	0	0	0
	1/4	1	0	1	1/4	1	0	0	1/4	1	0	1
	1/4	0	1	1	1/4	0	1	0	1/4	0	1	1
	1/4	1	1	1	1/4	1	1	1	1/4	1	1	0
$H(T)$	$2-0.75\times\log_2 3$				$2-0.75\times\log_2 3$				1			
$H(R1)=H(R2)$	1				1				1			
$H(T,R1)=H(T,R2)$	1.5				1.5				2			
$H(R1,R2)$	2				2				2			
$H(T,R1,R2)$	2				2				2			
$I(T;R1)=I(T;R2)= H(T)+ H(R1)- H(T,R1)$	$1.5-0.75\times\log_2 3$				$1.5-0.75\times\log_2 3$				0			
$I(T;R1,R2)= H(T)+ H(R1,R2) -H(T,R1,R2)$	$2-0.75\times\log_2 3$				$2-0.75\times\log_2 3$				1			
$I(T;R1,R2)- I(T;R1)-I(T;R2)$	$0.75\times\log_2 3-1$ $=0.189$				$0.75\times\log_2 3-1$ $=0.189$				1			

lower order interactions (20). For example, consider cases where two regulators effect a target via OR-, AND-, XOR-type relationships. The two regulators together account for much more in the target than they individually can (**Table 23.1**). Intuitively, such a non-additive effect can be described as coordination or synergy between parents (with respect to the target, more description in **Section 2.3**). On the other hand, confounding models commonly have zero or negative higher order interactions, such as redundant parents. We propose that when such high-order interaction is considered, we can better differentiate true causal models from confounding models.

We tested the MI3 algorithm using both synthetic and experimental data (22). In synthetic data experiment (**Figs. 23.1 and 23.2a**, detailed in (22)), MI3 achieved absolute sensitivity/precision of 0.77/0.83 and relative sensitivity/precision both of 0.99, and consistently and significantly outperformed the control methods including Bayesian networks, two-way mutual information, and a discrete version of MI3. We then used MI3 and control methods to infer a regulatory network centered at the MYC transcription factor from a published microarray data set (**Fig. 23.2b**, detailed in (22)). Unlike control methods, MI3 effectively differentiated true causal models from confounding models (22). MI3 recovered major MYC cofactors, and revealed major mechanisms involved in MYC-dependent transcriptional regulation, which are strongly supported by literature (22). The MI3 network (**Fig. 23.2b**) showed that limited sets of regulatory mechanisms are employed repeatedly to control the expression of large number of genes (22).

In this chapter, we describe the theoretic framework for learning GRN models using high-order mutual information via the MI3 method. We also discuss Bayesian network and classical mutual information-based methods, which are closely related to MI3. One other significant feature of MI3 is learning continuous probabilistic models for transcriptional regulation based on kernel density estimation (22). The same approach can be directly transplanted to learn more accurate regulatory models using Bayesian network and classical mutual information (22).

2. Theory

2.1. Mutual Information Definition, Extension, and Calculation

In information theory, for a discrete variable, X, Shannon entropy $H(X)$ is defined as (24)

$$H(X) = -\sum_{i=1}^{m_x} P(x_i) \log_2 P(x_i) \qquad [1]$$

where $X = x_i \, (i = 1, 2, m_x)$, corresponds to m_x different states of the variable X. The Shannon entropy is a measure for the randomness or unpredictability of variable distribution. Thus, the higher the Shannon entropy, the harder it is to predict the value of this variable. The corresponding definition for continuous variables is the same (17), except that the summation becomes integration.

The entropy of joint distribution of two discrete variables X and Y is similarly defined as (24)

$$H(X, Y) = -\sum_{i=1}^{m_x} \sum_{j=1}^{m_y} P(x_i, y_j) \log_2 P(x_i, y_j) \qquad [2]$$

where $Y = y_j \, (j = 1, 2, m_y)$ corresponds to m_y different states of the variable Y.

The mutual information between two variables X and Y, $I(X; Y)$ is defined based on Shannon entropy (24, 25):

$$I(X; Y) = H(X) + H(Y) - H(X, Y) \qquad [3]$$

Mutual information measures the difference in predictability when considering two variables together versus considering them independently. In other words, mutual information measures the interdependency between variables. High dependency or mutual information usually occurs when either there is causal relationship between variables or a common causal factor is influencing both variables. Therefore, mutual information can be used to identify correlative or even causal relationships.

For three variables X, Y, and Z, we can define three types of three-way mutual information measuring different types of dependency: total correlation $C(X; Y; Z)$ (different from linear correlation) (26), generalized two-way mutual information $I(X; Y, Z)$, and three-way interaction information $I(X; Y; Z)$ (19, 20):

$$C(X; Y; Z) = H(X) + H(Y) + H(Z) - H(X, Y, Z), \qquad [4]$$

$$I(X; Y, Z) = H(X) + H(Y, Z) - H(X, Y, Z) \qquad [5]$$

$$I(X; Y; Z) = H(X, Y) + H(Y, Z) + H(X, Z) - H(X) - H(Y)$$
$$- H(Z) - H(X, Y, Z)$$
$$[6]$$

These are all extended mutual information of order 3, different in lower order terms:

$$I(X; Y, Z) = C(X; Y; Z) - I(Y; Z) \qquad [7]$$

$$I(X;Y;Z) = I(X;Y,Z) - I(X;Y) - I(X;Z) \qquad [8]$$

Table 23.1 shows common examples where the relationships are high order and can only be fully captured by high-order (three-way) mutual information.

Conditional entropy and mutual information can also be defined based on conditional probability. A rearranged version of conditional mutual information can be derived by starting with the definition of conditional probability given Z:

$$I(X;Y|Z) = \frac{1}{n} \sum_{k=1}^{n} \log_2 \frac{P(x_k, y_k | z_k)}{P(x_k | z_k) P(y_k | z_k)} \qquad [9]$$

Next, apply Bayes' rule and rearrange to yield

$$I(X;Y|Z) = \frac{1}{n} \sum_{k=1}^{n} \log_2 \left[\frac{P(x_k, y_k, z_k)}{P(x_k) P(y_k, z_k)} \frac{P(y_k, z_k)}{P(y_k) P(z_k)} \right] \qquad [10]$$

Re-write into mutual information

$$I(X;Y|Z) = I(X;Y,Z) - I(X;Z) \qquad [11]$$

Apparently, this conditional mutual information is of order 3 and is closely related to all other types of three-way mutual information. So far, we have been focusing on three-way mutual information and entropy. Similarly, the concept of entropy and mutual information can be directly extended to arbitrary higher order to capture even complicated relationships among multiple variables or multiple sets of variables.

2.2. Comparison Between Mutual Information and Log-Based Local Conditional Probability

By substituting the entropy definition formulae [1] and [2] into equation [3], we get the expanded formula for mutual information based on probability:

$$I(X;Y) = \sum_{i=1}^{m_x} \sum_{j=1}^{m_y} P(x_i, y_j) \log_2 \frac{P(x_i, y_j)}{P(x_i) P(y_j)} = \frac{1}{n} \sum_{k=1}^{n} \log_2 \frac{P(x_k, y_k)}{P(x_k) P(y_k)} \qquad [12]$$

where $X = x_k$ (j $= 1, 2, n$), $Y = y_k$ (j $= 1, 2, n$), correspond to n data points of variable X or Y.

The counterpart to mutual information in Bayesian network (BN) terms is a log-based local conditional probability, or log likelihood (LL), which can be expanded as

$$LL(X|Y) = \log \prod_{k=1}^{n} P(x_k | y_k) = \sum_{k=1}^{n} \log \frac{P(x_k, y_k)}{P(y_k)} \qquad [13]$$

It can be seen that mutual information is similar to log likelihood, except with a weighted-averaging term $1/n$ and normalizing term $P(x_k)$, which minimize the effects of sample size and specific distribution of individual variables.

Given a particular data set and for fixed target node X, these two scores are equivalent:

$$I(X; Y) \propto \frac{1}{n} LL(X|Y) - \frac{1}{n} \sum_{k=1}^{n} \log P(x_k) \qquad [14]$$

The same equivalence holds for models with two parents or multiple parents. In other words, $I(\ ;\ )$ and $LL(\ |\ )$ scores are interchangeable when comparing the regulatory models for the same node.

2.3. MI3 Algorithm

2.3.1. Scoring Metric

The MI3 algorithm (22) uses three-way mutual information for local causal model inference. Our hypothesis is that gene expression regulation commonly involves more than two genes (i.e., more than one regulator genes) with higher order interaction, which can be faithfully captured by continuous higher order mutual information. The algorithm is limited to three-way mutual information (two regulators and one target), but the same method can be easily extended to higher order mutual information to model more complicated regulation mechanisms. Note that we call all types of mutual information involving three variables three-way mutual information (**Section 2.1**), while three-way interaction information refers to $I(T; R1; R2)$ only.

The MI3 scoring function has two parts, including correlative and coordinative information components.

Correlative component: $I(T; R1, R2)$
Coordinative component: $I(T; R1, R2) - I(T; R1) - I(T; R2)$
MI3 score: $2^*I(T; R1, R2) - I(T; R1) - I(T; R2) =$
$$I(T; R1\,|R2) + I(T; R2\,|R1)$$

Here T is the target gene, and $R1$ and $R2$ are the regulators. Mutual information definition and high-order extensions are described in the previous section.

The correlative component measures the correlation (both linear and nonlinear) between the target and the parent set. Pairs of regulators accurately describing the expression of the target gene will score well by the correlative component.

The coordinative component measures the coordination effect between the regulators with respect to the target. In other words, it describes how well pairs of regulators versus individual regulators predict the target (see the examples in **Table 23.1**). Confounding models commonly have a negative coordinative score because parents overlap in their correlation with the target. The coordinative component can be rearranged to $I(T; R1 | R2) - I(T; R1)$, suggesting that this component measures how much better $R1$ predicts T given $R2$ versus not given $R2$. Note this component is actually the third-order interaction information between T, $R1$, and $R2$, i.e., $I(T; R1; R2)$ (20), and is three-way symmetric.

The MI3 score is the sum of the correlative and the coordinative component. The symmetric coordinative component captures higher order interactions and differentiates causal relationships from confounding ones without telling the causal direction. The asymmetric correlative component determines the direction of the causal relationship. By merging these two components, the MI3 score considers connections between the regulators as well as dependency between the target and regulators. The MI3 score can be rearranged and simplified to $I(T; R1 | R2) + I(T; R2 | R1)$. This rearrangement can be interpreted as the conditional mutual information between the target gene and each regulator given the other regulator, which better shows the three-way nature of this score.

2.3.2. Network Inference Procedure

Instead of learning the global GRN all at once, we infer gene regulatory networks in two steps: learning and assembly (detailed rationale in **Section 7.1**). First, we learn the local regulatory network for each variable through an exhaustive search. Note that the local regulatory models for each target gene can have different number of parents despite that we started from two-parent models. Second, we can assemble local networks up into a unified network. Similar to Bayesian networks, the gene regulatory networks learned by using MI3 are directed and acyclic. In the assembly step, we may need to reconcile two-way edges and directed cycles in the network to create the required DAG structure. We solve conflicting local structures based on their scores (more details in **Section 4.4**).

Note that the key difference between MI3 and other methods is the scoring metric rather than the network construction procedure. The network construction procedure can be the same for all methods.

2.4. Nonparametric Probability Density Estimation for Continuous Variable

To avoid discretizing our data to calculate mutual information, we have adopted a continuous method for mutual information calculation based on a classical nonparametric Gaussian kernel method in probability density estimation (27, 28). To estimate the probability density at a specific location, we used all our data points. First we calculate the probability density at an interesting location based on a Gaussian distribution centered at each data point (kernel), and then take the average of all these densities using the following expression:

$$f_{\mathbf{H}}(x) = \frac{1}{n} \sum_{i=1}^{n} K_{\mathbf{H}}(\mathbf{x} - \mathbf{X}i) \qquad [15]$$

Here $\mathbf{x}$ is the position where probability density is to be estimated, and $\mathbf{X}i\,(i = 1, 2, \ldots, n)$ is the ith data point, both $\mathbf{x}$ and $\mathbf{X}i$ are d-dimension vectors. $K()$ is the kernel function, a symmetric probability density function, $\mathbf{H}$ is the bandwidth matrix which is symmetric and positive-definite, and $K_{\mathbf{H}}(\mathbf{x}) = |\mathbf{H}|^{-1/2} K(\mathbf{H}^{-1/2}\mathbf{x})$. The choice of kernel distribution makes little difference in probability estimation (27). We take $K(\mathbf{x}) = (2\pi)^{-d/2} \exp(-\mathbf{x}^{\mathrm{T}}\mathbf{x}/2)$, the multivariate Gaussian. On the other hand, the choice of bandwidth $\mathbf{H}$ is crucial, and we use optimal bandwidth described by Scott (28). Data are normally transformed into a uniform distribution (17) before the kernel density estimation to eliminate the potential effect of specific distributions.

Following our description above, to calculate entropy and mutual information for continuous variables, we calculated a probability density estimate at the positions of sample data points, then took the sample mean of log probability density (17) to approximate the full integration. The probability density estimation was the most computationally intensive step for this work.

Nonparametric probability density estimation for continuous variables effectively eliminates the inaccuracies introduced by discretizing data. However, this method is computationally demanding and requires a large sample size (n) (27–29). Due to these limitations, we limited our MI calculation to four or less variables. Notice that the sufficient sample only depends on the number of relevant dimensions of the local models (three nodes for two-parent models), and has nothing to do with the size of the total number of variables.

3. Software

MI3 is implemented in the open-source statistical computing language R as mi3 package. The package is available under the GNU GPL from online (http://sysbio.engin. umich.edu/~luow/downloads.php). **Figure 23.1** shows the workflow for GRN learning using mi3 package. A detailed procedure for mi3 package usage is described next in **Section 4**.

The same package also includes BN implementation, which is the same to MI3 procedure except that it uses a different scoring

1. Data preparation
   a. Synthetic data generation (*synData*) or microarray data processing (using R/Bioconductor)
   b. Data transformation (*rowunif, log2*)

2. Local two-parent model learning
   a. For each interesting node (*T*), exhaustively search
   b. Small dataset: *R1+R2* (*best2Pa*)
   c. Genome-scale dataset: *R2|R1* (*MI3r2*)

3. Local model refinement
   a. Solve conflicting models with two-way edges or forming directed circles (*twoPaModels*)
   b. Adjust for one-parent models (*onePaModels*)
   c. Adjust for three-parent models (*threePaModels*)

4. Network assembly and visualization
   a. Integrate and plot the global network (*netPlot*)

Fig. 23.1. The workflow for GRN learning using mi3 package. Relevant functions are given in parenthesis. The same workflow applied to both MI3 and BN methods, except that scoring metrics are different. Note that local model refinement and network assembly and visualization are currently only implemented for sub-genome-scale small data sets.

metric. As described above in formula [**14**], $I(\ ;\ )$ and $LL(\ |\ )$ scores are interchangeable when comparing the regulatory models for the same node. Therefore, we can learn BN models using mutual information score like $I(T; R1, R2)$, i.e., the correlative component of MI3 score, instead of $LL(T\ |\ R1, R2)$. This is the BN variant implemented in mi3 package. We will not describe BN learning procedure separately from MI3 in the following sections, as they are the same in all aspects except the scores being used in local model selection and improvement.

4. Procedure

4.1. Setup

Download mi3 package from http://sysbio.engin.uimich.edu/~luow/downloads. Three versions of this package are available: source package, binary for Mac OS *X*, and binary for Windows. Note that R needs to be pre-installed to set up and use mi3 package. Check http://www.r-project.org/ for R installation. Install mi3 package following the instructions in the tutorial. To use the visualization functionality of mi3, package graph is also needed.

4.2. Data Preparation

4.2.1. Generation of Synthetic Testing Data

We created a synthetic network structure with algebraic relationships between variables (**Fig. 23.2**) to validate MI3 method (22).

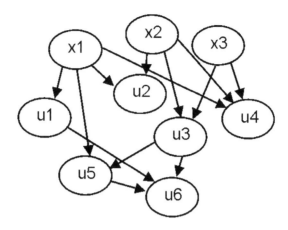

Variable	Algebraic formula	True parent set
x1	N(0,1)	
x2	N(10,5)	
x3	N(0,10)	
u1	$(x1)^3 + N(0,0.1*nl)$	x1
u2	$x1 + N(0,0.1*nl)$, $x1+10 \geq x2$ $x2/10 + N(0,0.1*nl)$, $x1+10 < x2$	x1, x2
u3	$(x2-x3)/(x2+10) + N(0,0.05*nl)$	x2, x3
u4	$x1-x3/10 + N(0,0.1*nl)$, $x1+10 \geq x2$ $x2/10+x3/10 + N(0,0.1*nl)$, $x1+10 < x2$	x1, x2, x3
u5	$\log(\exp(x1)+\exp(u3)) + N(0,0.1*nl)$	x1, u3
u6	$(u1+u5)*u3/2 + N(0,0.05*nl)$	u1, u3, u5

Fig. 23.2. Synthetic gene regulatory network. This synthetic model structure is designed to mimic a miniature gene regulatory network, with several major features. First the network contains nine variables in total, three of which are independent, and six dependent. Second, the variables are assembled into a hierarchy of regulatory relationships, with independent variable mimicking regulators and cofactors and dependent variables mimicking target genes. Third, the complexity of the network is controlled in that dependent variables have one to three parents, mostly two or three, and each regulator/cofactor controls a set of targets. Targets may share regulators and thus may have different levels of coregulation/coexpression, which can lead to confounding models. Fourth, a diverse set of continuous nonlinear and logical relationships among variables were encoded by the algebraic formulae in the table to describe a realistic, yet complicated regulatory network. Figure is modified, with permission, from (22).

MI3 and BN learning procedures can also be best demonstrated with this synthetic example of a completely known gene regulatory network. This model structure is designed (22) to mimic a miniature gene regulatory system, with regard to the network size, overall and local structure, and dependency relationships. To demonstrate the learning procedure, we randomly generate data sets of different sample sizes (data points) from this network first. Then we apply MI3 or BN to recover the network structure.

The user may customize their own synthetic examples by changing either the network structure or the algebraic relationships between variables or both (**Fig. 23.2**).

4.2.2. Gene Expression Data Processing

One key issue with learning regulatory model from gene expression microarray data is that there are usually multiple probe sets (as in Affymetrix GeneChip Data) for a single gene. This multiplicity interferes with gene network inference, where the nodes are individual genes instead of some part or transcript isoform of a gene. Dependency between different probe sets of a gene would normally be much higher than dependency between genes. A straightforward yet rough solution is to generate unique expression data entry for each gene by combining all probe sets for the same gene. A more rigorous strategy is to remap the probes to genes based on the latest genome annotation. This results in not only one data entry per gene naturally but also more accurate expression data as the Affymetrix probe sets were originally designed based on the outdated gene definition. However, the remapping of probes to genes is complex. Fortunately, this remapping is done routinely (30) and the results released online (http://brainarray.mhri.med.umich.edu) as customized CDF files ready to use. Please check the web site and our original MI3 paper (22) for details in processing raw expression data using these customized CDF files.

4.2.3. Data Transformation

The data should undergo uniform transformation in order to eliminate the effects of gene-specific distributions. For example, the distributions for individual nodes in the synthetic data are strongly divergent (**Fig. 23.2**). Note that microarray data are frequently log_2 transformed after pre-processing. Based on the author's experience, log_2 scaled data plus row-wise (gene-wise) $N(0, 1)$ normalization are satisfactory for most dependency/correlation analyses. This is default data transformation implemented in mi3 package. The uniform transformation became the new default in the new and future release for more statistical rigor and accuracy.

4.2.4. Data Filtering

For downstream analysis, all genes are included without a discriminative filtering process based on the magnitude of changes or based on the marginal dependency between candidate regulatory genes and the target gene. In other words, all other genes are included in the search for the best regulatory models of a target gene (details in **Section 7.2**). This inclusion is important as the

Reconstructing Transcriptional Regulatory Networks   413

filtering process could be arbitrary and misleading without considering the dependency, particularly the high-order dependency among genes.

4.3. Local Two-Parent Regulatory Model Learning

4.3.1. Rationale

MI3 assume two regulators for each target gene by default with the following reasons:

First, for a target gene which is governed by its regulator gene, there is often some other gene (a second regulator) which further affects the expression of the target, no matter what actual role it plays (such as being another transcription factor, a cofactor, a transporter for the first regulator) and no matter strong or weak this effect is.

Second, even if the first regulator plays a dominant role, when assuming only one regulator, it is impossible to tell the direction of regulation (i.e., which gene is the target and which is the regulator) from nonsequential observational data. By assuming two or more regulators per target, identifying the direction of regulation is no longer a problem as the models score differently when the regulators/targets switch position.

Note that the two-regulator models can be further refined when the number of regulators is different than two. This refinement process is available in mi3 package. Note that we limit the number of regulators to be 3 or less, because (1) a gene may be regulated by up to tens of other genes, but only limited number of them show significant regulatory effect in a specific experiment. Note that we are not trying to learn the inclusive and generally applicable regulatory models independent of any experimental condition; (2) the sample size and computing time become impractically large for learning models with more regulators. In our experience two-regulator models are good enough for most regular microarray studies.

4.3.2. Exhaustive Search for R1+R2 (Small Data Set)

As described above, exhaustive search is implemented and recommended by MI3 method (more details in **Section 7.2**). For small-to medium-sized data sets with up to 1,000 genes, it is possible to exhaustively search all two-parent models by computing of the full two-way and three-way mutual information arrays for all gene combinations. The actual computing time depends on the size of data matrix, i.e., number of genes/rows (v) and samples/columns (n), as $O(v^3 n^2)$. With this formula, you can estimate the compute time of your data on your computer based on the time for small synthetic data set.

MI3 learns the best two-parent regulatory models for all interesting genes in three steps. First, compute log-likelihood (negative entropy) for all single, pairs, and triplets of genes based on Gaussian kernel density estimation (details in **Section 2.4**). Second, compute the continuous two-way and three-way mutual information values by taking the mean of log probability density:

$$I(X; \Upsilon) = [ll(X, \Upsilon) - ll(X) - ll(\Upsilon)]/n$$
$$I(Z; X, \Upsilon) = [ll(X, \Upsilon, Z) - ll(Z) - ll(X, \Upsilon)]/n$$

where n is the number of data points, $I(Z; X, \Upsilon)$ is the correlative component of MI3 score. The coordinative component, $I(Z; X, \Upsilon) = I(Z; X, \Upsilon) - I(Z, X) - I(Z; \Upsilon)$, can be derived directly based on two-way and three-way mutual information arrays. Third, the best two-parent regulatory model for each gene or node selected by using MI3 score (**Section 2.3**).

Correspondingly, mi3 package has three major functions accomplishing by three computation steps. Function *ll* computes log-likelihood, function *compllCube* computes continuous two-way and three-way mutual information by calling *ll*, and function *best2Pa* selects the best two parents by calling *compllCube* in turn.

The procedure described above is for learning the best two-parent model-based continuous three-way mutual information – the default method implemented in the mi3 package. In addition, the mi3 package contains options for network learning using related scores (MI2, MI3, and BN) (22) based on both continuous and discrete probability densities.

4.3.3. Exhaustive Search for R2 Given R1 (Transcriptome-Scale Data Set)

For transcriptome-wide expression data, an exhaustive search for the best models for all genes is not tractable for MI3. While fast heuristics are possible in such case, exhaustive search is still required for optimal results. Here we introduce an application where genome-wide exhaustive search for regulatory models is feasible for all interesting target genes. Consider one regulator ($R1$) is known for a target gene (T), what is the second regulator gene ($R2$) that affects T's expression? Instead of searching all two-way or higher order combinations, we just need to search for a single $R2$ gene exhaustively. This is computationally tractable using MI3.

MI3 provides a routine (function *MI3r2*) to search for best $R2$'s for a list of known target genes (T's) of one common regulator ($R1$). First, compute log-likelihood hence two-way mutual information between all variables and $r1$. Then, pre-filter out T's from the *targList* with $I(T; R1) \geq mi0$ (some threshold), to ensure that $R1$ regulates T under the specific experimental condition. Finally, the two MI3 component scores, i.e., correlative and coordinative components are calculated based on Gaussian kernel density estimation. MI3 scores and top $R2$'s can be directly derived based on the results.

The function $MI3r2$ (with an auxiliary function as in **Section 6**) is the counterpart for function $best2Pa$ in this search for $R2$ given $R1$ scenario. In contrast to $best2Pa$, this function offers a way to incorporate prior knowledge and reduce the searching space by a factor of v (total number of variables/genes), hence make it tractable to learn the best two-parent regulatory models (given $R1$) from genome-wide microarray data.

4.4. Local Model Refinement

Two-parent regulatory models are frequently sufficient for learning a GRN from expression data as described above (**Section 4.3.2**). However, a GRN may be further improved by local model refinement. For example, the regulatory models for some genes may conflict with models for adjacent genes. In other words, edges in different models have opposite directions or form directed cycles. Such circular dependency is invalid in directed graphical modeling, hence need to be solved. Occasionally, models with different number of parents, i.e., one- or three-parents models, might be more consistent with the truth. We do not consider models with four or more regulators as described in **Section 4.3.2**. MI3 learns two-parent regulatory models as starting point and provides two utilities for further local model refinement: (1) reconciliation of conflicting local structures and (2) adjustment of the parent number. First, reconciliation of conflicting local structures is to evaluate and solve which model is the right one or the highest scoring one when there are multiple models with conflicting edge directions. Second, we adjust the parent number by first checking whether reducing or adding one parent significantly worsens or improves the model in terms of the dependency between the target and the parent set or $I(T; R$'s). The function $twoPaModels$ wraps the function $compI1Cube$ and the function $best2Pa$, and reconciles the conflicting models. The functions $onePaModels$ and $threePaModels$ further revise the number of parents of the models when needed.

4.5. Network Assembly and Visualization

When all the local regulatory models are ready, the GRN learning is essentially done. The last step is to assemble the models into a global network and present an integrated network view of the GRN. The function $netPlot$ uses the graph representation (graph package) and visualization (Rgraphviz package) facilities in R/Bioconductor.

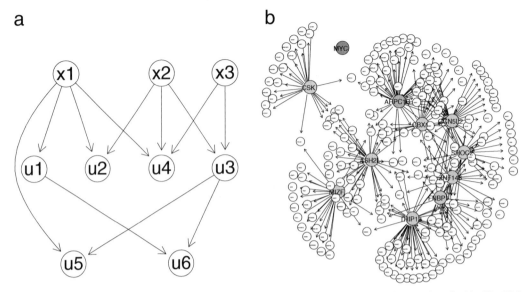

Fig. 23.3. Gene regulatory networks learned by using MI3 method: (**a**) from the synthetic data set described by **Fig. 23.2**; (**b**) from a published microarray data set, GSE2350, in Gene Expression Omnibus (http://www.ncbi.nlm.nih.gov/geo/). The synthetic network in (**b**) was inferred and visualized using mi3 package following the whole workflow described in **Fig. 23.1**, as an example for MI3 application to small data set. The gene regulatory network in (**b**) was inferred using *MI3r2* function of mi3 package and visualized using modified *netPlot* function as an example for MI3 application to genome-wide expression data set. Figure is modified, with permission, from (22).

5. A Sample Session

For the convenience of the users, script for a sample session is available online at http://sysbio.engin.umich.edu/~luow/downloads.php.

6. Conclusion

We have introduced a high-order mutual information-based statistical learning method, MI3, and demonstrated how to use MI3 to reconstruct networks from gene expression data. We also covered the theory and implementation of Bayesian network, a closely related classical method. Overall, we feel that a widespread adoption of statistical network reconstruction in bioinformatics would be of significant benefit for efficient knowledge discovery and representation.

7. Notes

7.1. Learning Local Regulatory Networks Versus the Global Network

When learning a GRN, our strategy is to learn local regulatory networks for individual target genes first, and then assemble the local networks into a single global network. Another strategy in graphical modeling is to learn the global network in one step. Compared to global network learning, local network learning is quicker and more tractable as the searching space is much bigger. The difference between these two strategies is not significant when the GRN is small, but local network learning became the only feasible strategy for learning genome-scale regulatory models from genome-scale expression data using exhaustive search.

7.2. Exhaustive Search for Optimal Local Models

In MI3, model learning was focused locally, i.e., we scored and compared all possible local regulatory models for specific target T. This target-centered model learning applied to both synthetic data and experimental data, even though biologically we are interested in constructing models centered at particular $R1$ in the latter case. It would be less appropriate to compare models across different T's because they are not mutually exclusive. Similarly, in Bayesian networks, $\log P(T \mid R1, R2)$ is only comparable for fixed T, where all other terms including $P(R1)P(R2)$ in the full product form of joint probability (10, 11) cancelled out. Therefore, we only searched for best $R1-R2$ pairs given T, but not the best $R2-T$ pairs given $R1$ when learning probabilistic models based on MI3 score or log conditional probability or any other established score. This local approach makes it possible for MI3 to conduct an exhaustive search, which leads to globally optimized models.

References

1. Lockhart, D.J., Dong, H., Byrne, M.C., Follettie, M.T., Gallo, M.V. et al. (1996) Expression monitoring by hybridization to high-density oligonucleotide arrays. *Nat Biotechnol* 14, 1675–1680.
2. Schena, M., Shalon, D., Davis, R.W., and Brown, P.O. (1995) Quantitative monitoring of gene expression patterns with a complementary DNA microarray. *Science* 270, 467–470.
3. Lee, W.P., and Tzou, W.S. (2009) Computational methods for discovering gene networks from expression data. *Brief Bioinform* 10, 408–23.
4. Eisen, M.B., Spellman, P.T., Brown, P.O., and Botstein, D. (1998) Cluster analysis and display of genome-wide expression patterns *Proc Natl Acad Sci USA* 95, 14863–14868.
5. Spellman, P.T., Sherlock, G., Zhang, M.Q., Iyer, V.R., Anders, K. et al. (1998) Comprehensive identification of cell cycle-regulated genes of the yeast Saccharomyces cerevisiae by microarray hybridization. *Mol Biol Cell* 9, 3273–3297.
6. Butte, A.J., Tamayo, P., Slonim, D., Golub, T.R., and Kohane, I.S. (2000) Discovering functional relationships between RNA expression and chemotherapeutic susceptibility using relevance networks. *Proc Natl Acad Sci USA* 97, 12182–12186.
7. Moriyama, M., Hoshida, Y., Otsuka, M., Nishimura, S., Kato, N. et al. (2003)

Relevance network between chemosensitivity and transcriptome in human hepatoma cells. *Mol Cancer Ther* 2, 199–205.

8. Schafer, J., and Strimmer, K. (2005) An empirical Bayes approach to inferring large-scale gene association networks. *Bioinformatics* 21, 754–764.

9. Alon, U. (2007) *An introduction to systems biology : design principles of biological circuits*, Chapman & Hall/CRC, Boca Raton, FL.

10. Hartemink, A.J., Gifford, D.K., Jaakkola, T.S., and Young, R.A. (2001) Using graphical models and genomic expression data to statistically validate models of genetic regulatory networks. *Pac Symp Biocomput* 6, 422–433.

11. Friedman, N., Linial, M., Nachman, I., and Pe'er, D. (2000) Using Bayesian networks to analyze expression data. *J Comput Biol* 7, 601–620.

12. Sachs, K., Perez, O., Pe'er, D., Lauffenburger, D.A., and Nolan, G.P. (2005) Causal protein-signaling networks derived from multiparameter single-cell data *Science* 308, 523–529.

13. Friedman, N. (2004) Inferring cellular networks using probabilistic graphical models. *Science* 303, 799–805.

14. Basso, K., Margolin, A.A., Stolovitzky, G., Klein, U., Dalla-Favera, R. et al. (2005) Reverse engineering of regulatory networks in human B cells. *Nat Genet* 37, 382–390.

15. Butte, A.J., and Kohane, I.S. (2000) Mutual information relevance networks: functional genomic clustering using pairwise entropy measurements. *Pac Symp Biocomput* 2000, 418–429.

16. Heckerman, D. (1995) Microsoft research.

17. Steuer, R., Kurths, J., Daub, C.O., Weise, J., and Selbig, J. (2002) The mutual information: detecting and evaluating dependencies between variables. *Bioinformatics* 18(Suppl. 2), S231–S240.

18. Friedman, N., Nachman, I., and Pe'er, D. (1999) In: *Proceedings of the 15th Annual Conference on Uncertainty in Artificial Intelligence (UAI-99).* pp. 206–215, Morgan Kaufmann, San Francisco, CA.

19. Mcgill, W.J. (1954) Multivariate information transmission. *Psychometrika* 19, 97–116.

20. Jakulin, A., and Bratko, I. (2004) Quantifying and visualizing attribute interactions: an approach based on entropy. *arXiv:cs.AI/0308002.*

21. Nemenman, I. (2004) Information theory, multivariate dependence, and genetic network inference. *arXiv:q-bio/0406015.*

22. Luo, W., Hankenson, K.D., and Woolf, P.J. (2008) Learning transcriptional regulatory networks from high throughput gene expression data using continuous three-way mutual information. *BMC Bioinformatics* 9, 467.

23. Luo, W. (2008) *Gene regulatory network reconstruction and pathway inference from high throughput gene expression data*, PhD thesis, University of Michigan, Ann Arbor, MI.

24. Shannon, C.E. (1948) A mathematical theory of communication. *Bell Sys Tech J* 27, 379–423.

25. Kolmogor.An. (1968) Logical basis for information theory and probability theory. *IEEE Trans Inform Theory* IT14, 662–664.

26. Watanabe, S. (1960) Information theoretical analysis of multivariate correlation. *IBM J Res Dev* 4, 66–82.

27. Silverman, B.W. (1986) *Density estimation for statistics and data analysis.* Chapman and Hall, London/New York, NY.

28. Scott, D.W. (1992) *Multivariate density estimation : theory, practice, and visualization.* Wiley, New York, NY.

29. Scott, D.W., and Wand, M.P. (1991) Feasibility of multivariate density estimates. *Biometrika* 78, 197–205.

30. Dai, M., Wang, P., Boyd, A.D., Kostov, G., Athey, B. et al. (2005) Evolving gene/transcript definitions significantly alter the interpretation of GeneChip data. *Nucleic Acids Res* 33, e175.

Chapter 24

Computational Methods for Analyzing Dynamic Regulatory Networks

Anthony Gitter, Yong Lu, and Ziv Bar-Joseph

Abstract

Regulatory and other networks in the cell change in a highly dynamic way over time and in response to internal and external stimuli. While several different types of high-throughput experimental procedures are available to study systems in the cell, most only measure static properties of such networks. Information derived from sequence data is inherently static, and most interaction data sets are measured in a static way as well. In this chapter we discuss one of the few abundant sources for temporal information, time series expression data. We provide an overview of the methods suggested for clustering this type of data to identify functionally related genes. We also discuss methods for inferring causality and interactions using lagged correlations and regression analysis. Finally, we present methods for combining time series expression data with static data to reconstruct dynamic regulatory networks. We point to software tools implementing the methods discussed in this chapter. As more temporal measurements become available, the importance of analyzing such data and of combining it with other types of data will greatly increase.

Key words: Gene expression, causality, clustering, data integration, time series.

1. Introduction

Biological systems are inherently dynamic in nature as they change in response to external and internal stimuli and over time (1–3). While several different types of high-throughput genomic data sets are being collected, most are static or measured in a static way. For example, DNA sequences are inherently static and do not change over years. Information derived from such data including DNA and miRNA binding motifs (4) is also static. Other types of data, including interaction data are often measured in a static way. For example, large-scale protein–DNA binding (5),

---

I. Ladunga (ed.), *Computational Biology of Transcription Factor Binding*, Methods in Molecular Biology 674,
DOI 10.1007/978-1-60761-854-6_24, © Springer Science+Business Media, LLC 2010

protein–protein interaction (6, 7), and miRNA–mRNA (8) interactions are all performed as a snapshot experiment and do not provide information about changes over time.

To understand the dynamics of regulatory networks, researchers must be able to obtain and analyze temporal data sets regarding the activity of such systems. So far the most abundant data source for this task has been time series gene expression data (*see* **Fig. 24.1**). This chapter will focus on how we can extract information about the regulation of biological systems from this type of data and how it can be integrated with other (mostly static) data sets to reconstruct models for the activity of these networks in the cell. Other types of dynamic data, most notably imaging data, are also becoming available for some of these systems. However, since they are so far limited to a small number of biological systems and relatively little computational work has been performed to use these data sets, we would not discuss them in this chapter.

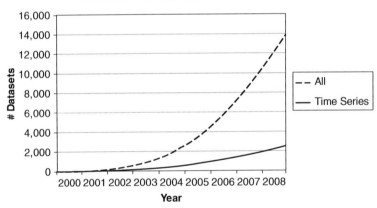

Fig. 24.1. Growth in the number of expression data sets deposited in GEO over the last decade. The *solid curve* represents the data sets from time series experiments. Over 2,000 time series data sets were deposited in GEO by the end of 2008, representing almost 20% of all gene expression experiments. Methods that can adequately analyze these data sets and that can combine them with other high-throughput data sets are required so that we can fully utilize the information obtained in these experiments.

Figure 24.2 presents four analysis levels that are often performed when using high-throughput data sets to study biological systems. While these should be considered for all high-throughput experiments, different types of data raise different sets of questions that should be addressed at each level. For example, while gene expression experiments can be used to study either static (snapshot) or dynamic processes, there are a number of unique issues when using time series expression data. When designing time series experiments, one needs to decide how many time points to use and what the sampling rates should be. When analyzing such data one should decide how to represent it

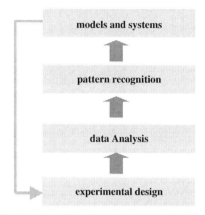

Fig. 24.2. Analysis levels for high-throughput experiments. Studies using high-throughput data sets often address issues related to these four analysis levels. Here we discuss issues that relate to the analysis of time series data sets and dynamic regulatory networks focusing on the top two levels (pattern recognition and modeling).

(a collection of points, some interpolated curve or a specific function). Similarly, while many clustering algorithms can be used for both static and time series data, these often do not make adequate use of the temporal information. Algorithms that are specifically designed for time series data may lead to better results for such data. In a previous review, we focused primarily on the lower two analysis levels: experimental design and data analysis (9). Here we focus on the upper two levels, clustering and systems biology. We discuss several new approaches for these tasks that have been developed since 2004. We also highlight the importance of data integration, specifically of time series and static data, which can lead to dynamic models while still relying on abundant static data.

The Methods section of this chapter is organized into three parts. **Section 3.1** discusses clustering methods for time series expression data. These methods are often used to obtain a first, global, view of the expression data. They are also used to infer groupings and large-scale organization and can provide functional information about unknown genes that are grouped with known genes. **Section 3.2** discusses methods that try to infer interactions and causality from temporal data sets. These methods rely on the observation that in many cases genes that are active at early time points will activate genes at later time points. Such an analysis can provide information about regulatory interactions between genes. While the first two sections focus on inference that can be drawn from expression data alone, **Section 3.3** describes methods that try to integrate additional data sets for this task. This helps to overcome the dimensionality problem that often exists in high-throughput experiments (relatively few samples and a large number of genes) by using additional data sources to constrain the set of parameters and their values. For each of these sections, we provide a list of software tools that implement some of the methods we discuss so that researchers may try these ideas on their own data. We conclude this chapter by discussing open problems and directions for future work.

2. Software

In this chapter we discuss several methods for analyzing dynamic regulatory networks. Some of these methods have been implemented and can be downloaded and used by researchers. Below we provide links for these software packages. The links are arranged according to the sections that discuss the methods they implement.

2.1. Clustering

The Eisen lab provides implementation for several general purpose clustering algorithms, including hierarchical clustering, self-organizing maps, and K-means. The software (Cluster) can be downloaded from http://rana.lbl.gov/EisenSoftware.htm. The STEM software for clustering short time series expression data is available from http://www.cs.cmu.edu/~jernst/stem/.

2.2. Regression and Dynamic Bayesian Networks (DBNs)

Kevin Murphy wrote a popular Bayesian Networks toolbox for MATLAB, which also contains DBN procedures. It can be downloaded from http://people.cs.ubc.ca/~murphyk/Software/BNT/bnt.html. A MATLAB implementation for aligning time series data sets studying the same system under different experimental conditions is available from http://www.sb.cs.cmu.edu/pages/software.html. An implementation of the method for combining experiments from different conditions can be found here: http://www.cs.cmu.edu/~yanxins/regulation_inference/Matlab.html.

2.3. Data-Integrated Methods

An R implementation of Inferelator, which combines time series and motif data, is available at http://err.bio.nyu.edu/inferelator. Perl scripts for the various components of the Statistical Analysis of Network Dynamics (SANDY) algorithm can be found under the "data download" section here: http://sandy.topnet.gersteinlab.org. The Network Component Analysis toolbox for MATLAB can be downloaded from http://www.seas.ucla.edu/~liaoj/download.htm. A MATLAB implementation of the post-transcriptional modification model, which combines sequence and expression data, can be downloaded at http://www.sb.cs.cmu.edu/PTMM/PTMM.html. The Dynamic Regulatory Events Miner (DREM) executable and Java source code can be obtained here: http://www.sb.cs.cmu.edu/drem.

3. Methods

3.1. Clustering Time Series Expression Data

In a gene expression time series experiment, one can observe the activity of thousands of genes over periods of time. To obtain a global view and facilitate further analysis, it is often useful to divide genes into smaller groups based on their temporal expression patterns. This can be done by cluster analysis, which assigns genes to subsets (clusters) such that genes in the same cluster exhibit similar expression patterns, while genes in distinct clusters do not. Cluster analysis has been used to discover co-regulated genes and genes sharing related functions (10).

There are a number of challenges when applying cluster analysis to gene expression time series. First, many general clustering methods do not take into account the order of time points. In contrast, methods specifically designed for time series may be better suited to the task. Second, most of the expression time series have relatively few (3–8) time points (11). As a result, methods that work well with long time series may tend to overfit when applied to such data sets. Third, because the number of genes observed is large, patterns may arise just by chance. One needs a way to distinguish between real patterns versus patterns arising by chance.

Several methods have been used to cluster time series data. However, general clustering algorithms do not take time into account. While we mention some of these here because of their popularity, our goal in this section is to focus on methods developed for clustering time series.

3.1.1. General Clustering Methods

The most popular methods for clustering expression data, which have also been used to cluster time series data, include hierarchical clustering, K-means, and projection-based methods. Hierarchical clustering groups genes either using a bottom-up (10) or a top-down (12) approach. In both cases, the resulting clusters are represented as a tree with leafs corresponding to genes and subtrees corresponding to clusters. K-means clustering (12) is an iterative algorithm that starts from k randomly selected clusters. In each iteration, genes are assigned to the cluster with the nearest mean, and the cluster-means are updated by calculating the average expression profile within each cluster. Projection-based methods, including principal component analysis (13, 14) and independent component analysis (15, 16), work by first mapping the gene expression to a new space and then clustering genes there.

Unlike the general clustering methods described above, several recent algorithms were developed specifically for clustering time series expression data. These methods utilize the dynamic information to improve the clustering results and often are shown to outperform the general clustering methods.

3.1.2. Clustering Using Continuous Representation

Time series experiments only generate snapshots at certain time points of the gene expression levels, which may be more naturally modeled by a continuous curve. One way is to represent the gene expression level by splines, piecewise polynomials with bounded constraints. Bar-Joseph et al. (17) proposed to use B-splines for this purpose, and the resulting model has fewer basis coefficients than the number of time points, which helps to avoid overfitting. The clustering method assumes a mixture model where each mixture component corresponds to a cluster, and the expression of each gene is generated through a noisy process from the model expression curve. Bar-Joseph et al. (17) describe a method that simultaneously estimates the parameters for the continuous representation and the assignment of genes to clusters.

Closely related methods have been proposed to represent expression time series by piece-wise linear (18), quadratic (19), or higher-order interpolation (20). In Magni et al. (18) and Liu et al. (19), the learned models are transformed into symbolic representations, which are in turn used to cluster genes, while Wang et al. (20) cluster genes directly based on the learned polynomials.

3.1.3. Clustering Using Dynamic Features

In these methods, features reflecting temporal patterns are extracted from expression time series and used for clustering. Kim and Kim (21) use first- and second-order differences between adjacent time points as temporal features. Genes are clustered based on the pattern determined by the sequence of features. One limitation of the method is that it requires several replicate experiments and most time series expression data sets are measured with very few or no replicates. Déjean et al. (22) represent genes by smoothing splines, and use the derivatives at some discretization points as features. Genes are clustered by applying hierarchical clustering to the extracted derivatives. Li et al. (23) convert expression time series to a sequence of slopes, and use an unsupervised conditional random field model to cluster the genes.

3.1.4. Clustering Using Hidden Markov Models

Schliep et al. (24) developed a hidden Markov model (HMM) method to model the dependency between observations of adjacent time points. An HMM is specified by a set of hidden states, the probability of starting at a given state, the probability of transition from one state to the other, and the probability of generating the gene expression level at each state. The clustering is modeled by a mixture of HMMs, where each HMM corresponds to a cluster. Gene assignment and model parameters are estimated by maximizing the likelihood of the observed expression time series using an Expectation Maximization (EM) style algorithm (*see* **Chapters 6** and **7**). The number of clusters is determined by a heuristic procedure that removes clusters with too few genes and splits clusters with too many genes.

In (24), a gene is assigned to the cluster corresponding to the most probable HMM. Schliep et al. (25) suggest a method to improve the assignment by grouping genes with ambiguous membership into a separate cluster, and show that the resulting groups are more robust to noise in the data. Schliep et al. (25) also propose an approach to incorporate prior biological knowledge to improve the clustering.

We note that while HMM-based methods works well for long time series, they require that the number of time points to be much larger than the number of states, which may be problematic for short time series.

3.1.5. Clustering Methods Based on Stochastic Processes

In the next section we discuss regression models for determining the effects genes have on other genes. These ideas can also be used for clustering by relying on an autoregressive model (26). An autoregressive model of order p assumes the expression level at a given time point is a linear function of the expression levels *of the same gene* in the previous p time points. The clustering algorithm uses an agglomerative procedure to search for the most probable set of clusters. It starts by assuming every expression time series is generated by a different process. In the next step, it computes the model likelihood for all possible pair-wise merges. The method then identifies the merge that results in the highest model likelihood, and, if it is higher than the current model likelihood, merges the two clusters. The procedure stops when the model likelihood cannot be improved by merging anymore. A closely related method by Zhou and Wakefield (27) models gene dynamics by a random walk, and uses birth-death MCMC to determine the number of clusters.

3.1.6. Clustering Using Model Profiles

Short Time-series Expression Miner (STEM) is designed specifically for short time series (28) (**Fig. 24.3**). The method starts by selecting a set of potential model profiles that can represent any expression profile. The number of potential profiles is controlled by a user parameter that determines the amount of change a gene can exhibit between two adjacent time points. A subset of the m profiles is selected by a procedure that maximizes the minimum distance between any two profiles. The rationale is to select a distinctive set of profiles that covers the entire space of possible expression profiles. Given a set of model profiles, each gene is assigned to the closest model profile. The significance of the model profiles is computed by hypothesis testing where the null hypothesis is that any profile observed is resulted from random fluctuation of the model profile.

Anand et al. (29) suggest a number of ways to improve this method by assigning genes to model profiles based on a fuzzy membership function, and selecting model profiles using an evolutionary algorithm that finds trade-off between minimizing quantization errors and minimizing the number of profiles. They show that in certain situations the proposed method improves the clustering quality.

3.2. Regression Analysis for Causal Inference in Time Series Data

There are two primary sources for inferring regulatory relationships from gene expression data. The first are perturbation experiments (either knockout or knockdown) that inactivate a gene or a pair of genes and study the downstream affects (30, 31). The second are time series experiments in which researchers use lagged correlations to search for regulatory relationships (32).

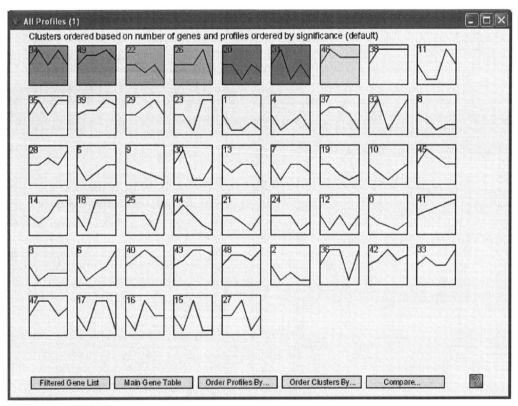

Fig. 24.3. Short time-series expression miner (STEM) example. The data are from experiments studying a serum response factor, SRF-VP16, of wild-type embryonic stem cells. Samples were taken at five time points: 0, 10, 30, 60, and 180 min (76). (**A**) Overview of clustering results. The number in the *top left-hand* corner of a profile box is the profile ID number. The shaded profiles (left part of the top row) had a statistically significant number of genes assigned, and those *non-white* profiles of the same color are similar and assigned to the same cluster of profiles. (**B**) Zoom in on profile 34. The figure displays all genes assigned to that profile. It also lists the expected and actual number of genes assigned to this profile and the *p*-value for having so many genes assigned to this profile.

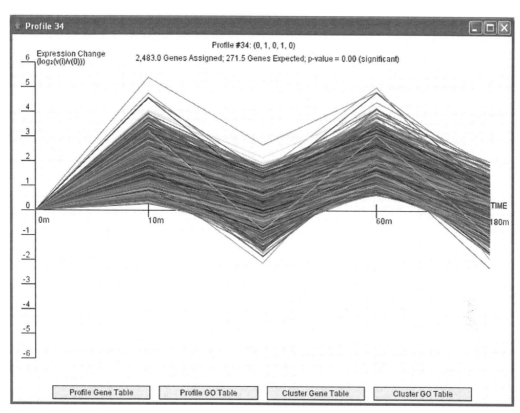

Fig. 24.3. (continued)

Unlike perturbation experiments, which usually start with a single perturbed gene, time series data allow researchers to study several different regulators at once. On the other hand, the application of methods for inferring lagged correlations to a data set containing measurements of thousands of genes over a relatively small number of time points may lead to a large number of false positives. In such a data set, many of the inferred regulatory relationships may result from noise or from unrelated sources (co-occurrence as opposed to activation). To address this problem, most algorithms developed for this task are focused on trying to limit overfitting by tightly controlling the algorithms used to learn the interaction parameters. Another possible solution to this problem is to combine different data sets (measuring the same set of genes under different experimental conditions) and search for regulatory relationships that are present in a subset of these data sets.

We divide the set of methods developed for lagged regulatory analysis into those that can only be applied to a single data set and those that can be applied to multiple data sets at once. Most of the work applied to a single data set relied on regression analysis to identify causal genes. Most of the work applied to multiple

428   Gitter, Lu, and Bar-Joseph

data sets used correlation coefficients (implicitly assuming a fixed delay of 0 in all experiments) though more recently researchers have used regression analysis for multiple data sets as well.

3.2.1. Time-Lagged Inference from a Single Data Set

Qian et al. (32) were among the first to use time series data for inferring interactions among genes. Their method relied on aligning the measured values for a pair of genes. To identify causal relationships they have used local alignment algorithms to find cases where a later expression of one gene matches an earlier expression of another gene and link these two genes. They have also looked at inverted relationships that could identify repression effects. Schmitt et al. (33) applied a similar analysis to a much larger data set from the photosynthetic cyanobacterium *Synechocystis* sp. The analysis identified networks of interactions and allowed inference of putative effects of light on this organism. The final network comprised 50 different groups containing 259 genes. Most of these gene groups possess known light-stimulated gene clusters while others represent novel findings in that work. Balasubramaniyan et al. (34) developed a tool called CLARITY (Clustering with Local shApe-based similaRITY). This tool uses the Spearman rank correlation as a shape-based similarity measure to compute the correlation between genes in a single time series expression data set. The method can also identify time-shifted (lagged) correlations, which can be used to infer causal relationships.

3.2.2. Dynamic Bayesian Networks (DBNs)

Another direction for determining causal relationships from time series data is the use of various graph theory-based methods also known as graphical models. These models include Bayesian networks (35) that have been successfully applied to study static expression data. An extension of Bayesian networks, Dynamic Bayesian Networks (DBNs) can be used to determine regulatory relationships from time series data, often improving on the static version for this type of data (36). For example, Ong et al. (37) applied DBNs to study *Escherichia coli* time series data. While Ong et al. used discretized expression values (up or down regulation) in their DBNs, most follow-up studies worked with continuous values and regression analysis. To illustrate the general concept of using DBNs for time series expression data, consider the graph in **Fig. 24.4**. This figure presents five genes in two consecutive time points. Genes A and C on the left are connected to gene B indicating that their expression levels in a previous time point affect the expression of B in the next time point. This could represent transcription factors (TFs) regulating some of their targets (note that A also regulates C so the figure represents a temporal feed-forward loop). We denote the node representing A on the left as a parent of B and C. The exact nature of this effect is indicated by the Conditional Probability for B, which is specified in the figure. This function assumes that B is expressed as a linear

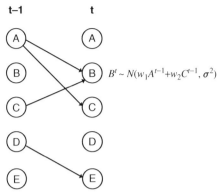

Fig. 24.4. Dynamic Bayesian network (DBN) example. Five genes are represented using a DBN. Edges represent conditional dependence between levels of genes in two consecutive time points. The probability distribution for gene B is specified. Since both A and C connect to B, this probability is a function of the level of these two genes at the previous time point plus some noise term. Note that in the DBN in the picture we only allow connections between time points. In general, DBNs connections are also allowed within a time point, but most applications using DBNs for time series expression data only use the between time points connections.

function of A and C plus some noise. More generally, we usually use functions of the form:

$$x_i^t = \sum_{j \in p(i)} w_{i,j} f(x_i^{t-1}) + \varepsilon$$

where at time t the x_i^t expression of gene i is dependent on the expression of its parents $(p(i))$ at the previous time point $(t-1)$. $w_{i,j}$ represents the strength of the influence of parent j on gene i. ε accounts for experimental and other types of noise and is assumed to be independent of the measurements and distributed as a Gaussian with 0 mean ($\varepsilon \sim N(0, \sigma^2)$). The dependency between parents and genes can either be linear (in which case f is the identity function, as in **Fig. 24.4**) or nonlinear (in which case f can take an arbitrary form). More generally, given a vector of expression levels at time $t - 1(X^{t-1})$, we can predict the levels at time t using the following equation:

$$X^t = Wf(X^{t-1}) + \mathbf{e}$$

where W is a sparse matrix and $\mathbf{e}$ is a vector of normally distributed random noise variables.

The major challenge associated with learning such networks from data is the condition that we need to estimate a large number of parameters from a relatively small number of data points. Time series experiments are very short (often no longer than 8 time points, (11)), which means that we face the curse of dimensionality problem when trying to infer DBNs from a single time series data set. To address this problem, researchers have used a number of regularization methods. These include *L1* penalty terms on the values of the weight parameters (effectively minimizing the number of parents for each node) (38), information theoretic penalty terms on the number of parameters (39, 40), limiting possible parents to a coherent subset of the genes (41), and constraints on the time distance between the activation of the regulator and the activation of its target (42).

Recently, Ahmed and Xing (43) developed a method termed TESLA for learning time-varying networks. This method combines ideas from both static and dynamic network analyses. A regression network is constructed for each time point. However, the networks for consecutive time points are linked so that the presence of edges in one time point depends on their presence in the previous time point allowing TESLA to uncover the evolvability of the networks over time.

3.2.3. Interaction Inference from Multiple Temporal Data Sets

Combining multiple time series data sets when learning DBNs or other time lag models may help in overcoming the skewed dimensionality problem. However, combining multiple data sets for this task is a non-trivial problem. First, sampling rates differ between different data sets, making it hard to determine a common temporal unit for DBNs. Second, for a specific interaction pair (a TF and its target gene) the actual time lag may differ between different experiments since the *time scale* of the series data may change. For example, using different arrest methods leads to very different cell cycle durations (44). These different cell cycle durations translate to differences on the molecular level, which affect the time it takes a TF to activate the genes it regulates. Finally, even for a pair of genes displaying time lagged regulation, this relationship might exist in only a subset of the data sets since different pathways may be activated under different conditions.

A possible way to combine multiple data sets is to ignore the time lag and rely instead on correlation between the profiles of genes in the data set. This effectively assumes a time lag of 0 for all pairs. For example, Lee et al. (45) used the correlation method to combine a large number of human expression data sets to search for correlated pairs. Another way to address this issue, which is appropriate for combining experiments that study the same system under different conditions (for example, different cell cycle arrest methods) is to align the data sets assuming that genes behave in the same way in all experiments though with different time units. The alignment process determines the appropriate transformation from one time series to another. Once the alignment is determined, we can transform the different data sets into a common temporal representation and they can then be used to infer DBNs and other lagged models as discussed above. Various

Computational Methods for Analyzing Dynamic Regulatory Networks 431

alignment techniques have been suggested including methods based on dynamic programming (46), methods based on continuous representations and alignment of curves (47), a combination of dynamic programming and continuous representation (48), and HMM-based methods (49). There have been a few attempts at using aligned data sets for reconstructing DBNs, most notably for cell cycle studies in yeast and human (50). Note however, that all of these alignment methods assume that the data sets measure the same system. However, when combining more diverse experiments (for example, cell cycle and stress experiments), such an assumption cannot be expected to hold anymore. Thus, unlike static BNs that have been used to combine a large number of data sets from a wide range of experimental conditions, DBNs have so far been limited to modeling individual data sets or similar data sets for the same biological system.

Shi et al. (51) presented method that may overcome this problem and allow researchers to combine experiments from different conditions in a single DBN. These authors presented an algorithm that uses a set of known interacting pairs to compute a temporal transformation between every two data sets, regardless of the condition they study. The underlying idea is that some interactions would be present in both data sets and these can be used to learn the temporal transformation between the two data sets. Using an EM algorithm, they align all time series data sets to a common reference data set (usually the longest) and use the aligned experiments to search for additional regulatory interactions, not used in the learning phase, that are present in multiple data sets. From 16 yeast time series data sets from cell cycle, various stress conditions, and DNA damage response experiments, the method was able to greatly improve upon the accuracy of models constructed from a single data set.

3.3. Integrating Additional Data for Improved Network Inference

As discussed in the previous section, network inference techniques that rely solely on time series gene expression data often suffer because there are many more parameters to fit than time points. To remedy this problem, inference algorithms can incorporate other data sources to impose additional constraints and reduce the number of feasible models.

Adding new types of data to existing models gives rise to its own set of challenges. Such information can be used in a pre- and/or post-processing step to eliminate inconsistent networks or can be tightly coupled with the network inference algorithm, which may require a fundamentally different computational framework. Furthermore, not all types of data are prevalent in certain species. For instance, whereas sequence data are readily available for many species of interest, genome-wide protein–DNA binding studies have only been performed for a few species (*see* **Chapters 11**, **12**, and **20** (52)). In addition, as noted earlier, sequence data are inherently static and protein–DNA binding, protein–protein interactions (PPI), and miRNA–mRNA interactions are generally measured at a single time point in a single condition. Thus, it is not always straightforward to use this information to provide additional insight into dynamic regulatory processes.

The data-integrative methods we present here are broadly grouped by the types of additional information they utilize: sequence and motif data, protein–DNA binding interactions, and/or other types of interactions.

3.3.1. Combining Time Series Expression and Sequence Data

In the context of inferring dynamic regulatory networks, sequence data are most often used to predict protein–DNA binding by identifying TF binding site motifs in genes' promoter regions. Kundaje et al. (53) combined time series gene expression profiles and occurrence counts of known motifs to learn transcriptional modules. Splines were used to model the dynamic expression data, and the modules were learned by using Expectation Maximization to optimize a generative probabilistic graph model. Ramsey et al. (54) extended the time-lagged correlation method discussed in the previous section to include a motif scanning step. Differentially expressed genes were clustered, a time lagged correlation procedure calculated significance for TF-gene pairs, and the significance scores were combined to yield TF-cluster scores. Position-weight matrices were used to scan the promoter regions of the differentially expressed genes and motif enrichments were computed for each cluster. Inferelator (55) first formed biclusters based on gene expression data, regulatory motifs in promoter regions, and a network of functional associations. Kinetic equations were then fit to determine the regulatory impacts between predictor variables, TFs and external stimuli, and the biclusters. This method also models pairwise combinatorial interactions between predictors. An extension to Inferelator (56) adopts a Bayesian approach to improve predictions under long time scales.

3.3.2. Utilizing Protein–DNA Binding Interactions

While incorporating sequence data is appealing due to its prevalence in many species, motif-based binding predictions are not as informative as experimental protein–DNA binding interaction data. The availability of genome-wide ChIP–chip data in model organisms such as *Saccharomyces cerevisiae* (5) has given rise to techniques that make use of such information, sometimes in conjunction with sequence data.

Luscombe et al. (57) presented Statistical Analysis of Network Dynamics (SANDY), a tool for calculating network statistics for dynamic systems. Differentially expressed genes were assigned to a stage in the cell cycle, and an iterative trace-back algorithm was applied to isolate the active TFs and sub-network at that stage. Sub-networks were subsequently compared based on graph statistics such as topology, presence of network motifs, and TF usage. A rule-based method by Chawade et al. (58) clustered genes such that their promoter regions were enriched for a common set of motifs that are known to be bound by a TF. In addition, each gene in a cluster must be first significantly expressed at the same time point or immediately after the TF regulating that cluster is first expressed, and the expression profiles of the clustered genes must be correlated. We previously discussed integrating time lagged correlation models with motifs (54), and Wu and Li (59) extended this class of models by incorporating TF binding and deletion gene expression data as well. Lin et al. (60) employed a first-order nonlinear differential equation to combine cell cycle TF binding data and dynamic gene expression data and extract dynamic interactions among the TFs. Network Component Analysis (61) decomposes a data matrix containing dynamic gene expression levels into a connectivity matrix and a signal matrix and provides criteria for doing so uniquely. The signal matrix corresponds to the activity levels of TFs over time, and the connectivity matrix quantifies how strongly TFs regulate their target genes. Protein–DNA binding data are used to constrain the connectivity matrix so that TFs cannot regulate genes they do not bind, and extensions to Network Component Analysis (62, 63) use gene knockout data to further constrain the signal matrix.

Several regression-based methods also include protein–DNA binding data to guide the estimation of model parameters. Cokus et al. (64) applied linear regression to time series gene expression data and binding interaction data to estimate dynamic TF activity levels at each time point. The authors then used least squares to estimate a transition matrix that specifies how TFs affect each other's activity levels over time. Multivariate Random Forests, developed by Xiao and Segal (65), consist of a random forest of multivariate regression trees that use protein–DNA binding and motif data as input and temporal gene expression levels as outcomes. The resulting proximity matrix specifies pairwise gene similarity based on both time series expression and binding information. The authors used the proximity matrix as input to a guided clustering method to identify regulatory cliques.

Probabilistic graphical models have also benefitted from the integration of TF binding information. Dynamic Bayesian Networks, discussed in the previous section, were adapted to include TF binding data as a prior by Bernard and Hartemink (66). The strength of the prior for the presence of an edge is greater for binding interactions with lower p-values, and the prior is factorable in order to enable efficient computation. Sanguinetti et al. (67) incorporated ChIP–chip data in their Kalman filter model to represent network connectivity. They applied a variational Expectation Maximization inference algorithm to learn TFs' dynamic protein concentration levels and regulatory influences on their

3.3.3. Dynamic Regulatory Events Miner

target genes. The post-transcriptional modification model presented by Shi et al. (68) learns temporal TF activity levels via a switching model that determines whether a TF is regulated transcriptionally or post-transcriptionally. TFs' activity levels can then be respectively inferred from either their own gene expression levels or the expression levels of their regulatory targets. Protein–DNA binding data are incorporated as prior in the log-likelihood score function to penalize TF-gene regulatory interactions in the model that disagree with the ChIP–chip data. Below we present one type of graphical model applied to this problem, an extension of HMMs, in greater detail.

Dynamic Regulatory Events Miner (DREM) (69) takes a unique approach by focusing its modeling of temporal regulatory interactions on bifurcation points. Bifurcation events occur when a set of genes share a similar expression trajectory up to a certain time point and then diverge (**Fig. 24.5**). To identify these splits, groups of genes are assigned to the hidden states of an input–output hidden Markov model (IOHMM). IOHMM is an extension of hidden Markov models that allows static input, in this case TF-gene interaction data, to influence the state transition probabilities. At each state that has more than one child state, an $L1$-penalized logistic regression classifier maps the subsets of TFs that are potentially active at that state to transition probabilities for the genes assigned to that state. After DREM assigns each gene's expression profile to one of the paths along hidden states in the model, it uses protein–DNA binding or motif data to determine which TFs are responsible for the bifurcations by calculating enrichment scores based on the hypergeometric distribution.

Because DREM infers the times at which TFs regulate their targets, it can differentiate master regulators that control the immediate response to a stimulus from secondary regulators that are active later. Determining the time at which a TF is most active also enables DREM to identify the best time to conduct binding experiments for particular TFs to experimentally verify their role in the stimulus response. DREM was so far applied to yeast and *E. coli* (70) and in both cases led to specific temporal predictions, which were experimentally verified.

Fig. 24.5. (continued) that pass through the corresponding hidden state (represented by a *light shaded* node) are shown. These genes share a common response through the first two time steps, exhibiting relatively little response to the stimulus. However, after the second time point this set of genes diverges into two distinct groups, one of which is significantly repressed. DREM identifies Fhl1, Rap1, Gat3, Yap5, Pd1, Leu3 and Smp1 as the TFs responsible for down regulated genes and Adr1 as controlling the up regulated genes for this bifurcation point. Several predictions made by DREM were experimentally validated as discussed in the main text.

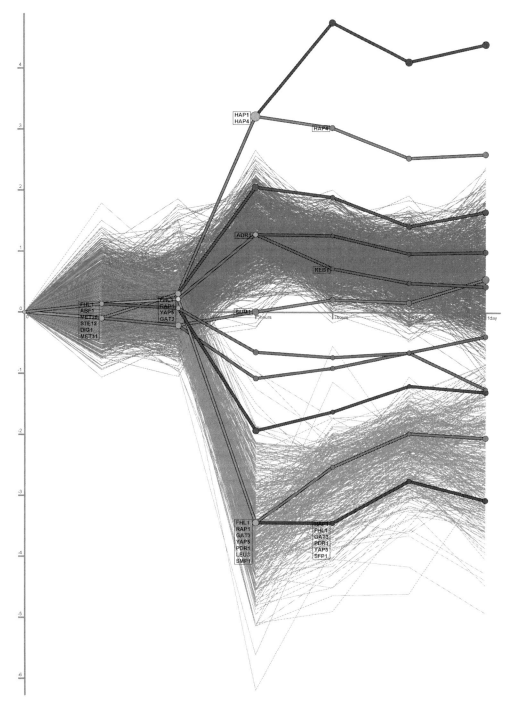

Fig. 24.5. Dynamic regulatory events miner (DREM) example. Here DREM has been applied to stationary phase expression data from a *Saccharomyces cerevisiae* strain in which the gene encoding for the TF Ypl230w had been deleted (77). Because protein–DNA binding data specific to this condition are not available, general binding interactions (5) were used. *Thick lines* show the paths between states in the input–output hidden Markov model and *thin lines* indicate the expression levels of individual genes. A bifurcation event at the second time point has been selected so that only genes

3.3.4. Integrating Additional Interactions Beyond Protein–DNA Binding

As techniques for integrating static sequence and protein–DNA binding data with time series gene expression data continue to become more commonplace, it becomes increasingly interesting to explore how other types of interactions that influence regulatory networks can be incorporated into dynamic models. Protein–protein interaction data are widely available for the most commonly studied species, but it is not obvious how such protein interactions dynamically control gene regulation. One preliminary approach by Vu and Vohradsky (71) applied a neural network-based ordinary differential equation model that combined PPI data with time series gene expression, sequence, and ChIP–chip binding data. PPI were used to model the regulation of a gene by a protein complex in the case where not all members of the complex directly bind the promoter region. However, this method only allowed for complexes of two proteins, and general integration of PPI data remains an open problem.

Post-translational modifications (PTMs) of histones and TFs are another promising source of interaction data. TSAP (72) focused on integrating dynamic histone PTMs to infer regulatory states and model the trajectories of genes between those regulatory states over time, where a regulatory state was defined by a joint gene expression and histone PTM pattern. It extended the Affinity Propagation (73) algorithm to cluster genes at each interval in a time series gene expression experiment such that each cluster is dependent on clusters at other time points. Histone PTMs near genes' start sites were incorporated as additional features in the clustering similarity measure between genes. Using TSAP on mouse data provided new insights into the regulation of Hox genes, which play an important role in motor neuron development.

4. Discussion

High-throughput data about the activity in the cell are rapidly accumulating. However, most of these data are static. To infer the dynamic activation of regulatory programs within the cell, researchers need temporal data. The most abundant source for such data is time series expression experiments, which now account for more than 20% of all expression studies.

When analyzing time series data, researchers need to use tools specifically designed for such data sets. These tools take advantage of the temporal ordering in the data sets and use these to infer groupings and causal relationships. A number of clustering methods have been specifically designed for these data sets

Computational Methods for Analyzing Dynamic Regulatory Networks 437

and many provide an easy to use interface. Other methods use regression-based analysis to infer causality and interaction from time series data. Both types of methods have been applied to time series data sets from a large number of species.

While methods that analyze expression data provide useful information, they are limited by the large dimensionality of the data and the relatively few time points that are sampled in each study. To overcome this, researchers have been developing methods for integrating time series data with other data sources. This helps limit the parameter space and allows for temporal predictions regarding data that were measured in static way. These methods often lead to detailed hypotheses regarding the time of specific interactions, some of which have been experimentally tested and shown to be accurate.

4.1. Evaluation

Of the methods discussed in this chapter, clustering approaches are so far the most widely used. Almost every high-throughput study uses clustering to visualize and organize the results. While general clustering algorithms are still the most popular methods for analyzing time series data, the temporal clustering methods have been gaining ground recently and have been increasingly used when analyzing data from species ranging from bacteria to yeast to mammals.

The use of methods for causality inference in time series data, discussed in **Section 3.2**, has so far been primarily limited to computational reanalysis of expression and other types of biological data and has not been widely used by experimentalists. This may be because of the more complex nature of these methods or because, as discussed above, of the skewed dimensionality problem that leads to many false positives. However, even though it has not been very popular on its own, this types of analysis may prove to be very useful as part of the larger analysis framework, specifically when combined with other types of biological data.

Data integration methods for inferring dynamic networks are increasingly used by experimentalists. Many recent high profile papers (74, 75) use various forms of data integration to improve the inference of dynamic regulatory networks. Unlike clustering, this area is still in its infancy and so it remains to be seen which computational methods would prove the most beneficial. However, this direction has already led to important findings and is likely to remain an important research direction as we discuss below.

4.2. Future Work

While we believe that more temporal data will become available (including temporal data regarding binding of TFs to genes and protein interaction), some data sets are inherently static (most notably DNA). Thus, a major challenge remains to develop methods for combining these temporal data sets with static data sets for reconstructing dynamic modules for regulatory networks. Many methods can already successfully integrate time series expression data with protein–DNA interactions, but the important goal of connecting regulatory, signaling, and metabolic networks would require the use of other types of interactions including protein interactions and protein modification information. Methods for integrating all of these data sources would provide a much needed view for the dynamic activation of signaling and regulatory networks, which lie at the heart of any response system.

References

1. Gasch, A.P., Spellman, P.T., and Kao, C.M. et al. (2000) Genomic expression programs in the response of yeast cells to environmental changes. *Mol Biol Cell* 11, 4241–4257.
2. Nau, G.J., Richmond, J.F.L., Schlesinger, A. et al. (2002) Human macrophage activation programs induced by bacterial pathogens. *Proc Natl Acad Sci USA* 99, 1503–1508.
3. Bar-Joseph, Z., Siegfried, Z., Brandeis, M. et al. (2008) Genome-wide transcriptional analysis of the human cell cycle identifies genes differentially regulated in normal and cancer cells. *Proc Natl Acad Sci USA* 105, 955–960.
4. Xie, X., Lu, J., Kulbokas, E.J. et al. (2005) Systematic discovery of regulatory motifs in human promoters and 3ʹ UTRs by comparison of several mammals, *Nature* 434, 338–345.
5. Harbison, C.T., Gordon, D.B., Lee, T.I. et al. (2004) Transcriptional regulatory code of a eukaryotic genome. *Nature* 431, 99–104.
6. Krogan, N.J., Cagney, G., Yu, H. et al. (2006) Global landscape of protein complexes in the yeast *Saccharomyces cerevisiae*. *Nature* 440, 637–643.
7. Gavin, A., Aloy, P., Grandi, P. et al. (2006) Proteome survey reveals modularity of the yeast cell machinery. *Nature* 440, 631–636.
8. Tan, L.P., Seinen, E., Duns, G. et al. (2009) A high throughput experimental approach to identify miRNA targets in human cells. *Nucleic Acids Res* 2009, gkp715.
9. Bar-Joseph, Z. (2004) Analyzing time series gene expression data. *Bioinformatics* 20, 2493–2503.
10. Eisen, M.B., Spellman, P.T., Brown, P.O. et al. (1998) Cluster analysis and display of genome-wide expression patterns. *Proc Natl Acad Sci USA* 95, 14863–14868.
11. Ernst, J., Nau, G.J., and Bar-Joseph, Z. (2005) Clustering short time series gene expression data. *Bioinformatics* 21, i159–i168.
12. Tavazoie, S., Hughes, J.D., Campbell, M.J. et al. (1999) Systematic determination of genetic network architecture. *Nat Genet* 22, 281–285.
13. Alter, O., Brown, P.O., and Botstein, D. (2000) Singular value decomposition for genome-wide expression data processing and modeling. *Proc Natl Acad Sci USA* 97, 10101–10106.
14. Holter, N.S., Mitra, M., Maritan, A. et al. (2000) Fundamental patterns underlying gene expression profiles: simplicity from complexity. *Proc Natl Acad Sci USA* 97, 8409–8414.
15. Lee, S., and Batzoglou, S. (2003) Application of independent component analysis to microarrays. *Genome Biol* 4, R76.
16. Frigyesi, A., Veerla, S., Lindgren, D. et al. (2006) Independent component analysis reveals new and biologically significant structures in micro array data. *BMC Bioinformatics* 7, 290.
17. Bar-Joseph, Z., Gerber, G., Simon, I. et al. (2003) Comparing the continuous representation of time-series expression profiles to identify differentially expressed genes. *Proc Natl Acad Sci USA* 100, 10146–10151.
18. Magni, P., Ferrazzi, F., Sacchi, L. et al. (2008) TimeClust: a clustering tool for gene expression time series. *Bioinformatics* 24, 430–432.
19. Liu, H., Tarima, S., Borders, A. et al. (2005) Quadratic regression analysis for gene discovery and pattern recognition for non-cyclic short time-course microarray experiments. *BMC Bioinformatics* 6, 106.
20. Wang, L., Ramoni, M., and Sebastiani, P. (2006) Clustering short gene expression profiles. In: *Research in Computational Molecular Biology*. pp. 60–68.
21. Kim, J., and Kim, J.H. (2007) Difference-based clustering of short time-course

microarray data with replicates. *BMC Bioinformatics* 8, 253.

22. Déjean, S., Martin, P.G.P., Baccini, A. et al. (2007) Clustering time-series gene expression data using smoothing spline derivatives. *EURASIP J Bioinform Syst Biol* 2007, 70561.

23. Li, C., Yuan, Y., and Wilson, R. (2008) An unsupervised conditional random fields approach for clustering gene expression time series. *Bioinformatics* 24, 2467–2473.

24. Schliep, A., Schonhuth, A., and Steinhoff, C. (2003) Using hidden Markov models to analyze gene expression time course data. *Bioinformatics* 19, i255–i263.

25. Schliep, A., Steinhoff, C., and Schonhuth, A. (2004) Robust inference of groups in gene expression time-courses using mixtures of HMMs. *Bioinformatics* 20, i283–i289.

26. Ramoni, M.F., Sebastiani, P., and Kohane, I.S. (2002) Cluster analysis of gene expression dynamics. *Proc Natl Acad Sci USA* 99, 9121–9126.

27. Zhou, C., and Wakefield, J. (2006) A Bayesian mixture model for partitioning gene expression data. *Biometrics* 62, 515–525.

28. Ernst, J., and Bar-Joseph, Z. (2006) STEM: a tool for the analysis of short time series gene expression data. *BMC Bioinformatics* 7, 191.

29. Anand, A., Suganthan, P., and Deb, K. (2007) A novel fuzzy and multiobjective evolutionary algorithm based gene assignment for clustering short time series expression data. In: *IEEE Congress on Evolutionary Computation 2007.* pp. 297–304.

30. Workman, C.T., Mak, H.C., McCuine, S. et al. (2006) A systems approach to mapping DNA damage response pathways. *Science* 312, 1054–1059.

31. Yeang, C., Mak, H.C., McCuine, S. et al. (2005) Validation and refinement of gene-regulatory pathways on a network of physical interactions. *Genome Biol* 6, R62.

32. Qian, J., Dolled-Filhart, M., Lin, J. et al. (2001) Beyond synexpression relationships: local clustering of time-shifted and inverted gene expression profiles identifies new, biologically relevant interactions. *J Mol Biol* 314, 1053–1066.

33. Schmitt, W.A., Raab, R.M., and Stephanopoulos, G. (2004) Elucidation of gene interaction networks through time-lagged correlation analysis of transcriptional data. *Genome Res* 14, 1654–1663.

34. Balasubramaniyan, R., Hullermeier, E., Weskamp, N. et al. (2005) Clustering of gene expression data using a local shape-based similarity measure. *Bioinformatics* 21, 1069–1077.

35. Pe'er, D., Regev, A., Elidan, G. et al. (2001) Inferring subnetworks from perturbed expression profiles. *Bioinformatics* 17, S215–S224.

36. Hartemink, A.J. (2005) Reverse engineering gene regulatory networks. *Nat Biotechnol* 23, 554–555.

37. Ong, I.M., Glasner, J.D., and Page, D. (2002) Modelling regulatory pathways in *E. coli* from time series expression profiles. *Bioinformatics* 18, S241–S248.

38. Perrin, B., Ralaivola, L., Mazurie, A. et al. (2003) Gene networks inference using dynamic Bayesian networks. *Bioinformatics* 19, ii138–ii148.

39. Kim, S., Imoto, S., and Miyano, S. (2004) Dynamic Bayesian network and nonparametric regression for nonlinear modeling of gene networks from time series gene expression data. *Biosystems* 75, 57–65.

40. de Hoon, M., Imoto, S., and Miyano, S. (2009) Inferring gene regulatory networks from time-ordered gene expression data using differential equations. In: *Discovery Science.* pp. 283–288. Springer, Berlin/Heidelberg.

41. Shermin, A., and Orgun, M.A. (2009) Using dynamic Bayesian networks to infer gene regulatory networks from expression profiles. In: *Proceedings of the 2009 ACM Symposium on Applied Computing.* pp. 799–803. ACM, Honolulu, Hawaii.

42. Zou, M., and Conzen, S.D. (2005) A new dynamic Bayesian network (DBN) approach for identifying gene regulatory networks from time course microarray data. *Bioinformatics* 21, 71–79.

43. Ahmed, A., and Xing, E.P. (2009) Recovering time-varying networks of dependencies in social and biological studies. *Proc Natl Acad Sci USA* 106, 11878–11883.

44. Spellman, P.T., Sherlock, G., Zhang, M.Q. et al. (1998) Comprehensive identification of cell cycle-regulated genes of the yeast *Saccharomyces cerevisiae* by microarray hybridization. *Mol Biol Cell* 9, 3273–3297.

45. Lee, H.K., Hsu, A.K., Sajdak, J. et al. (2004) Coexpression analysis of human genes across many microarray data sets. *Genome Res* 14, 1085–1094.

46. Aach, J., and Church, G.M. (2001) Aligning gene expression time series with time warping algorithms. *Bioinformatics* 17, 495–508.

47. Bar-Joseph, Z., Gerber, G.K., Gifford, D.K. et al. (2003) Continuous representations of time-series gene expression data. *J Comput Biol* 10, 341–356.

48. Smith, A.A., Vollrath, A., Bradfield, C.A. et al. (2008) Similarity queries for temporal

toxicogenomic expression profiles. *PLoS Comput Biol* 4, e1000116.

49. Lin, T., Kaminski, N., and Bar-Joseph, Z. (2008) Alignment and classification of time series gene expression in clinical studies. *Bioinformatics* 24, i147–i155.

50. Wilczynski, B., and Tiuryn, J. (2007) Reconstruction of mammalian cell cycle regulatory network from microarray data using stochastic logical networks. In: *Computational Methods in Systems Biology*. pp. 121–135.

51. Shi, Y., Mitchell, T., and Bar-Joseph, Z. (2007) Inferring pairwise regulatory relationships from multiple time series datasets. *Bioinformatics* 23, 755–763.

52. The ENCODE Project Consortium. (2007) Identification and analysis of functional elements in 1% of the human genome by the ENCODE pilot project. *Nature* 447, 799–816.

53. Kundaje, A., Middendorf, M., Gao, F. et al. (2005) Combining sequence and time series expression data to learn transcriptional modules. *IEEE ACM Trans Comput Biol Bioinform* 2, 194–202.

54. Ramsey, S.A., Klemm, S.L., Zak, D.E. et al. (2008) Uncovering a macrophage transcriptional program by integrating evidence from motif scanning and expression dynamics. *PLoS Comput Biol* 4, e1000021.

55. Bonneau, R., Reiss, D., Shannon, P. et al. (2006) The Inferelator: an algorithm for learning parsimonious regulatory networks from systems-biology data sets de novo. *Genome Biol* 7, R36.

56. Madar, A., Greenfield, A., Oster, H. et al. (2009) The Inferelator 2.0: a scalable framework for reconstruction of dynamic regulatory network models. In: *Proceedings of the 31st Annual International Conference of the IEEE EMBS*. Minneapolis, MN.

57. Luscombe, N.M., Madan Babu, M., Yu, H. et al. (2004) Genomic analysis of regulatory network dynamics reveals large topological changes. *Nature* 431, 308–312.

58. Chawade, A., Brautigam, M., Lindlof, A. et al. (2007) Putative cold acclimation pathways in Arabidopsis thaliana identified by a combined analysis of mRNA co-expression patterns, promoter motifs and transcription factors. *BMC Genomics* 8, 304.

59. Wu, W., and Li, W. (2008) Systematic identification of yeast cell cycle transcription factors using multiple data sources. *BMC Bioinformatics* 9, 522.

60. Lin, L., Lee, H., Li, W. et al. (2005) Dynamic modeling of cis-regulatory circuits and gene expression prediction via cross-gene identification. *BMC Bioinformatics* 6, 258.

61. Liao, J.C., Boscolo, R., Yang, Y. et al. (2003) Network component analysis: reconstruction of regulatory signals in biological systems. *Proc Natl Acad Sci USA* 100, 15522–15527.

62. Tran, L.M., Brynildsen, M.P., Kao, K.C. et al. (2005) gNCA: a framework for determining transcription factor activity based on transcriptome: identifiability and numerical implementation. *Metab Eng* 7, 128–141.

63. Galbraith, S.J., Tran, L.M., and Liao, J.C. (2006) Transcriptome network component analysis with limited microarray data. *Bioinformatics* 22, 1886–1894.

64. Cokus, S., Rose, S., Haynor, D. et al. (2006) Modelling the network of cell cycle transcription factors in the yeast *Saccharomyces cerevisiae*. *BMC Bioinformatics* 7, 381.

65. Xiao, Y., and Segal, M.R. (2009) Identification of yeast transcriptional regulation networks using multivariate random forests. *PLoS Comput Biol* 5, e1000414.

66. Bernard, A., and Hartemink, A.J. (2005) Informative structure priors: joint learning of dynamic regulatory networks from multiple types of data. *Pac Symp Biocomput* 2005, 459–470.

67. Sanguinetti, G., Lawrence, N.D., and Rattray, M. (2006) Probabilistic inference of transcription factor concentrations and gene-specific regulatory activities. *Bioinformatics* 22, 2775–2781.

68. Shi, Y., Klutstein, M., Simon, I. et al. (2009) A combined expression-interaction model for inferring the temporal activity of transcription factors. *J Comput Biol* 16, 1035–1049.

69. Ernst, J., Vainas, O., Harbison, C.T. et al. (2007) Reconstructing dynamic regulatory maps. *Mol Syst Biol* 3, 74.

70. Ernst, J., Beg, Q.K., Kay, K.A. et al. (2008) A semi-supervised method for predicting transcription factor-gene interactions in *Escherichia coli*. *PLoS Comput Biol* 4, e1000044.

71. Vu, T.T., and Vohradsky, J. (2009) Inference of active transcriptional networks by integration of gene expression kinetics modeling and multisource data. *Genomics* 93, 426–433.

72. Reeder, C.C. (2008) A novel computational method for inferring dynamic genetic regulatory trajectories. Thesis, Massachusetts Institute of Technology.

73. Frey, B.J., and Dueck, D. (2007) Clustering by passing messages between data points. *Science* 315, 972–976.

74. Amit, I., Garber, M., Chevrier, N. et al. (2009) Unbiased reconstruction of a mammalian transcriptional network

mediating pathogen responses. *Science* 326, 257–263.

75. Lu, R., Markowetz, F., Unwin, R.D. et al. (2009) Systems-level dynamic analyses of fate change in murine embryonic stem cells. *Nature* 462, 358–362.

76. Philippar, U., Schratt, G., Dieterich, C. et al. (2004) The SRF target gene Fhl2 antag-onizes RhoA/MAL-dependent activation of SRF. *Mol Cell* 16, 867–880.

77. Segal, E., Shapira, M., Regev, A. et al. (2003) Module networks: identifying regu-latory modules and their condition-specific regulators from gene expression data. *Nat Genet* 34, 166–176.

Subject Index

A

AbstractClassifier 99, 105, 111
AbstractModel 99
Access paradigm 369
Acetylation 298–299
Acetyltransferase 306–307
Activation
    mechanism 242
    -repression mechanism 279
Acyclic graph 17, 402
Adaptive selection 289, 292
Aetiology 314, 335
Affinity 2, 8–10, 25, 34, 46, 48, 50, 180–181, 191,
        195–196, 198, 200–203, 243, 271, 277, 285,
        287, 292, 372, 436
Affymetrix 412
AGAMOUS 217, 361
Agar 134–135
Agarose 253–254
Agglomerative clustering 425
AGRIS database 29, 319, 352
AhdI 270, 277
Algorithm evaluation 122–123, 127
AlignACE 12, 131–133, 135, 183
Aligning time series datasets 422
Alignment-free prediction 11
Allele-specific protein-binding 145
Allosteric mechanism 51
Alpha 243, 246–247, 270, 376
Alta-Cyclic base caller 5
Alternative promoters 14, 226
Alzheimer's disease 247
AmnSINE1, 228
Amphioxus 342
Amplification 5–6, 168, 196–197, 208, 253–255,
       263
Amplification bias 5
Androgen receptors 100, 244
Androgen-responsive 245
Annotation 15–16, 65–66, 68, 74, 125–126,
        128, 133, 136, 145, 155, 221, 256, 262, 280,
        300, 304, 315–339, 353, 355, 364–365, 367,
        370–372, 375, 378, 380–385, 387–388, 391,
        393, 412
Annotation View 385, 387–388
ANN-Spec 132–133, 135
Anthocyanin 367
Anthocyanin pathway 367
Antibody 4, 6, 8, 143, 162–164, 171–172, 180,
        196, 220, 330
Anti-GST antibodies 9
Antitoxin 270

Apache 356, 372, 384
Apache Derby 372, 384
Apoptosis 377
Applied Biosystems (Life Technologies) 5, 164
Arabidopsis gene regulatory information server
        (AGRIS) 29, 319, 352
Arabidopsis Information Resource (TAIR) 352,
        354–355
Arabidopsis thaliana 216, 354–355, 363–364
Arabidopsis transcription factor database
        (RARTF) 352–353
Archiving annotations 318–323
Area under curve (AUC) 110, 112–113, 130
Argonaute 319
Artificial neural network 60, 75, 436
Athena Database 59
Atherosclerosis 245
AtTFDB 352–353
Autoimmune disease 2
Autoinhibition 52
Autoregressive model 425
Auto regulatory 280
Autoregulatory binding site 280
Auxin 359, 362, 367
Auxin response 359, 367
Average binding affinity 198, 201
Awk 234

B

Bacillus subtilis 25, 124
Background correction 5–7, 165
Background model 92, 104–105, 107, 146–153, 189
Background noise 3, 6–7, 146–150, 152–153, 164–165,
       237, 258
Bacterial promoters 73–74
Bacteriophage 34, 270
Baker's yeast 36
Bandwidth selection 168
Barrier insulator 39
Basal transcription . . .26, 34, 36, 57–58, 373–374, 377–378
Base calling 5
Base interaction 46
Bayesian model 307
Bayesian network 16–17, 104, 115–116, 401–417, 422,
       428–430, 433
Bayesian Networks toolbox for MATLAB 422
Bayesian statistics 115
Bayesian tree 115
Bed files (for Genome Browser) 234
Benchmark 13, 127, 129, 132, 134, 136, 157, 163,
       300–301
Beta-globin 38, 262

I. Ladunga (ed.), *Computational Biology of Transcription Factor Binding*, Methods in Molecular Biology 674,
DOI 10.1007/978-1-60761-854-6, © Springer Science+Business Media, LLC 2010

444 COMPUTATIONAL BIOLOGY OF TRANSCRIPTION FACTOR BINDING
Subject Index

Beta-globin cluster 262
Beta-scaffold factor 6
BglII 254, 272
Bias . . . 3, 5–6, 11–13, 91, 93, 110, 116, 136, 145, 149, 157,
      196, 292, 294, 298
Bicluster 432
Bicoid 284, 288
Bidentate interactions 45
Bifurcation 434–435
Bimodality 303
Binary classification 70, 294
Binding dissociation constant 201
Binding energy 94, 199, 201, 204–207
Binding probability 92, 94, 201–202, 204, 206, 209
Binding site 1–17, 85–91, 97–118, 121–140, 151,
      154–155, 161–175, 181–184, 190, 199–201,
      204–206, 214–215, 225–238, 241–248,
      272–279, 285–294, 315–316, 319, 329,
      338–339
Binomial distribution 6, 76, 146–150, 152, 165
BioJavaAdapter 101
BioPERL 16, 137
BioProspector 12
BioPython 137
BioTapestry 370, 372, 388–389
Birth-death MCMC 425
BLAST 5, 136, 138, 230, 257, 273, 280, 320, 329,
      338–339, 352, 356, 358, 364–365, 367, 387
Blastoderm 287
BLAT 5, 230, 328, 338–339
Blender 372, 394
Blood disease 38
Bonferroni correction 188
Bowtie 5, 15, 151–152, 167, 175, 230–231, 233–234,
      236–237
B-splines 16, 424
Burroughs-Wheeler transformation 231
BZIP family 49, 365, 367

C

Caenorhabditis elegans 2
CAMP-response 298
CAMP-Response Element Binding Protein (CREB) . 164,
      298–299
Cancer 2, 315
Canonical binding site 243, 336
CAP binding site 34
5CArrayBuilder 255–258, 262
Causal relationships 16, 402–403, 405, 408, 428, 436
C-box 272–279
C/C++ 123
CC (carbon copy) 253–255
CCCTC-binding factor (CTCF) 15, 40, 233–236,
      238, 321
3C-Carbon Copy (5C) 254–255
5C3D 255–256, 259–262, 264–265
CD95 326–331
Celera 371, 380–381, 383, 385, 393
Celera Genome Browser 371, 380–381, 383, 385, 393
Cell cycle 319, 360, 376–377, 430–433
Cell migration 377
Charles Lawrence 12, 87
Chemical potential (μ) 201–202,
      204–206, 286–287
Chimpanzee evolution 314

ChIP-chip 4, 6, 10, 14–15, 17, 28, 86, 113, 123,
      144–145, 157, 162–164, 180, 182–184, 186,
      188, 190, 229–230, 246–248, 294, 307, 321,
      352, 432–434, 436
Chip design 4
ChIP enrichment 155, 188
ChIP environment 170–171
ChIPOTle 183
ChIP sample 143, 145–146, 148–149, 151, 154–157
ChIP-seq 3–7, 10, 15, 17, 28, 86, 98, 109, 113, 123,
      143–158, 161–174, 180, 183–190, 225–238,
      294, 316, 320–321, 335–336
ChIP-seq, resolution 4–5, 10, 145,
      157, 162, 164–165, 172, 184, 188, 231–232
Chi-square distribution 189
Cholesterol 241, 245–247
Cholesterol biosynthesis 247
Chordate 342
Chromatin architecture 252, 262
Chromatin conformation signatures (CCSs) 262
Chromatin free DNA 172
Chromatin immunoprecipitation (ChIP) 3–8, 28,
      86, 110, 143–144, 162–164, 179–192, 214–215,
      218, 220, 229, 246, 352
Chromatin-mediated silencing 39
Chromatin modification 33, 144
Chromatin remodeling 6, 51, 260, 359
Chromatin remodeling enzyme 6
Chromosomal map 151
Chromosome conformation 38, 251–266
Chromosome Conformation Capture (3C) 38, 253
Chromosome rearrangement 38
Cis-Browser 369–371, 373, 380–388, 393–394
Cis-evolution 292
CisGenome 7, 16, 151–155, 165, 183, 232
CisGRN-browser 16, 369–394
CisGRN-Lexicon 369–394
Cis-Lexicon 15, 369–380, 383–385, 387–388, 391–392
Cis-Lexicon search engine 370, 391
Cis-regulatory element 24, 26, 39, 297, 352, 356,
      364, 368, 370, 374
Cis-regulatory module (CRM) 14, 17, 162,
      214–216, 219, 283–285, 287–288, 291–293,
      319, 370–372, 374–375, 380–381, 385,
      387–389, 390–391
Cis-regulatory module evolution 292
Classifier 98–100, 103, 105, 107–113, 116–117,
      182, 187, 189, 216, 218–221, 306, 434
CLAVATA3 217
CLOSE Project (*cis*-Lexicon Search Engine) 391
ClustalW 356
Clustering
    time series expression data 423–426
    using continuous representation 424
    using Hidden Markov models 424–425
Clustering with Local shApe-based similaRITY
      (CLARITY) 428
ClusterTree-RS procedure 275
C-Myb oncogene 46, 51
C-Myc oncogene 51
Co-Bind algorithm (Cooperative BINDing) 75
Co-bound genes 133
COCAS 183
Co-expressed genes 14, 214, 302, 304
Co-expression 3, 10, 16–17, 302–305, 404
Co-factor interactions 298

Subject Index

Co-localization analysis 14
Colorspace (SOLiD, Life Technologies)..............152
Combinatorial regulation 34, 75
Combining strand densities 169
Comparative sequence analysis 228
Compartmentalization.............................242
Compartmentalization signal.......................242
COMPELL database of composite regulatory
    elements 75
Compendia of gene expression experiments 3, 10, 16
Competitor assay..................................203
Complete information 89
Composite elements (CE)............................60
Computational pipeline............................162
Conditional distribution........................17, 402
Conditional entropy................................406
Conditional mutual information 406, 408
Conditional probability 402–403, 406–407, 417, 428
Confidence intervals 76–77, 259–261
Conformation capture..........................38, 253
Confounding models 404, 408
Confusion matrix 108–110, 117
Connectivity matrix 433
Connectivity networks..............................433
CONREAL...293
Consensus binder............................196, 201
Consensus sequence 9–10, 36–37, 39, 75, 87,
    101–102, 114, 180, 191, 199, 245, 272,
    387–388
Consensus Sequence View 387–388
Conserved blocks 65
Conserved non-coding elements (CNEs) 228, 321
Controlled ontology................................323
Controlled vocabulary 323, 372, 375, 389
Control library 7, 164–165, 169–170, 174
Cooperative binding 51, 75, 248, 291
Cooperative recognition 51
Coordinative component 407–408, 414–415
Copy number 173, 226
Core evaluation statistics 125–128
Co-regulation 216, 368, 411
Core promoter 26, 34–38, 57–58, 217, 319
CORE-SINE family of transposable elements 228
CORG Database....................................59
Corona Lite 151–152, 167
Corpus...316–317, 323
Correlation networks 402
Correlative component................407–408, 410, 414
Correlative metric.................................402
Covariance matrix 62
Coverage 5, 16, 145, 164, 258, 263–264, 317, 366
5CPrimer.....................253, 255–258, 262–264
CpG deserts......................................118
CpG islands...........................36, 58, 118, 190
C-protein 2, 269–281
CRACR algorithm 9
Cross-hybridization 4, 145, 164, 229, 257
Cross-link 4, 6, 86, 143, 163–164, 170–172,
    180, 253–254
Cross-linked chromatin 86, 143, 180, 253
Cross-species analysis 133
Cross-species comparisons 138, 284, 314
Cross-validation ... 110–112, 117, 129, 187–188, 219, 246
Crystal structure 45
CTCF 15, 40, 233–236, 321
Cumulative density function (CDF) 412

Curatability 316
Curation..........273, 308, 316–321, 323, 327, 334, 336,
    340, 357, 387
Custom microarrays........................253, 255, 258
Custom track, Genome Browser....16, 171, 175, 233–234
'Cut and paste' mechanism.........................226
Cutter enzyme....................................256
CWINNOWER.....................................11
Cyanobacterium 428
Cyclic AMP.......................................34
CyIR2 271
Cyrene 370–372, 380, 385–389, 395
CYRENE Project...................................395

D

Danio rerio...............................343–344, 346
Database of composite elements (COMPEL)........60, 75
Database of tunicate gene regulation 319
Davidson Criterion................................369
DbSNP database...................................320
DBTSS database...................................59
Degenerate binding 213
Degradation 270, 298, 308, 377
Denoising 173
De novo motif discovery 10–13, 98–99, 155, 158,
    183–187
Density bandwidth 167
Density distribution 3, 165–166
Density estimate 7, 261, 409
Density function 147, 152, 167, 409
Density profile....................................157
Density threshold 169
Dependency/correlation analysis 412
Depression 315
Derived evaluation statistics 128–129
Diabetes 242, 245
Diagram problem 378
Differential expression 302
Dimeric transcription factor 10
Dimerization domain.............................. 50
Directed acyclic graph (DAG)...............17, 402, 408
Directed causal networks..........................402
Direct interaction.............................303, 306
Direct readout 44
Dirichlet prior 104, 116
Discrete probability 414
Discretizing data.................................409
Discriminant analysis 61–62, 74
Distance function 122
Distance matrix 259
Distance measure 214
Distant enhancer 38–39
DNA
    damage 431
    duplex destabilization 75
    helicase 330
    ligase 253, 255, 258
    looping 35, 51, 373–374, 378
    methylation............................3, 16, 30, 161
    motif..........................12–13, 34, 38
    packaging 34, 252
    polymerase 208
    Sample Class 99, 101
From DNA to diversity..........................284
DNase 3, 14–15, 318

COMPUTATIONAL BIOLOGY OF TRANSCRIPTION FACTOR BINDING
446 Subject Index

DNase hypersensitive regions 15
DNase hypersensitivity regions 14
Domain 27–28, 45–50, 186, 205, 242, 263, 270,
        280, 352, 355, 357, 359, 362–363, 365,
        375–376, 394, 402
DoOP Database 59
Dorsal-ventral patterning 124
Double palindromes 277
DPE element 35–37, 58
DPInteract database 74
DPTF project 323, 354
Driver TFBS 374
Drosophila
    melanogaster . . . 284, 287–290, 293–294, 314, 377, 380
    pseudoobscura 293–294
    yakuba 288
Drug discovery 241
Duodenal homeobox 50
Dynamic activation 436, 438
Dynamic Bayesian networks (DBNs) 16, 422,
    428–430, 433
Dynamic model 421, 436
Dynamic network 16, 252, 430, 437
Dynamic programming 431
Dynamic programming algorithms 431
Dynamic range 145
Dynamic Regulatory Events Miner (DREM) 422,
    434–435

E

EcoRI 254
EcoRV 271–272, 275
Ectopic 373
EEL 293–294
E2F factor 28, 360
EGFP reporter 342
Eland 151–152, 167, 175
Electrophoretic mobility shift assays (EMSA) 28, 326,
    330–331
Embryo 16, 124, 360, 371–372, 388–389, 394
EMMA 293–294
Empirical p-value 188
ENCODE project 233, 314, 353
Encyclopedia of DNA Elements (ENCODE) . . . 4, 27, 233
Endoderm 377
Endogenous retroviral elements 235
Endonuclease 269–271
Energy
    matrix 204–207
    model 94
    score 288
Enhanceosome 51
Enhancer
    blocking 39–40, 233
    'boosting', 228
    -promoter interactions 39
Enrichment calculation 186
Enrichment score 9, 434
Ensembl
    algorithm 13
    browser 16
    database 338, 343–344
Entrez 320, 327–328, 337, 343, 346, 381–382, 394
Entropy 45, 404–406, 409
Epigenetic control 40

Epigenomics 15
Epistatic interactions 288
Epitope structure 6, 171
Equilibrium distribution of nucleotides 286–287
ERANGE 165
Escherichia coli 8, 13, 25, 34, 48–49, 73,
    124–125, 131–132, 139, 180, 200, 317, 319,
    428, 434
*Esp*1396I 271
ESR1 321
Estrogen receptor binding sites 164, 242
Euclidean distance 191, 259
Eukaryotic Promoter Database (EPD) 59, 246
E-value 280–281, 365–366
Even-skipped 292
Even-skipped gene 292
Everted repeat (ER) 243, 245
EVOC cell-type ontology 329, 340, 345
Evolutionary algorithm 426
Evolutionary constraint 291, 314
Evolutionary models of transcription factor binding
    sites 285–289
Evolutionary signatures 284
Evolutionary simulation 288
Exaptation events 226–227
Exaptation of transposable elements 226
Exclusive or (XOR) relationships 16
Exhaustively search 304, 413
Exhaustive search 408, 413–415, 417
Exon 15, 35, 38–39, 58, 60–61, 78, 118, 154, 162,
    167, 217, 220, 380–381
Expectation maximization . . 11, 85–94, 162, 172, 186, 204,
    425, 432–433
Experimental design 202, 220, 255–256, 258, 263, 421
Experimental perturbations 370
Experimental protocol 196, 323
Experimental validation 245, 294, 300–301, 318
Experimental verification 277
Exponential enrichment 3, 8–9, 15, 28, 195, 198, 208
Expressed sequence tag (EST) 27, 59, 67, 78, 353, 356,
    364–366
Extended *k*-mer 11
External database 320, 332
Extracellular matrix 377
Extracellular space 377
Extreme diversity 269
Eye morphogenesis 377

F

False discovery rate (FDR) 6–8, 145, 153–154,
    156–157, 161–174, 188–189
False negatives 2–3, 7, 73, 109, 125, 146, 189, 205,
    232, 367
False positives 3, 6–7, 10, 60, 63, 72–74, 109, 117, 123,
    125, 128, 130, 133, 146, 151, 164, 169–172,
    188–189, 200, 205, 229, 232, 237, 265, 277,
    280, 284, 294, 306, 364, 367, 427, 437
Fasta-files 101, 139, 175, 218, 233
Fasta format 79
FASTQ file 233, 237
Feed-forward loop 428
Fermi-Dirac binding energies 201, 204
Fermi-Dirac binding probability 201, 204, 209
Fgenesh 60, 78
FindPeaks 7, 157, 165

Finite temperature corrections 209
First exon 35, 58, 380
Fitness 286–287, 292
Fitness function 292
Fitting motif models to ChIP data 183
Fixed stringency 201–207
Fluid cells 5
Fluorescence signal 229
FlyTF project 323
F81 model 285–286
Fold enrichment 150, 153–154
Formaldehyde 163, 171–172, 253
FOXA2 . . . . 321, 377
Fprom promoter prediction program 61
Framefinder program 356
Functional polymorphism 340
Functional validation 340
Fuzzy membership 426

G

GABP 171
Galaxy analysis workbench for genomic data 218
Gaussian kernel density estimation 414–415
Gaussian models 402
GBrowse 16, 140
Gel shift assay experiments 243–244
Gene Expression Omnibus (GEO) 302, 355, 418
Gene Ontology (GO) 155, 355–356, 365, 375
Generalized two-way mutual information 405
Generative learning principles 102–105
Gene Regulation Ontology 317
Gene regulatory network (GRN) 16, 28–29, 140,
            369–394, 401–402, 408–411, 415–417
Genetic manipulations 307
Genetic medicine, viii
Genetic network 356
Genetic variation 315
Genome Browser 15–16, 155, 167, 171–172, 230,
            233–234, 238, 320, 322, 329, 371, 380–381,
            383, 385, 393
Genomic windows 148–149, 156
Genotypic variations 287
Genscan 60, 78
Gibbs Motif Sampler 209
Gibbs sampler 216, 219
Gibbs' sampling 12, 85–94, 162, 214, 309
Gibbs tool 14
Glam 131–135
GLAM2 134
GLITR 165
Global regulatory network 17
Globin imbalance 39
Glucocorticoid receptor (GR) 28, 60, 100–103, 108,
            110, 112–114, 242, 244–245
Glucose depletion 34
Glucose metabolism 245
Glut4 gene 68
Glycine max 353–354, 366
Glycosylation 298–299
GNF Atlas II 14
G-protein binding protein CRFG (GTPBP4) 171
Graph model 432
Graph representation 415
Graph statistics 432–433
Graph theory 140, 428

Green alga 354–355, 365–366
GRN models 401, 404
Growth-associated binding protein (GABP-alpha) 163
Gypsy element 39–40

H

Haar wavelet 173
Hairpin 49
Hairy-wing 39
Halpern-Bruno model (HB) 286–289
Havana annotation 65
Helicos 5, 229
Helix-loop-helix 6, 48
Helix-turn-helix (HTH) 46, 48–50, 270, 280, 375
Hemoglobin 67, 360
Heritability 313
Hermaphrodite 2
Herpes simplex 27
Heterochromatin 39
Heterodimer formation 50
Heuristic motif 13
Heuristic procedure 425
Hidden Markov models (HMMs) 16, 27, 75, 78,
            245–248, 275, 281, 356–358, 364, 367,
            424–425, 431, 434–435
Hidden partition 304
Hierarchical clustering 422–424
High-affinity binders 8
High order mutual information 403–404, 416
HindIII 254
Hinge region 242
Histone
    methylation 335
    modification 3, 16, 30, 161, 221, 252, 302–303
Hmmcalibrate 281
HMMER search 357, 364
Holdout sampling 110–113, 117
Homeobox 50
Homeobox-binding domain 91
Homeodomain 10, 50, 91, 362–363, 375–376
Homeostasis 241, 313
Homolog 361
Housekeeping genes 25, 58, 375–377
Hox genes 436
HTH motif 46, 48
HTPSELEX Database 9, 15, 203
Human-chimpanzee divergence 226
Human-mouse divergence 227
Hydrogen bond 44–45, 198
Hypergeometric distribution 186, 434
Hypergeometric test 291

I

IFCalculator 255, 258–260, 262, 264
Illumina 5, 151, 164, 166–167, 229–231, 237
Immunoglobulin 47, 50
Immunoprecipitation 3–8, 28, 86, 110,
            143–144, 162–164, 175, 179–192, 214–215,
            218, 220, 229, 246, 352
INCLUSive MotifSampler 215
Independent component analysis 423
Indirect readout 44
Inferelator 422, 432
Inferring causality 419
Inferring causality from temporal datasets 421

COMPUTATIONAL BIOLOGY OF TRANSCRIPTION FACTOR BINDING
448 Subject Index

Inferring interactions 428
Information content 11, 87, 93, 134, 166, 168, 284
Information measures 403, 405
Information theory 204–205, 207, 402, 404
Information transmission 315
Inhomogeneous Markov models . . . 102, 104, 107, 114, 116
Initiator element 35–37
Input-output hidden Markov model (IOHMM) . . 434–435
Insulator 33–40, 233, 374
Insulator element 39
Integrative informatics 241
Integrative Modeling of MFG Triplets and
    TF-PTMs 305–307
Interferon 46, 51, 315
Intergenic regions 9, 154, 217
Intermolecular ligation of cross-linked DNA
    fragments 253–254
InterProScan domains 364–365
Interspersed repeat 228
Intragenic regions 154
Intron 27, 38–39, 61, 69, 72, 154, 217, 220, 247, 331,
    388
In vitro affinity 8–9
In vitro binding 219
In vitro selection 195
In vitro substrates 307
In vivo binding 8–9, 180
In vivo transcription 188, 315
Ion binding 377
Ionic interactions 45–46
IUPAC ambiguous nucleotide code 75, 243

J

JAK-STAT signaling 305
JASPAR database 190, 243–244
Java 12, 98–99, 101, 123, 137–138, 256, 323,
    371, 381, 384, 393–394, 423
JBD 183
Jstacs 12, 98–101, 104–108, 111–113

K

Kalman filter 433
Kernel-based identification of regulatory modules in
    eukaryotic sequences (KIRMES) 14,
    213–214, 216–219, 221–222
Kernel density . . . 7, 157, 165, 167–169, 404, 409, 414–415
Kernel density estimation 157, 165, 167–168, 404, 409,
    414–415
Kernel methods 408
Kernel smoother 173
Kinetic equations 432
K-means clustering 423
K-mer 8–9, 11, 122, 215
Knockdown experiment 220, 247
Knockout experiment 9
Knockout mutant 17
Kruppel sites 291
Kuhn-Tucker condition 205
Kullback-Leibler (KL) divergence 191

L

Lac operon 25–26, 34
Lagrangian multiplier 71
Lambda phage 34

Lambda repressor 34
Lanosterol 247
Learning-based ontology 323
Learning classifiers 306
Learning machines 69
LegumeTFDB 353
Leucine zipper 6, 48–49, 359
Leukemia 260
Library 7–9, 137, 156, 165–166, 168–170,
    195–199, 207, 245–246, 253–258, 262–264,
    266, 356, 370
LiftOver tool 338
Ligation 253–255, 258
Ligation-mediated amplification (LMA) 254–255
Light-stimulated 428
Likelihood ratio classifier 100
Limma package 155–156
Lineage-specific TFBS 228
Linear combination 61
Linear correlation 401–402, 405
Linear discriminant function (LDF) 61–63, 74
Linear network 16
Linker TFBS 374
LINUX 5, 74, 136–138, 233, 356
Lipid raft 247
Liquid association 303
Literature curation 319
Literature extraction 370, 392
Literature search 320, 324
Literature survey 357, 364
Liver carcinoma 328
Liver Specific Gene Promoter Database (LSPD) 319
Liver-specific promoters 319
Liver X receptor 242, 245–246
Local two-parent regulatory model learning 413–415
and Logic 374
Logic functions 370
Logistic regression 182, 434
Log-odds matrix motif models 181
Long-range electrostatic contacts 50
Lotus japonicus 353–354, 366
Low-affinity regulatory site 2
Luciferase 331, 345
Luciferase assay 331, 345
Lupus erythematosus 315
LXRE.HMM algorithm 245–246, 248
Lymphoblastoma 171

M

Machine-learning methods 75
Macromolecular crystallography 45
Macrophages 260
MACS 6–7, 157, 165, 183
Mahalonobis distance (D2) 62–63
Major groove 44, 46–50
Mapped reads 146, 148, 152–153
Markov chain 215, 285–286
Mass spectrometry 306
Master regulator 434
Mat-2 homeodomain 10
Match score 188
MatInspector 75, 243–245
MatInspector Database 243
MATLAB 205, 422
MATRIX search 75

Matthews correlation 127–128
Maximum conditional likelihood (MCL)
    principle 105–106, 108
Maximum likelihood (ML) principle 102–103,
    105–106, 113
Maximum a posteriori (MAP)
    principle 103–104, 106, 113, 115–116
Maximum supervised posterior
    (MSP) 106–108, 113, 116
McPromoter promoter prediction program 65
MDScan 183
Medicago truncatula 353–354
Melting curve 253–254
Melting point (Tm) 257
Metabolic diseases 245–246
Metabolic networks 438
Methyltransferase 270, 306
MI3 algorithm 404, 407–408
Micrococcal nuclease (MNase) 4
Microcosm program 256, 260, 264
MicroRNA (miRNA) 2–3, 16, 28, 30, 78, 161, 319,
    376–377, 419–420, 432
Mimosa: A Mixture Model of Co-Expression Data for
    Detecting TF Modulators 304–305
Minor groove 6, 44–50
MIPS 352
MIR elements 228
Mismatch Tree Algorithm (MITRA) 11, 14
MLAGANS 136
Model-based Analysis of ChIP-Seq (MACS) 6–7, 157,
    165, 183
ModENCODE project 314
Modification enzyme 302–303
Modular organization 60
Modulator gene 299, 301–304
Modulators of transcription factor activity 297–309
Molecular clock 289
Molecular switchboard 298
MolQuest package 74
Molscript 356
MONKEY 236, 294
Monte-Carlo simulation 7, 165
Morphs 293
Mother function 173
Mother wavelet 173
Motif
    alignment 10
    -based classifier 187
    databases 136, 139, 184
    discovery 4, 8–14, 17, 85–94, 98–99, 127, 133–134,
    136, 155, 158, 183–187, 190
    finder 2, 215
    -finding tools 172
Motif elicitation by maximizing expectation
    (MEME) 11, 13, 91, 131–135, 172, 183
MotifSampler 132–133, 135, 209,
    215
Motif Tool Assessment Platform (MTAP) ... 13, 121–140,
    183
Motiftool pipeline 139
MS/MS experiments 306
MSP classifier 107–108
Multiple alignment 11, 27, 87, 280, 293
Multiple annotation 316
Multivariate Gaussian distribution 409
Multivariate Random Forests 433

Multiz alignment 235
MuMRescueLite 15, 230, 232, 234
MUSCLE 65, 274–275, 318, 321
Mutation selection cycle 292
Mutual information 17, 302–303, 401–417
Mutual information networks 402
MYB family of transcription factors 27
Myeloblastocys 27
Myelomonocytes 259
MySQL 16, 322, 356, 368

N

NCBI Entrez Programming Utilities 381, 394
Nearest-neighbor algorithm 257
Negative binomial 6–7, 146–149, 152, 165
Negative binomial distribution 6, 146–148,
    152, 165
NetPlot 415–416
Network
    analysis 352
    assembly 408, 410, 415–416
    building 388
    reconstruction 4, 307, 416
    simulator 372
    statistics 432
    structure 412
    toolbox 422
    topology 125
    visualization 410, 415–416
Network Component Analysis (NCA) 422, 433
Network Component Analysis toolbox for
    MATLAB 422
Neutral selection 226
Neutral sequences 285
Neutral theory 289
Neutral theory of evolution 289
Newton-Raphson method 309
Next-generation sequencing 3–9, 15, 162, 164, 366
NFIRegulomeDB 321
NHR-scan model 245
Nickel-induced transcriptional repressor 46
NMR 9, 51, 242
Non-coding DNA 225, 236, 336–337
Non-coding polymorphisms 315
Non-coding RNA 29, 228, 315
Non-DNA-binding factor 51
Nonlinear regulatory relationships 16, 402
Non palindromic motif 279
Nonparametric Gaussian distribution 408
Nonparametric probability 408–409
Nonparametric probability density estimation 408–409
Nonparametric statistics 167, 408–409
Non-redundant motifs 183, 191
Non-specific binders 198, 200–201, 203, 209
Nonstoichiometric quantities 38
Normalization 150–151, 165, 254, 412
N-scan 65
NsiteH program 77–79
NUBIScan 244–245
Nuclear compartmentalization signal 242
Nuclear factor 242
Nuclear hormone 243–245
Nuclear lamina 39–40
Nuclear receptor 241–248, 375–376
Nuclear receptor binding site prediction 241–248

COMPUTATIONAL BIOLOGY OF TRANSCRIPTION FACTOR BINDING
450 Subject Index

Nucleosome
-depleted regions 3, 5–6
depletion 6
-rich regions 6
Null distribution 188
Numerical optimization 107–108

O

Obesity 315
Object-oriented database 368
Object-oriented design 98
Object-oriented Java framework 12, 98
OgreMax 372
Oligomer counting 215, 219
Oligonucleotide arrays 338
Oligonucleotide competitions 338
Oligonucleotide library 195, 207
Oligonucleotide preferences 62
OMGProm Database 59
One-dimensional browsers 16
Ontology analysis 155
Ontology annotations 365
Open-access 319, 323
Open Reading Frame (ORF) 74, 78, 300, 356
Operon 25–26, 34, 74, 78
Optimal selection 168
Optimization problem 71
Optimization toolbox 205
ORegAnno Database 321–323, 326
ORFeome 353
OR logic 374
Orphan receptors 242–243
Ortholog 138, 228, 354, 365–367, 389–390
Orthologous promoters 13, 59, 98, 113
Oryza japonica 353–354
Oryza sativa 353–354, 363, 366
Osiris Database 59
Osteoporosis 242
Overfitting 103–104, 106, 115, 187, 424, 427
Overrepresented oligomers 214
Over-selection 8, 200–201
Oxysterols 246–247

P

P50 50–51
P53 . . . . 2, 14, 47, 165, 326, 328–331, 343, 375–376
P65 50–51
Pairwise alignment 293
Pairwise entropy 274
Palindrome 10, 47, 272–278
binding 244
consensus 277
DNA 47, 50
inverted repeat 243
motif 277, 279
Panther classification 375
Paralog 390
Pathway . . 26, 124, 155, 241, 243, 245–248, 251–252, 298,
        300–302, 305, 315, 360, 362, 367, 372, 377, 430
Pathway databases 300
Pattern analysis 162
Pattern extraction 174
Pattern Extraction and Regular Expression Language
        (PERL) 15, 174

Pattern matching 243, 245
Pattern recognition 8, 421
Pax genes 380
PAZAR Database 15, 29, 316–319
PCR amplification 208, 254–255, 263
Peak calling 6–7, 16, 161–175, 232, 234
Peak calling windows 168
Peak characteristics 15
Peak detection 156
PeakFind 237
Peak saturation 171
Peak shape 150–151
Peak shift 168–169
Performance characteristics 128
Performance evaluation 11, 128
Performance graph 129–130, 134
Perinatal HIV-1 transmission 315
Perl DBI uploader 320
Peroxisome proliferator 185, 187, 242
Peroxisome proliferator-activated 242
Peroxisome Proliferator-Activated Receptor Gamma
        (PPARG) 2, 242–244
Peroxisome proliferator response element 185, 187
Perturbation experiment 4, 371, 426–427
PFAM Database 27, 356–358, 364
Pharmacology, viii
Phenol-chloroform DNA extraction 254
Phosphorylation 298–299, 306, 373
Photomicrograph 388
Photomorphogenesis 365
PHP 356
PHYLIP package 274, 356
Phylocon 14
Phylogenetic footprinting 13–14, 98, 228, 236, 280
Phylogenetic screen 228
Phylogenetic shadowing 98
Phylogenetic tree 138, 274, 280–281, 285, 289, 292
PhyloGibbs 14, 133–135
PhyloGibbs-MP 133–134
PhyloMEME 133–135
PhyloP 236
Phylo-Weeder 133, 135
PhyME 14, 133–135
Physical connectivity networks 251
Piecewise polynomials 424
Pilot-ENCODE regions 164
PLACE Database 59
PlantCARE Database 59
PlantGDB 354–356
PlantProm Database 59, 72
Plant promoter identification 69–73
PlantTFDB 29, 59, 351–368
Plant transcription factors 351–368
Poisson distribution 6, 146–147, 149, 165
Poly-cistronic RNAs 26
Polyclonal antibodies 172
Polycomb 164, 362
Polycomb group 164, 362
Polydactyly 39
Polymerase detachment 7
Polymorphisms 15, 315, 320–321, 326, 336–337,
        340, 343, 345
Population genetics 286–287
Population genetics theory 286
Populus trichocarpa, 354, 363, 366

Subject Index

Positional oligomer importance matrices (POIMs) 117, 219, 221–222
Positional scoring matrix 122, 229
Positional weight matrix 12
Position-Specific Scoring Matrix (PSSM) . . . 122, 229, 238
Positive amplification 6
Posterior probability 62, 100, 181
Post-transcriptional modification model
    (PTMM) 422, 434
Post-transcriptional regulation 30
Post-translational modification 33, 298, 436
Power-law 322
PPARG 2, 242–244
Preaxial polydactyly 39
Prediction
    accuracy 246
    performance 122–123
Pre-initiation complex (PIC) 26–27, 35–36, 38–39
P53-responsive 331
Primer
    design 263–264
    pair 253–254
Principal component analysis 423
Prior
    density 103, 116
    information 91, 93
    probability 182
PRIORITY 215
Probabilistic
    assignments 15
    base 5
    graph 402, 432–433
    graph model 432
    model 11, 75, 87, 284–285, 402, 404, 417
    peak calling 161–175
Probability
    of fixation 286
    score 402
ProbCons 293–294
Prodoric Database 125
Profile 27–28, 133, 157, 175, 180, 243–245,
    275, 278, 281, 293, 301, 303–307, 315, 319,
    356–357, 364–365, 367, 401, 423, 425–426,
    432–434, 437
Profile-based method 244
Progesterone receptors 100
Programming API 319
Programming language 15, 256–257
Proliferation 260, 359–362
Proliferin 60
PromH promoter prediction program 61
Promiscuous binding 220
PROML procedure 274
PromoSer Database 59
Promoter
    bashing 329, 344
    context 298
    -distal site 277
    microarray 4, 164
    model 73
    prediction 60–61, 64–66, 78–79
    region . . 2, 4, 10–11, 13, 25–26, 34–35, 38, 57–79, 154,
    162, 164, 171, 190, 220, 246, 248, 302, 352,
    432–433, 436
Promoterh 66–67
Property Inspector View 385, 387–388

Protein
    -binding microarray 3–4, 8–9
    -binding regions 154
    -coding region 281
    cross-links 254
    degradation 298
    /DNA complex 330
    engineering, vii–ix
    kinase 298
    levels 24
    microarray 307
    -protein interaction 17, 25, 28, 51–52, 306–307,
    373–374, 418, 432, 436
    stability 24, 345
    substrates 307
    transport 24
Proteinase k 254
Proximal
    palindrome 278
    promoter 34–35, 38, 58–59
Proximity matrix 433
Pseudo-ChIP library 168
Pseudo-ChIP-seq library 170
Pseudocount 12, 92
Pseudo-peak 170
PSI-BLAST 352
PTM-Switchboard 299–301, 305, 308
PTM-Switchboard Database 299–301
Pure algorithm 3
Purifying selection 228
P-values 72, 165, 187, 189, 433
PvuII system 270
PyMol 47
Python 123–124, 139, 218, 320

Q

QPMEME 205
Quadratic programming 205–206, 209
Quantitative Enrichment of Sequencing Tags
    (QuEST) . . . 7, 16, 157, 165–168, 171–175, 232

R

Raloxifene 242
Random binders 203
Random drift 289
Random fluctuation 426
Random forests 433
Random spike 168
Random walk 425
R/Bioconductor 415
Read density distribution 3, 165–166
Read rescue 231–232, 234
ReBase Database 272–275, 277, 279
Receiver-Operator Characteristics (ROC) . . 110, 112, 117,
    129–130, 133, 136
Receptor binding 241–248, 377
Receptor isoforms 242
Recognition helix 48, 50
Recognition sequence 242–243
REDfly 318–319, 321
Regression model 307
Regression network 430
Regression tree 433
RegsiteDB Database 59
REGTRANSBASE 125

COMPUTATIONAL BIOLOGY OF TRANSCRIPTION FACTOR BINDING
Subject Index

Regular expression 174, 243
Regular expression-type pattern matching methods 243
Regularization
    constant 151
    methods 430
Regulating the regulators 297–309
Regulatory
    architecture 370–372, 375–376, 378–379, 389–390
    complexity 315
    corpus 316–317, 323
    haplotype 326, 336–337
    interaction 14, 304, 307, 317, 421, 431, 434
    logic 313
    map 124–125
    model 403–404, 408, 410, 412–415, 417
    modules (RM) kernel 213–222
    motif 38–39, 59, 66–68, 74–75, 77, 79, 132, 432
    network 1–2, 14, 16–17, 28–29, 124–125, 162, 164,
        216, 218, 226, 230, 294, 317, 319, 323, 352,
        367, 369–395, 401–417, 419–438
    ontologies 323
    pathways 245, 248, 315
    polymorphism 320, 326, 336–337, 340, 343, 345
    programs 179, 299, 436
    triplet 299, 301, 304–307
Regulon 128
RegulonDB 13, 125, 131–135, 139, 316–317, 319
Relational database 15, 356, 368, 383–384
Relative accuracy 65
Relative performance graph 130, 134
RepeatMasker 235
Repeat sequence 235
Replicate experiments 424
Reporter assay 331
Repulsive forces 45–46
Rescue ratio 170
REST 228–229, 321
Restriction-modification 269
Retroelement 225
Retrovirus 235
Rgraphviz 415
β ribbon/hairpin proteins 48–49
Ribbon/hairpin 49
Ribosome 377
RIKEN BioResource Center 352–353
RNA polymerase II 57–59, 298, 377
RNA polymerase IV 24
ROC analysis 130
ROC curve 110, 112, 117, 130, 133–134
Roche/454 pyrosequencing 164
Roche 5, 164
Rosiglitazone 242
Royal Holloway 74
RSNP_DB 321
Rule-based learning 317
Rule-based methods 433
RVISTA 294

S

Saccharomyces cerevisiae 14, 36, 299, 317, 432, 435
Saccharomyces Genome Database (SGD) 300, 337
S-adenosyl-methionine (SAM) 46
Salmonella phage P22, 49
Sampling algorithm 92
Sampling techniques 12

Sanger sequencing 203
Saturation analysis 166, 171–172
Saturation curve 171
Scaffolding proteins 6
Scalability 254
Scalar product 219
SCAN2 .... 68, 78–79
Scoring matrices 306
Scoring methods 186
Scoring metric 403, 407–410
Sea urchin .. 15–16, 124, 314, 370–372, 378, 388–389, 394
Sea urchin database 15
Selective Alzheimer's Disease Indicator-1
    (Seladin-1/DHCR24) gene 247
Selective Androgen Receptor Elements (sAREs) 244
Selective Evolution of Ligands by EXponential (SELEX)
    _DB 9, 15, 203–204
    assay 217
    library 198–199, 207
    procedure 197, 199–204, 206, 208–209
    protocol 196, 199–201
    -SAGE 203
Selectivity 6, 13, 162, 164–165, 243, 245–247
Self-organizing maps 422
Sensitivity 13–14, 65, 74, 109, 112, 117, 127–128, 130,
    157, 162, 164–165, 173, 184–185, 189, 200,
    229, 245–247, 404
Sensitivity/specificity tradeoff 128, 130
F-Seq 165
SeqFinder 387–388
SeqMap 151–152
Sequence extraction 184–185
Sequence logo . 15, 101–102, 114, 214, 216, 219, 221–222,
    235, 244, 273, 278, 284
Sequence Tag Analysis of Genomic Enrichment (STAGE)
    164
Sequence windows 215–216, 219, 221
Serial Analysis of Gene Expression (SAGE) 203
Serum response 166, 426
Shannon entropy 404–405
Shearing 4
β-sheet group 49
Shift between the peaks on opposing strands 7, 165
SHOGUN 218
Short-read assembler 229–230, 232, 236
Short Time-series Expression Miner (STEM) 425–426
SHRiMP SHort Read MaPper 152
Sigma70 subunit 73
Sigma factor 131
Sigmoid function 205
Signal
    cutoff 264
    intensity 258
    matrix 433
    response 373
    saturation 255
Signal-to-noise ratio 7, 158, 184
SIGNAL SCAN 75
Significance criteria 145
Silencer 33–40, 51, 166, 172, 247, 252
SimAnn 293
Similarity search 356
Single-stranded oligos 196
SiRNA-mediated transcriptional gene silencing
    (TGS) 24
SiRNA(small interfering RNA) 24, 30

Subject Index

SISSRs 7, 15, 157, 165, 232–235, 238
Site-level Selection (SS) model 287–288
Site level statistics 125, 127
Skeletal muscle 65
Skewed dimensionality problem................430, 437
Skin development 377
Sliding window 150, 153–154, 157, 168
Smoothed density 167
Smoothing splines 424
SOAP 322
Softberry, Inc 74
SOLiD 5, 152, 164, 229, 231, 237
Sonication 4
Sonic hedgehog (SHH) 39
Spacer DNA 242–243
Sparse matrix 429
Spatial
  analysis 388–389
  chromatin 259–260
  correlations 14–15
  resolution 157
Spatio-temporal expression 371
SPBASE 380
Species-specific exaptation 227
Specificity profiles.................................. 315
Squalene synthase................................... 247
Ssake tool 167
Starting library 9, 198
Statistical Analysis of Network Dynamics
      (SANDY)..........................422, 432
Statistical computing 409
Statistical learning 401, 403, 416
Statistical mechanics 201
Statistical power 14, 156, 157
Statistical significance......12, 75–77, 145, 156, 183–184,
      187–188, 285
Steady state..270
STEM software for clustering short time series
      expression data 422
Step function 205–206
Steroid receptors 242
Stimulus response 434
Stochastic processes 425
Strand bias 11
β-strands 46–47, 49
Strongest binders 199
Strongylocentrotus purpuratus 314, 379, 388
Structural models............................260, 265
Structural proteins 28–29
Structured Query Language (SQL) 15
Substitution rate 286–287, 294
Subview Container View 385, 387
Supervised learning 100–101, 129
Supervised machine 214
Supervised posterior................................106
Support vector...........................14, 69–71, 116
Support vector machine.......................14, 69–70
Surface accessibility 45
SWISS-PROT Database.......................358–359
Switch position 413
Symbolic representations............................424
Synechocystis, 428
Synthetic gene...................................... 411
Synthetic network 410, 416
Systems biology 315, 371, 401, 421

T

Takifugu rubripes.............................337–338
Tamoxifen..242
Tandem DNA-binding 46
Tandem mass spectrometry (MS/MS)...............306
Tandem repeat....................................... 50
Target gene 2, 8, 29, 38, 241–248, 298–300, 305,
      307–308, 320, 327–328, 337, 339, 382, 402,
      407–408, 413–414, 430
TATA-binding protein (TBP).............26, 36, 45, 47
TATA-box........................57–59, 63, 66, 68
TATA promoters..................62–63, 66, 68, 72–73
Temperature corrections 209
Temporal clustering 437
Temporal expression pattern 283, 423
Temporal patterns 28, 424
TESLA for learning time-varying networks 430
Testing 62, 69, 72, 111, 128–129, 184, 301, 304, 314,
      410–412, 425
Text mining 15, 301, 308, 317–318, 320,
      323–324
TFcat database 316
TFDB project 323
TFIIB complex 36
Thalassemia 38
Thermal energy..................................... 201
Three-dimensional sea urchin embryonic models......372
Three-way interaction information..................405
Three-way mutual information (3MI) 17, 401–417
TiMAT 183
Time-lagged correlation 432
Time-series experiments 218, 220
Time-varying networks 430
Tissue specific 248
TNF receptor 327–328
Tobacco Transcription Factor Database
      (TOBFAC)...........................352–353
Topoisomerase-i-interacting protein 39
Total correlation.............................134, 405
Tracks, Genome Browser........................171–172
Training 2, 11, 61–62, 70–72, 100–101, 103,
      105, 109–111, 115–116, 118, 128–129, 182,
      214–222, 301, 316
Trait aetiology...................................... 314
Transactivation 242, 331
Transcriptional
  activation domain (TAD)..........................27
  apparatus 58
  complex...36
  factor families 121
  modules 432
  network 28–29, 254, 319
  pre-initiation complex (PIC) 26–27, 35–36,
      38–39
Transcription factor post-translational modifications
      (TF-PTMs) 298–299, 301,
      305–307
Transductive Confidence Machine 69
TRANSFAC Database..........................59, 63
Transfection 331, 345
Transient interactions 180
Transposable DNA elements 13
Transposable-element-derived regulatory networks 226
Transposable elements 14–15, 17, 225–227, 238
Tree algorithm...................................... 11

454 COMPUTATIONAL BIOLOGY OF TRANSCRIPTION FACTOR BINDING
Subject Index

Tree representation 394
Trp operon 34
TRRD Database 59, 319–320
Tryptophan 34, 375–376
TSSG promoter prediction program 61
TSSP promoter prediction program 61
TSSP-TCM promoter prediction program 61
TSSS 61
TSSW promoter prediction program 61
Tubulin 377
Turnover rate 289–291
Twin peaks 7, 165, 168
Two-parent model 408, 413
Two-sample analysis 150, 153–154

U

Ubiquitin ligase 39
Ubuntu LINUX 136
UCSC Table Browser 230
UCSC Genome Browser 16, 171, 175, 233–234, 238, 320, 322
UNIX 5, 174, 233
Unspecific oligonucleotides 330
Unsupervised learning 129
Upstream activating sequences (UAS) 35, 39
Upstream activator 34
USeq method 157

V

Van der Waals interaction 44–45, 198
Vector features 221
Virtual Sea Urchin embryo modeling 394
Virtual Sea Urchin software system 369

Vista Enhancer 321
Visualization 15–16, 78, 136, 171–172, 370–372, 388, 410, 415–416
Vitamin D Receptor (VDR) 242
Vocabulary 316, 323, 372, 375, 389
VOMBAT web-server 115
Voting-based methods 345
VSU model 388

W

Wavelet method 173
Weak binding site 13
WebMOTIFS 183
Window size 152, 157–158, 185, 234
Window switch 157
Winged helixturnhelix (HTH) motif 46, 48
WINNOWER 11
WUSCHEL 216–217

X

Xenobiotics 241, 246
XML format 341, 383, 393
XML language 394
XML parser 393
X-ray crystallography 51

Z

Zero temperature approximation 205, 209
Zinc-finger 242, 375
Z-score threshold 245